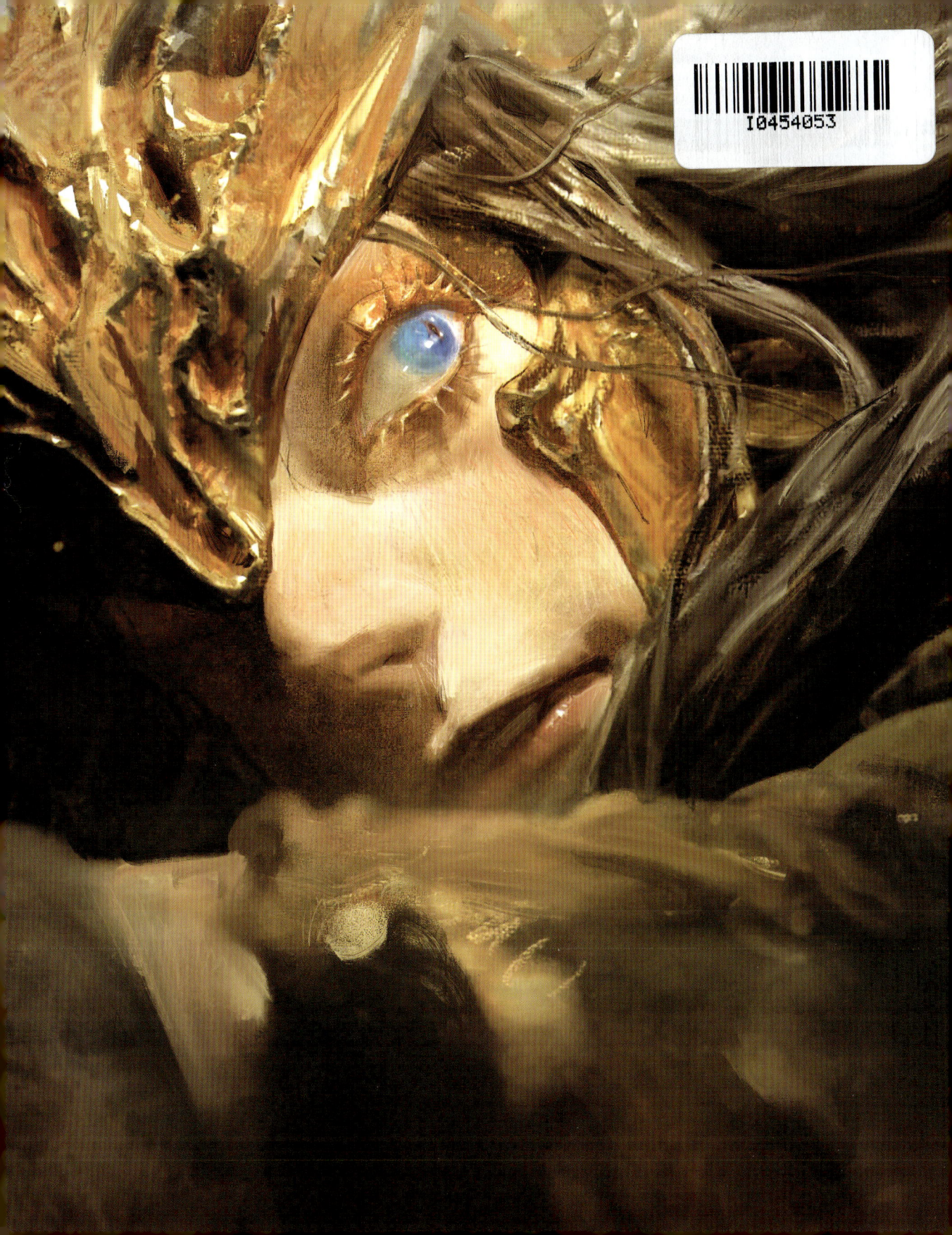

# ART WORKS

# Entei Ryu

## CHIMERA

エンテイ・リュウ アートワークス キメラ

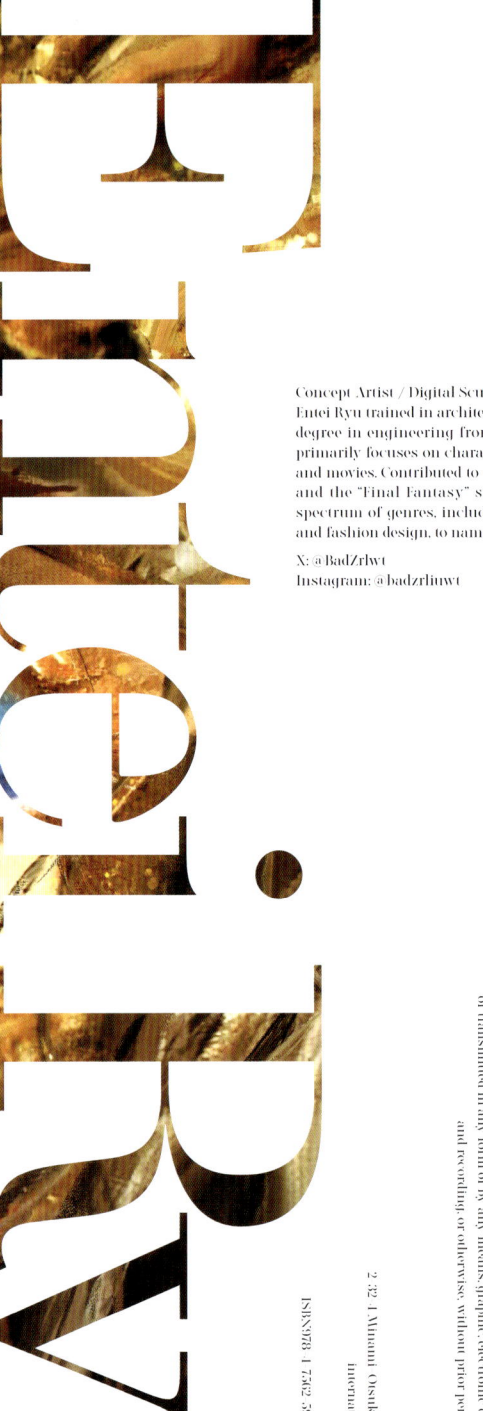

**Concept Artist / Digital Sculptor**
Entei Ryu trained in architecture and earned a Master's degree in engineering from the University of Tokyo, primarily focuses on character design for video games and movies. Contributed to titles like "Assassin's Creed" and the "Final Fantasy" series. Style spans a broad spectrum of genres, including sculpting, illustration, and fashion design, to name a few.

X: @BadZrlwt
Instagram: @badzrliuwt

2024.7.11 first printing of the first edition

Author — Entei Ryu

Facilitator — W's Co., Ltd.

Designer — Minoru Manata (Bonmachi Kunihan)

Editor — Yoshiyuki Oba

Editorial collaborator — Rina Tanaka (Hyoko sha)

Acknowledgements — KOJIMA PRODUCTIONS Co., Ltd.
Underverse Co., Ltd.
Saporani
Tokuma Shoten Publishing Co., Ltd.
Akita Publishing Co., Ltd.

Printed and Bound in Japan by Shinano Co., Ltd.

© 2024 Entei Ryu  PIE International

PIE International Inc.
2-32-4 Minami Otsuka, Toshima-ku, Tokyo 170-0005 JAPAN
international@pie.co.jp www.pie.co.jp/english
ISBN978-4-7562-5903-3 C0 outside Japan  Printed in Japan

Chimera, the fiery monster born of Greek myth,
with a lion's head, a goat's body,
and a serpent's tail. In biology, it refers to a genetic hybrid
formed by the fusion of cells from two different species.
It is a symbol of the artist who, in the creative process,
employs new technologies such as digital sculpture and 3D printing,
as well as traditional media like rock painting tools.
The artist deftly moves between the digital and traditional,
the two-dimensional and three-dimensional worlds,
creating works that are naturally hybrid, blending
the "virtual" and "real" genes into one.
Within Entei Ryu's imaginative world,
where CHIMERA represents countless fantasy
beings whose forms dance and intertwine
in a dazzling display of light and vision.
Here, the Chant of the Maiden and the Rock Armour of the Giant Beast
symbolize the balance of individual and nature.
Different faces and totems are woven with light and shadow
to create a stunning decoration and figuration.
flowing energy is engraved in an abstractive monument,
where organic and inorganic intermingled.
And life, exuberantly bursts forth in the chimerism of CHIMERA.

"キメラ（Chimera）"とは、ギリシャ神話に登場する
「ライオンの頭にヤギの体と蛇の尻尾を持ち、火を吐く異形の怪物（Xíμαιρα）」であり、
生物学においては「異なる種の細胞が共存する遺伝子嵌合体」を指す。
創作フローにおいてデジタル彫刻や3Dプリントなどの「新しいテクノロジー」と
岩絵具などの「伝統的な画材や描画方法」を同時に使用することの象徴でもあり、
「デジタルとトラディショナル」「平面と立体」の世界を往来し相互に影響させることで
「虚」と「実」2つの遺伝子を作品にごく自然に"混血"させる。
また同時に、それは作者の無尽のファンタジーに棲む数多の生物の姿でもある。
少女たちの歌声と巨獣の岩鎧 ― 個体と自然。
異なる顔と記号を光と影で織り上げ ― 装飾と造形。
流れるエネルギーを無機質なモニュメントとして綴る ― 有機と無機。
キメラ。生命は嵌合によって前進する。

# 01

CHIMERA

CHIMERA is not only the title of this book but also the name of a series of my creations themed around mythical creatures. Just like the meaning of the word itself, they are chimeric beings, possessing the vast bodies of beasts and the delicate torsos of young girls, displaying the forces of nature and humanity within the same entity. I have also endowed each individual with cultural symbols from different regions of the real world, and you can see decorative styles from Nordic, East Asian, African, and other regions.

I often reminisce about an afternoon when I was four years old, alone at home, watching 'Fantasia 1940' on television. As Stravinsky's "The Rite of Spring" and Beethoven's "Pastoral Symphony" played, the animation brought to life a panorama featuring Greek gods, centaurs, and the extinction of dinosaurs. These vivid images were etched deeply into my heart, and I believe they may very well have been the initial sources of my inspiration.

「CHIMERA」は本書の表題であるだけでなく、神話上の生物をテーマにした一連の作品のタイトルでもあります。言葉自体が示すとおり、それは巨大な獣の身体と繊細な少女の上半身を持ち、同じ実体の内に自然の力と人間の力を宿す、嵌合体的存在です。現実世界における多様な文化的シンボルを与えた個々の作品からは、北欧、東アジア、アフリカ、その他の地域の装飾様式を見て取れるでしょう。

私は4歳のときにテレビで放映されていた1940年の映画「ファンタジア」を、家でひとり観ていた午後のことをいつも思い出します。ストラヴィンスキーの「春の祭典」とベートーヴェンの「交響曲 第6番 田園」が流れる中、ギリシャ神話の神々、ケンタウロス、恐竜の絶滅をテーマにしたパノラマが、アニメーションで生き生きと表現されていました。心に深く刻まれたこうした鮮烈なイメージが、私のインスピレーションの最初の源泉になったのだろうと思います。

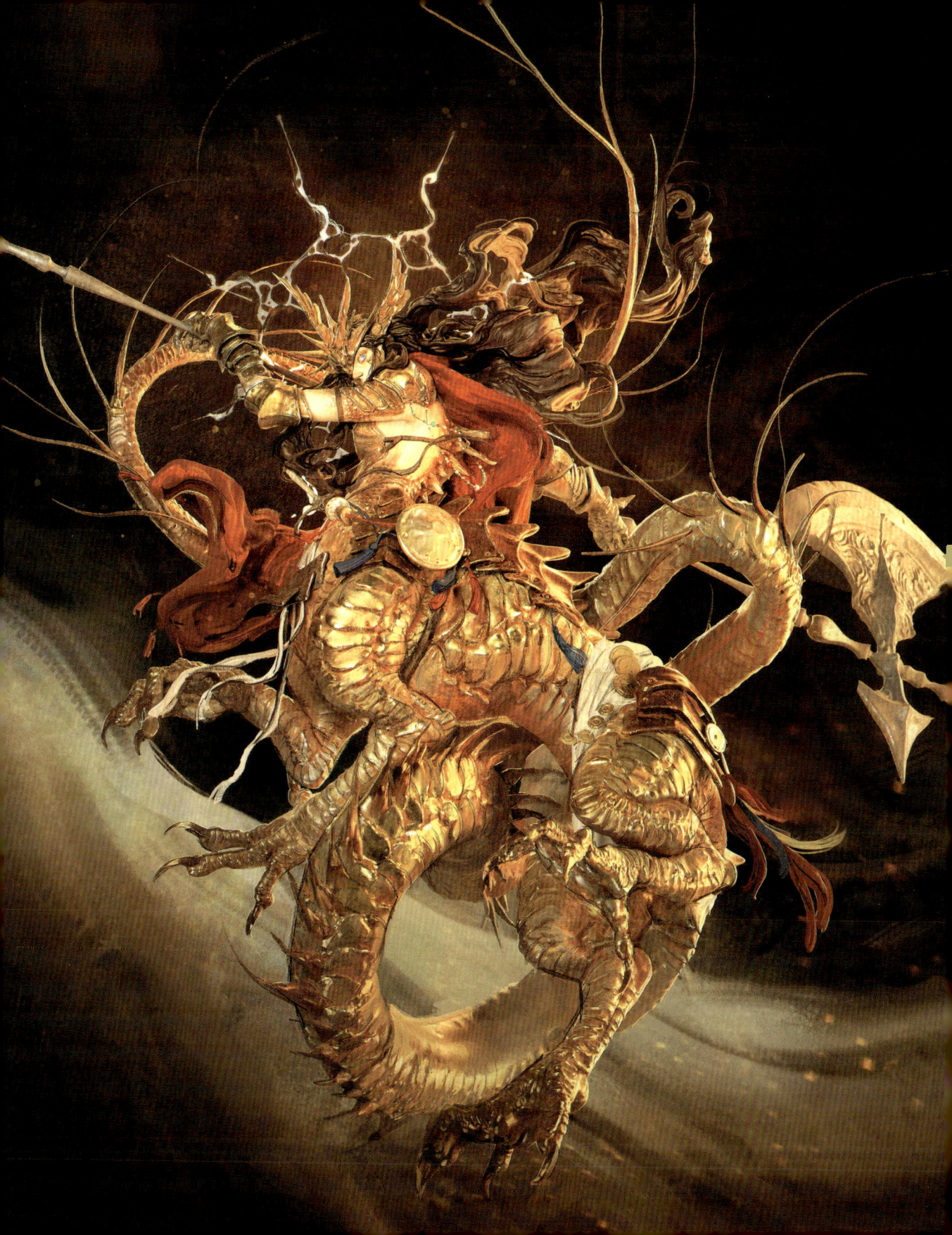

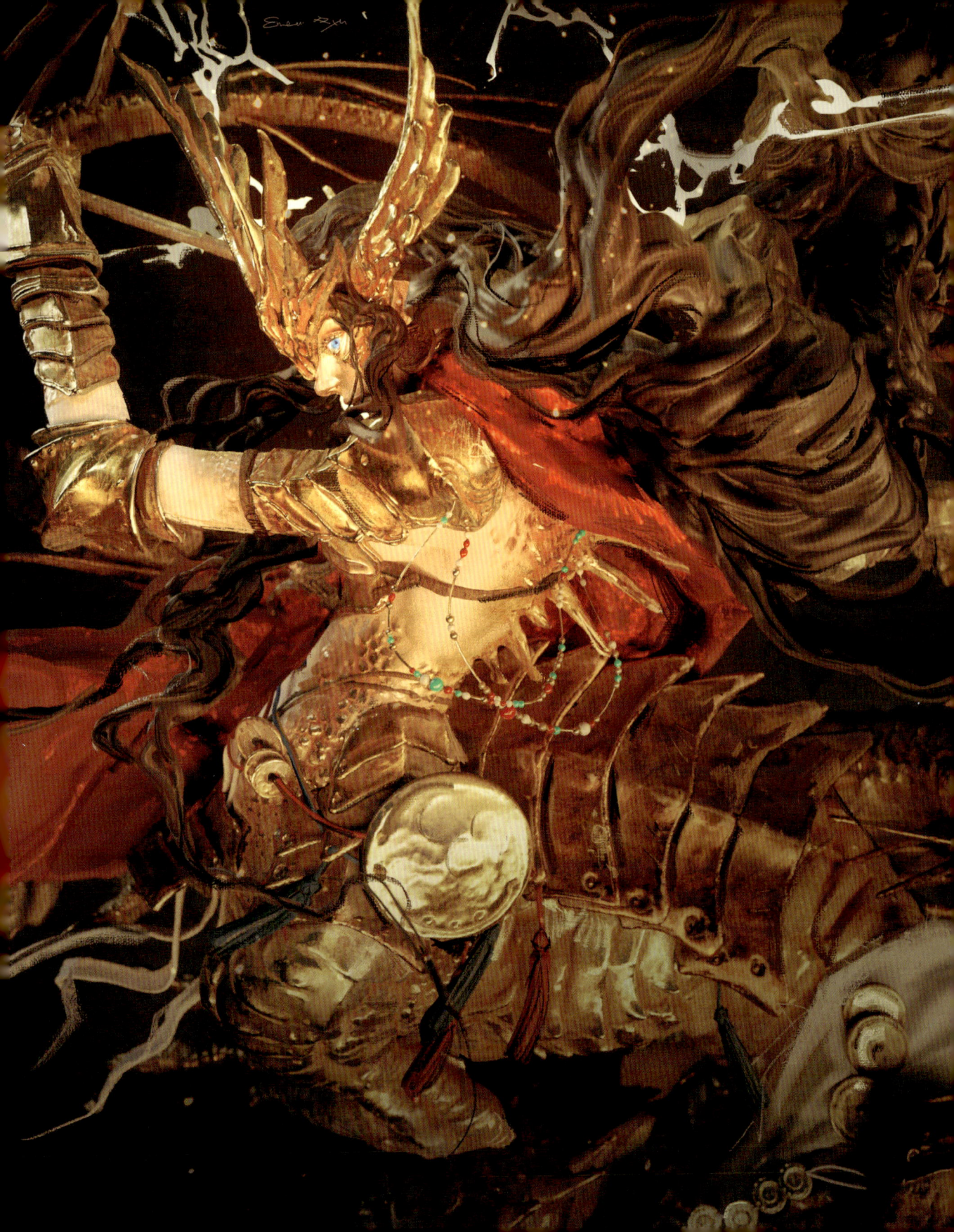

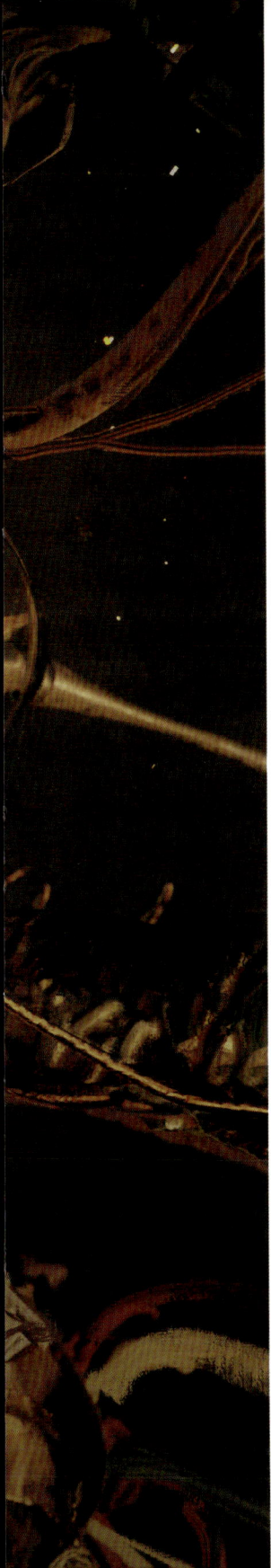

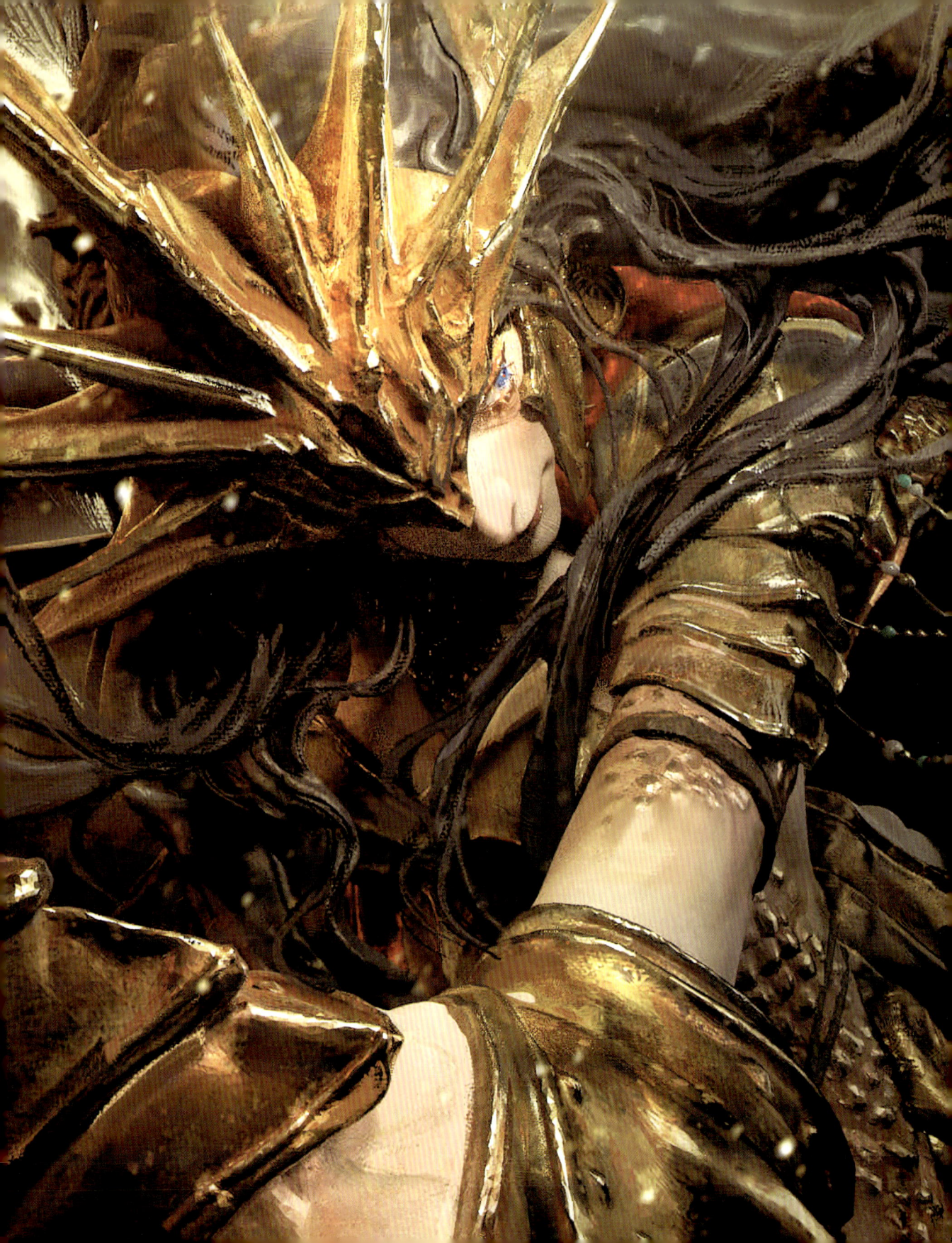

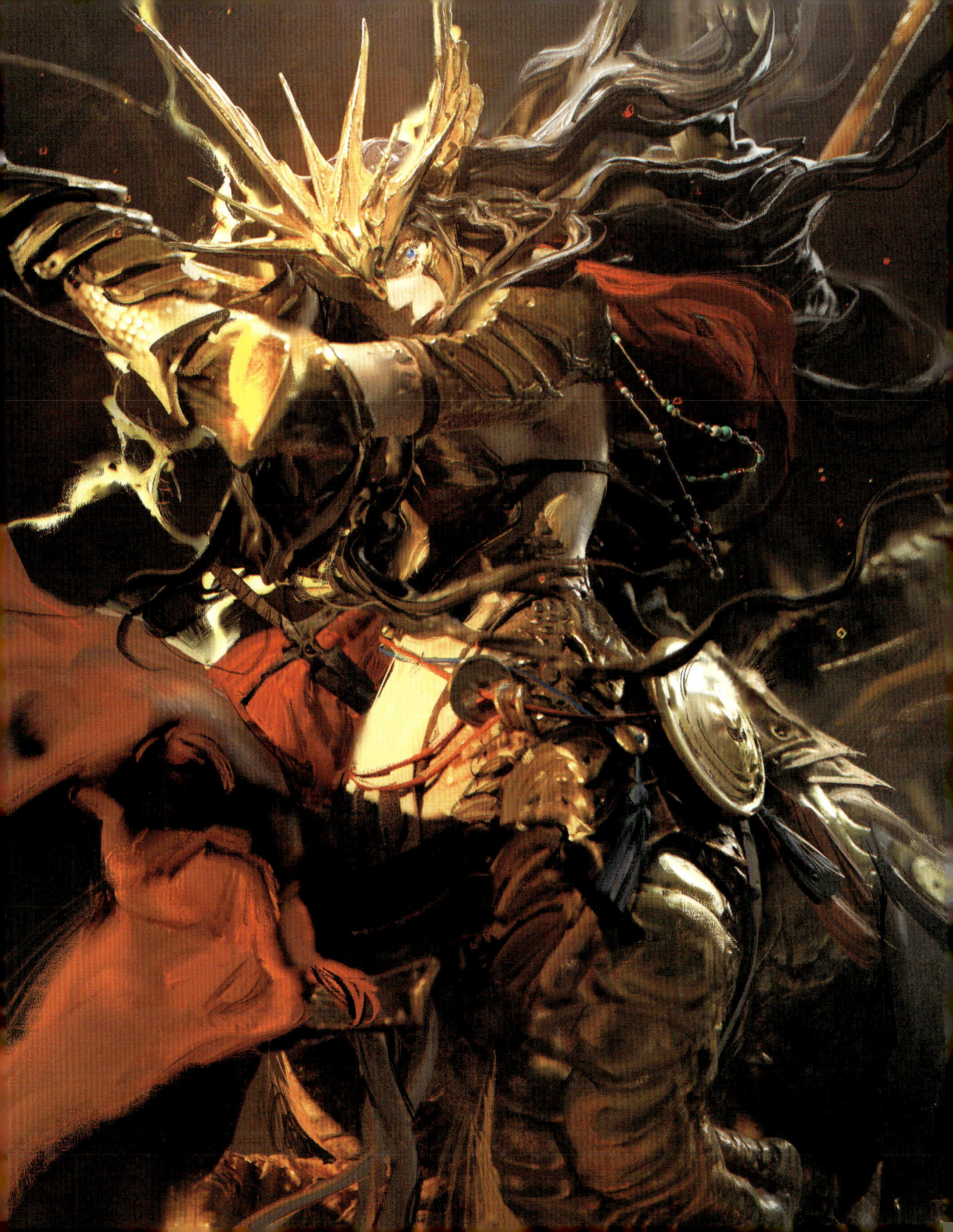

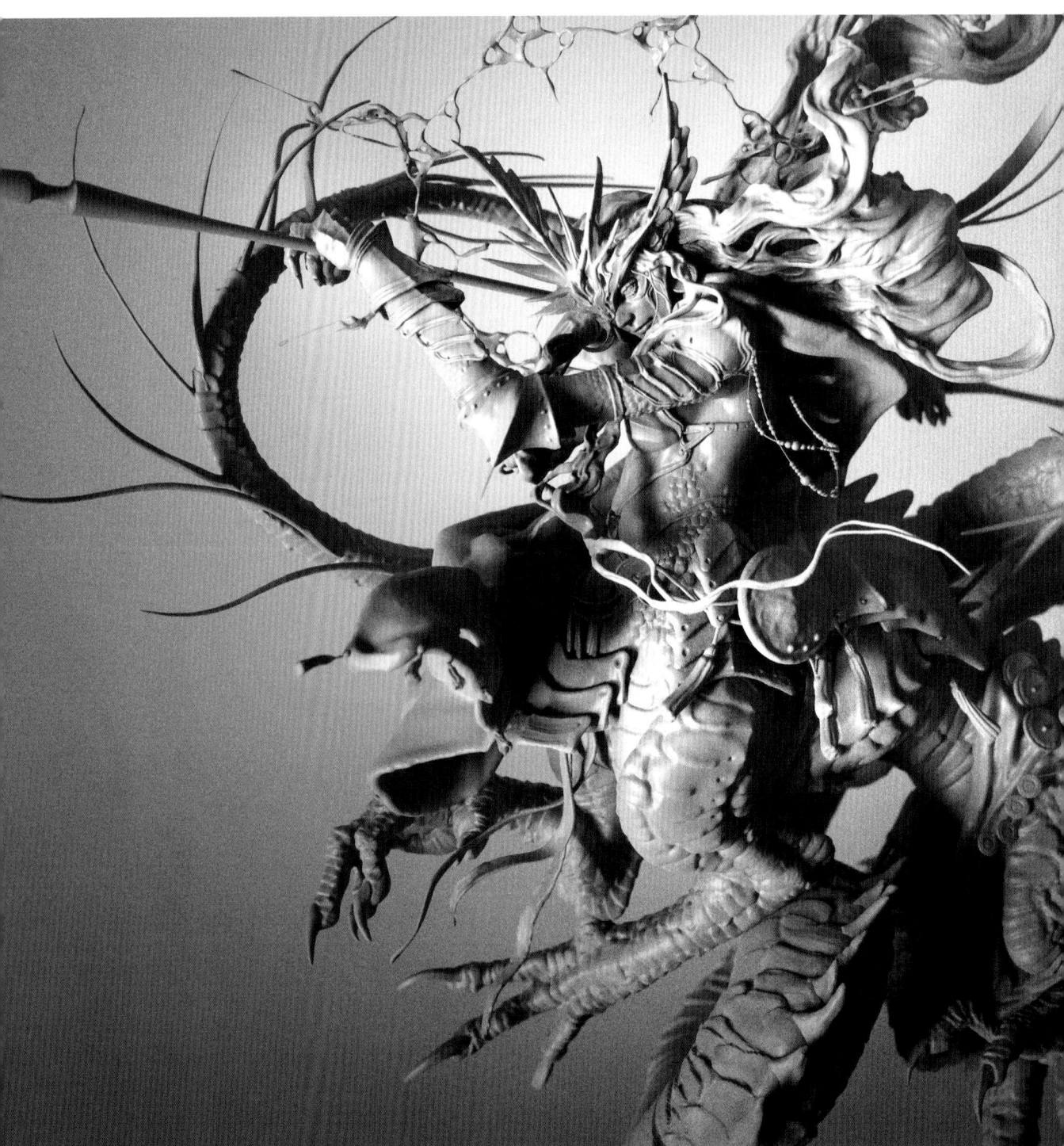

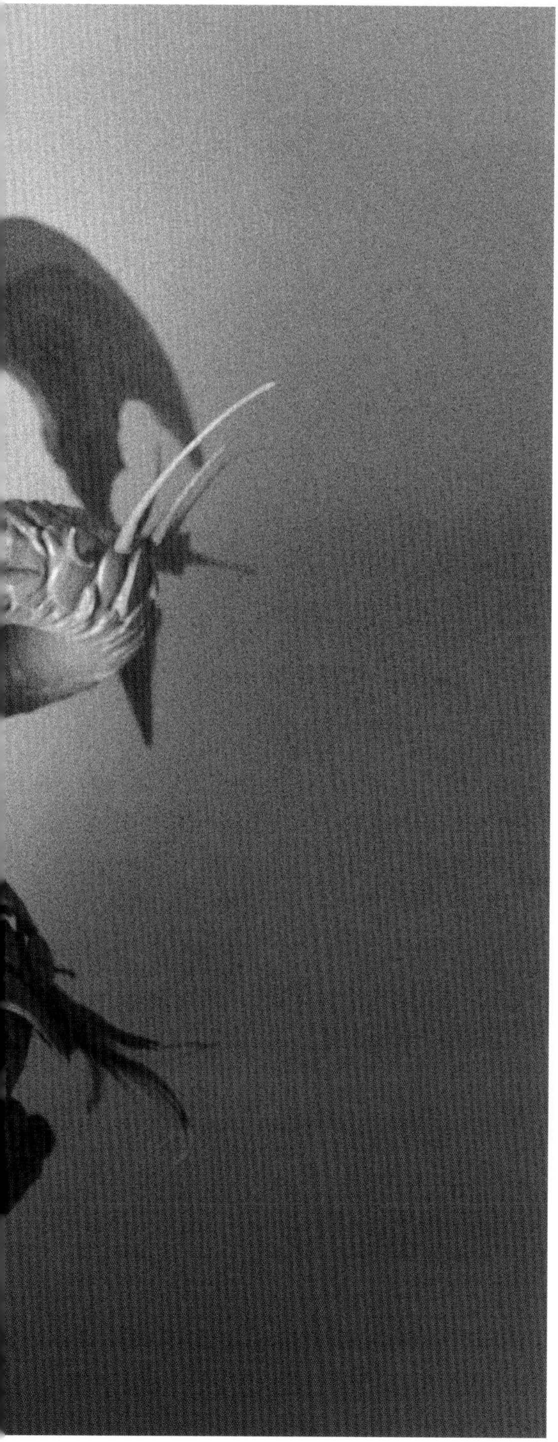

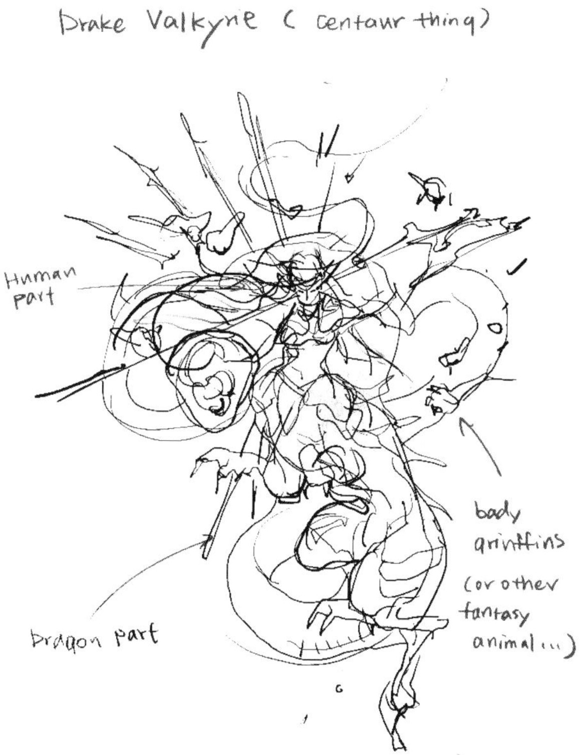

Drake Valkyne ( centaur thing)

Human
Part

Dragon part

body
griffins
(or other
fantasy
animal...)

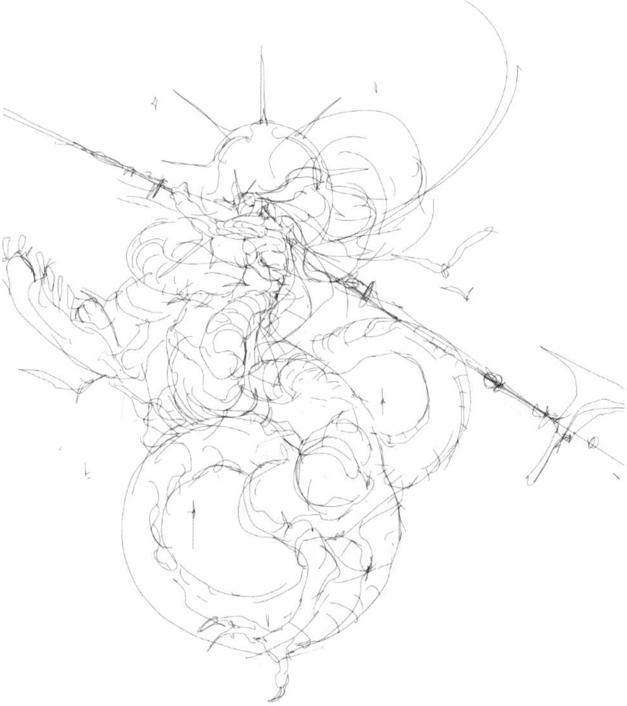

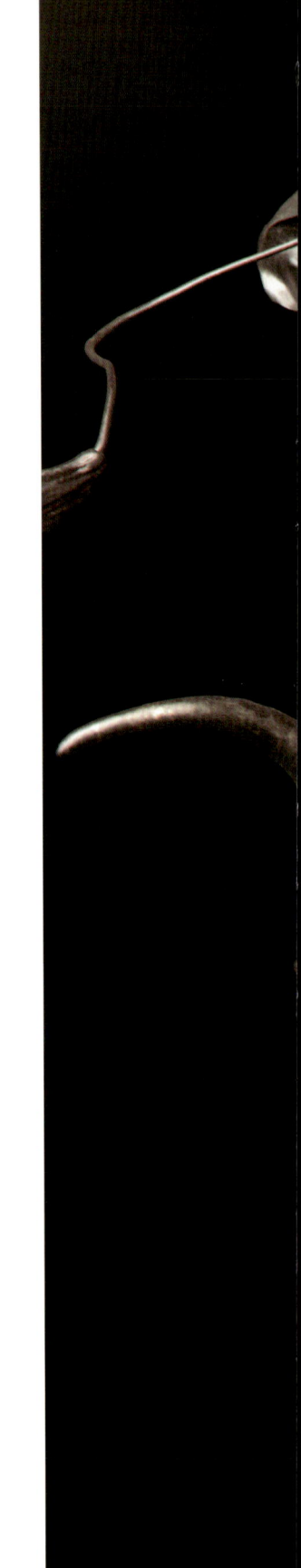

# Killing Angel 01

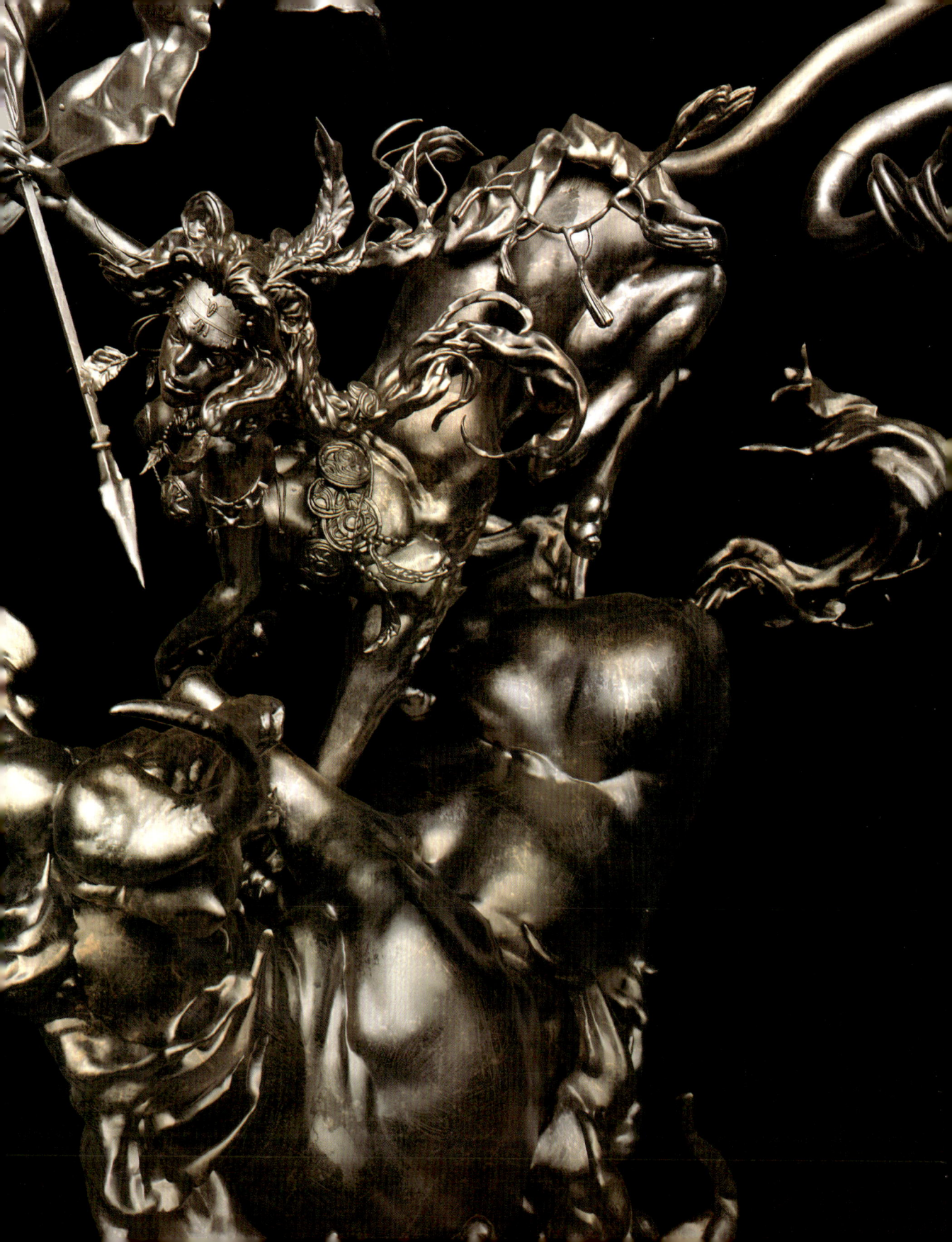

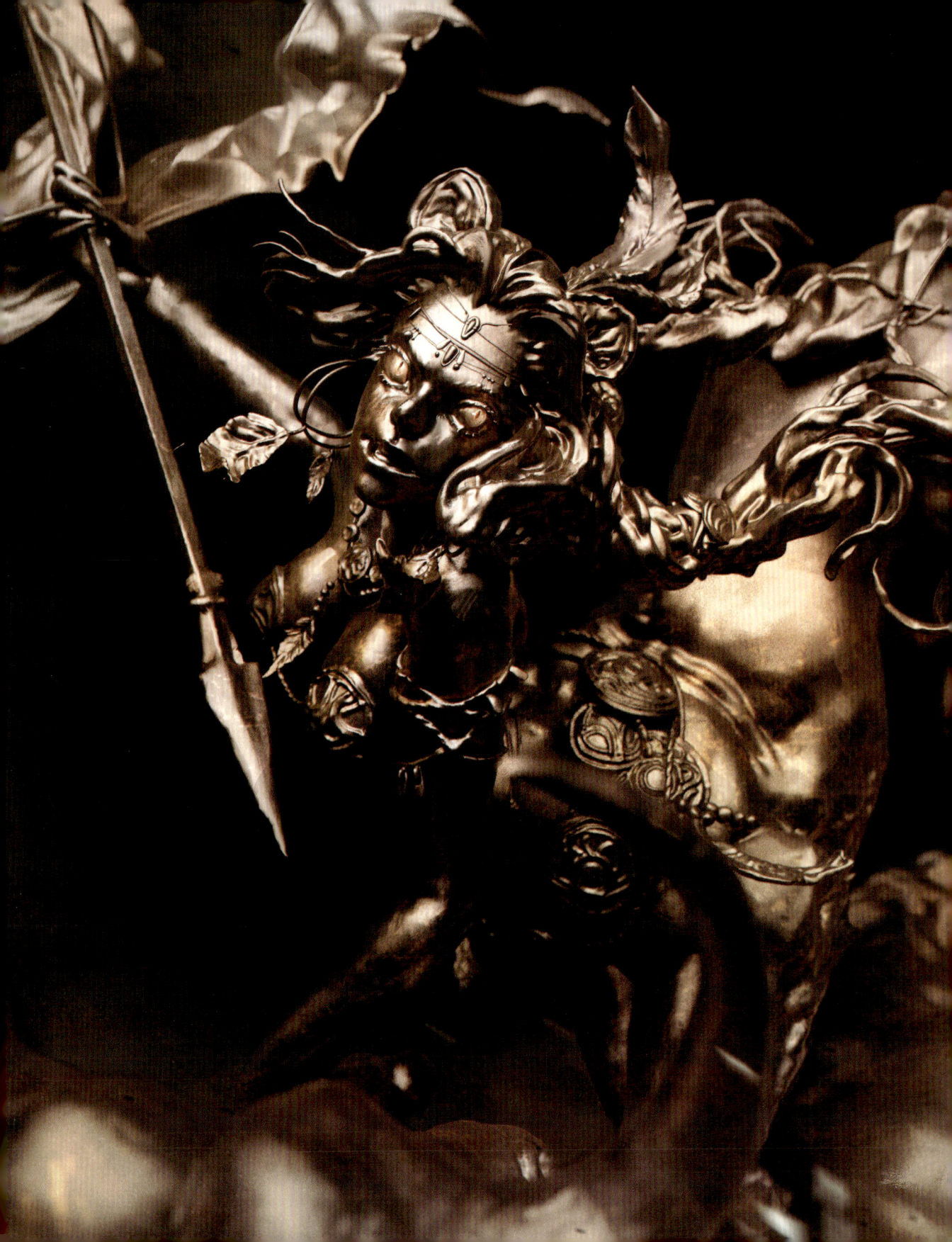

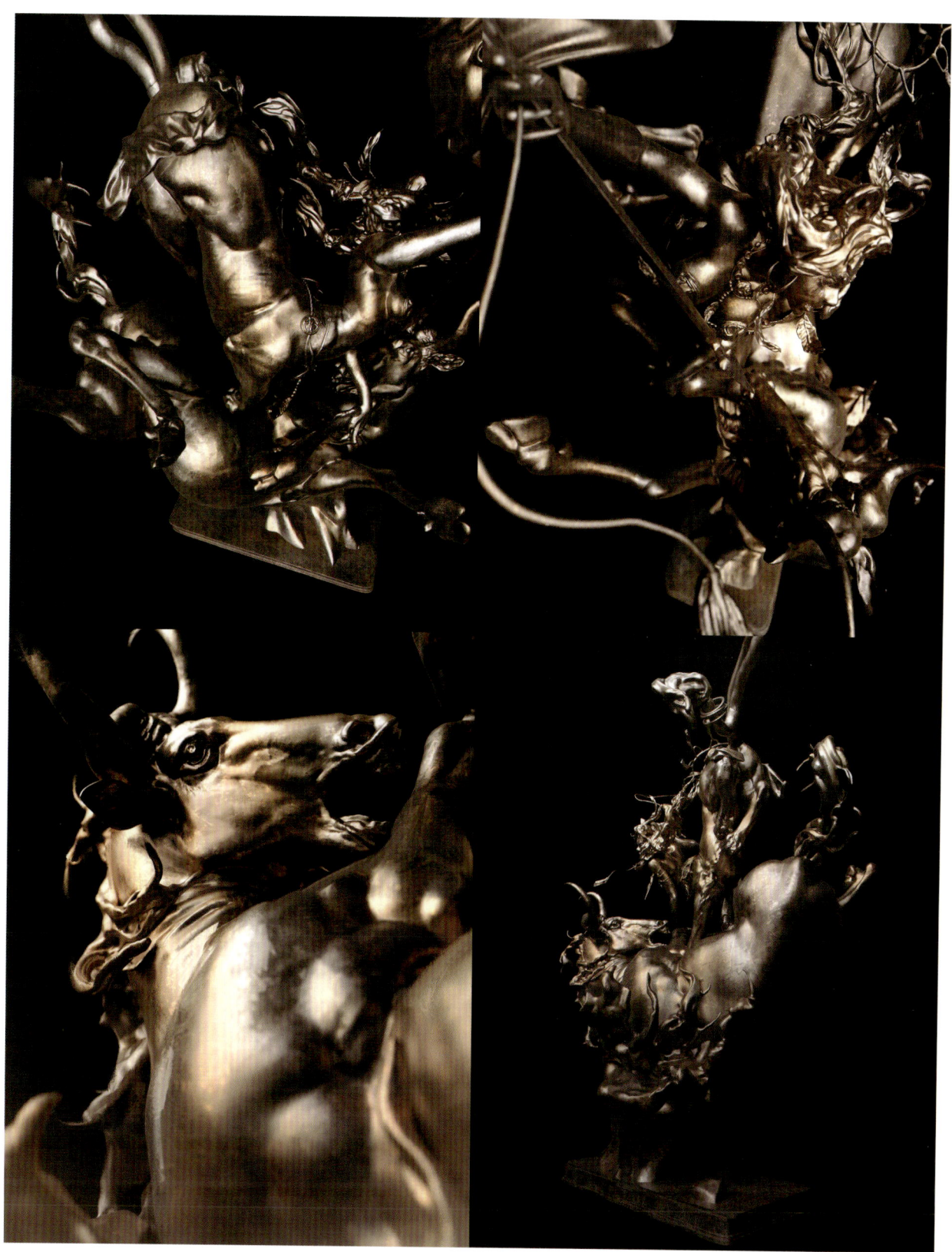

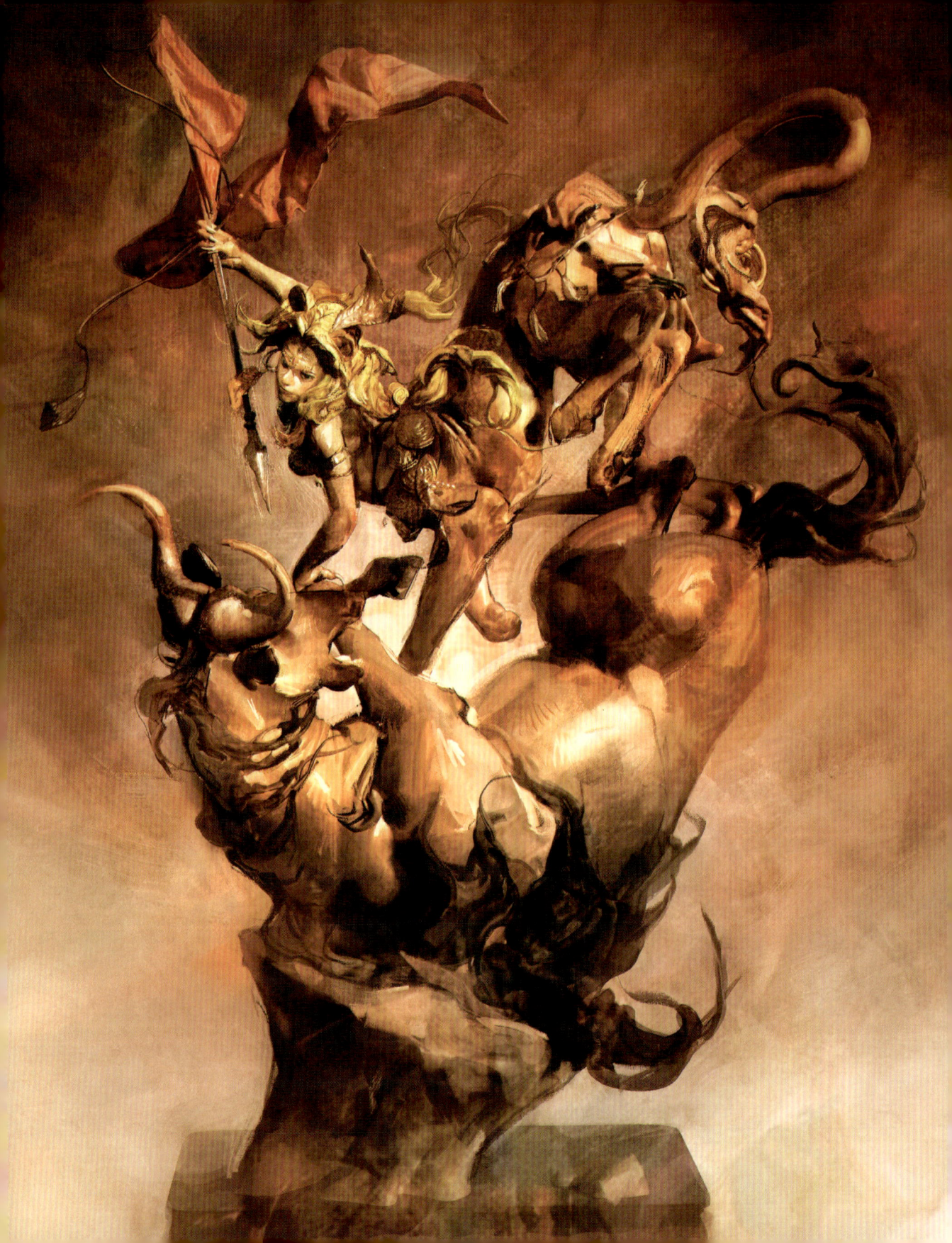

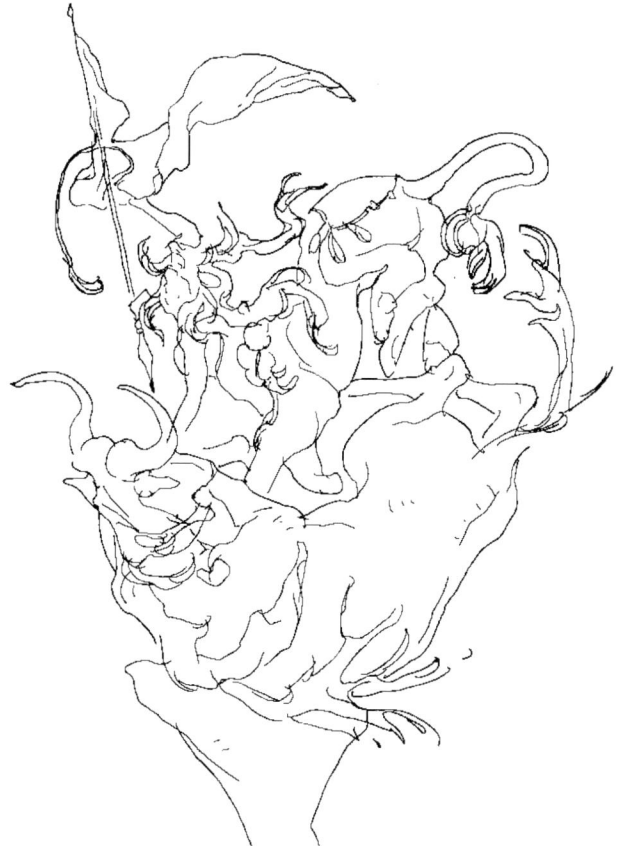

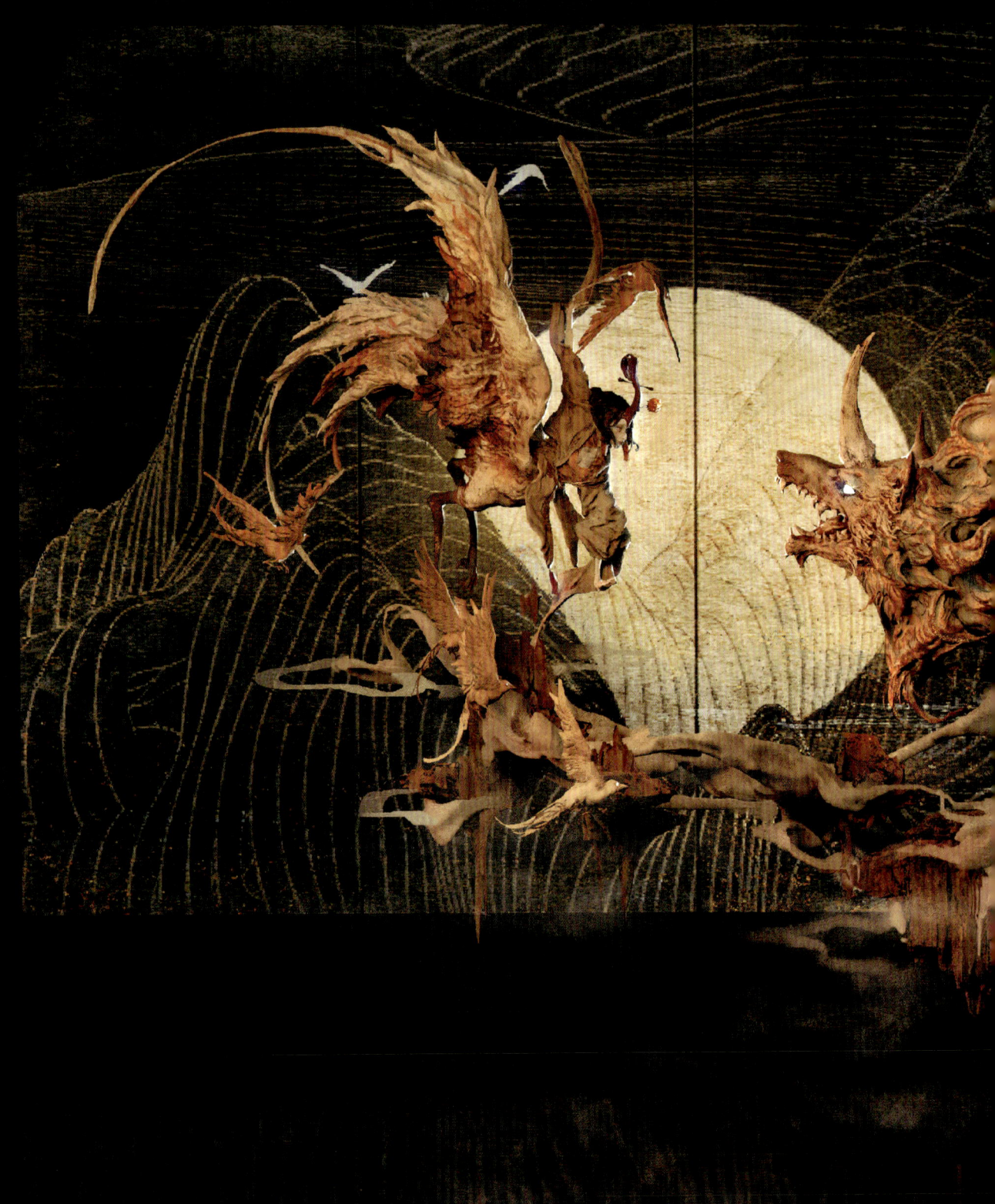

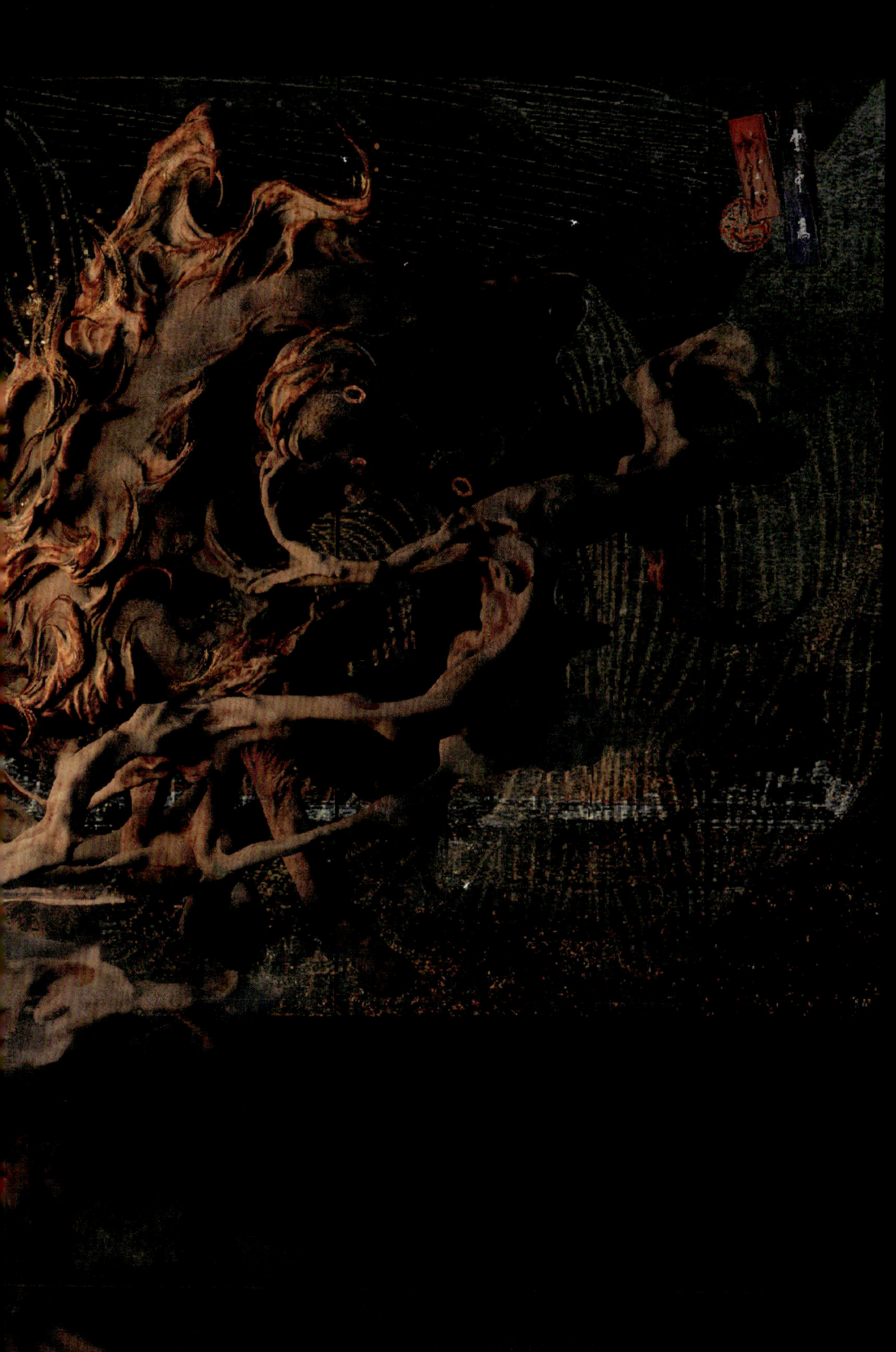

Moon Dancer 01

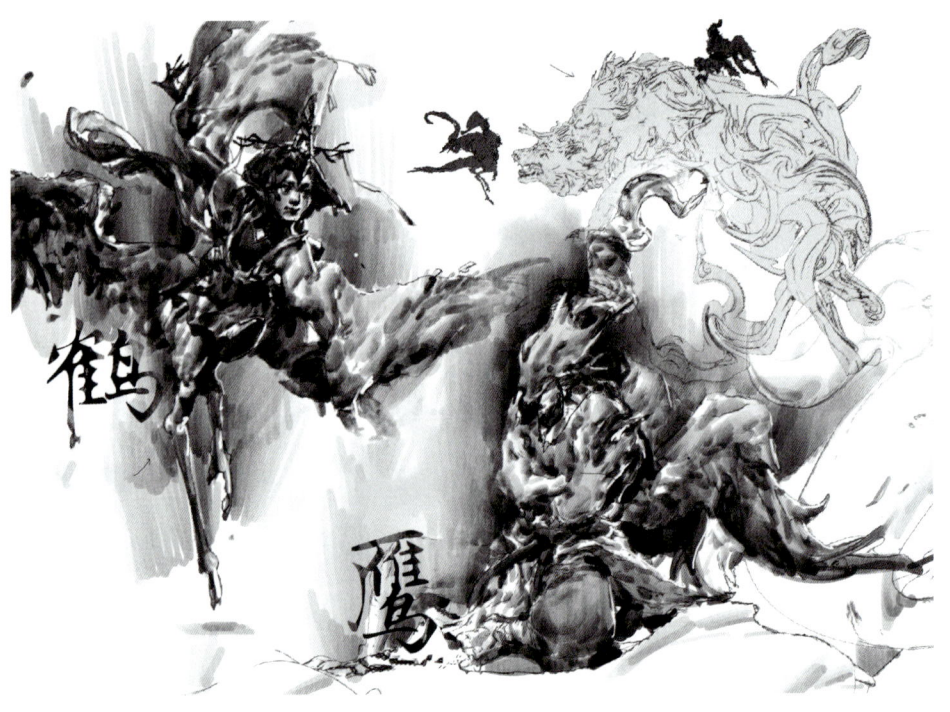

雀鳥

鷹

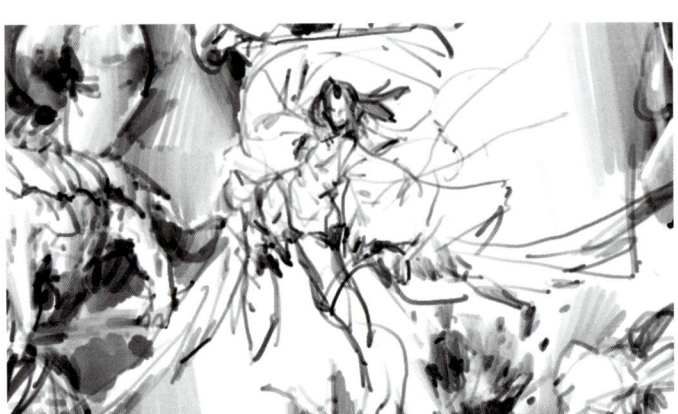

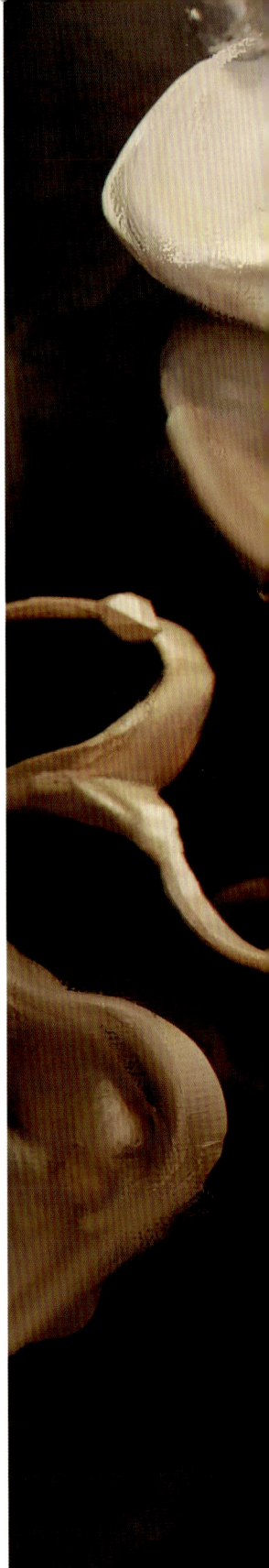

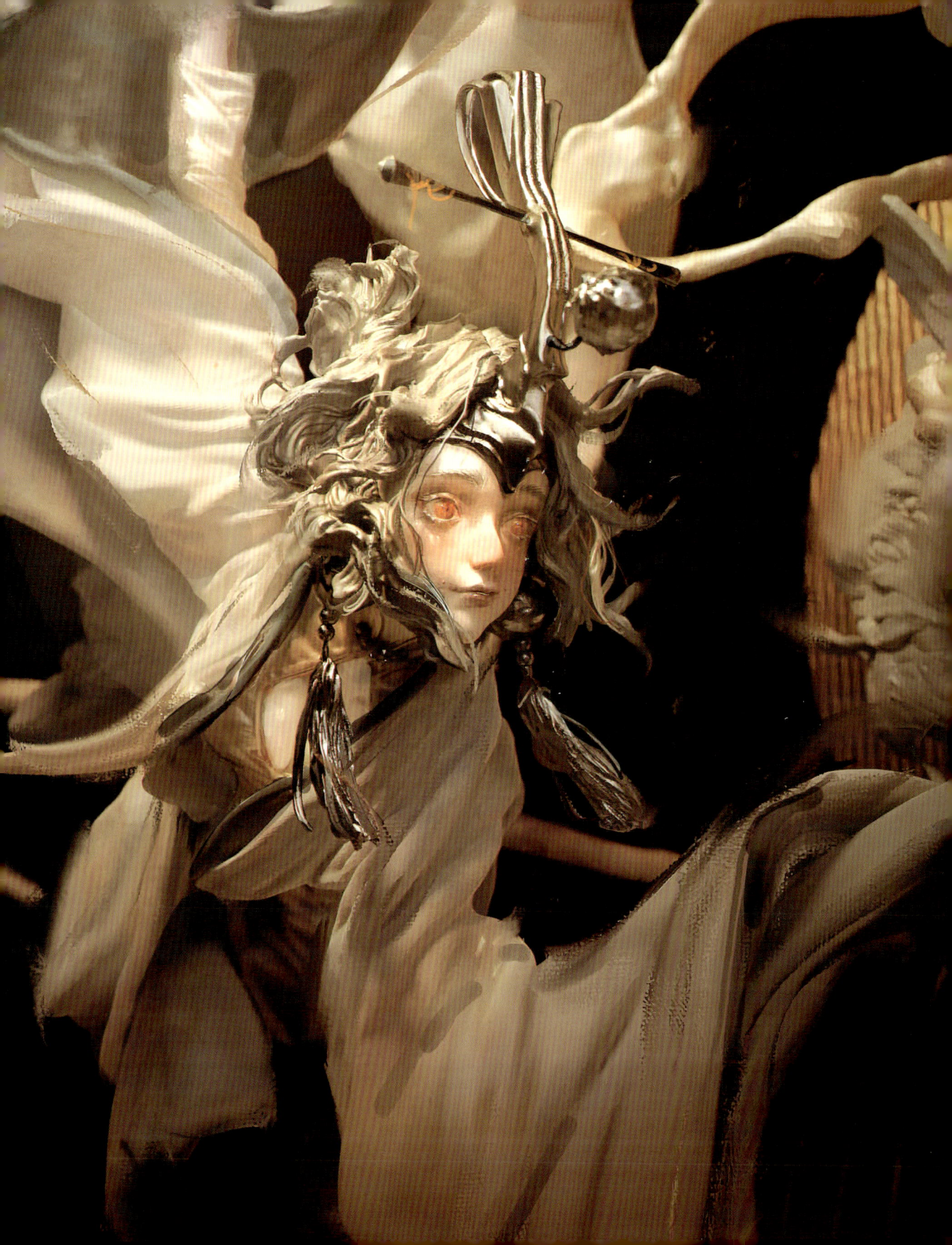

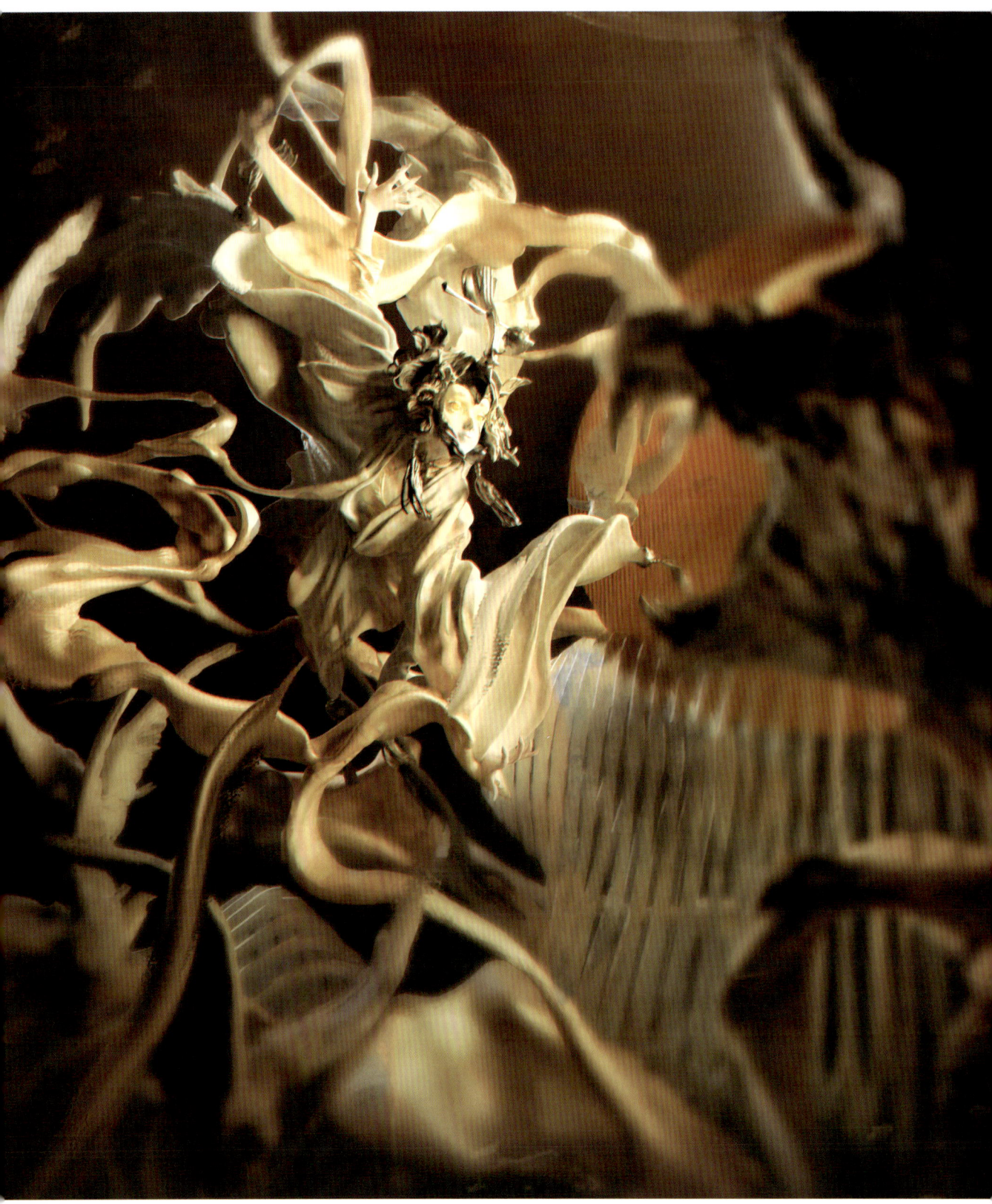

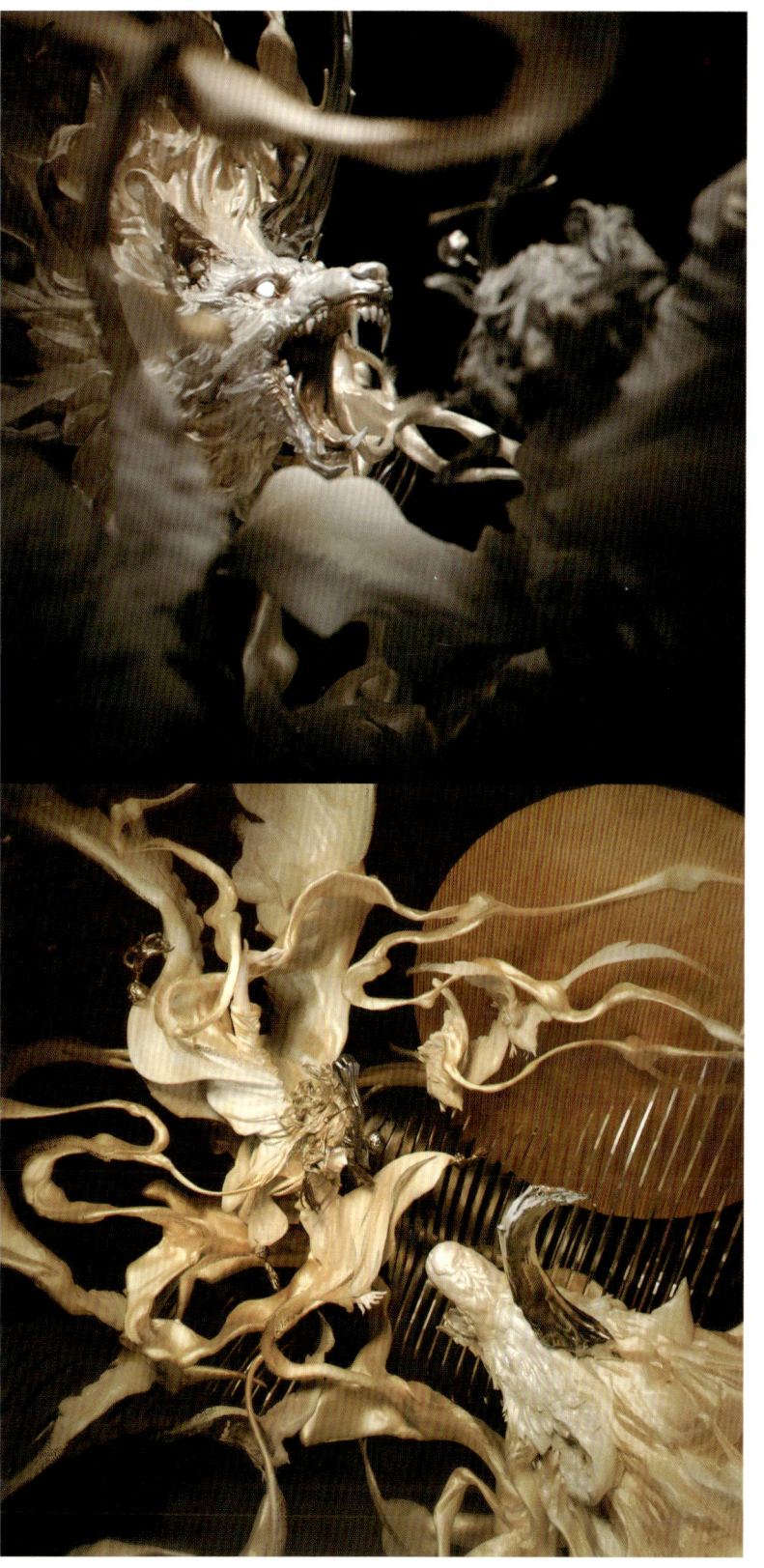

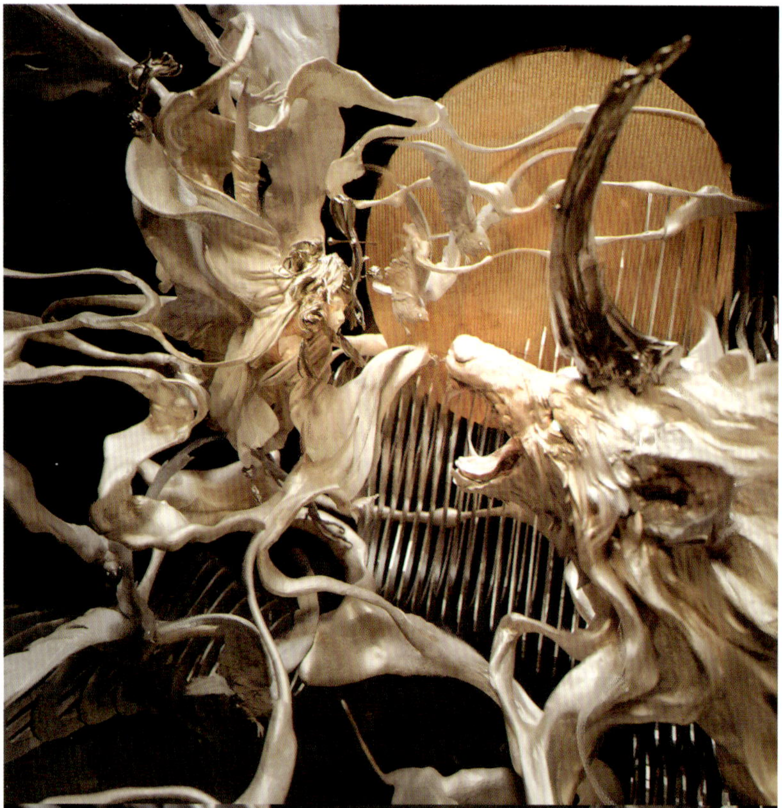

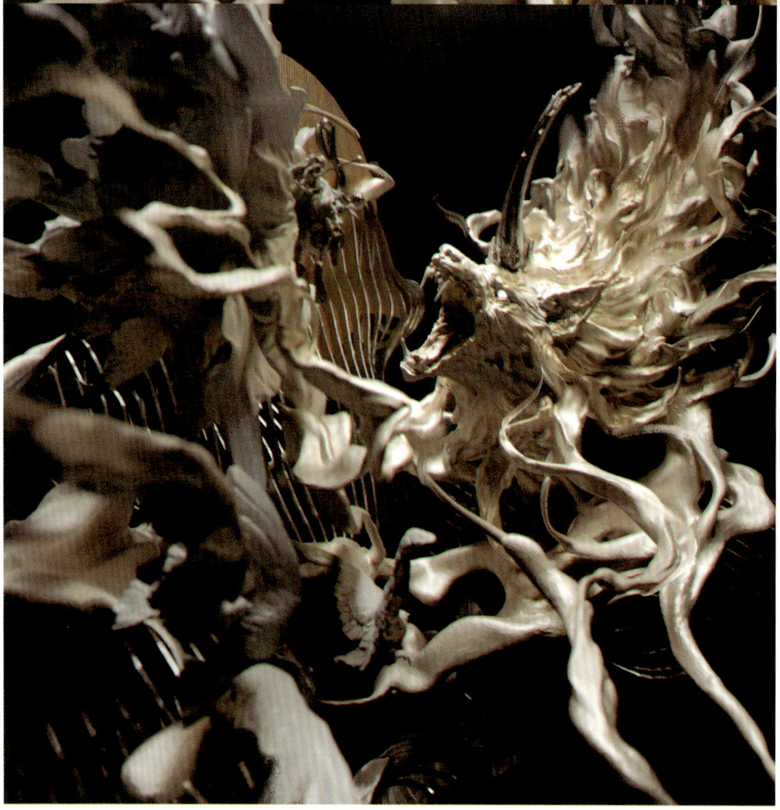

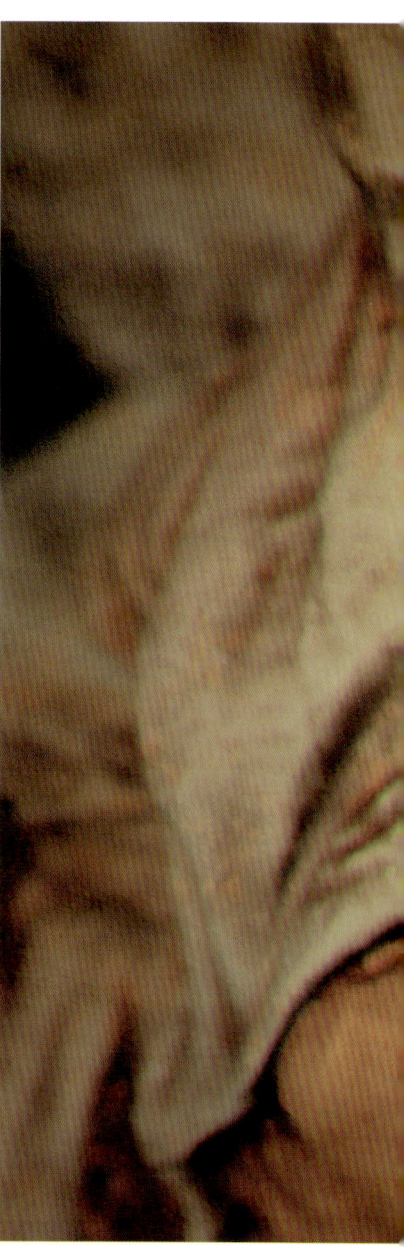

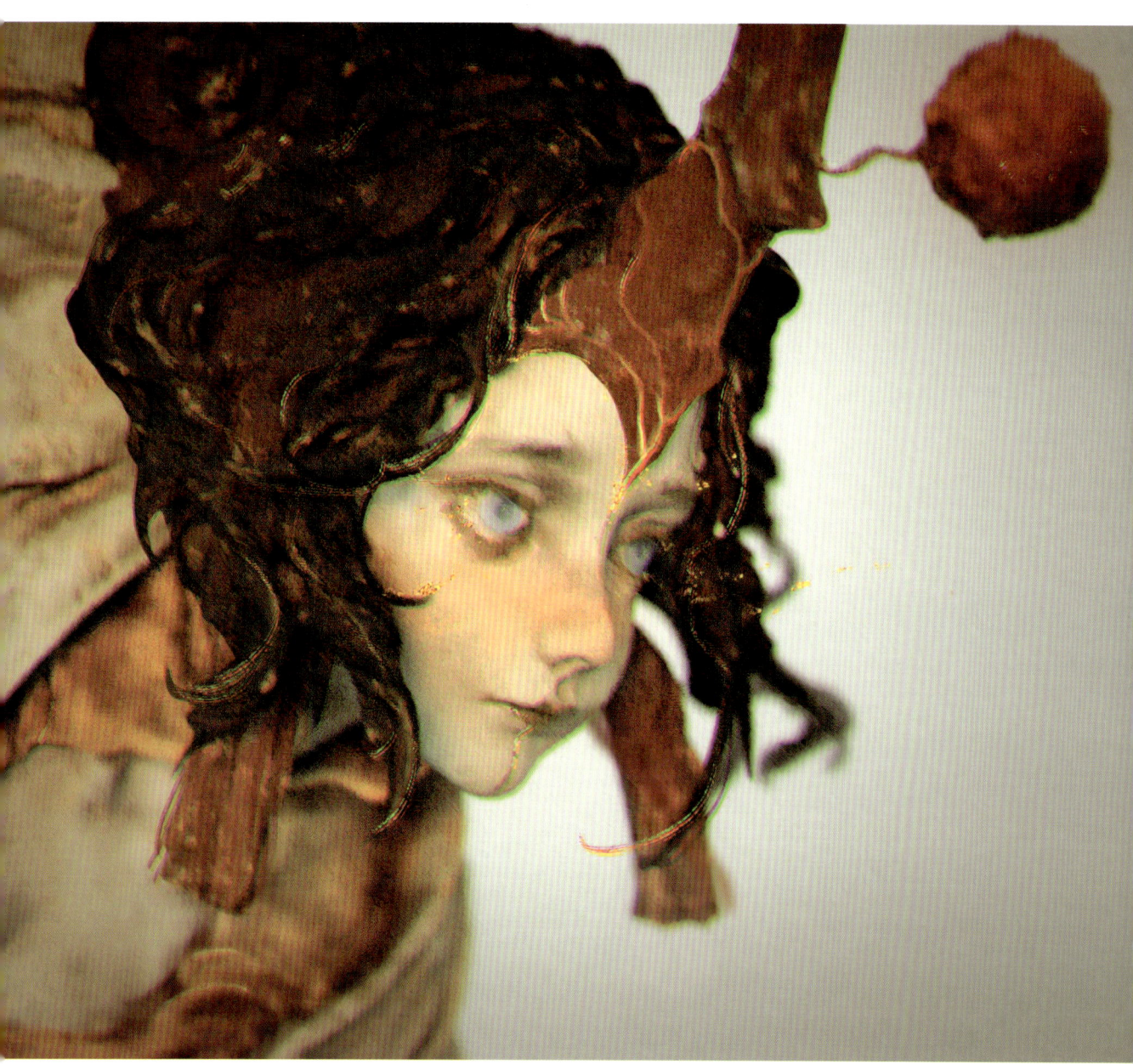

# Dragon Mermaid 020

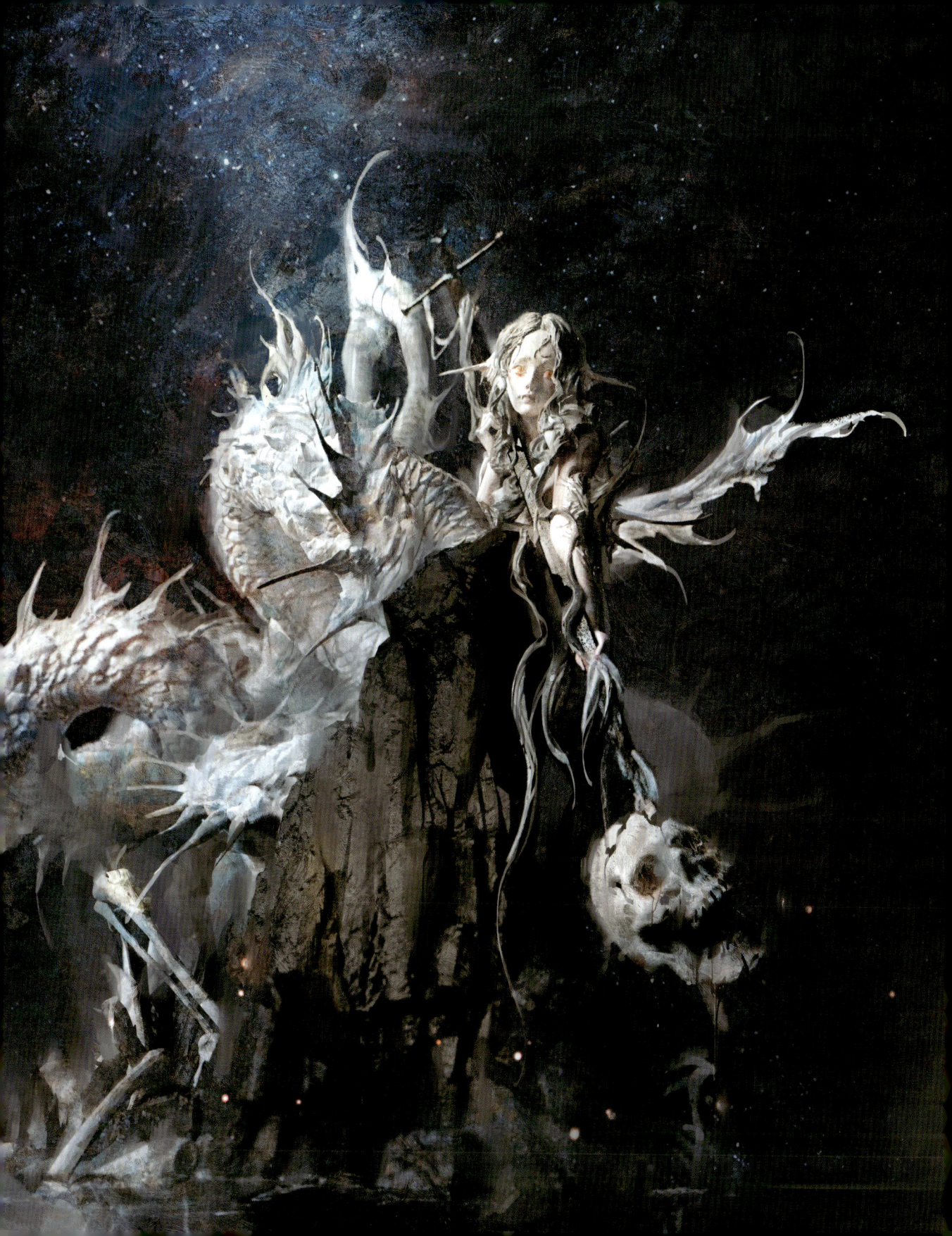

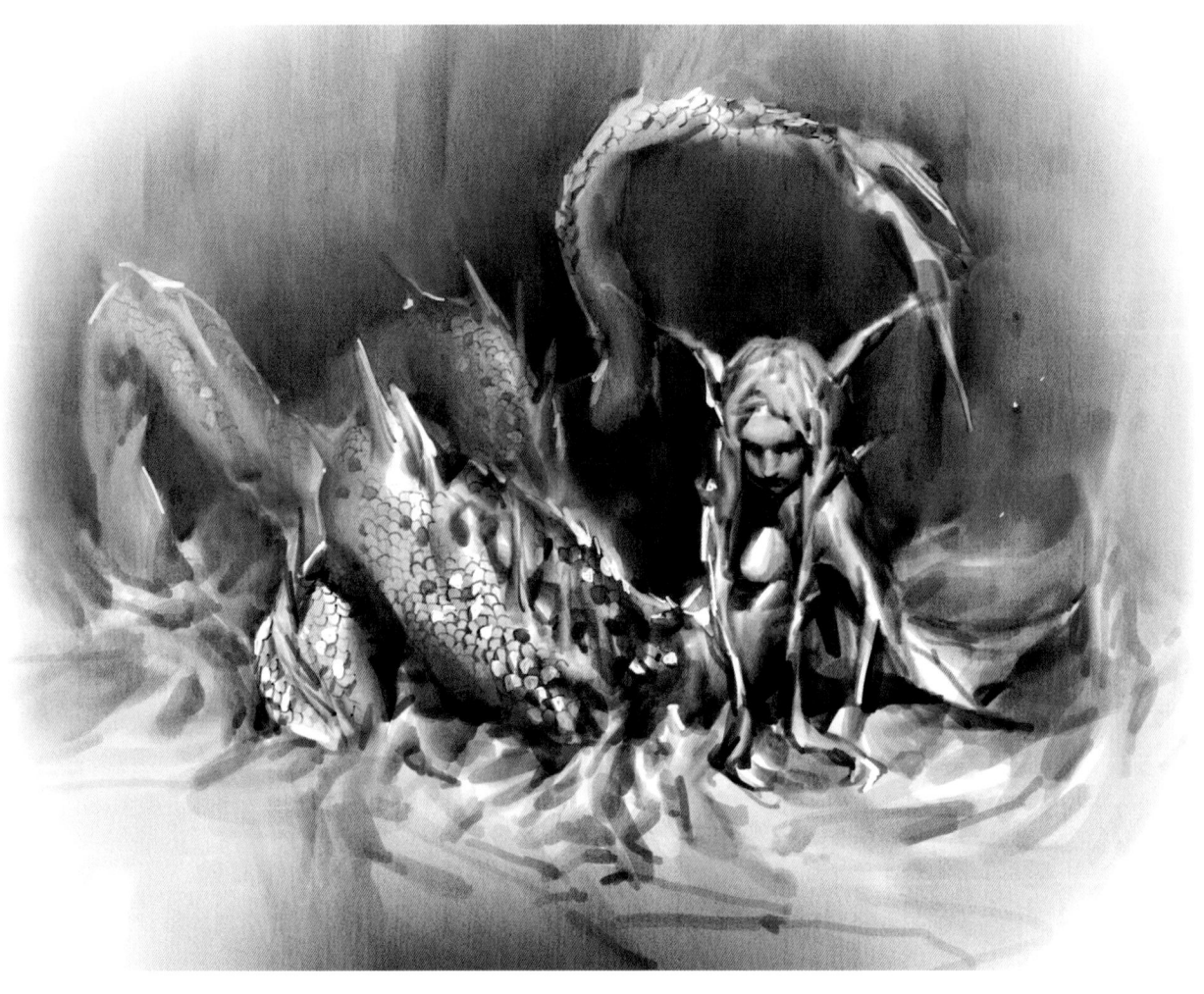

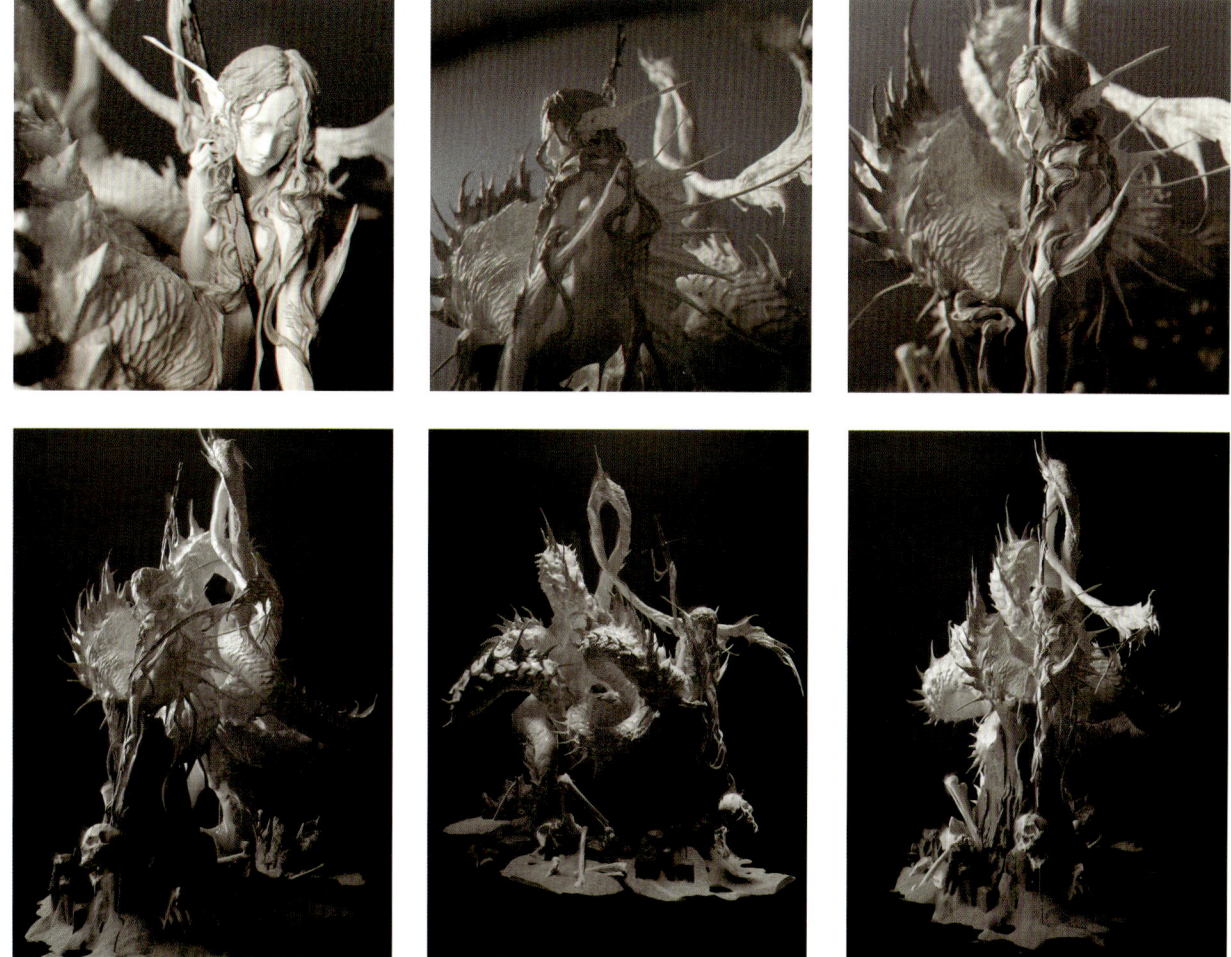

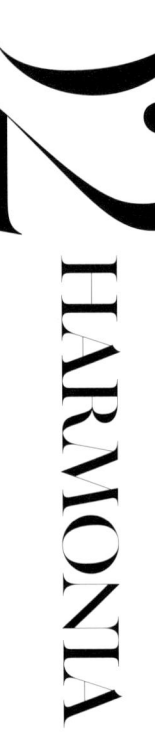

"We are not alone,"
 as long as love resides in our hearts. The lute can
transform wind and sand into wondrous melodies,
cherry blossoms can bloom upon the bowstrings, and
even the mountain tigers can be melted into lilies by
tenderness. The creatures of nature are gentle
companions as well as steadfast guardians. Only when
we are in HARMONIA with nature is our world
complete. In this chapter, I depict the girls and their
animal companions.

「私たちは独りではない」
心に愛がある限り。リュートは風と砂を不思議なメロディーに変え、桜は
弦の上に花開き、山の虎たちさえも優しく百合の花に溶けてゆきます。
自然界の生き物たちは、優しい仲間であると同時に揺るぎない守護者
でもあります。自然とのHARMONIA（ハルモニア、調和）の中にあってこ
そ、私たちの世界は完成します。この章では少女たち、そしてその動物
の仲間たちを描きます。

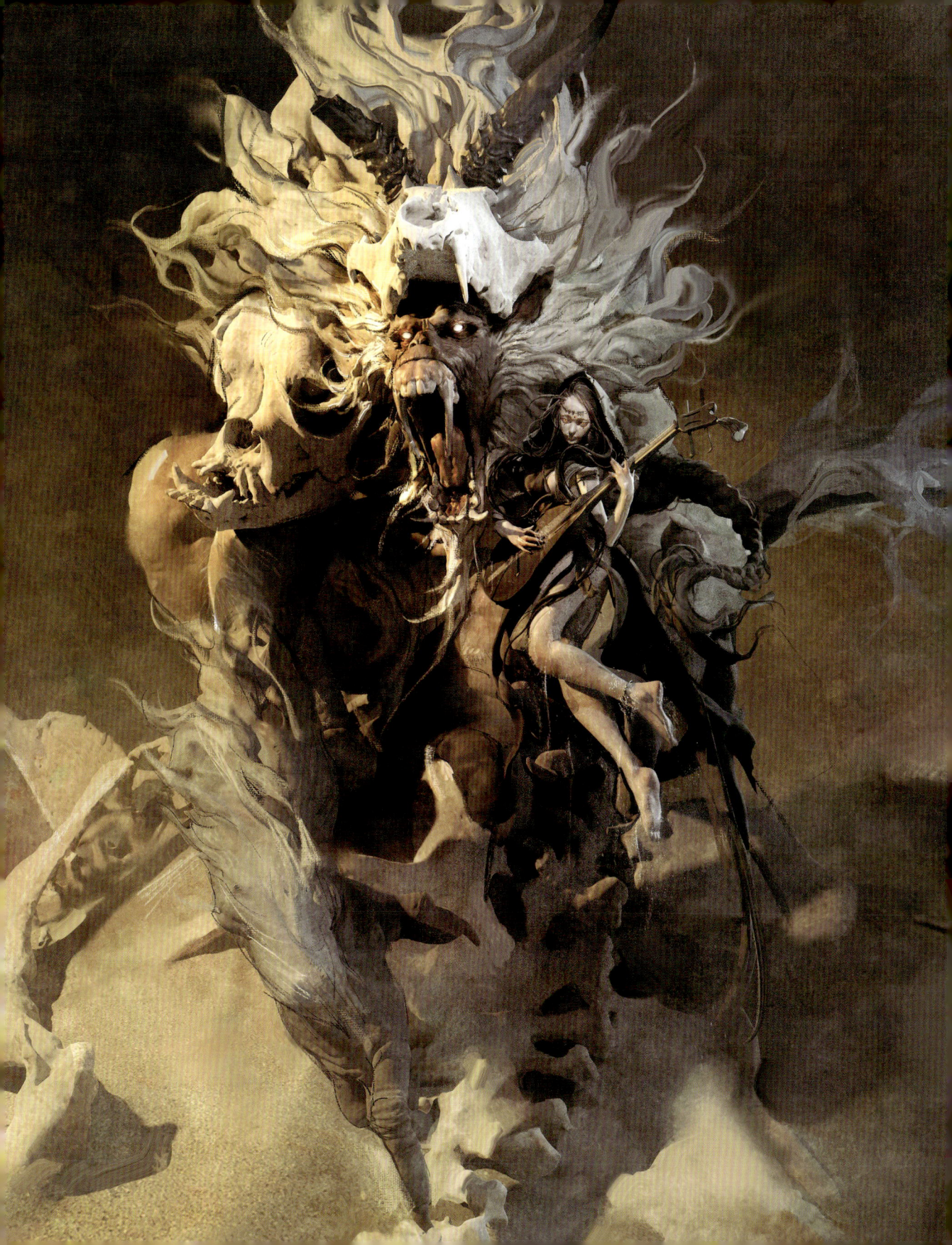

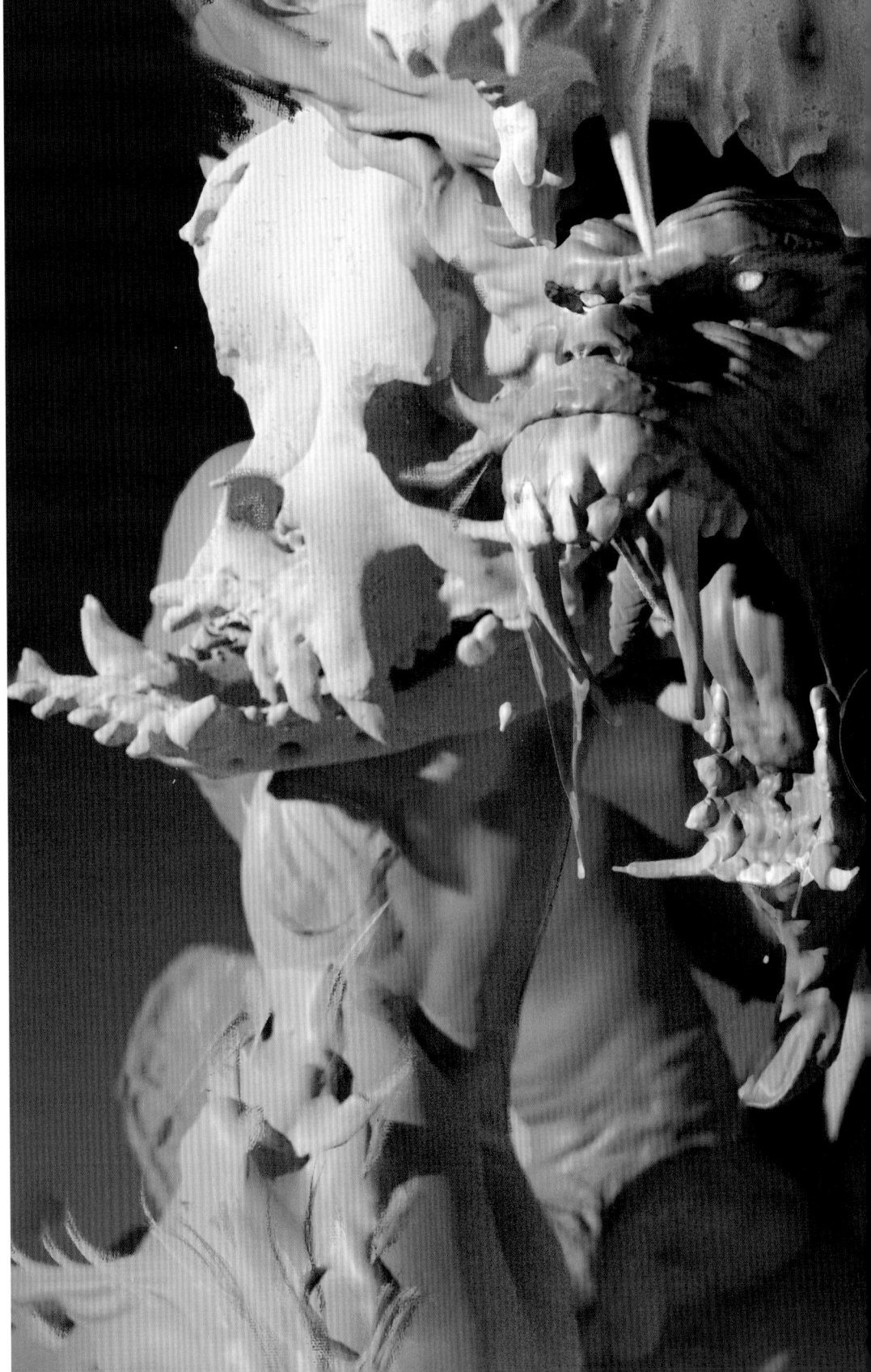

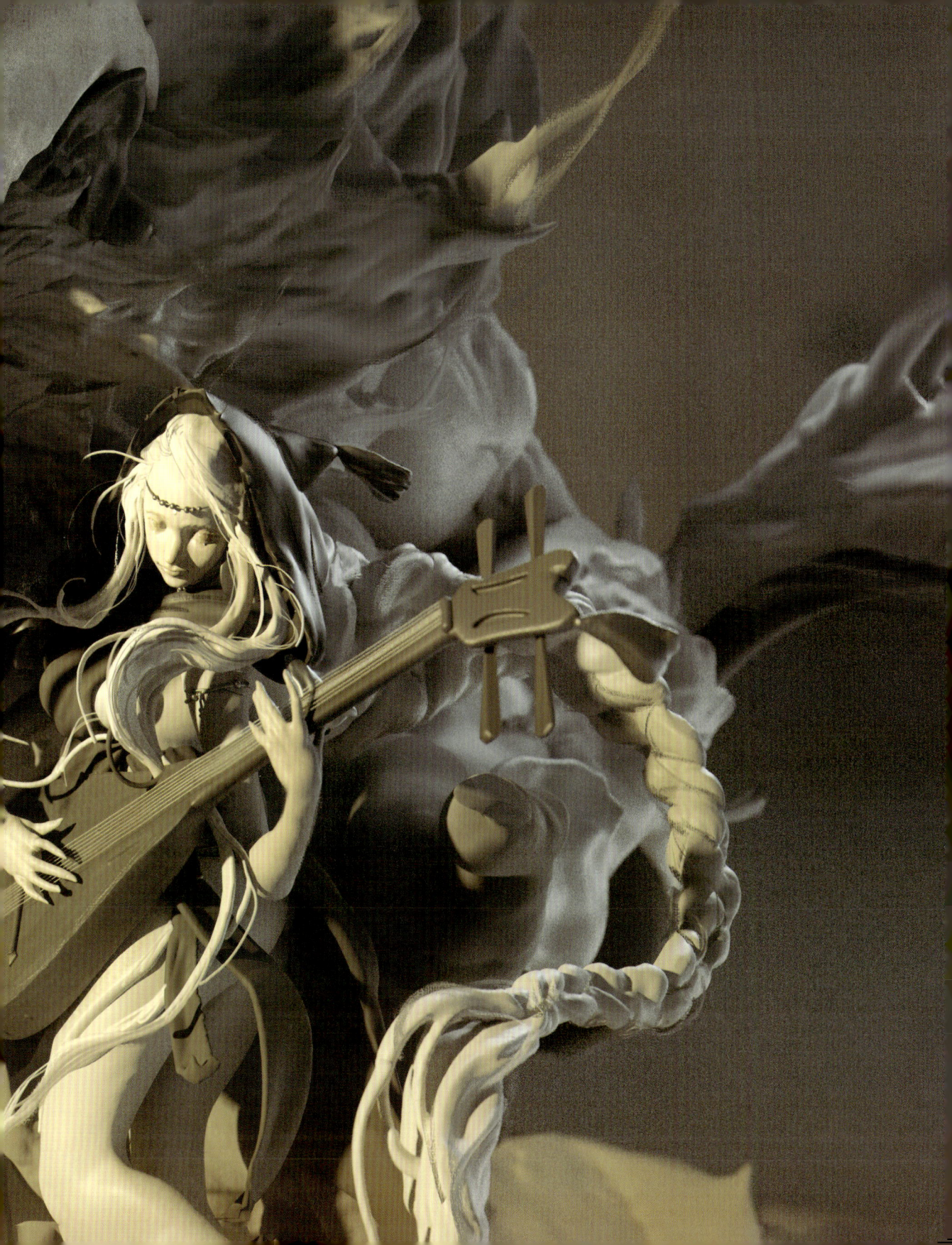

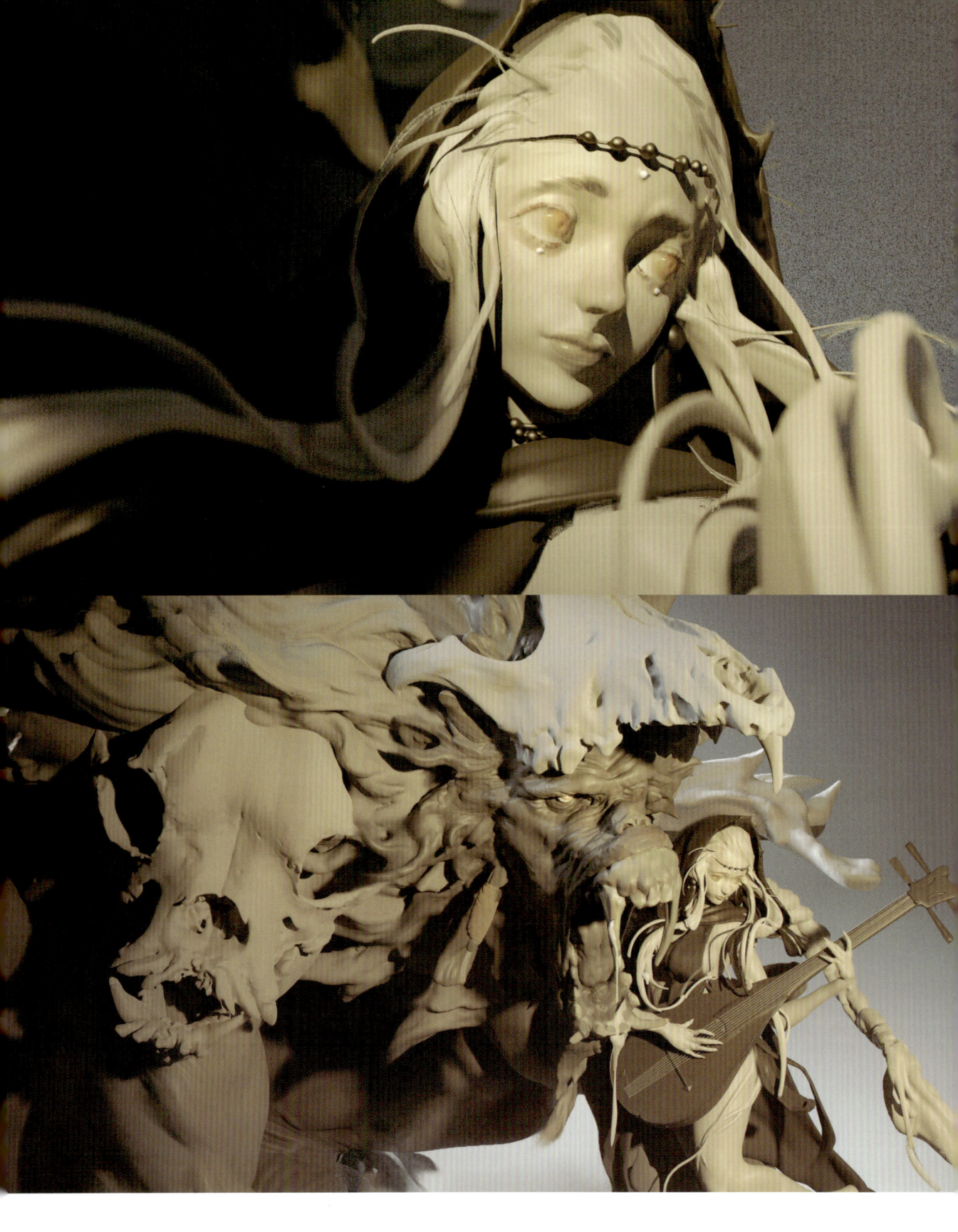

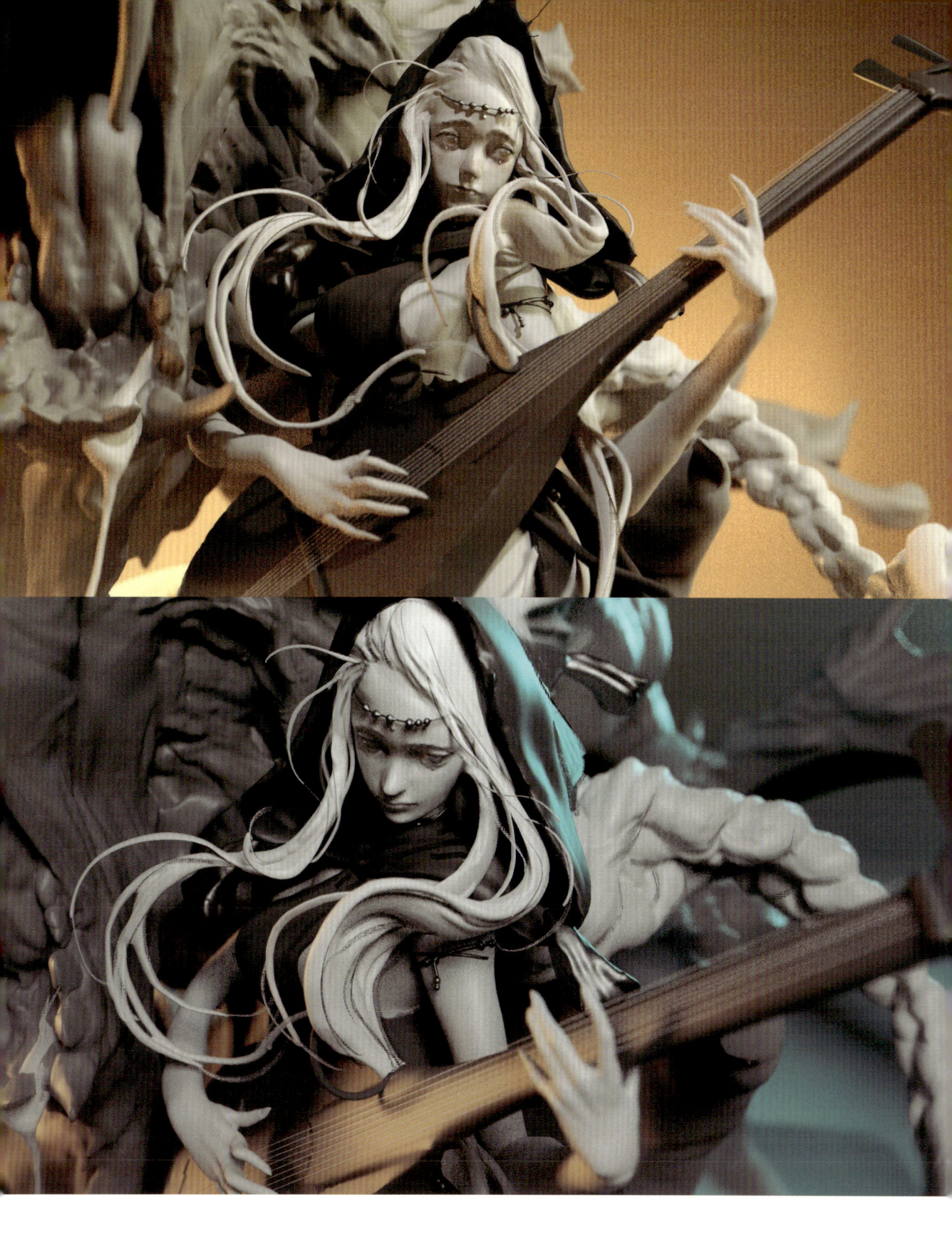

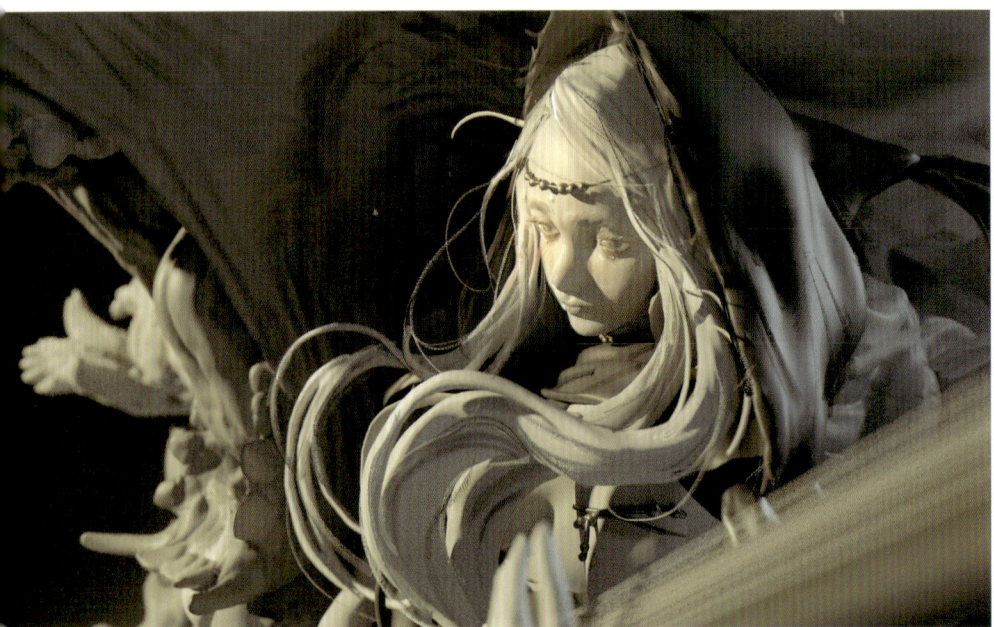

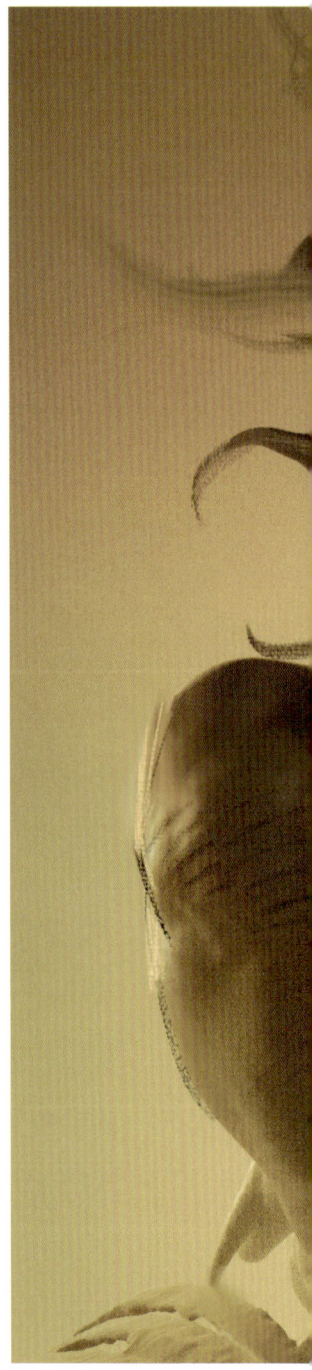

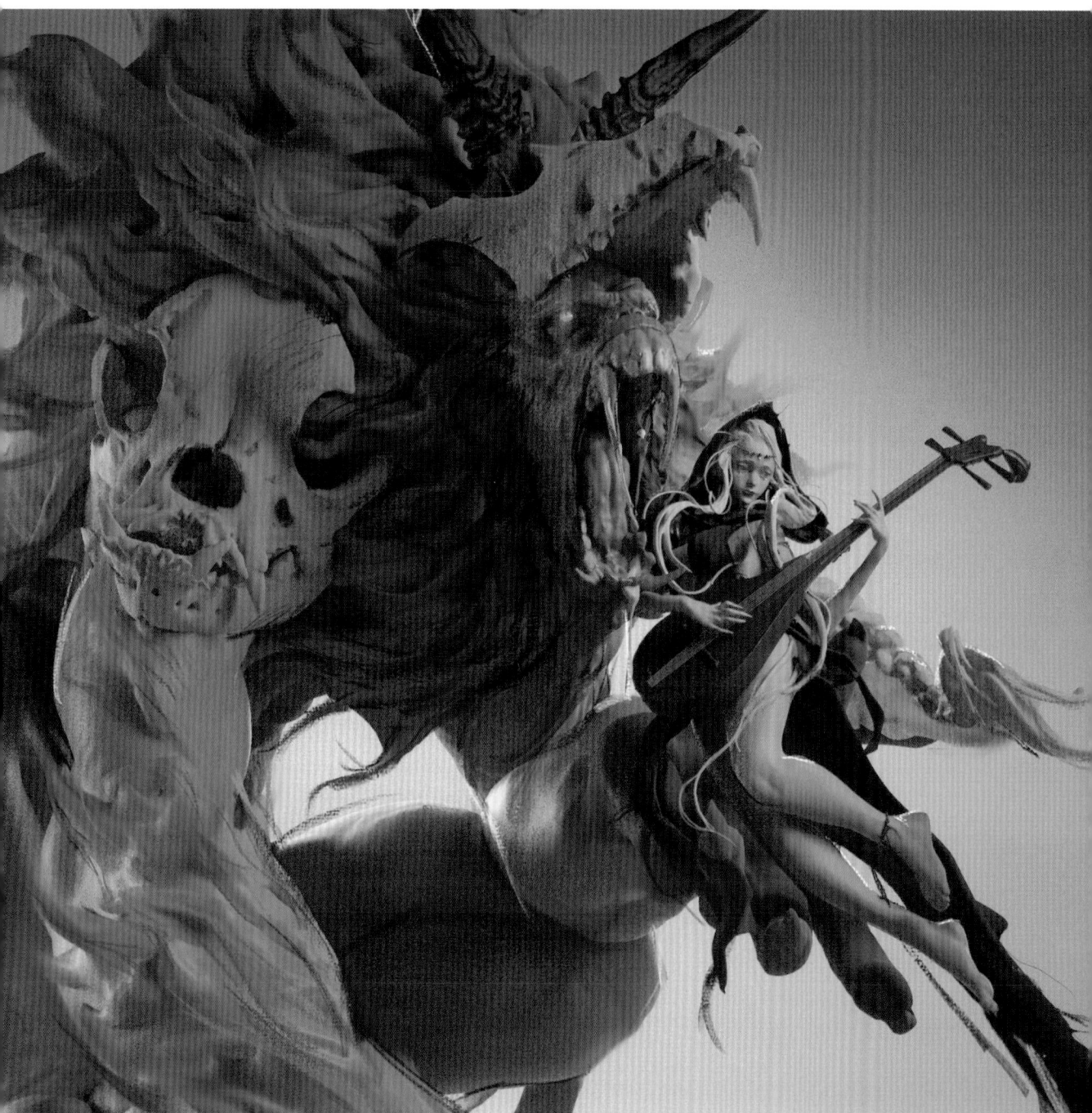

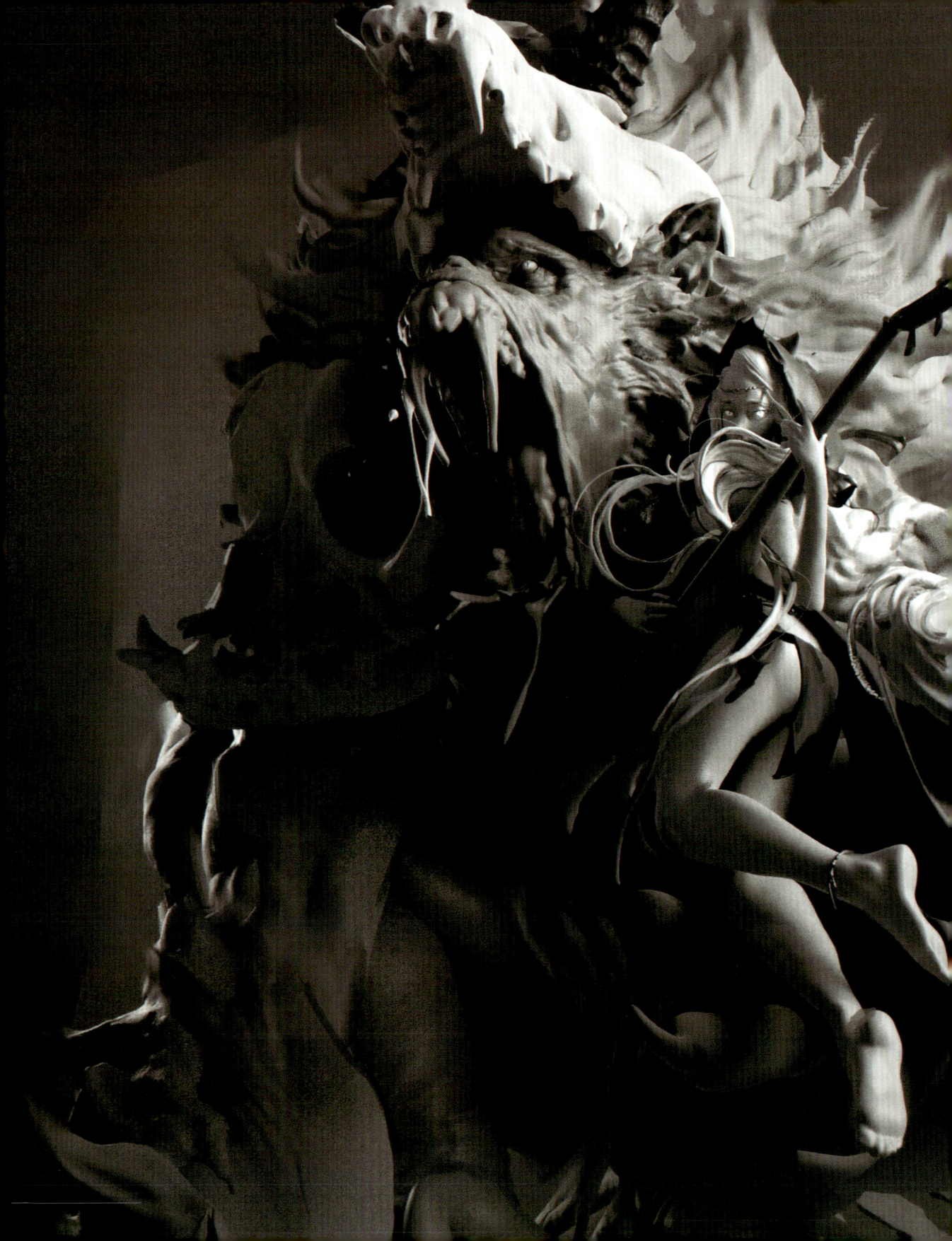

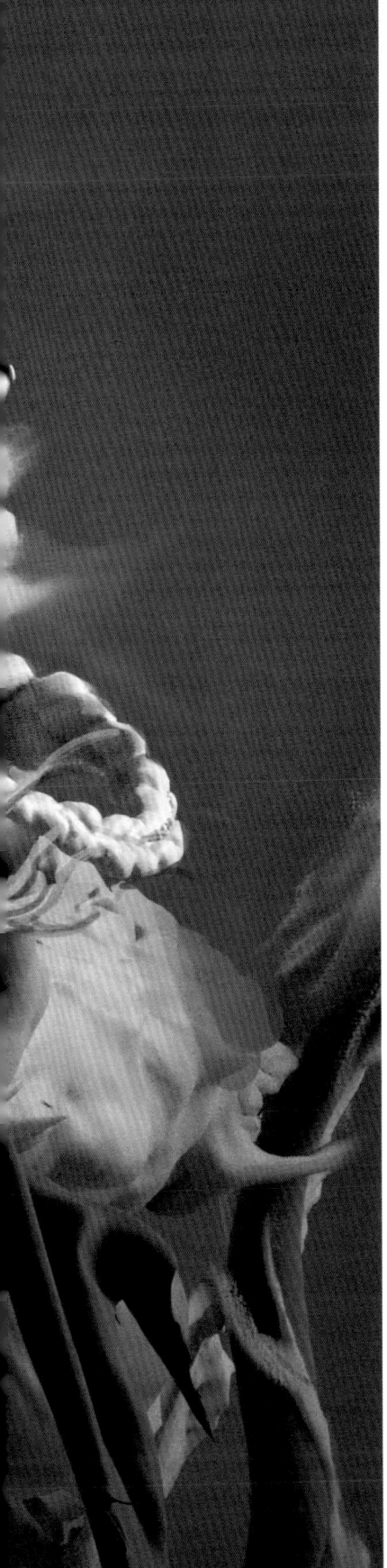

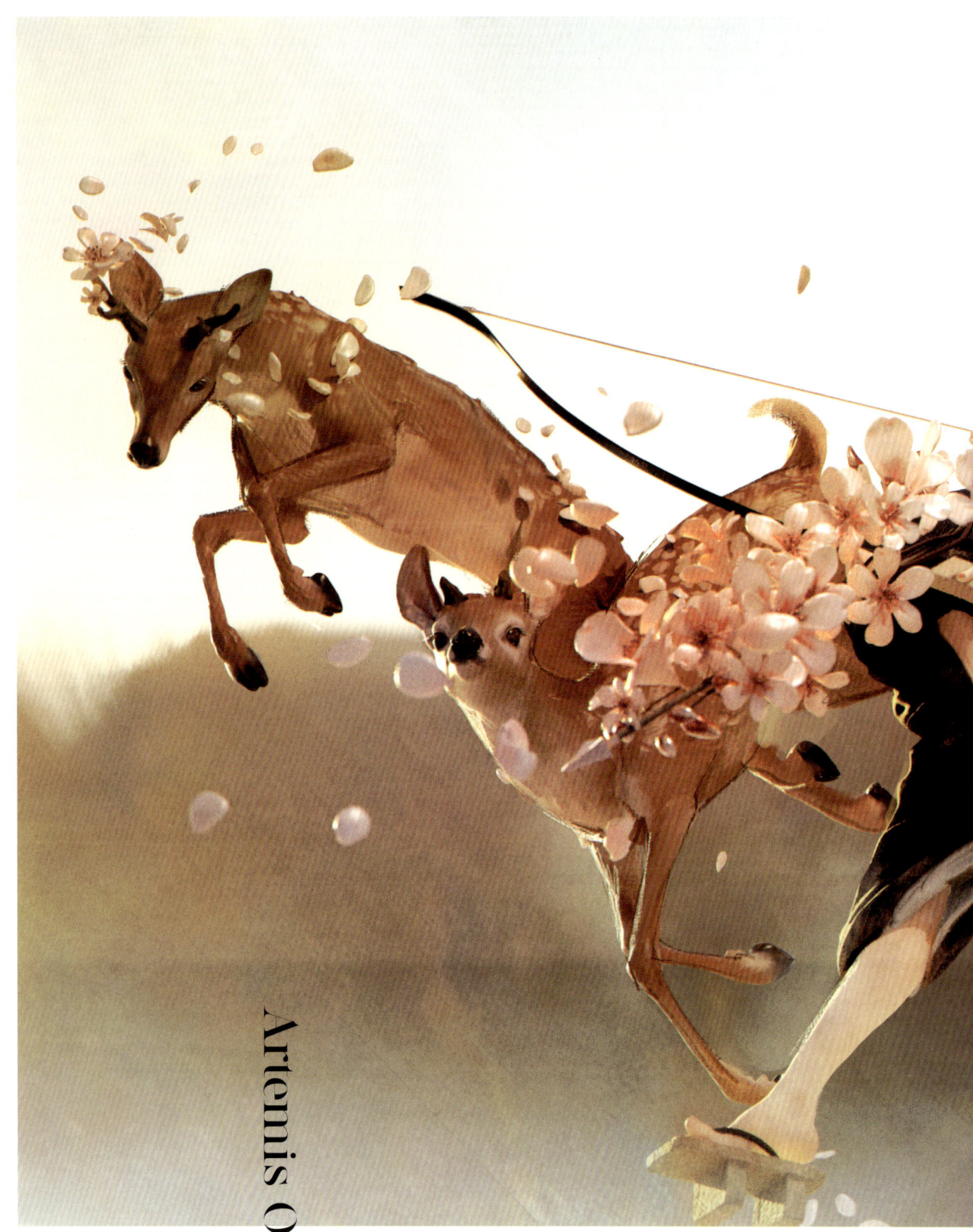

Artemis 040

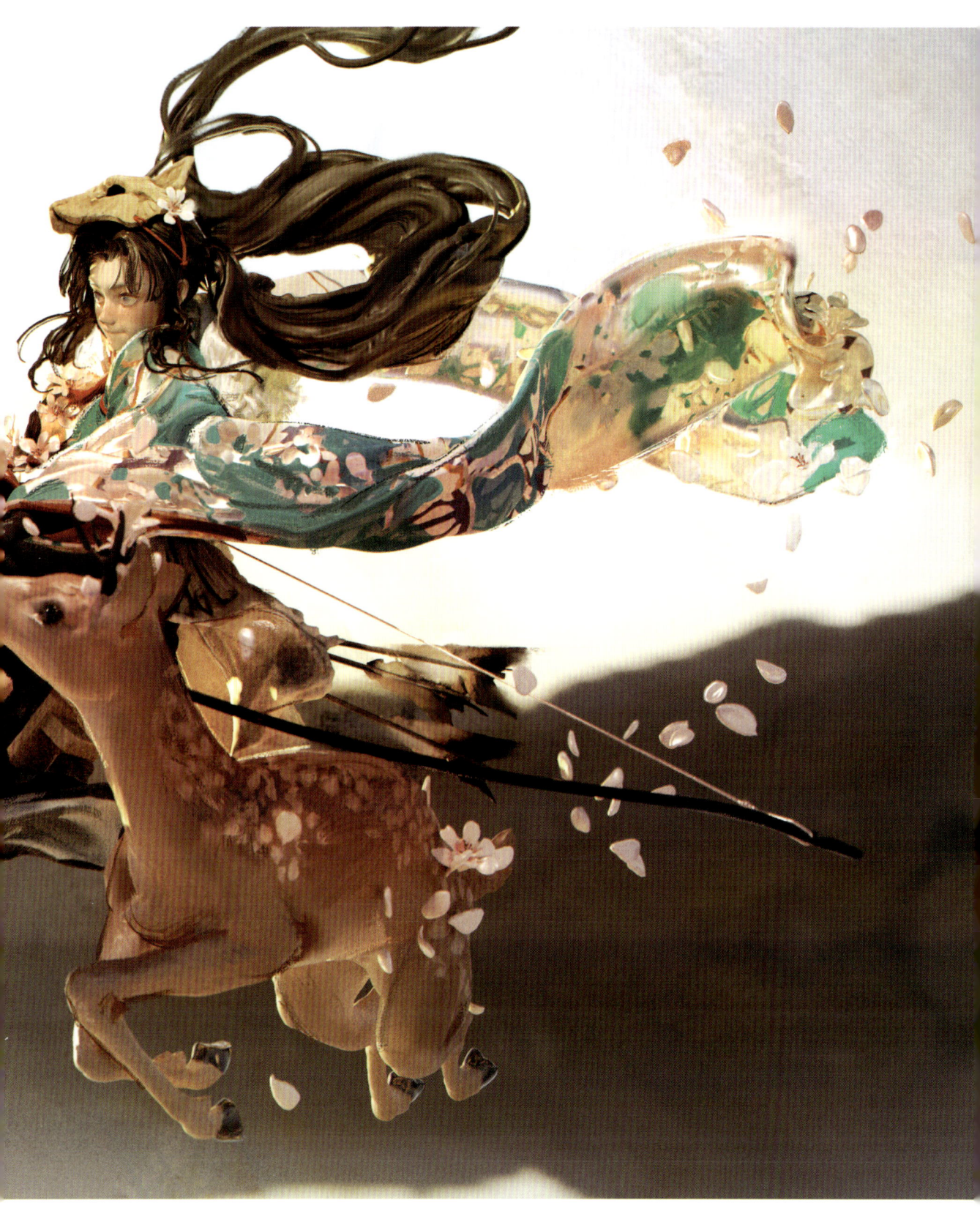

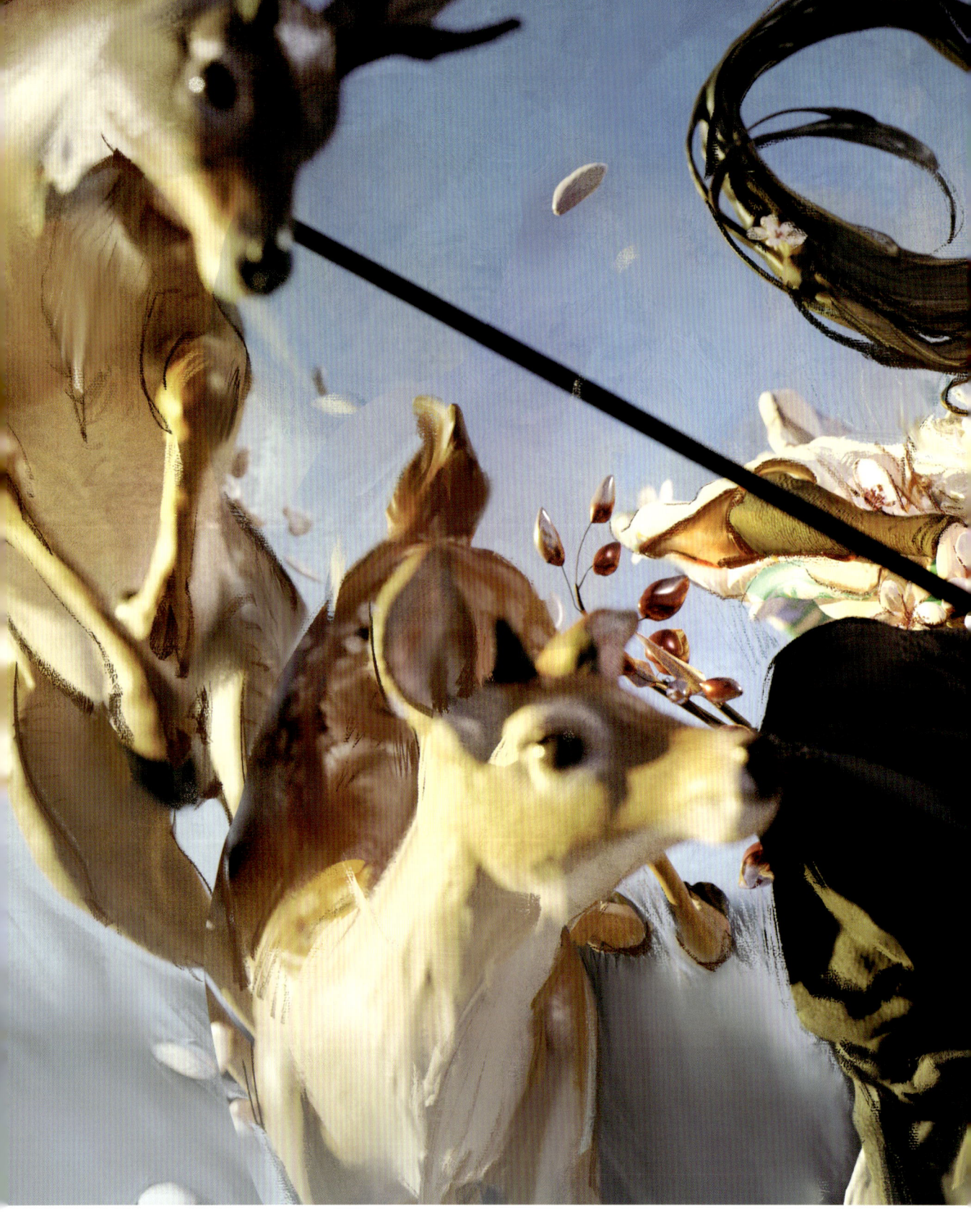

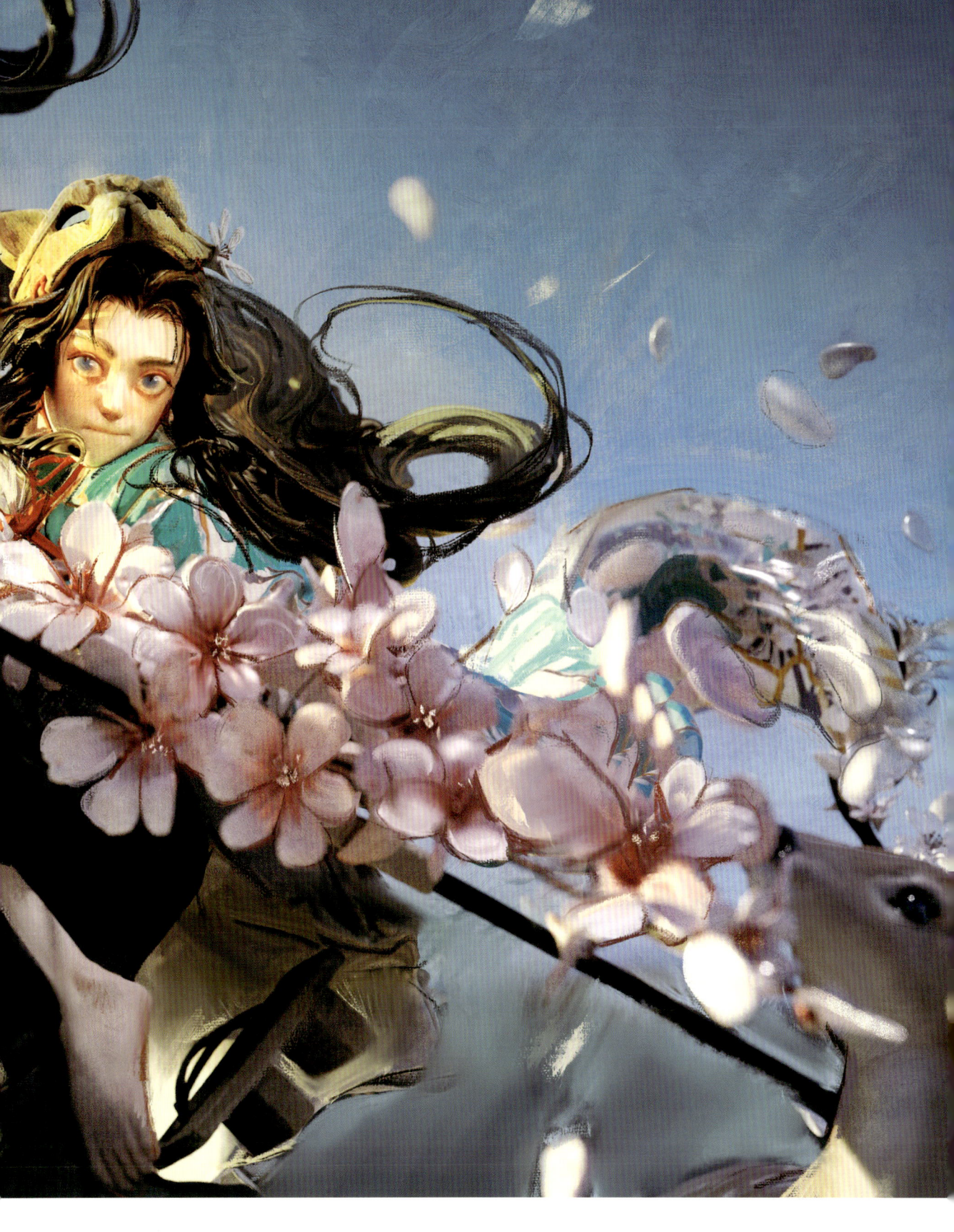

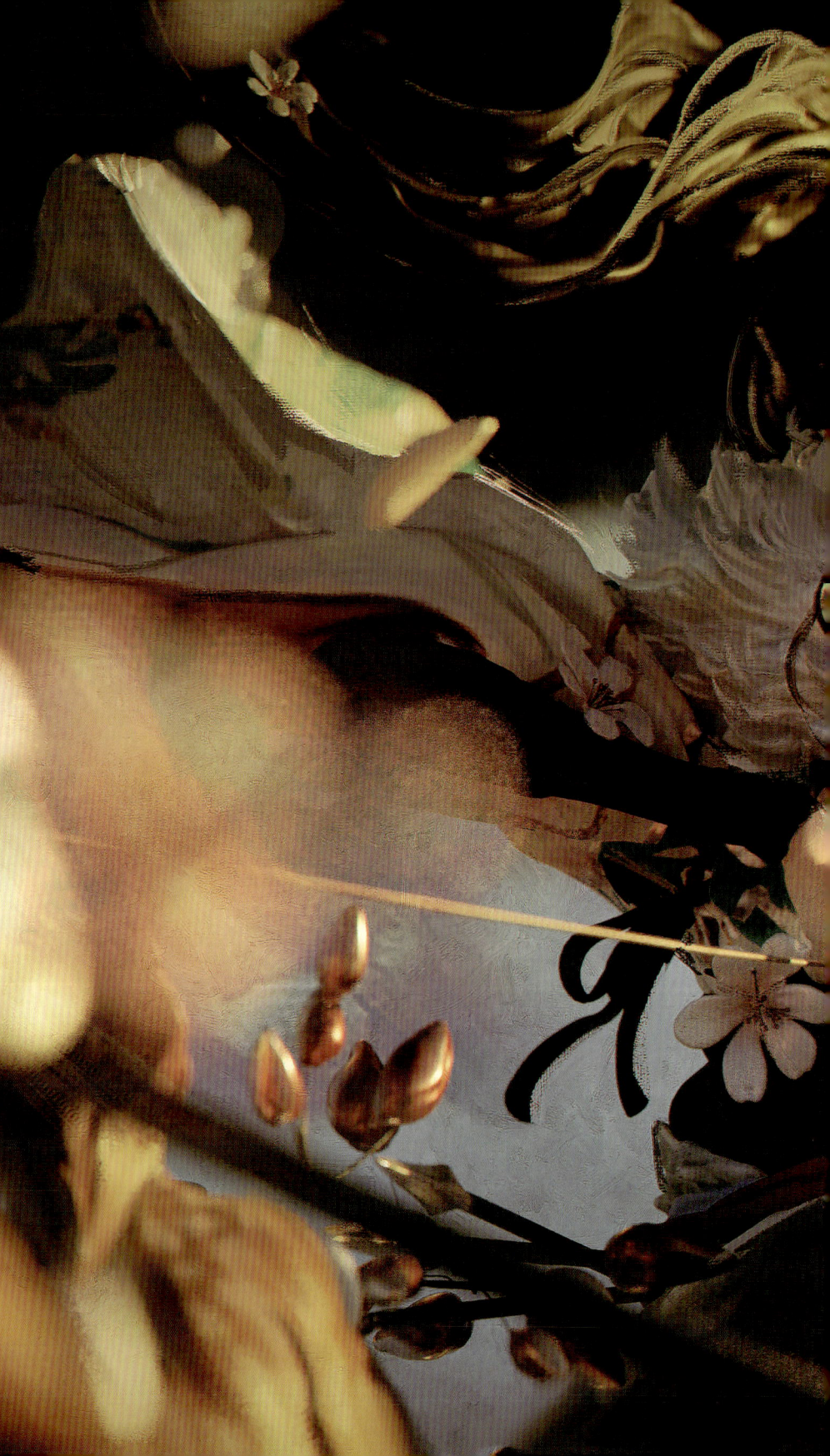

04

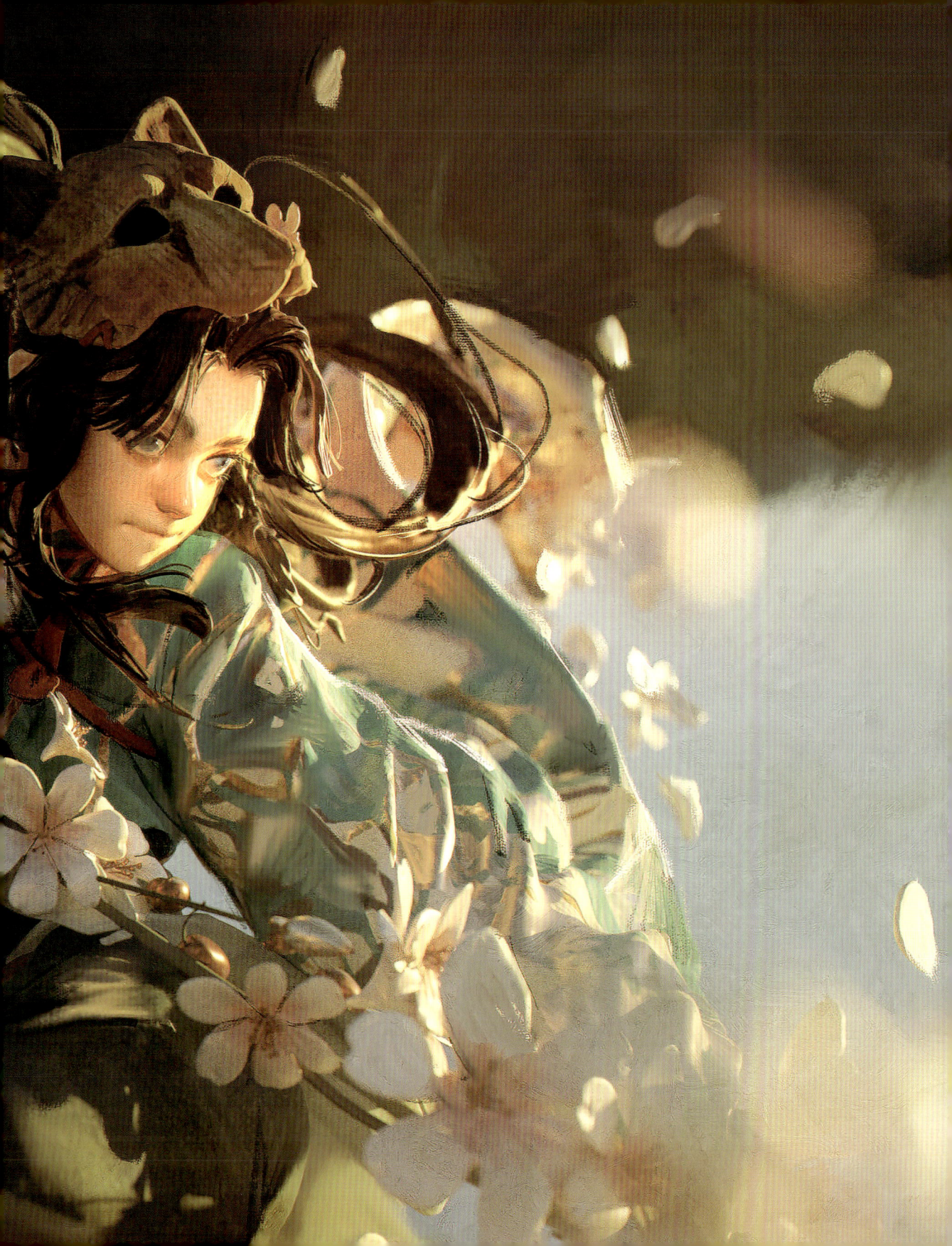

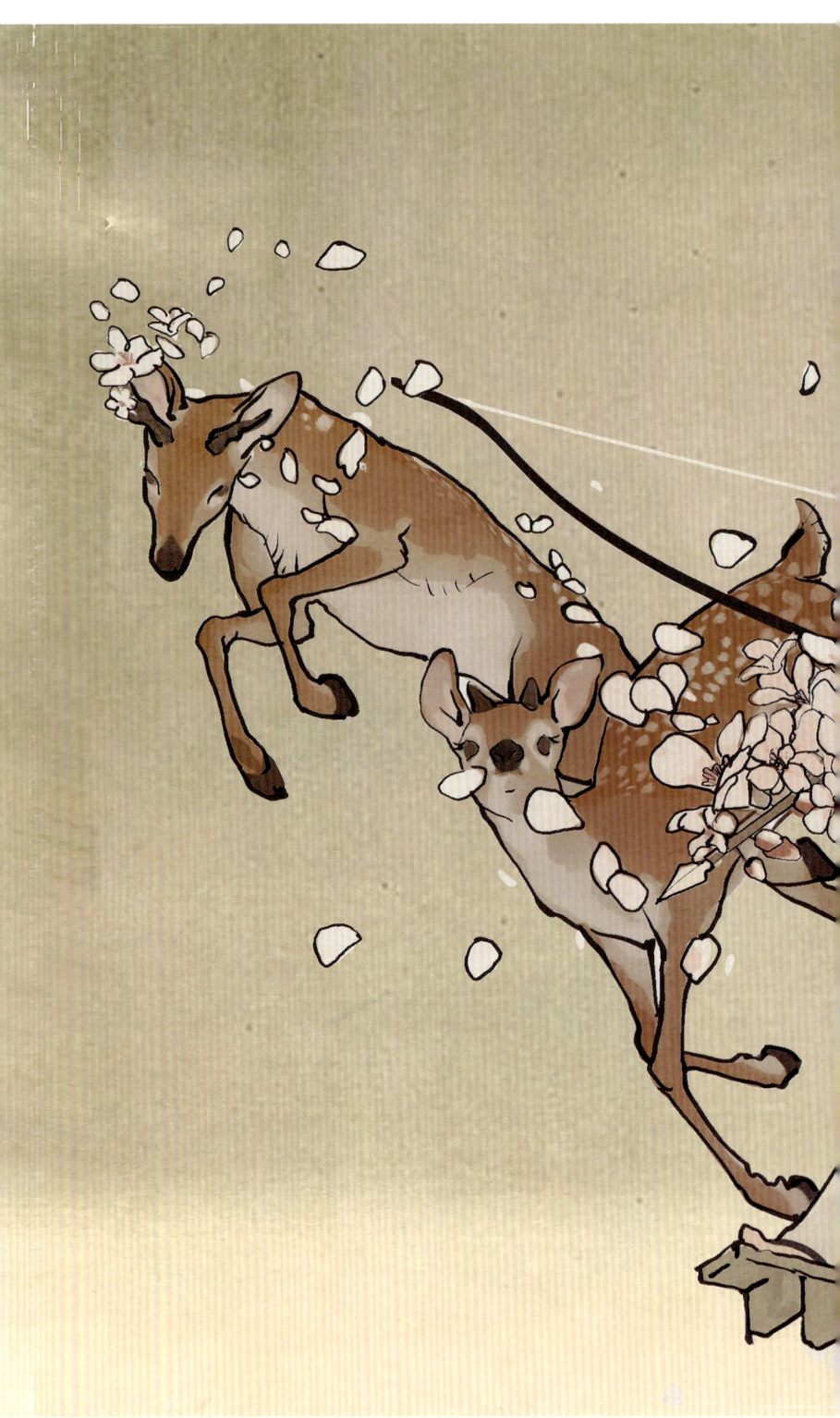

046

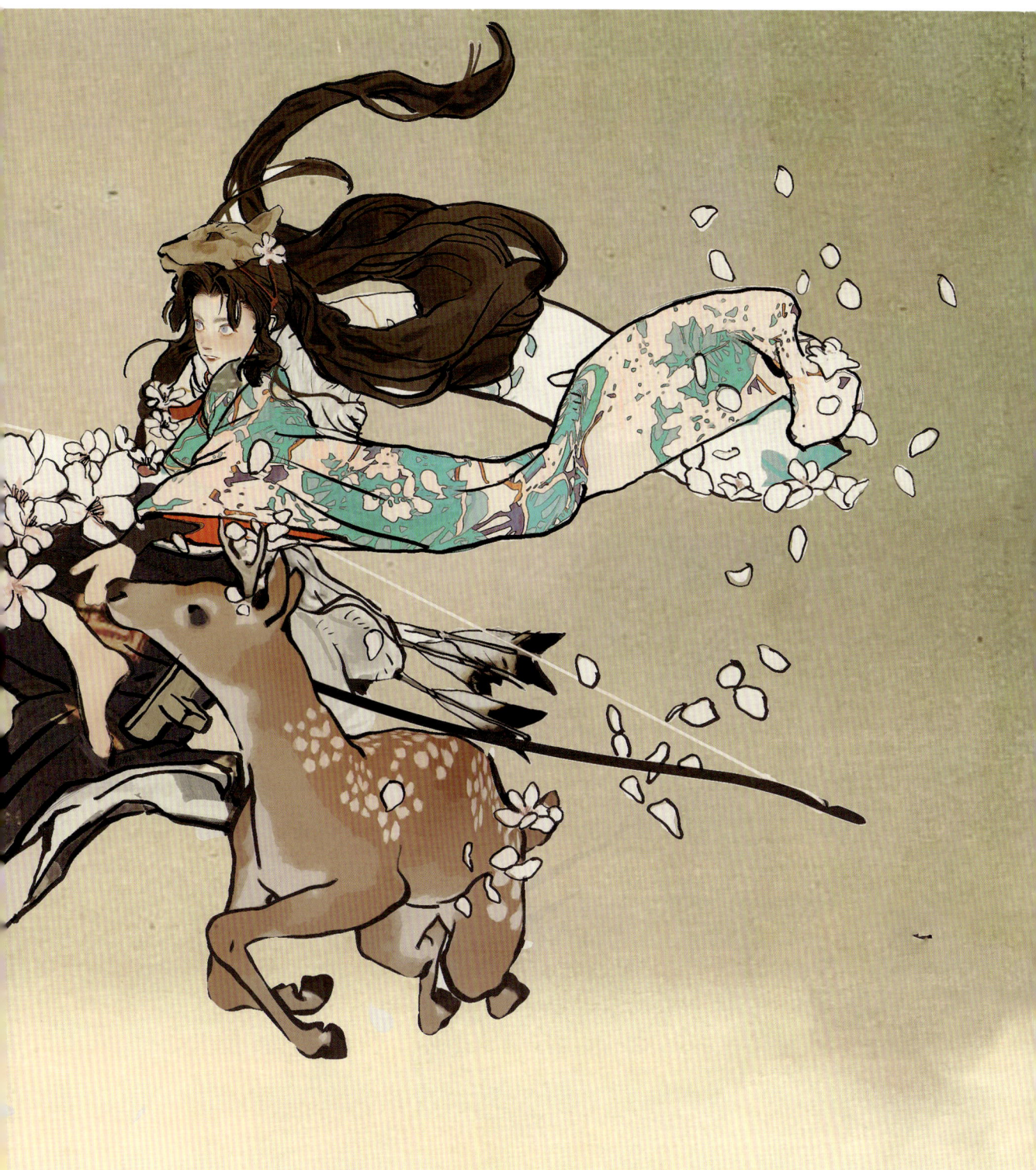

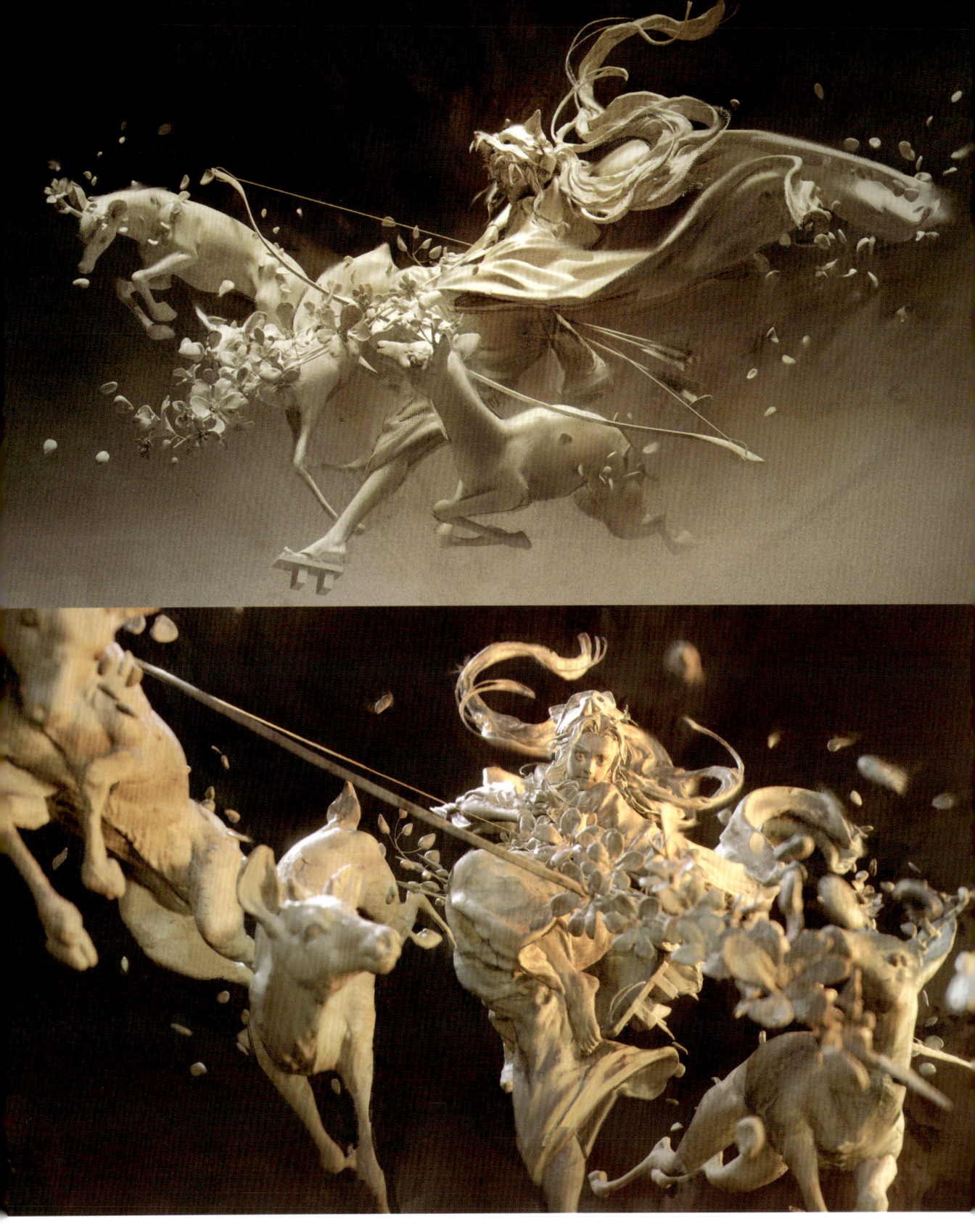

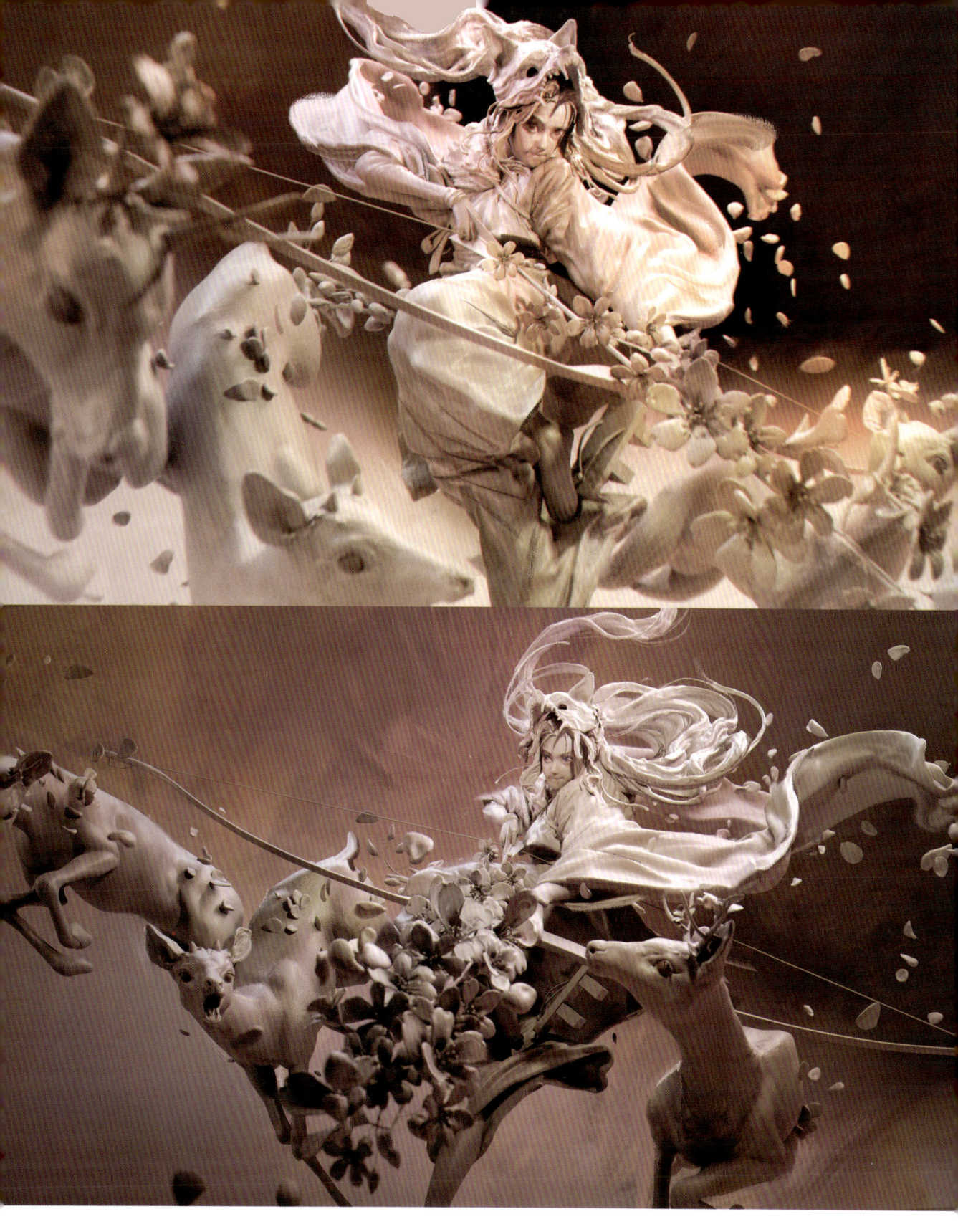

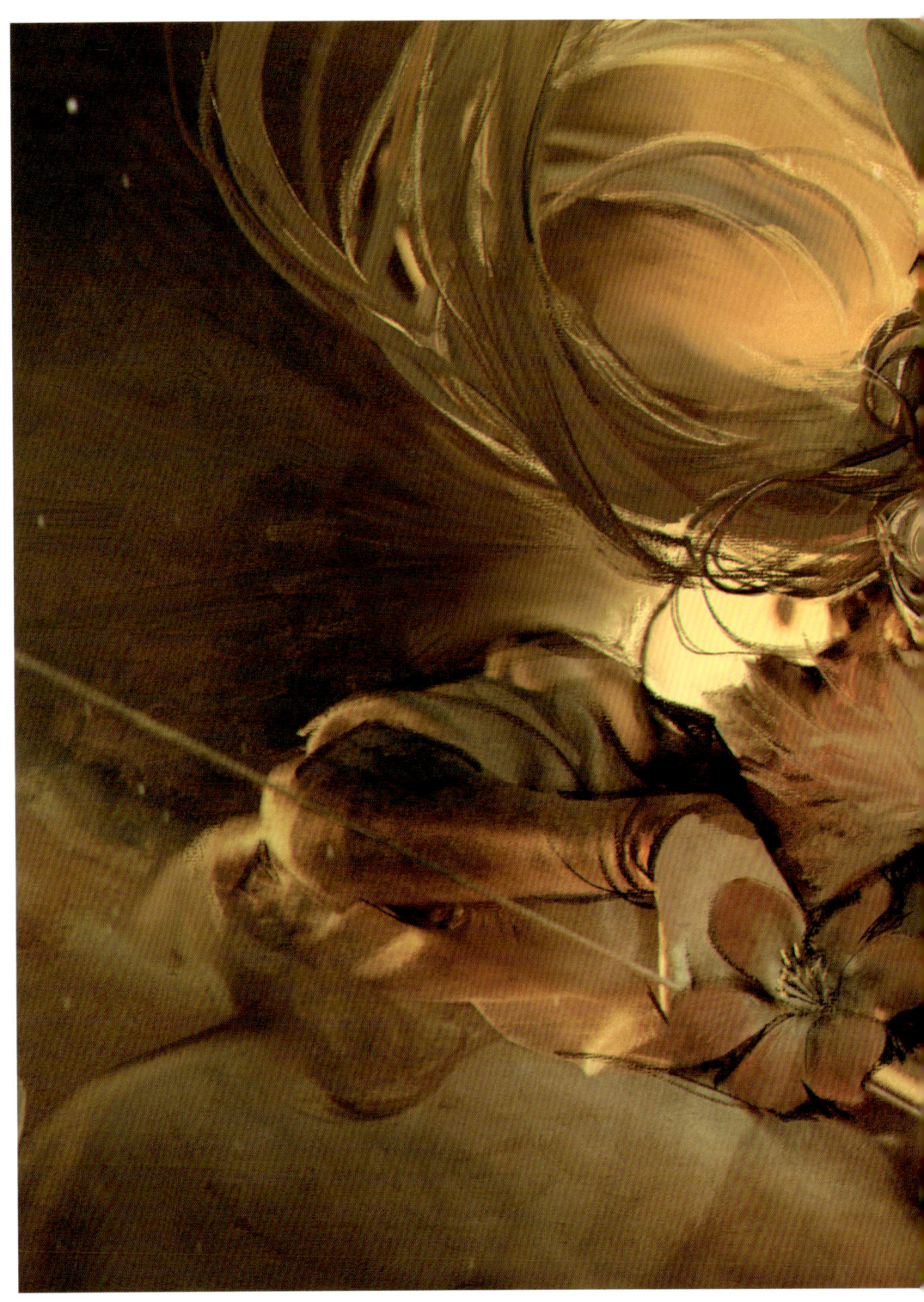

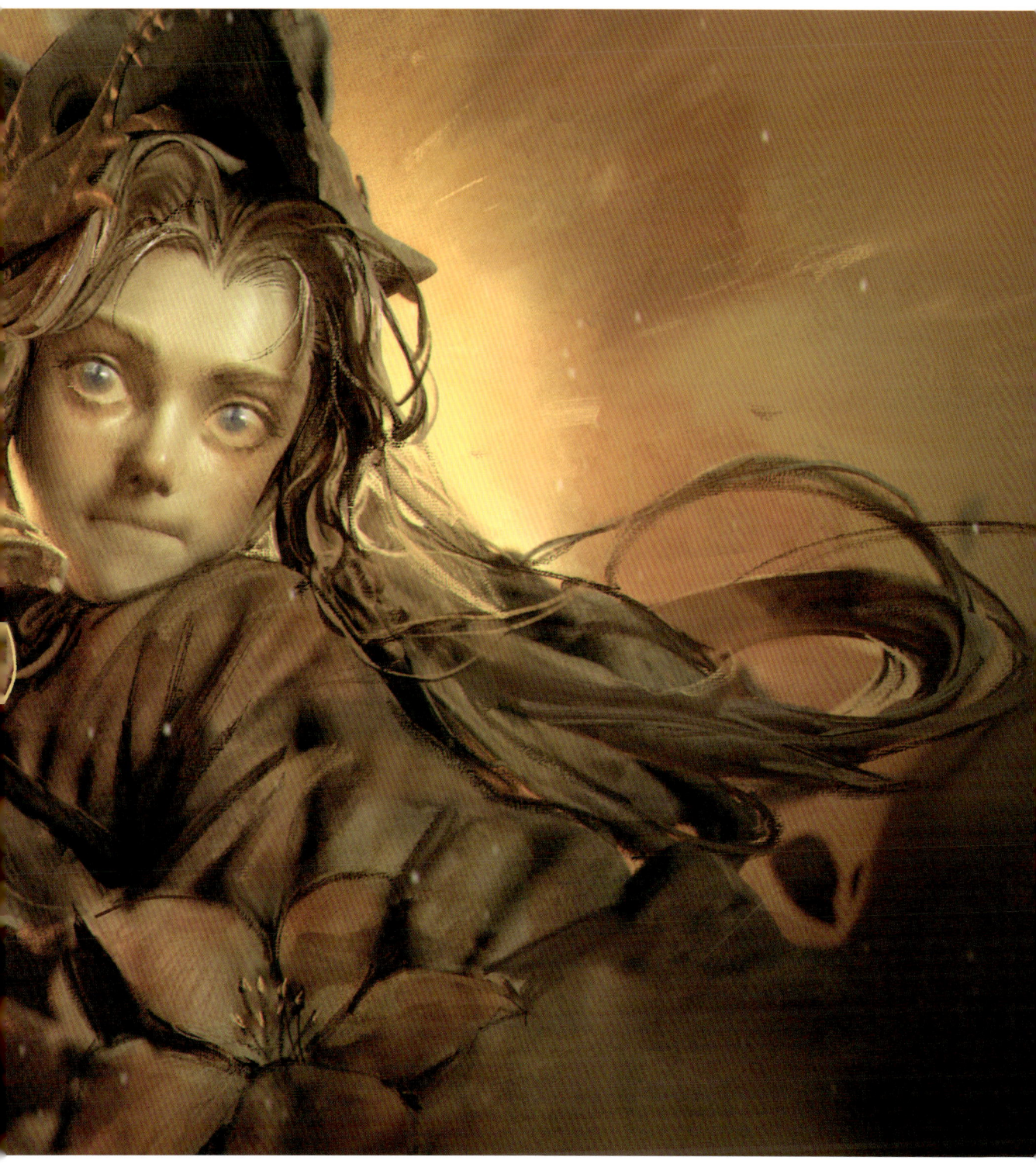

# Tiger and Flower 052

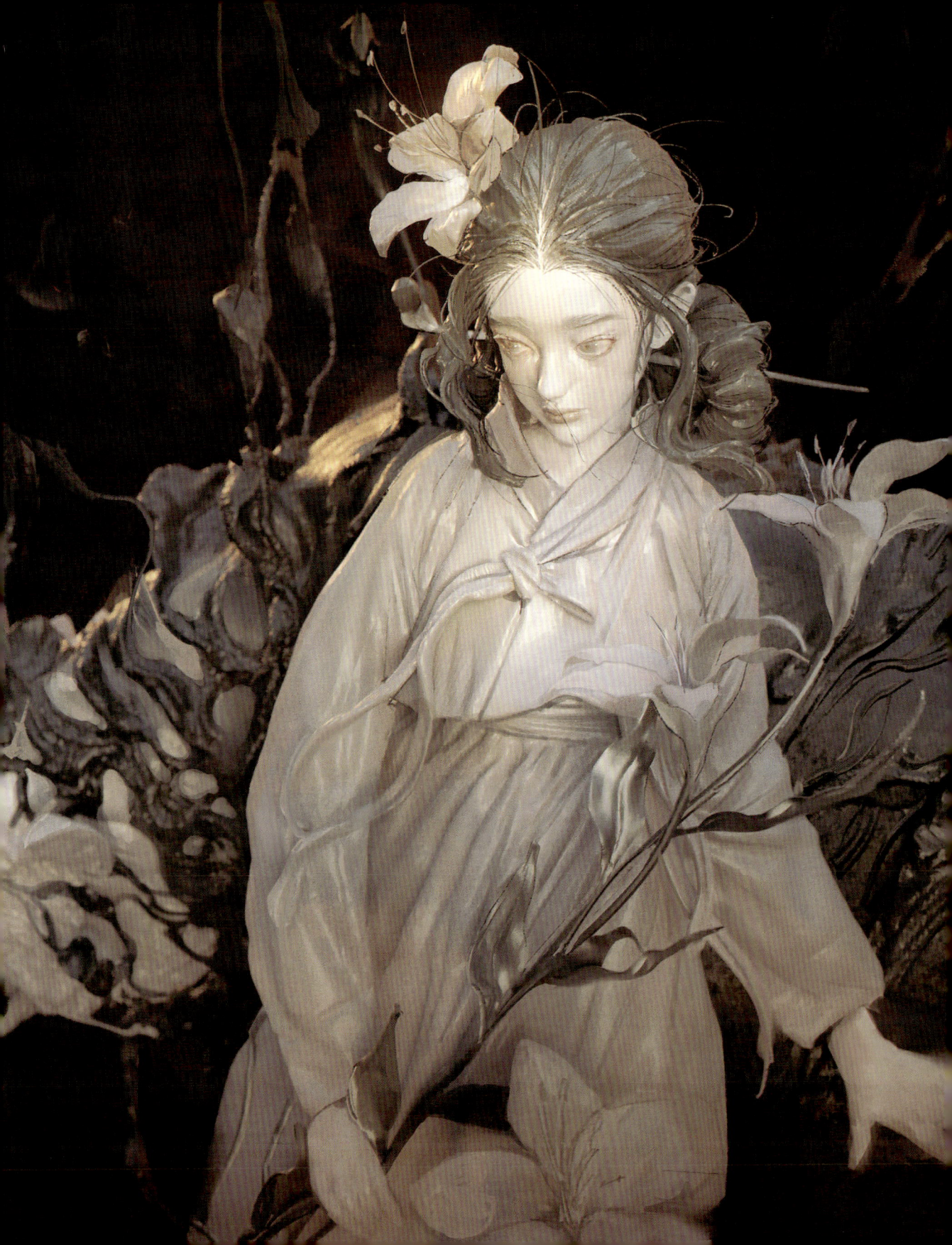

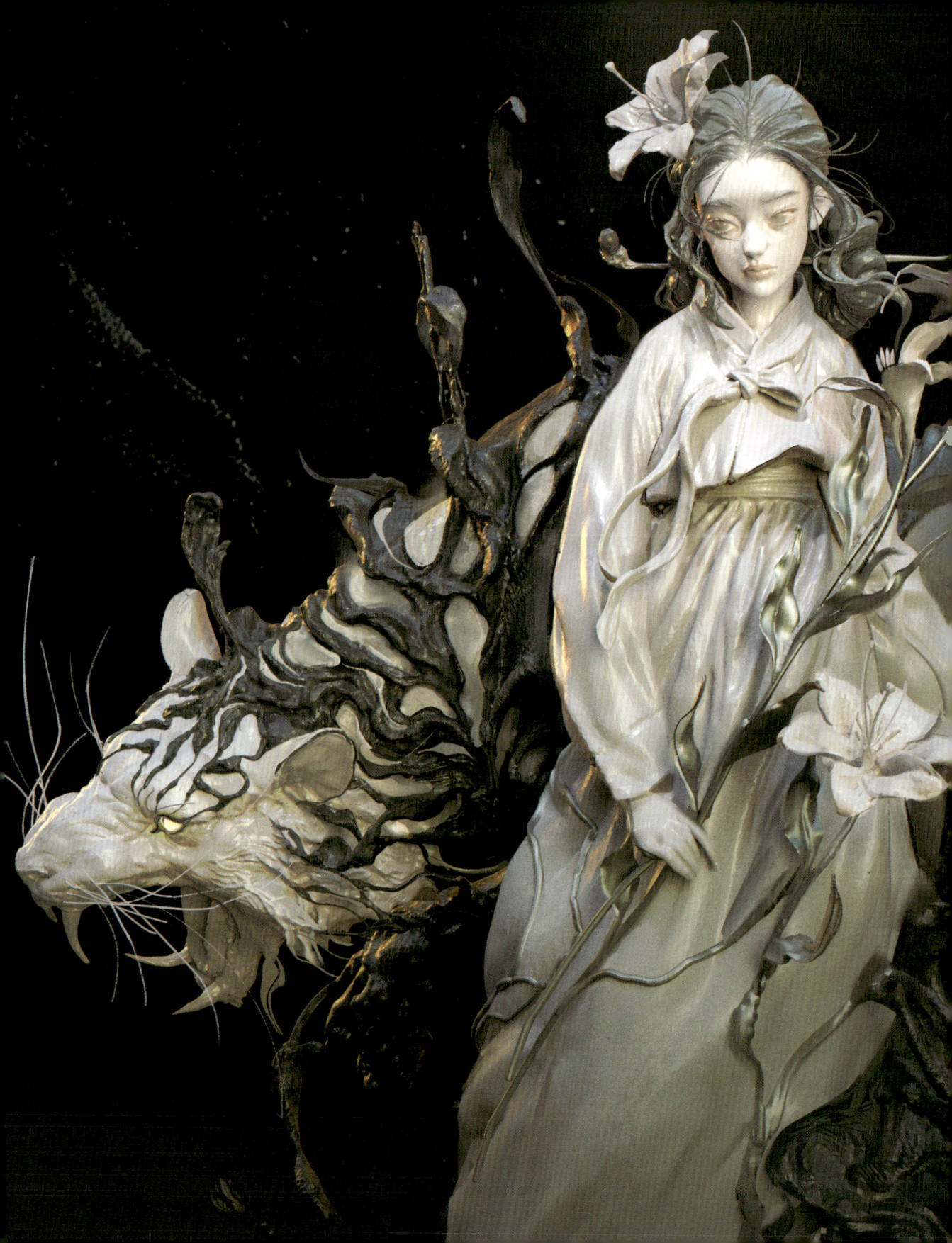

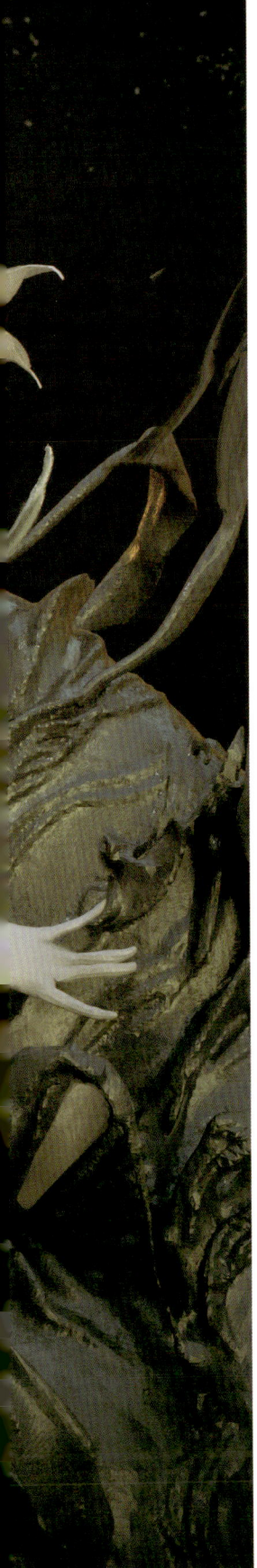

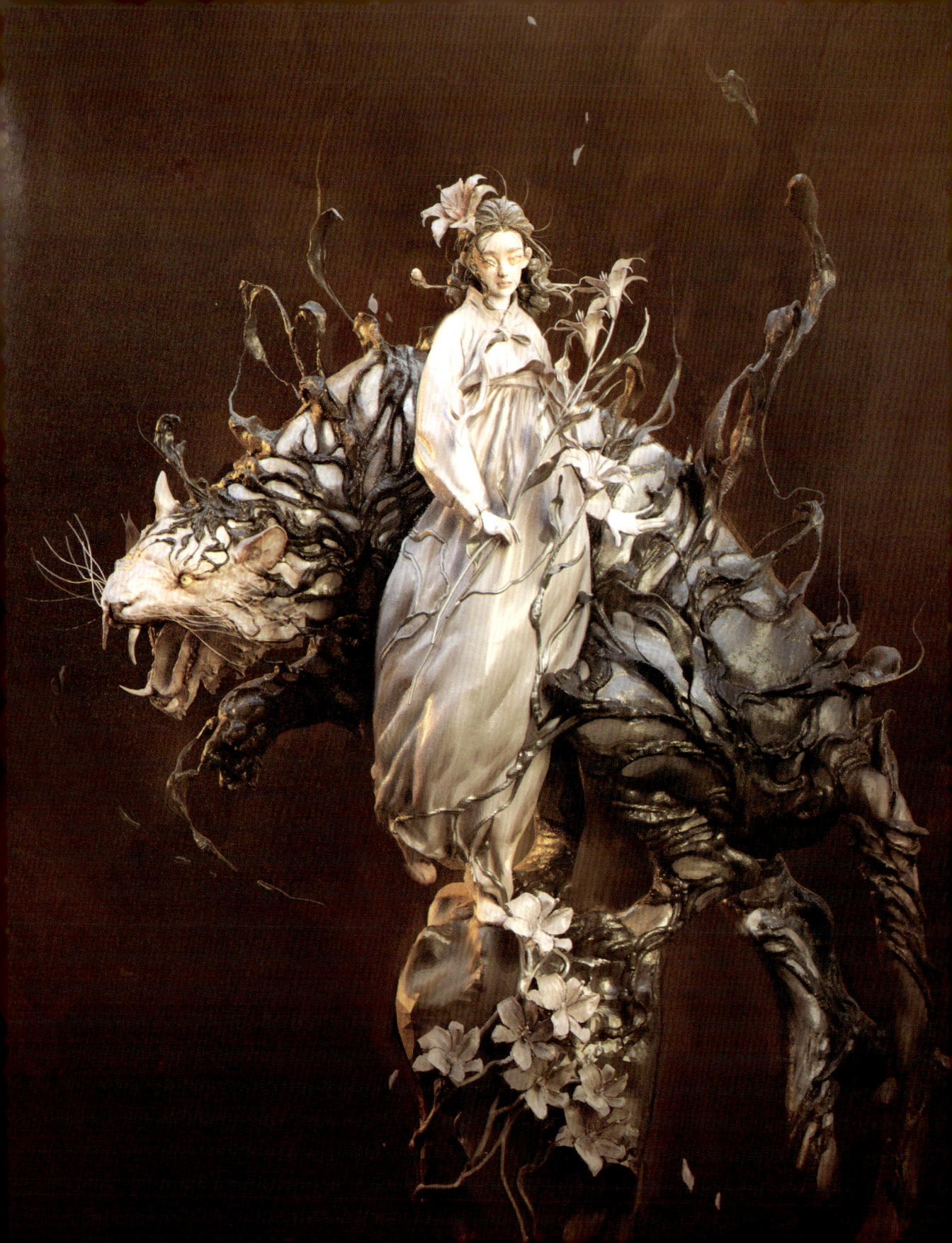

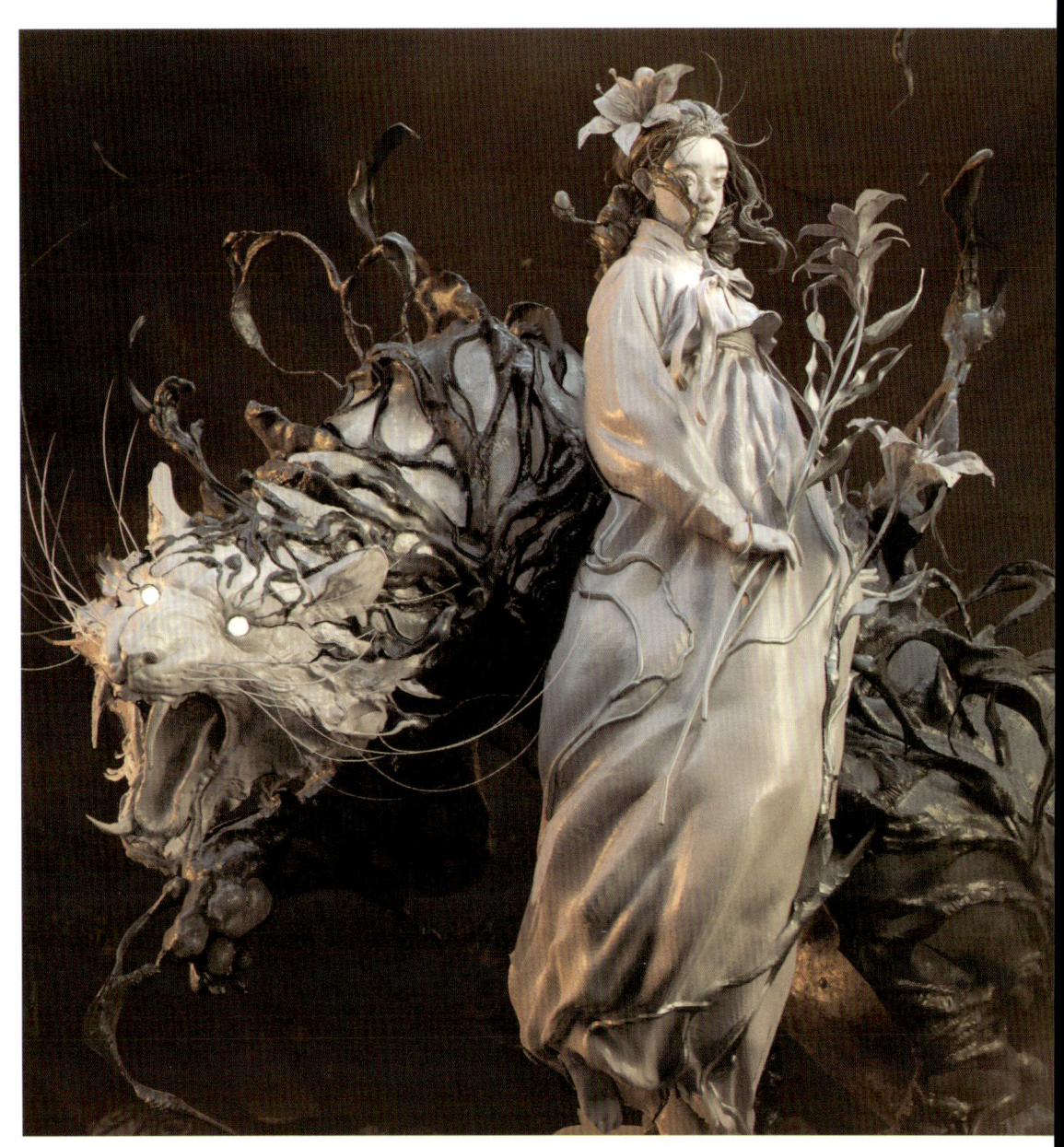

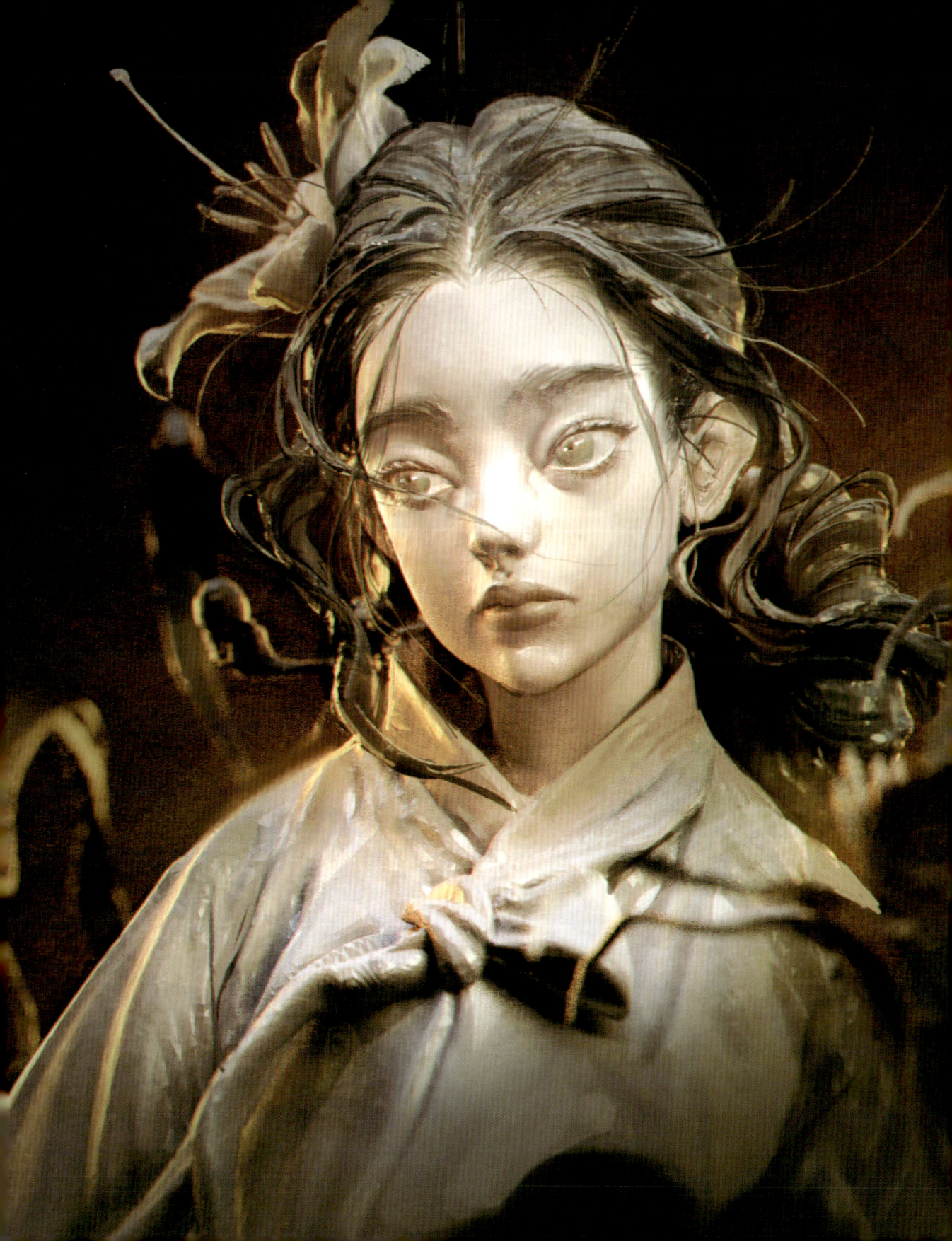

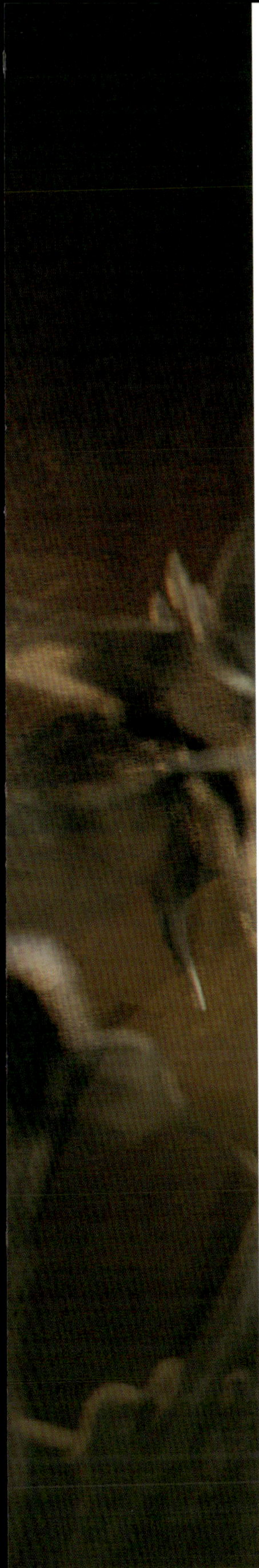

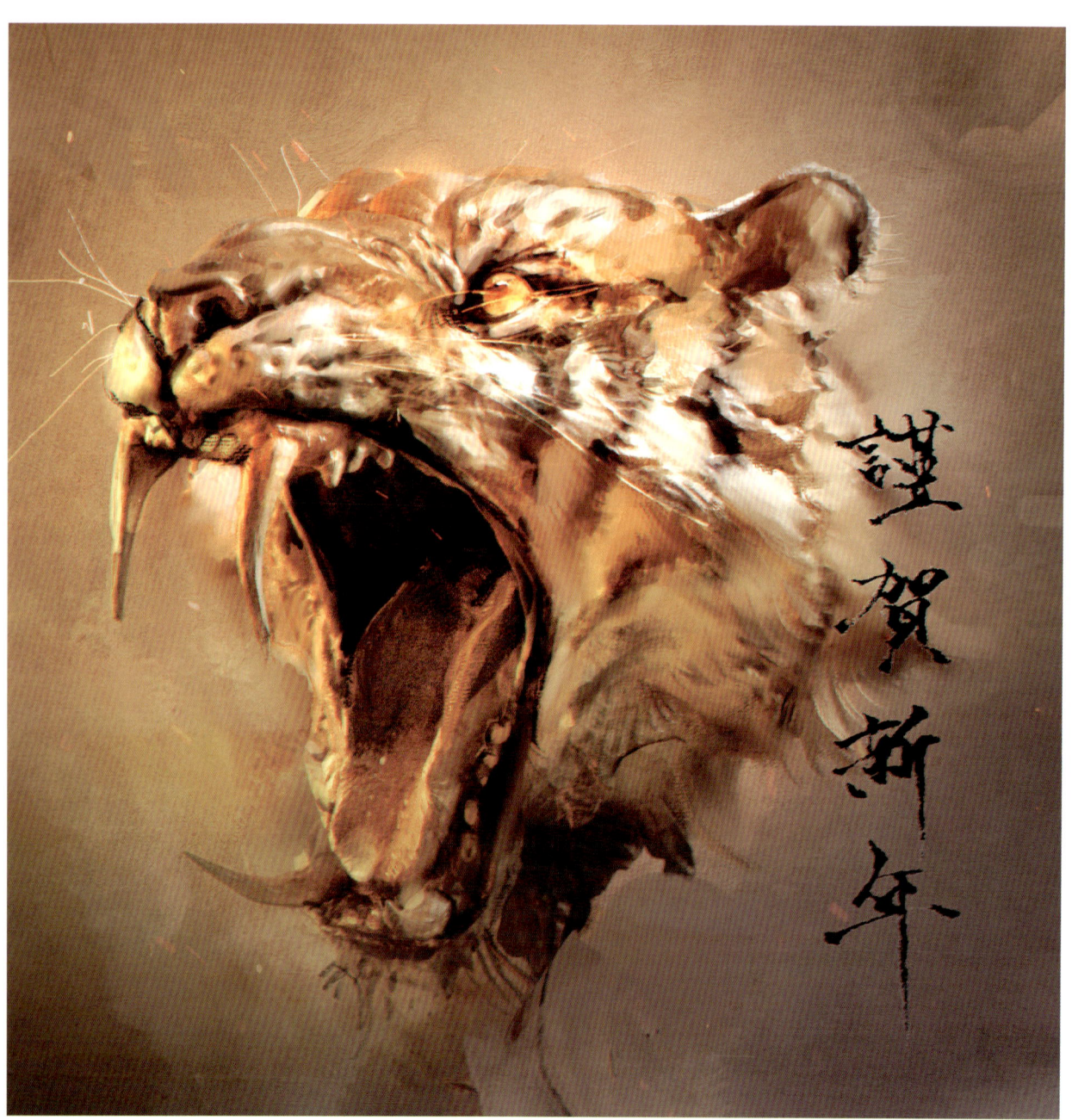

謹賀新年

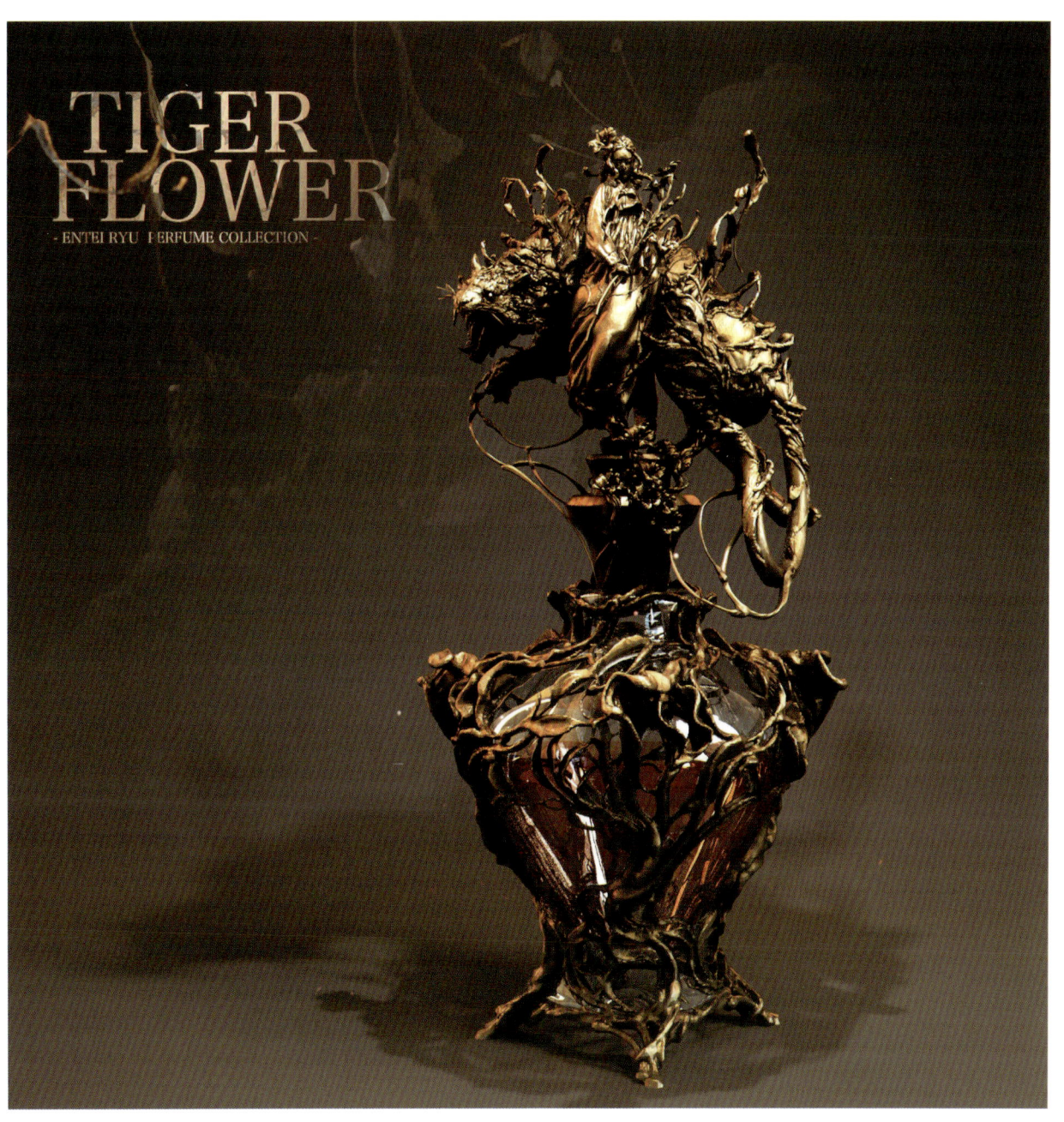

TIGER
FLOWER

- ENTEI RYU PERFUME COLLECTION -

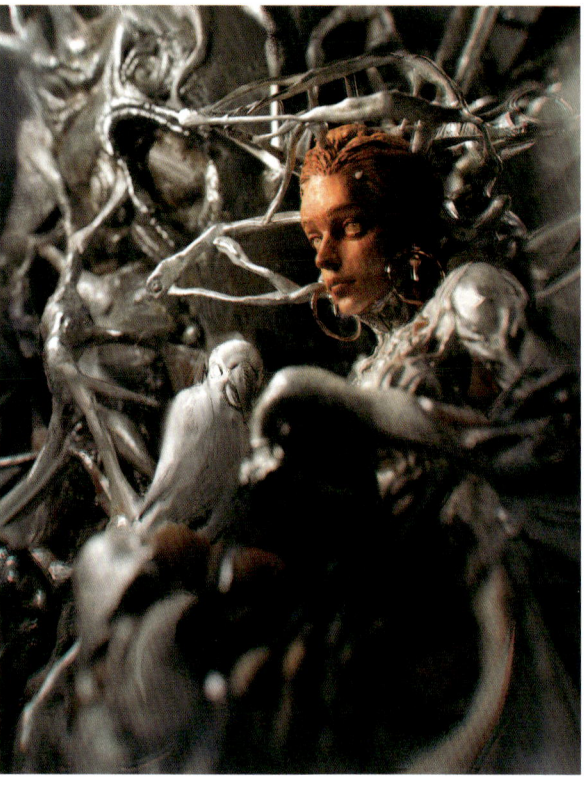

Mercury, the youngest sibling in our solar system, is a shape-shifting liquid metal with an atomic weight of 200.592 u, and represents the god of wisdom and communication in "The Iliad."

This chapter features some of the science fiction-themed works I have created over the past few years. Notably, it includes the artificiall woman who sacrificed herself for love, THE BRIDE; the broken yet resilient lovers supporting each other in a wasteland, THE LOVERS; and the white-haired youth and his loyal red military dog, THE FRIENDS. These three works belong to the same series, centering on the dynamics of relationships.

I weave in design details from sculpture and painting to give depth to the characters, using these unique sets of relationships to express my worldview.

Mercury（水星）は太陽系で一番小さい惑星であり、形状を変化させる原子量200.592uの液体金属（水銀）であり、「イーリアス」では知恵とコミュニケーションの神（ヘルメス）の象徴です。

この章では、過去数年間にSFをテーマに創作した作品のいくつかを紹介します。その筆頭は、愛のために身を捧げる人工の女性「The Bride」、荒廃した地で傷つきながらも逞しく支え合う恋人たち「The Lovers」、白髪の青年とその忠実な赤い軍用犬「The Friends」です。これら3つの作品は、関係性の力学に焦点を合わせた同じシリーズに属しています。

造形と絵画の詳細なデザインを織り込んでキャラクターに深みを与えるとともに、これら唯一無二の関係性を用いて私自身の世界観を表現しています。

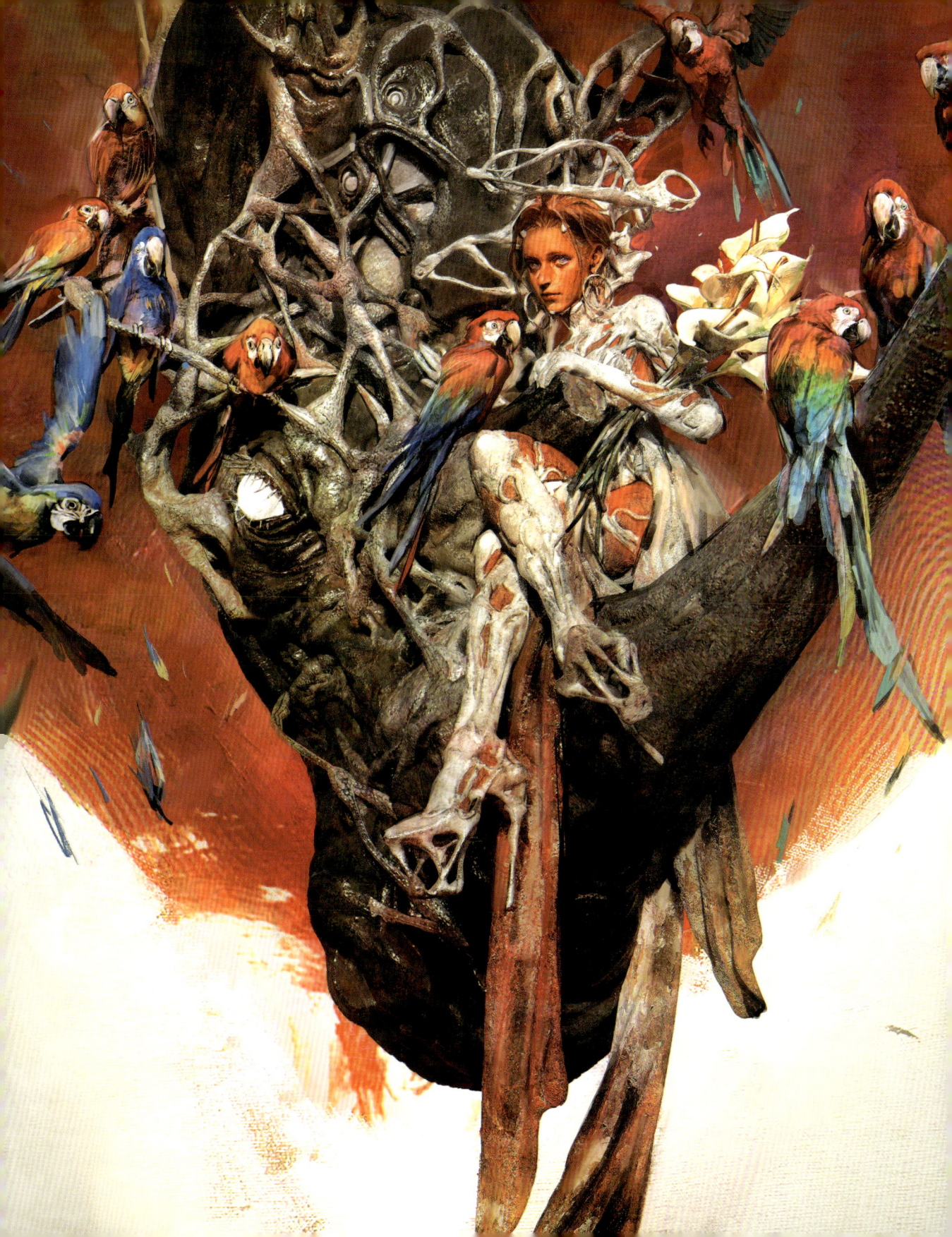

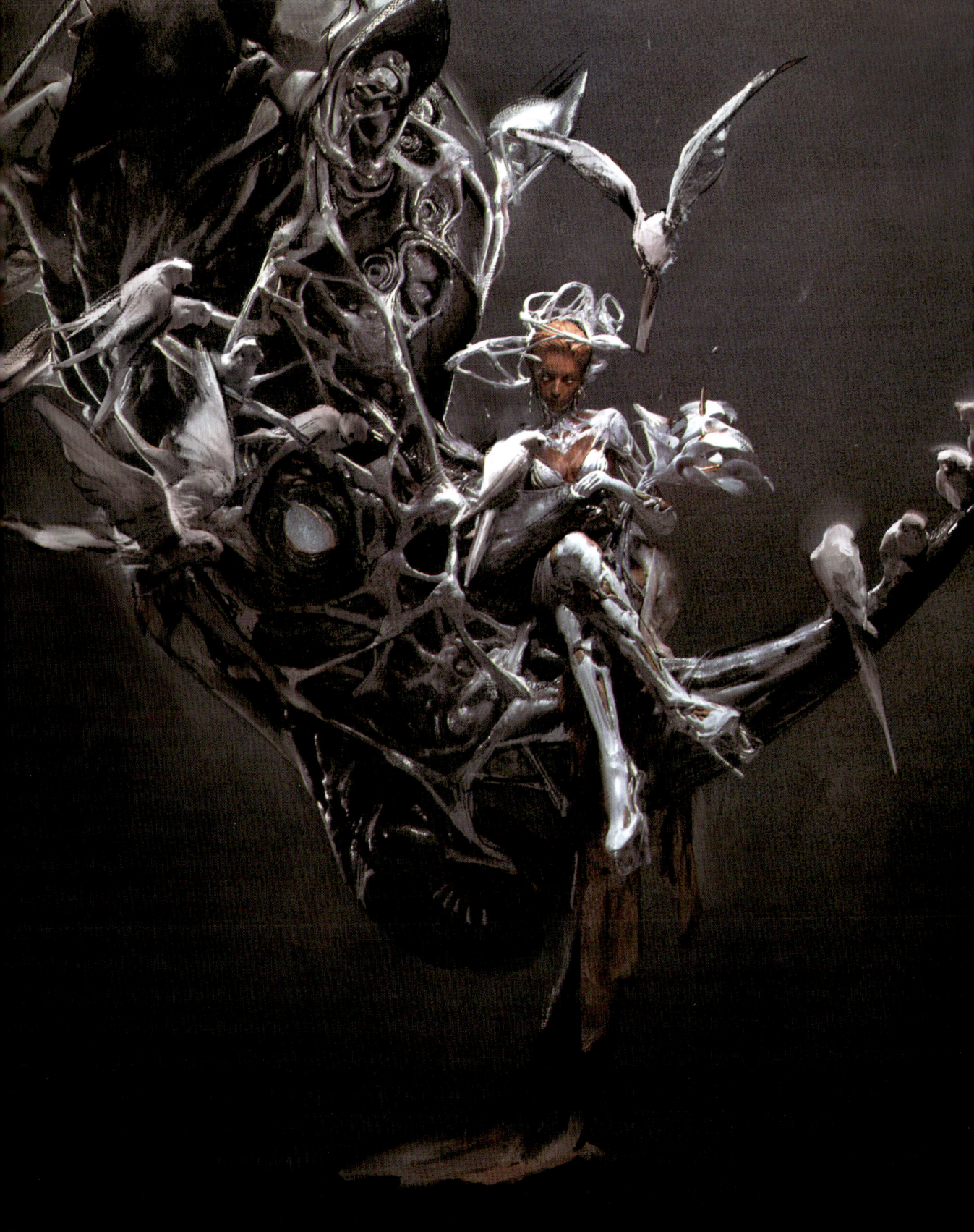

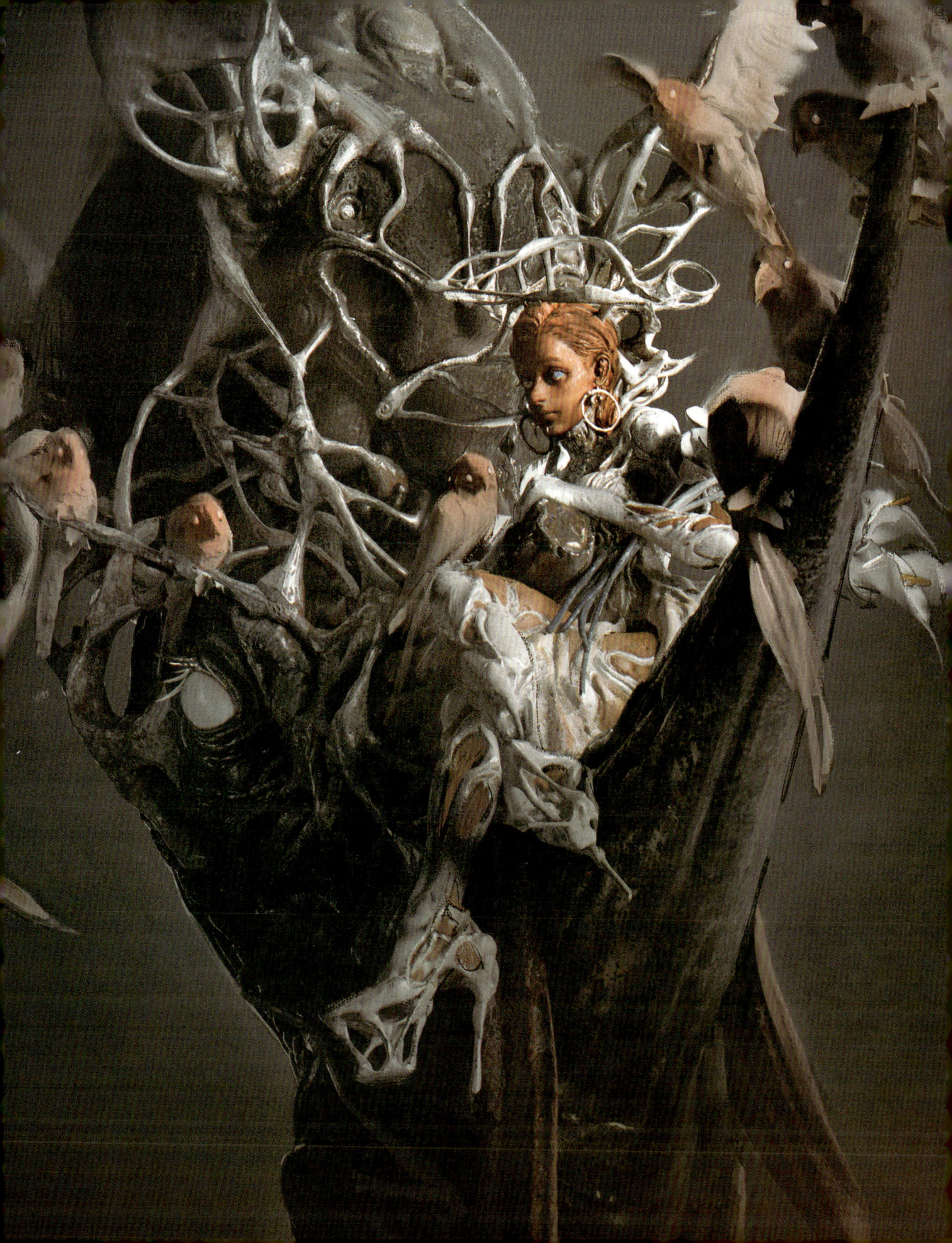

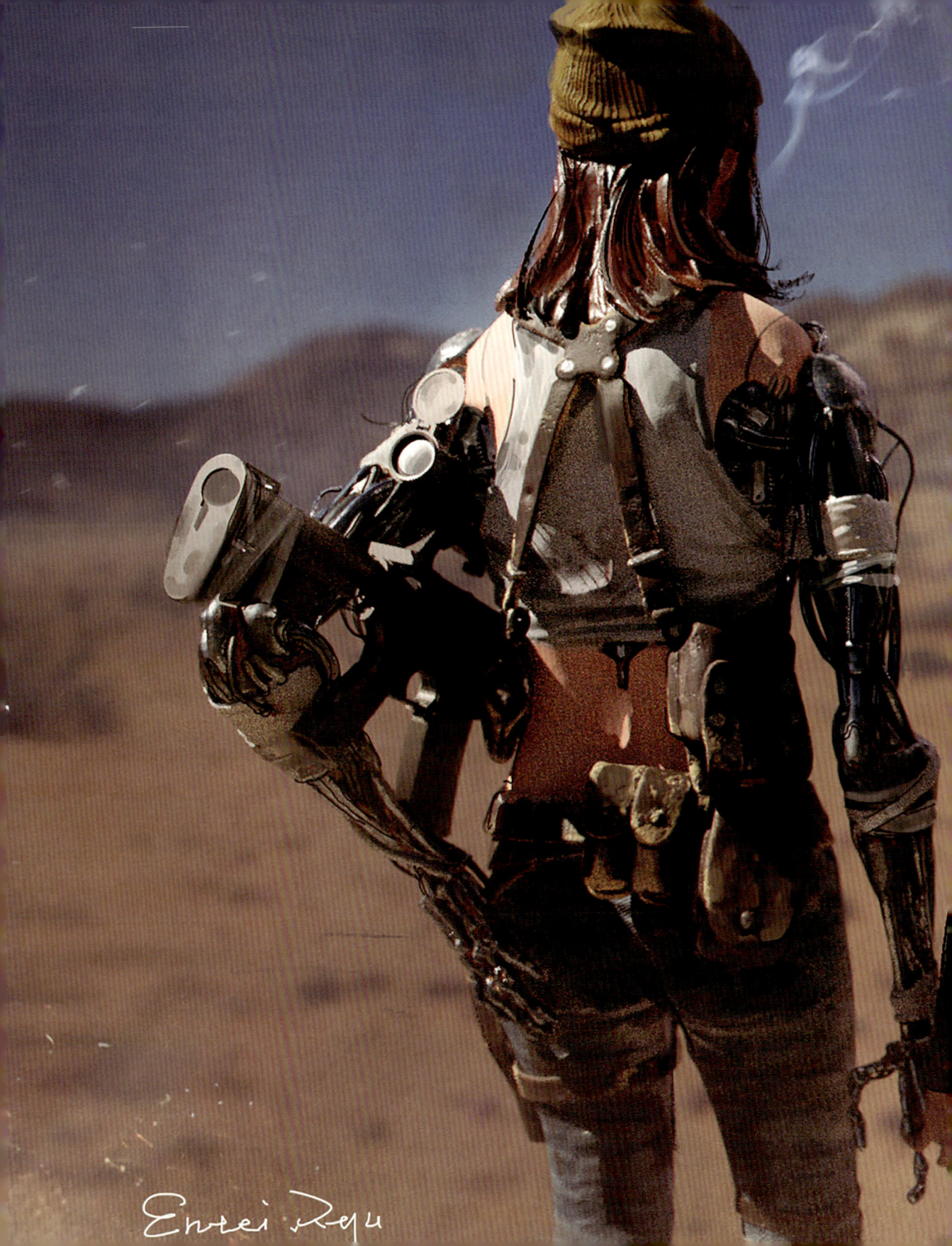

Enrei Ryu

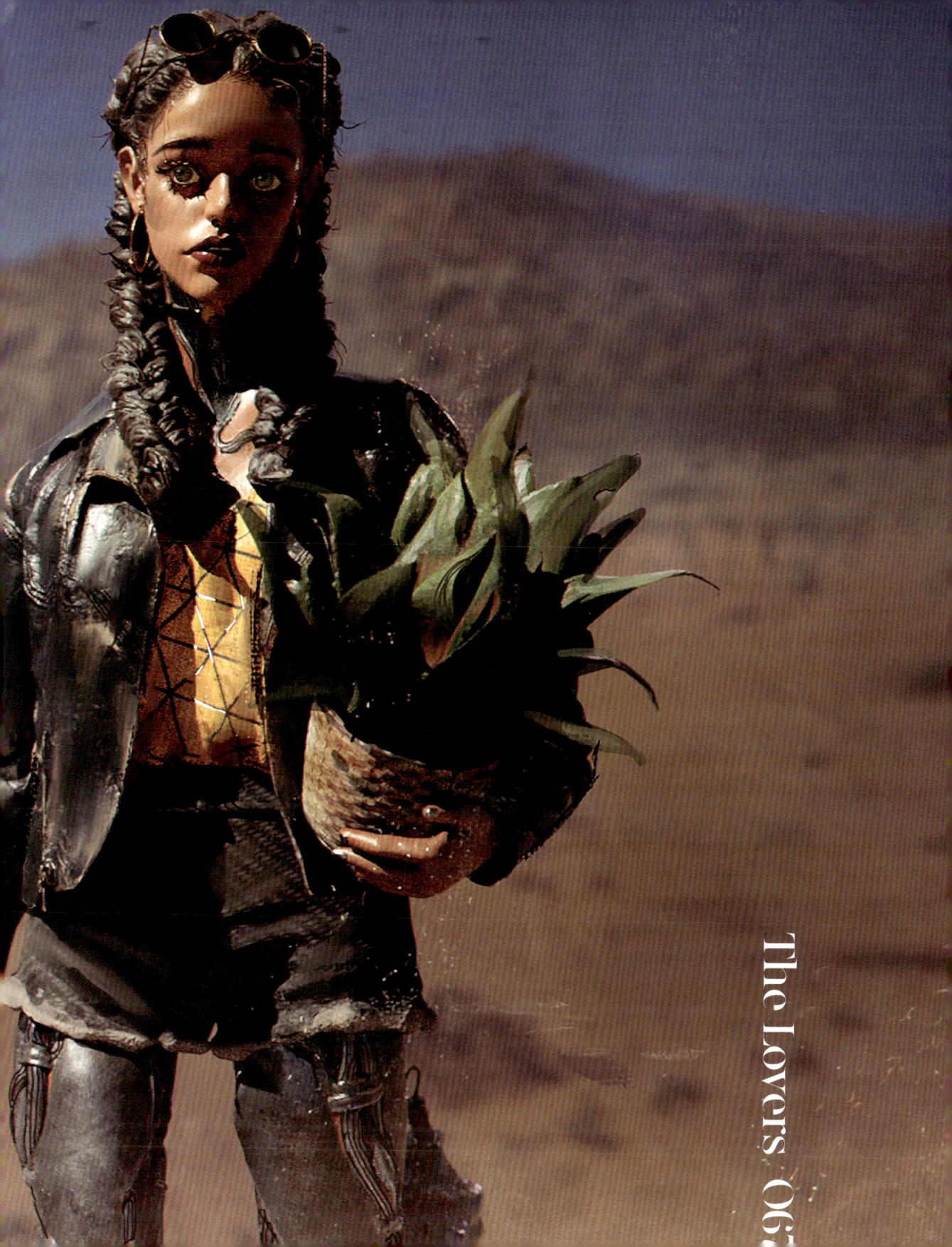

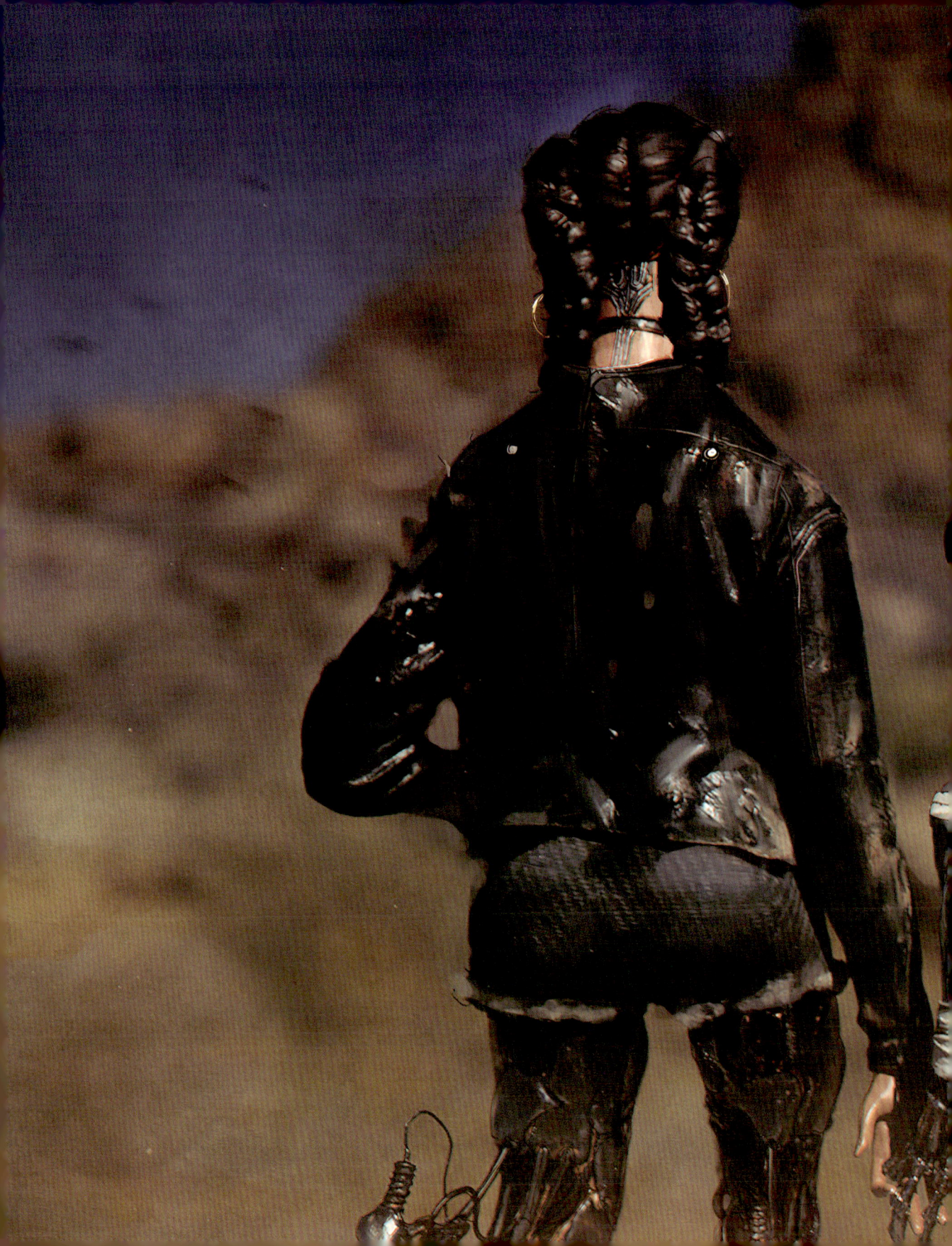

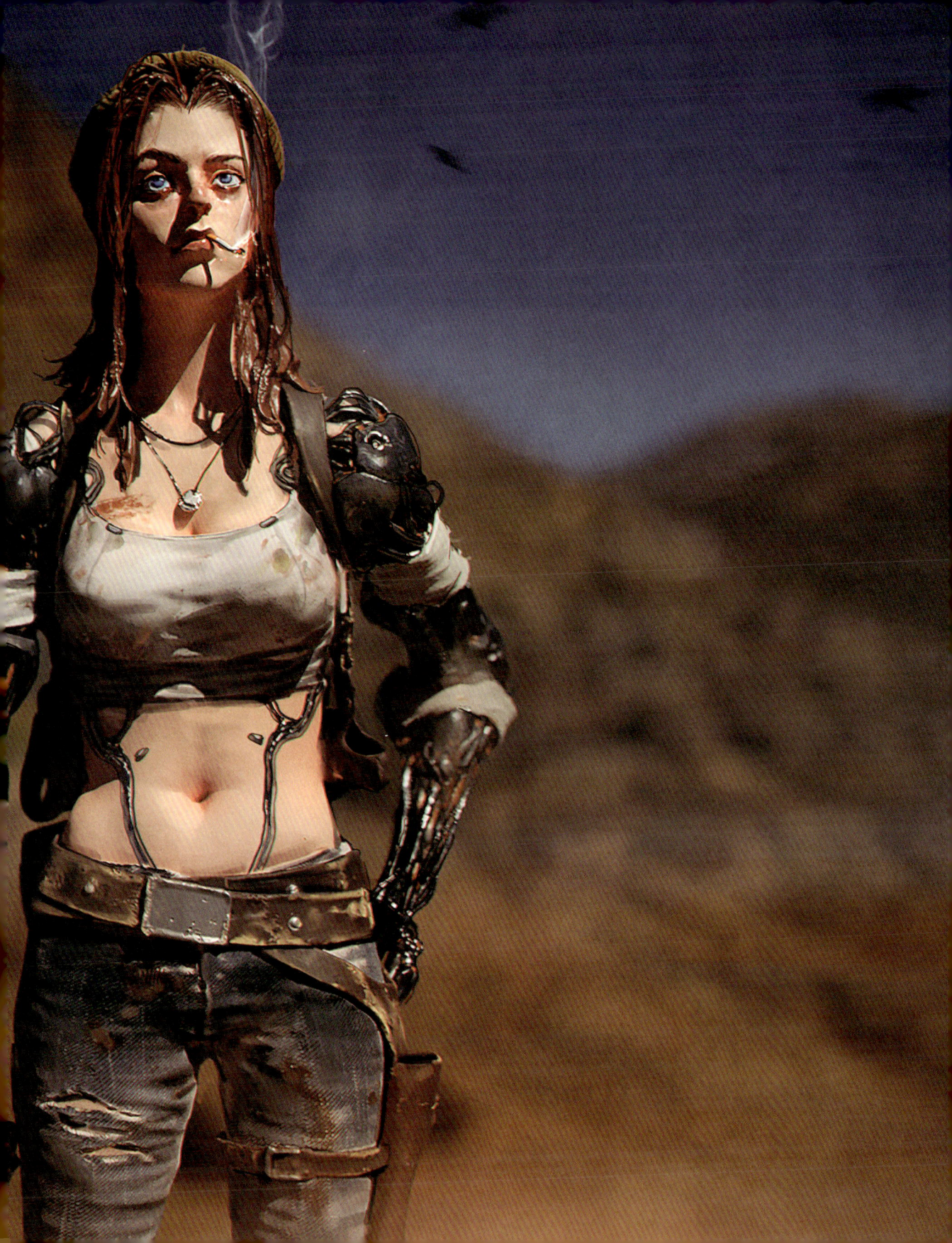

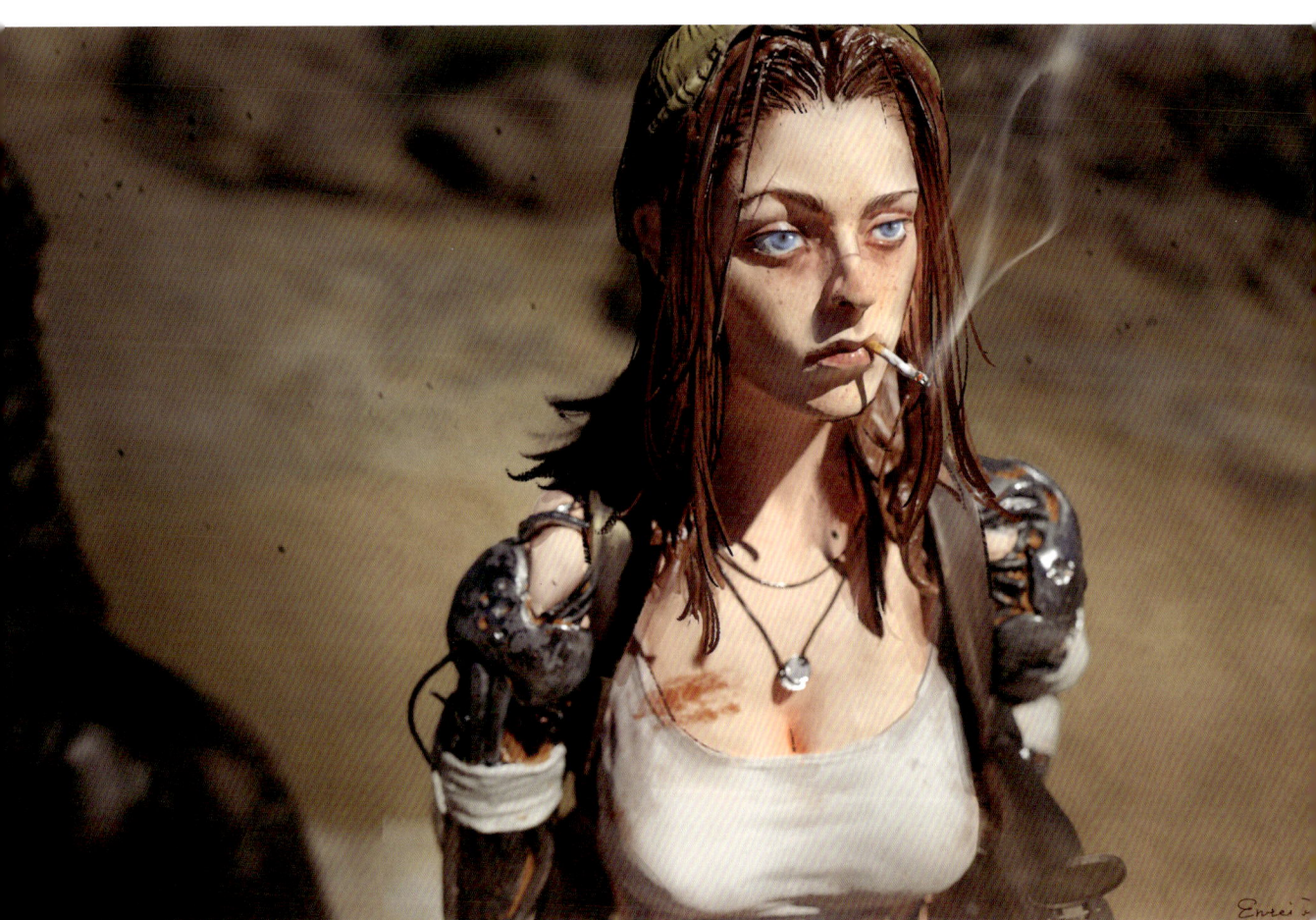

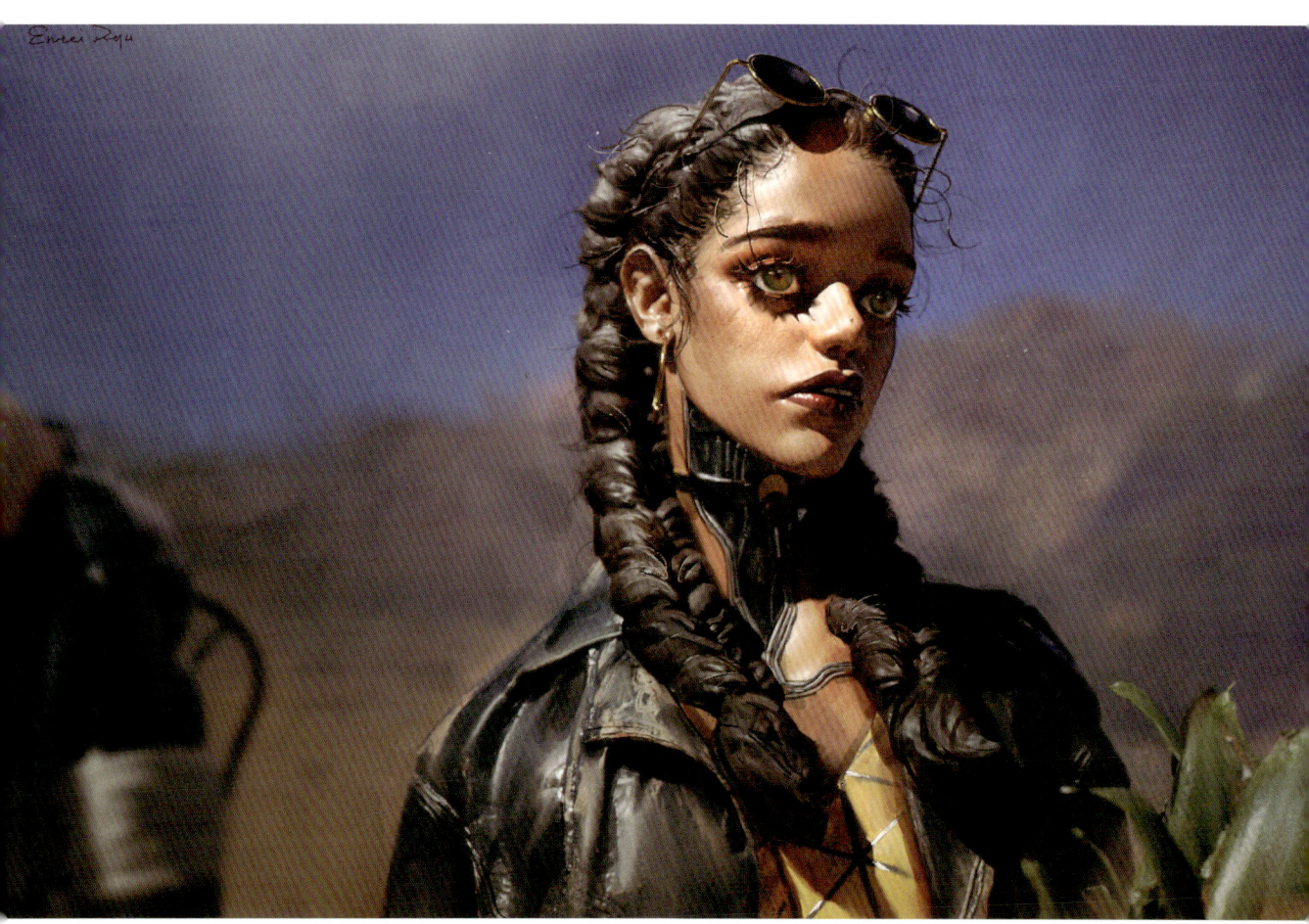

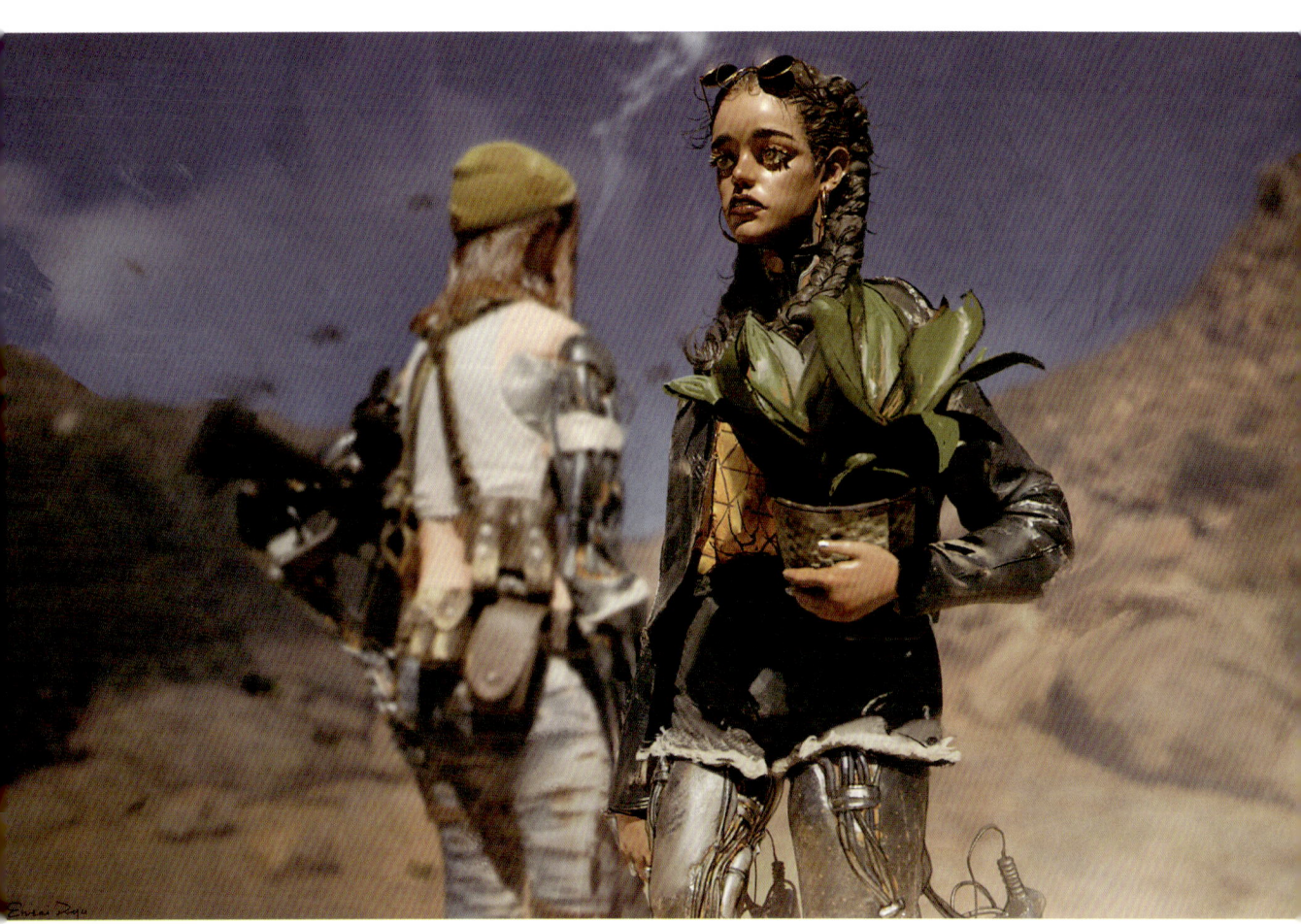

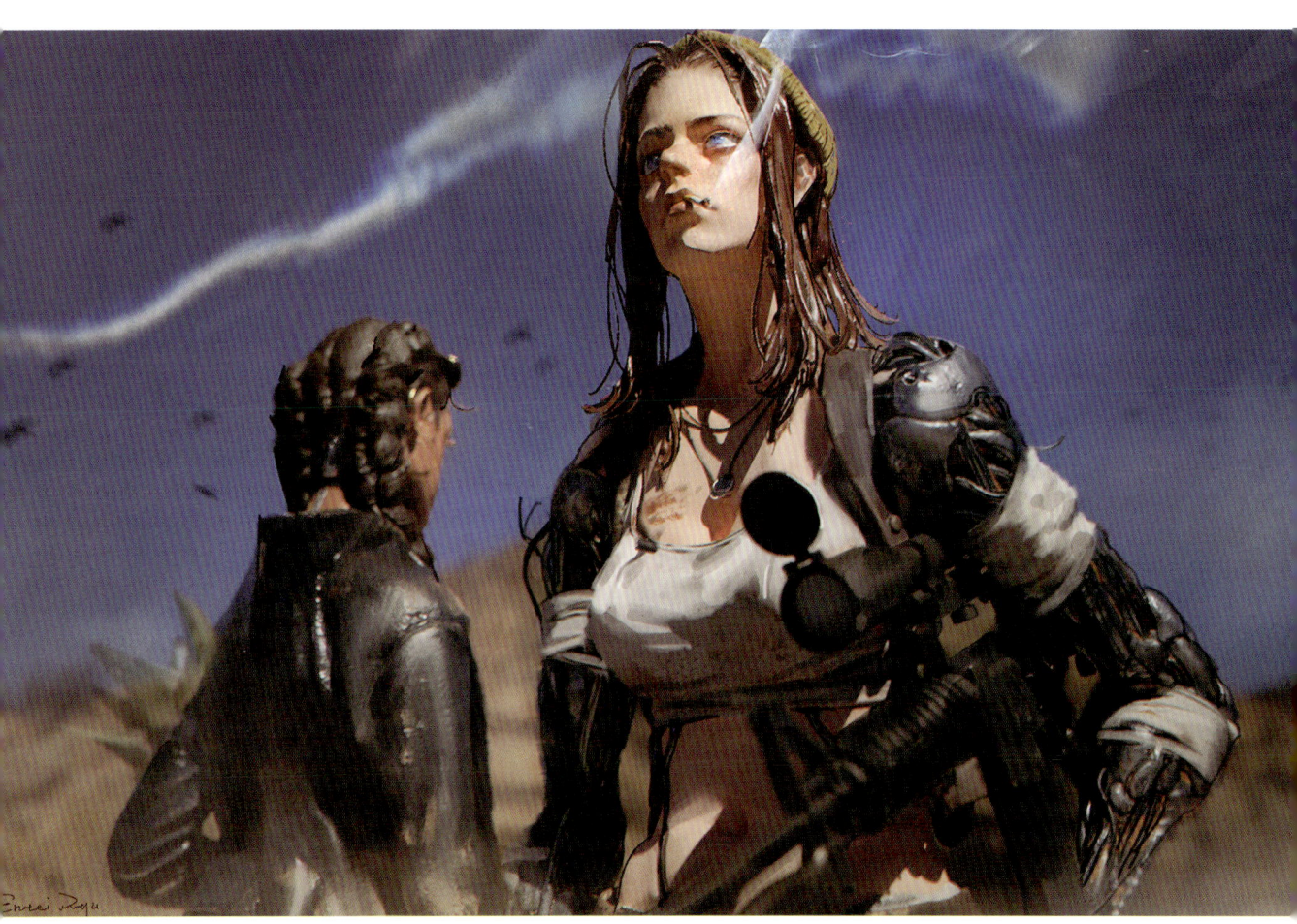

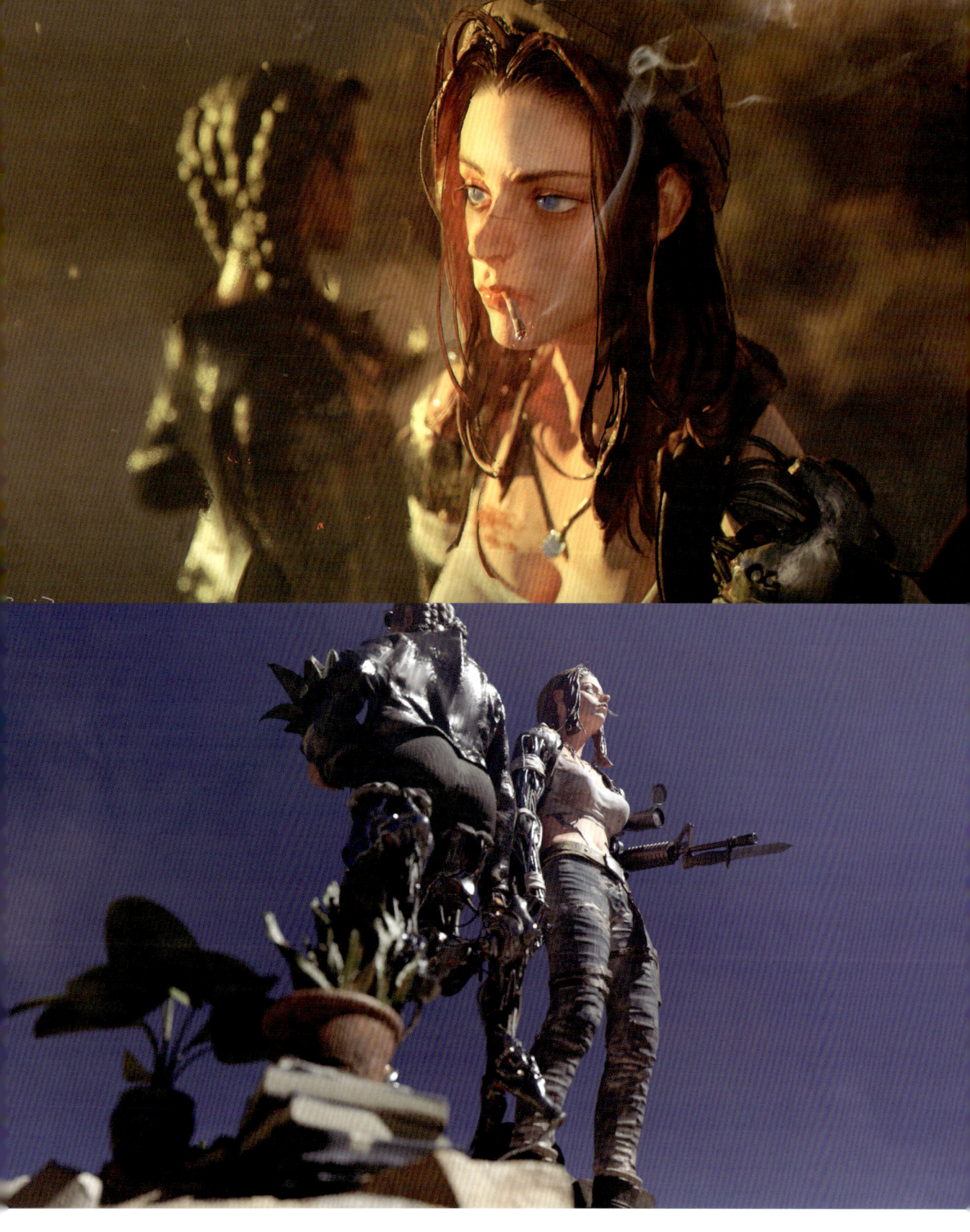

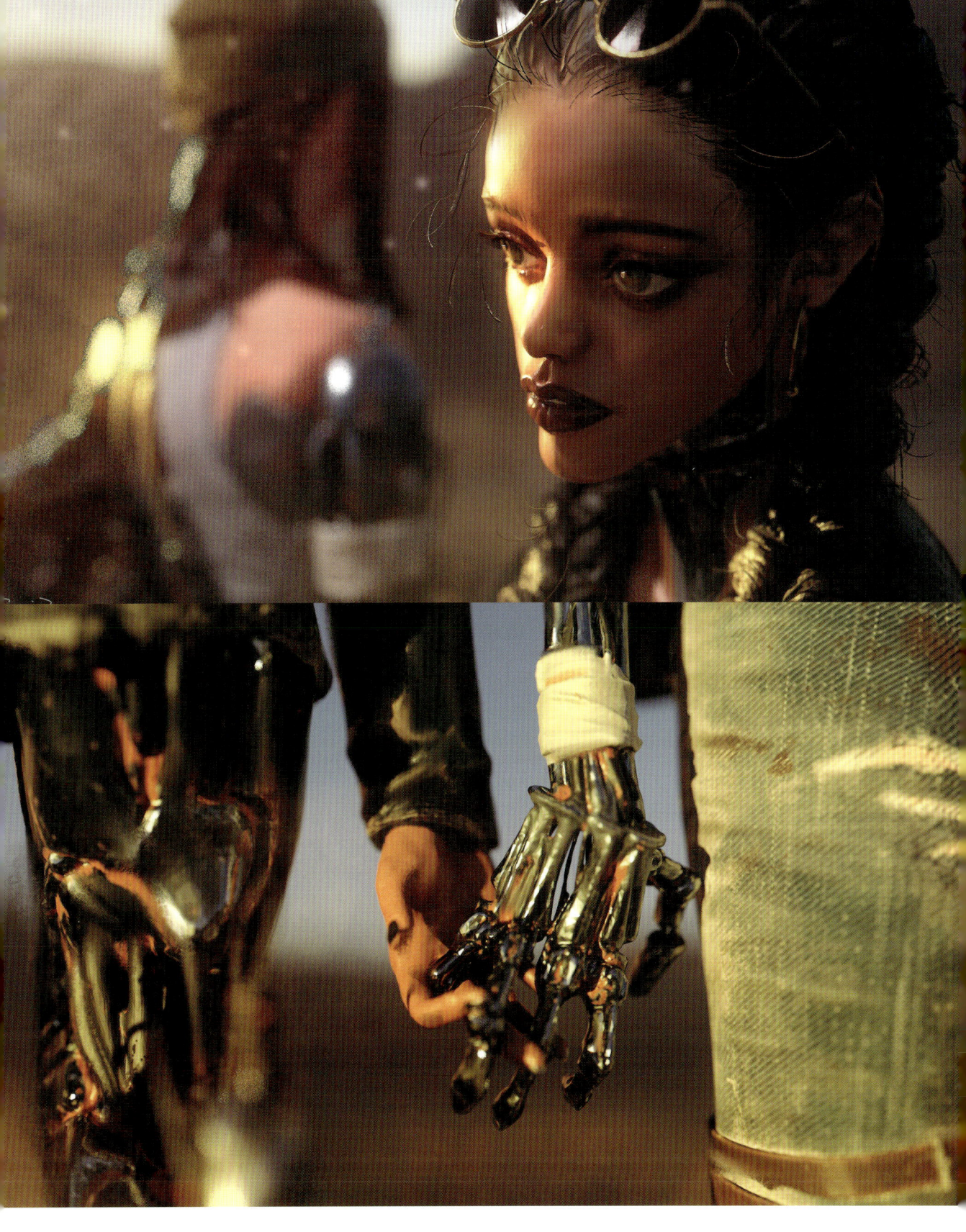

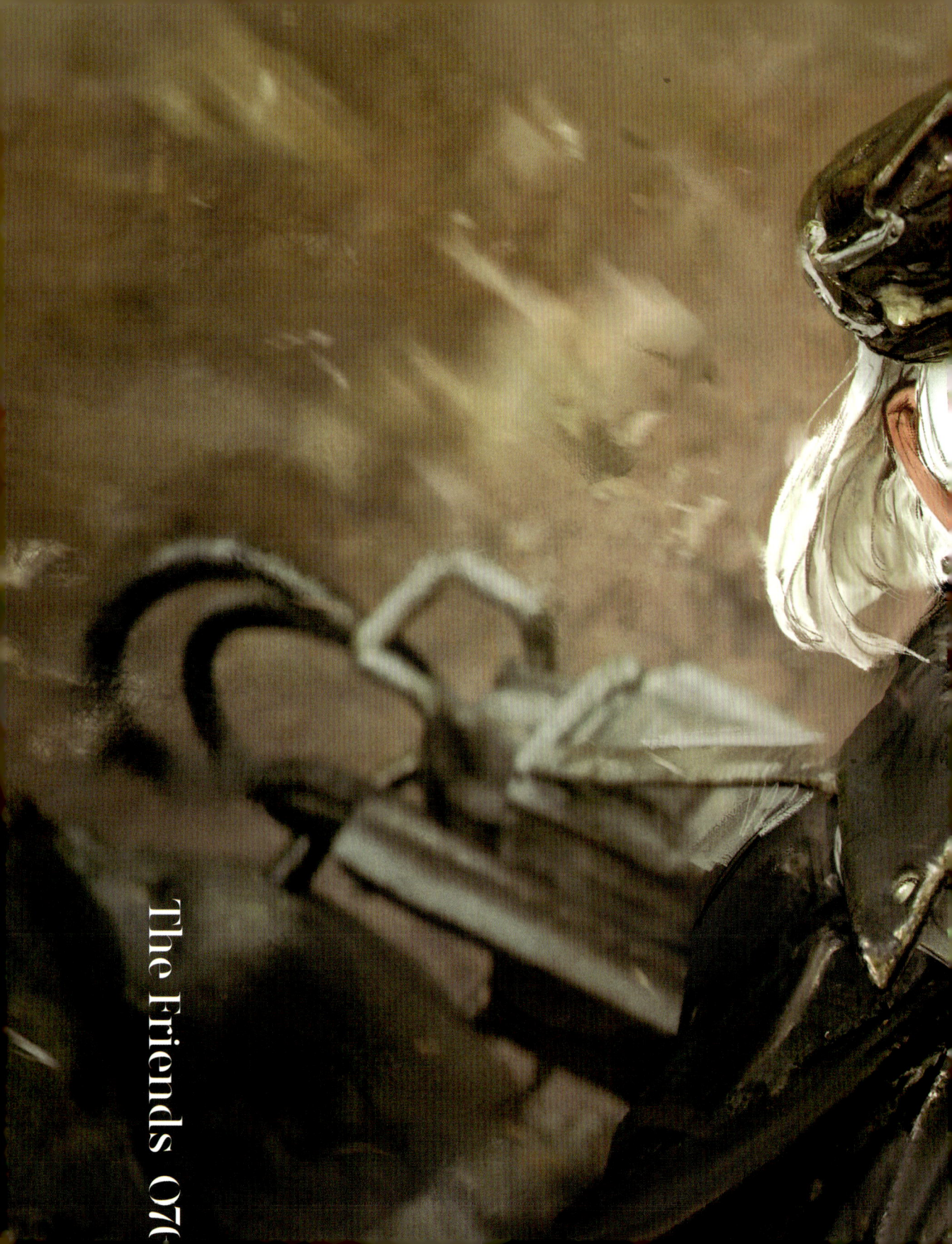

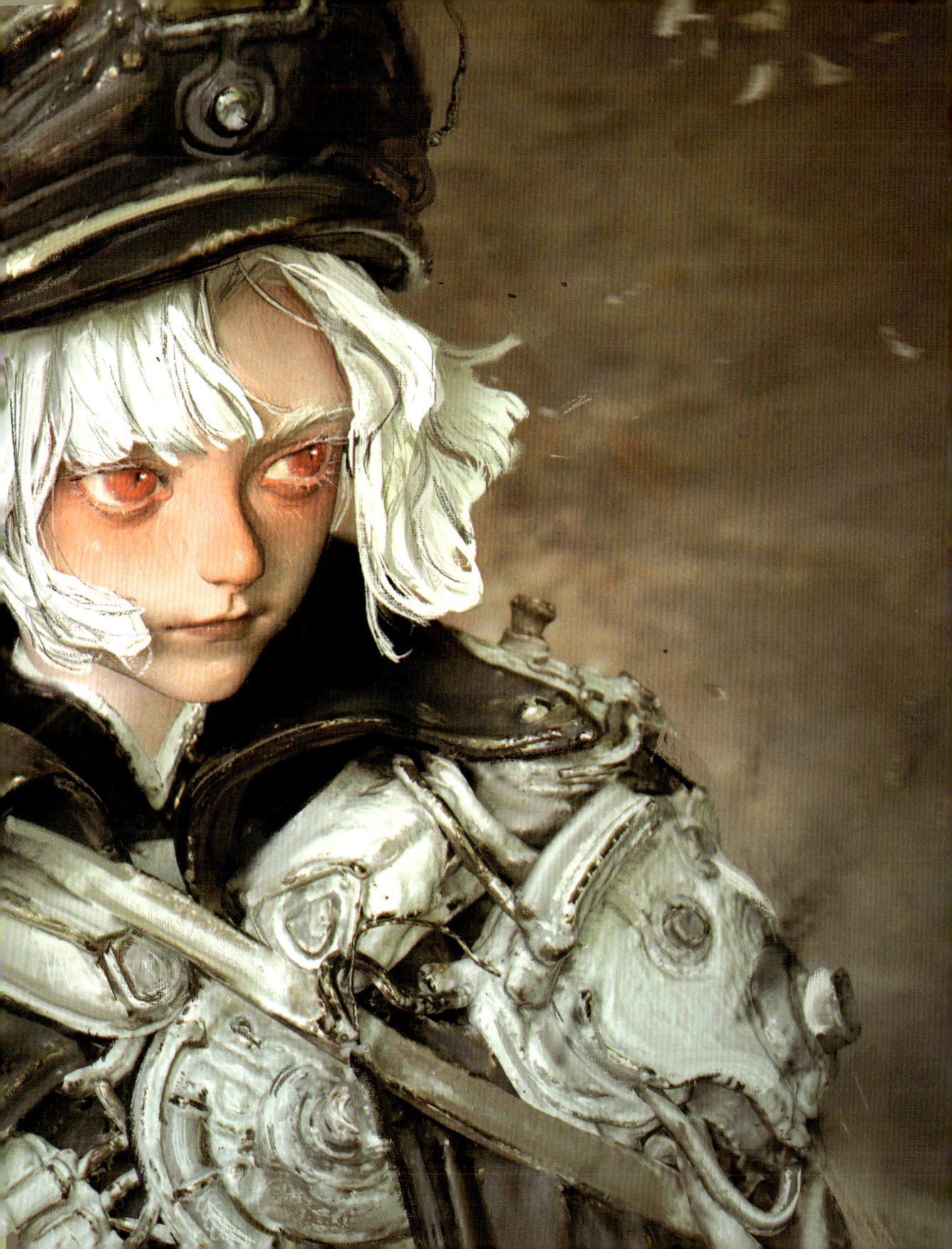

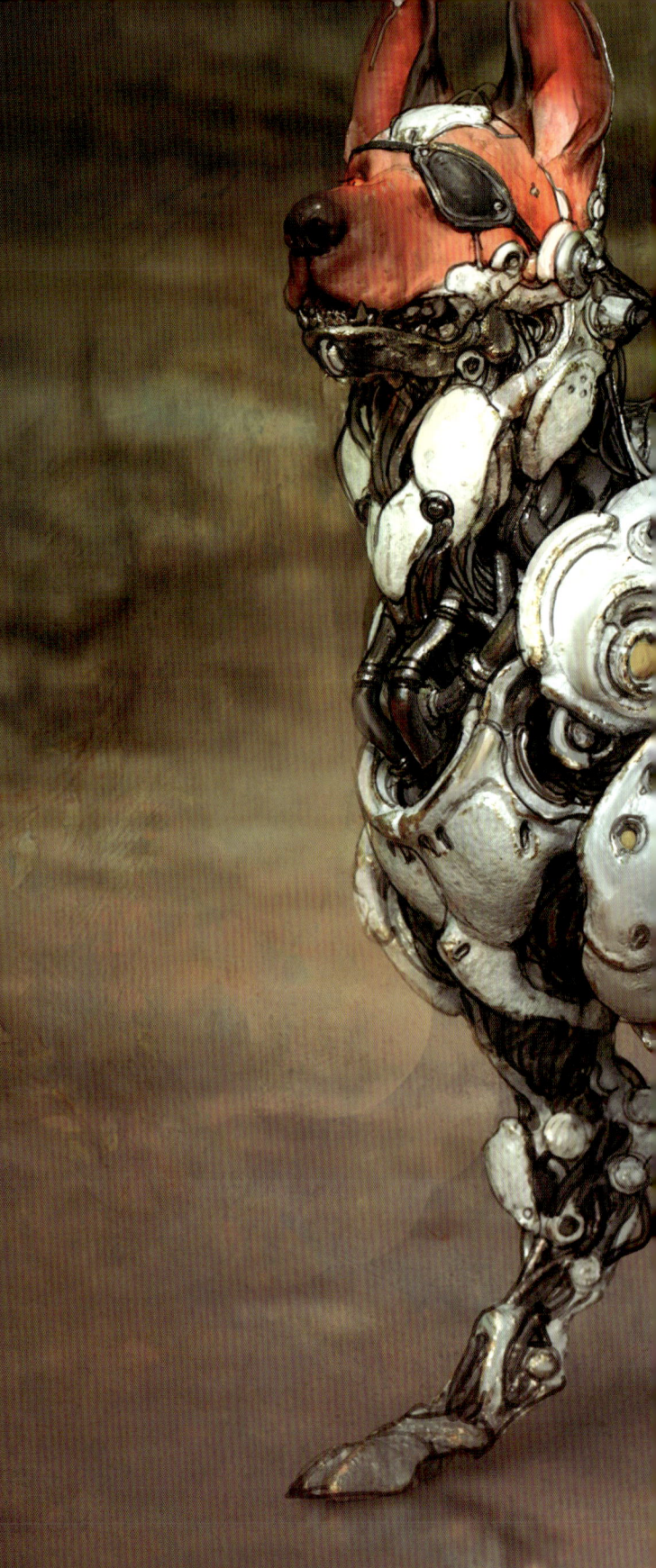

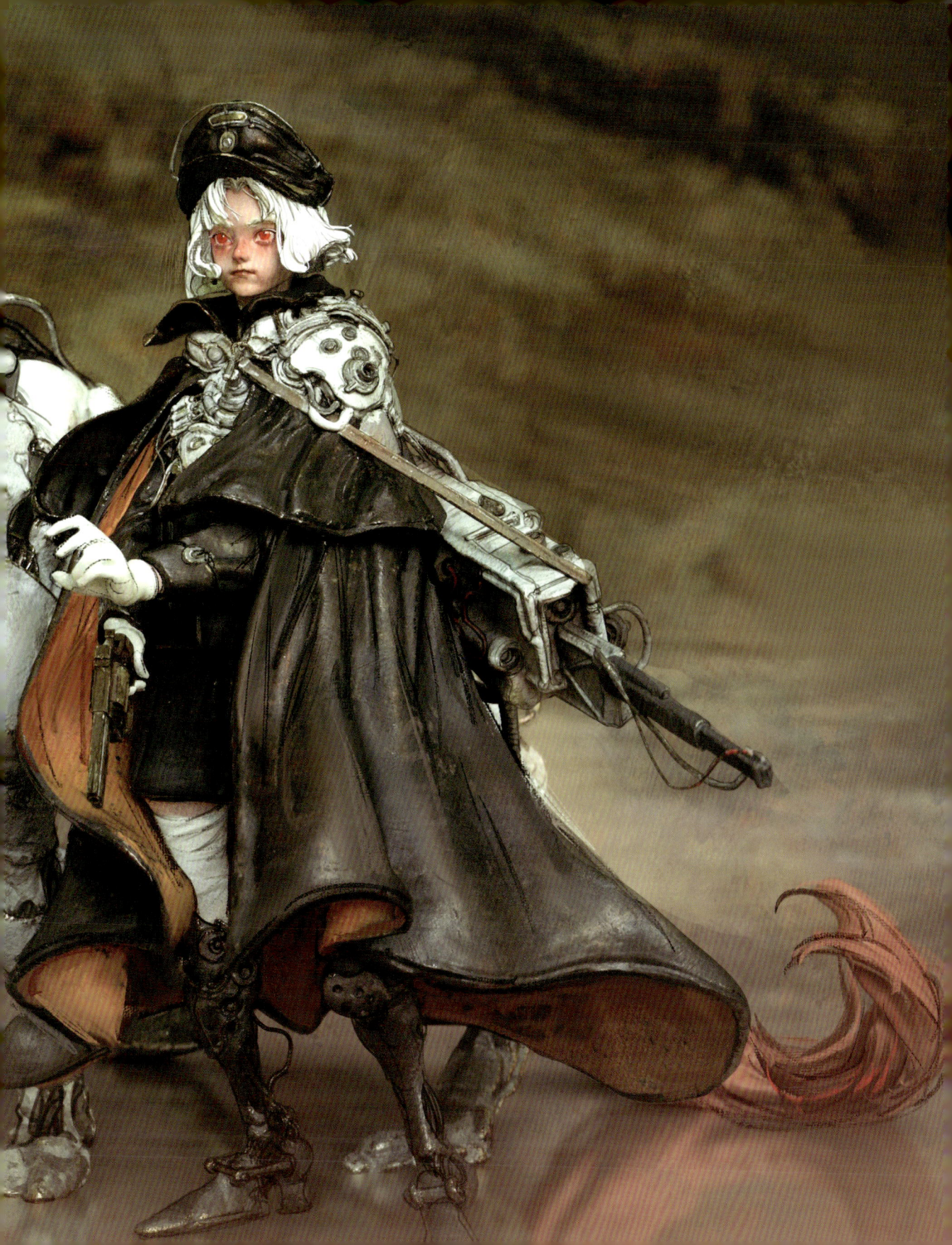

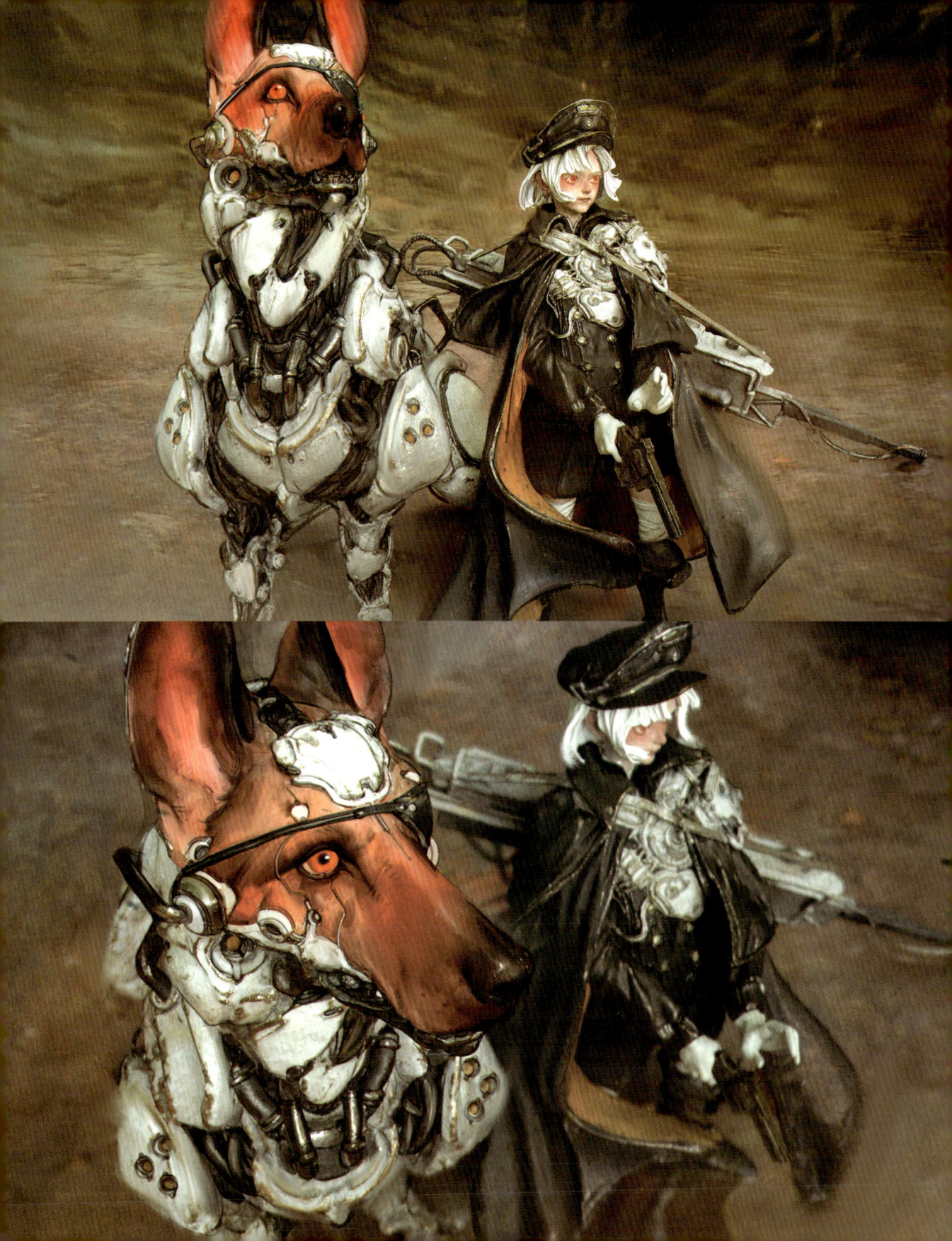

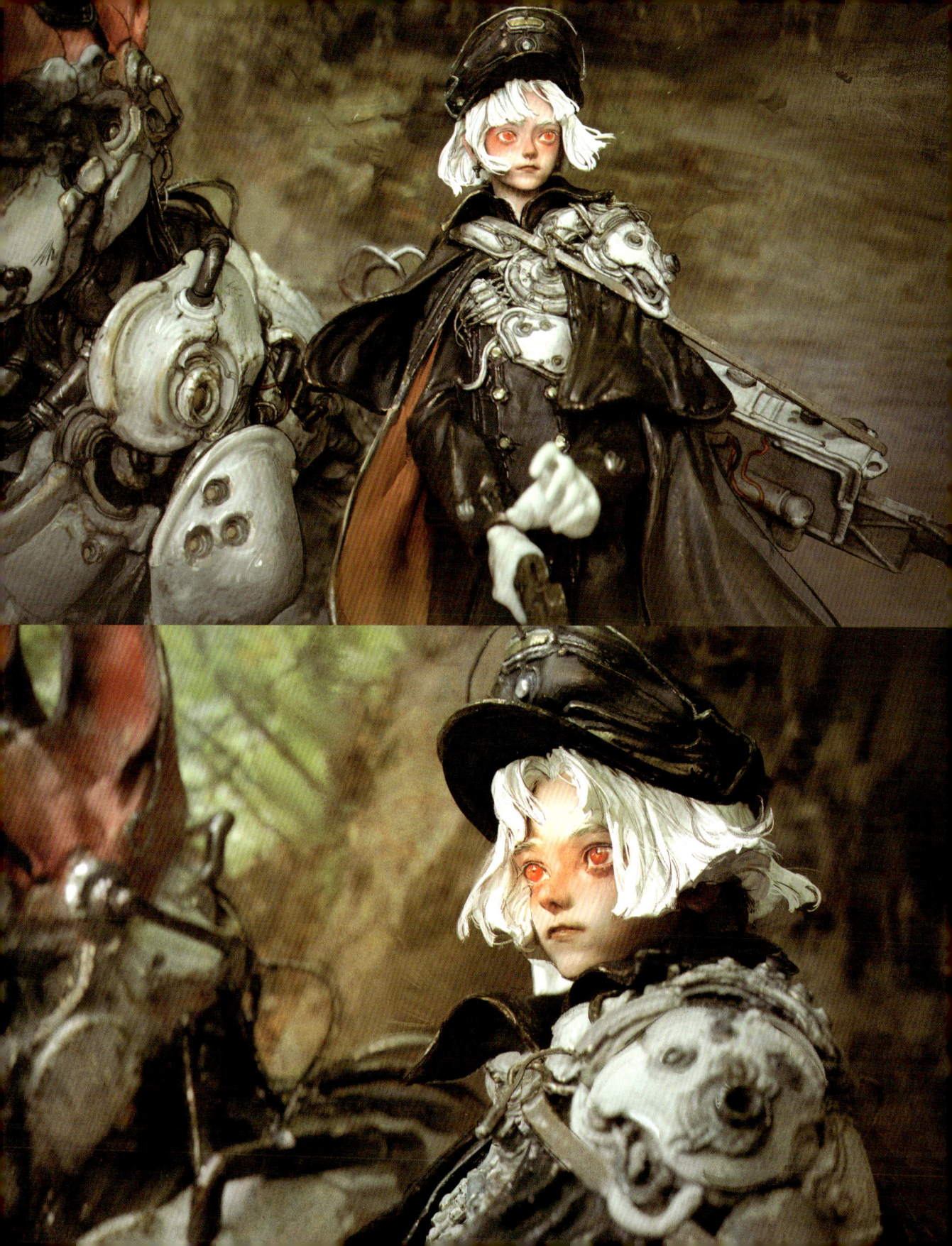

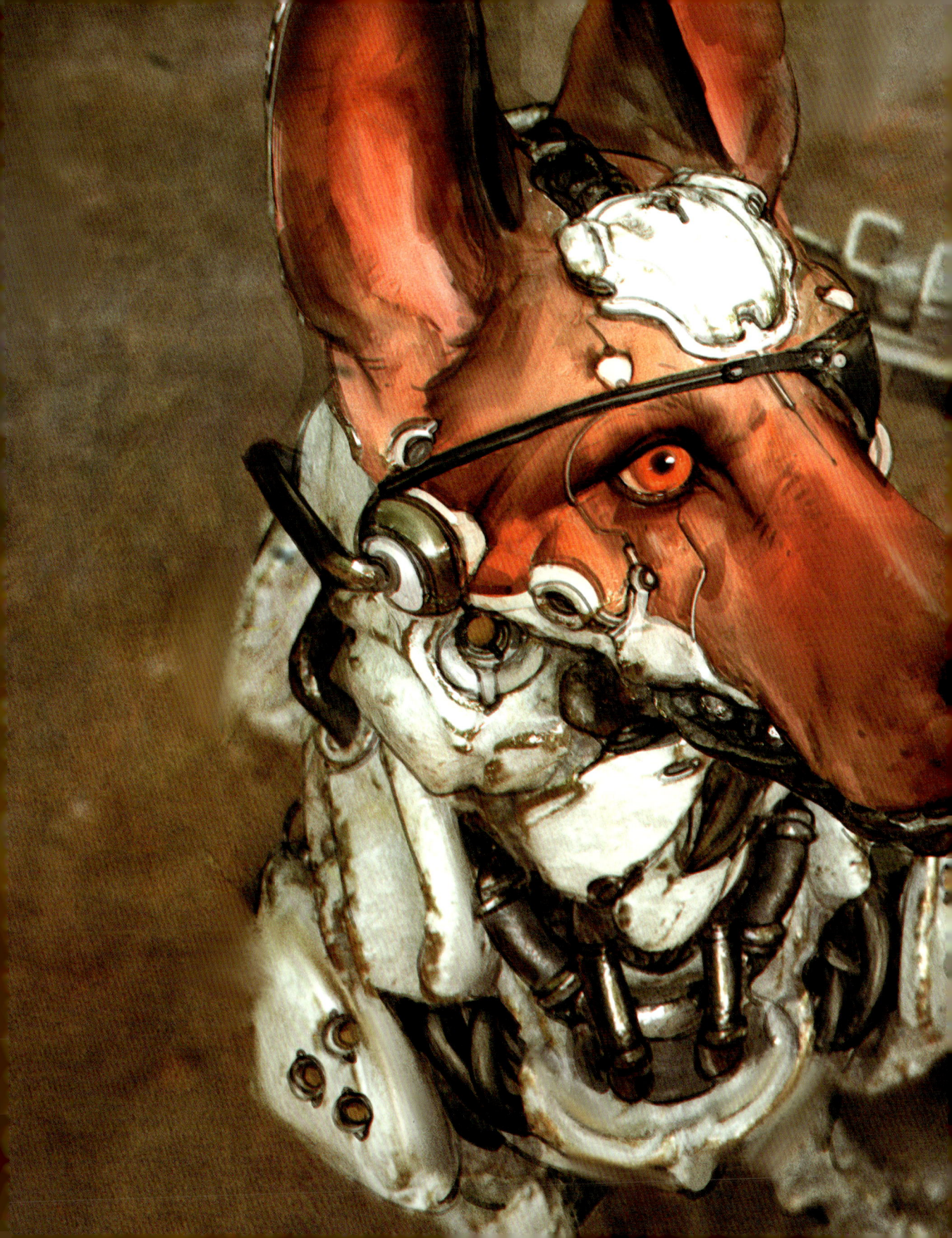

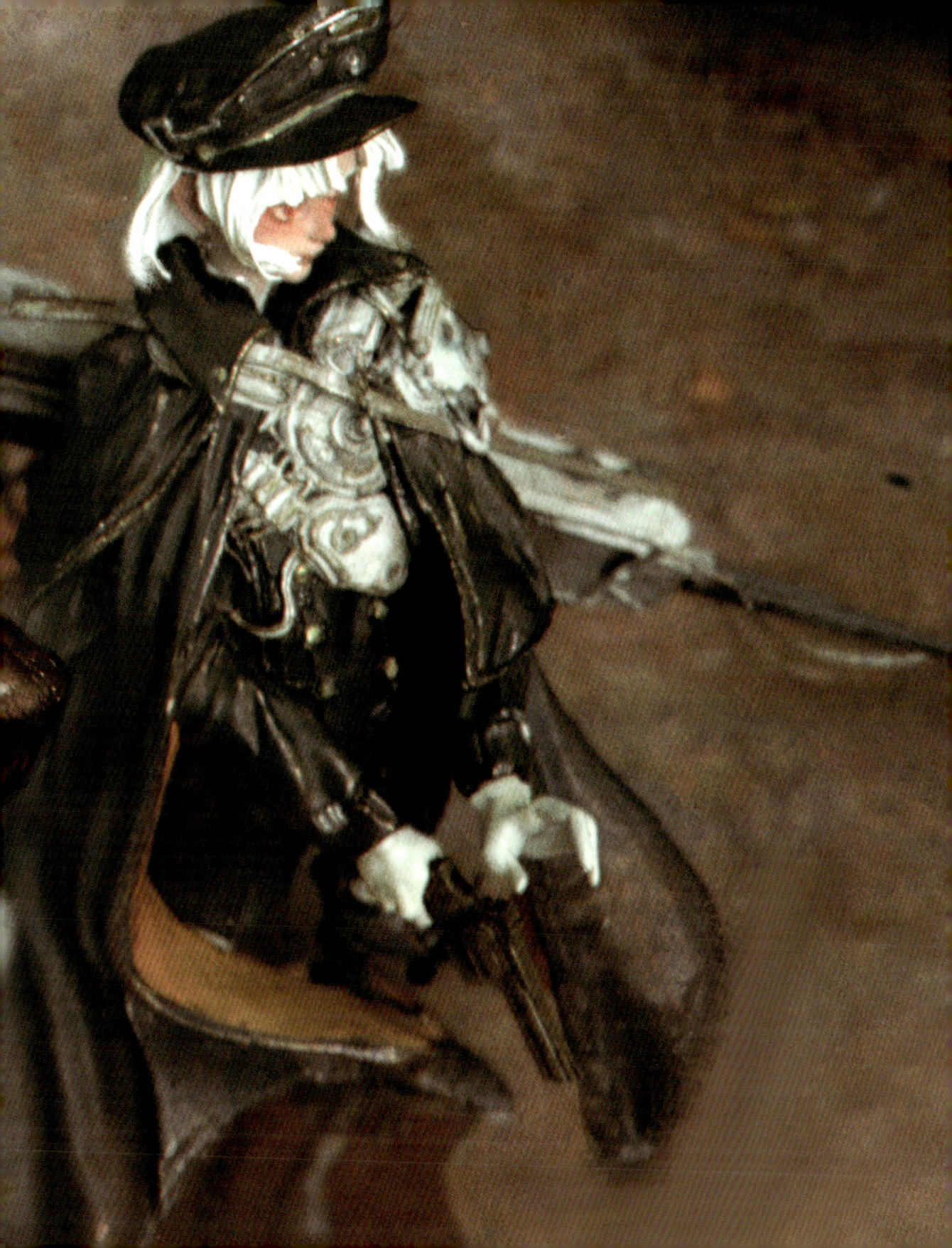

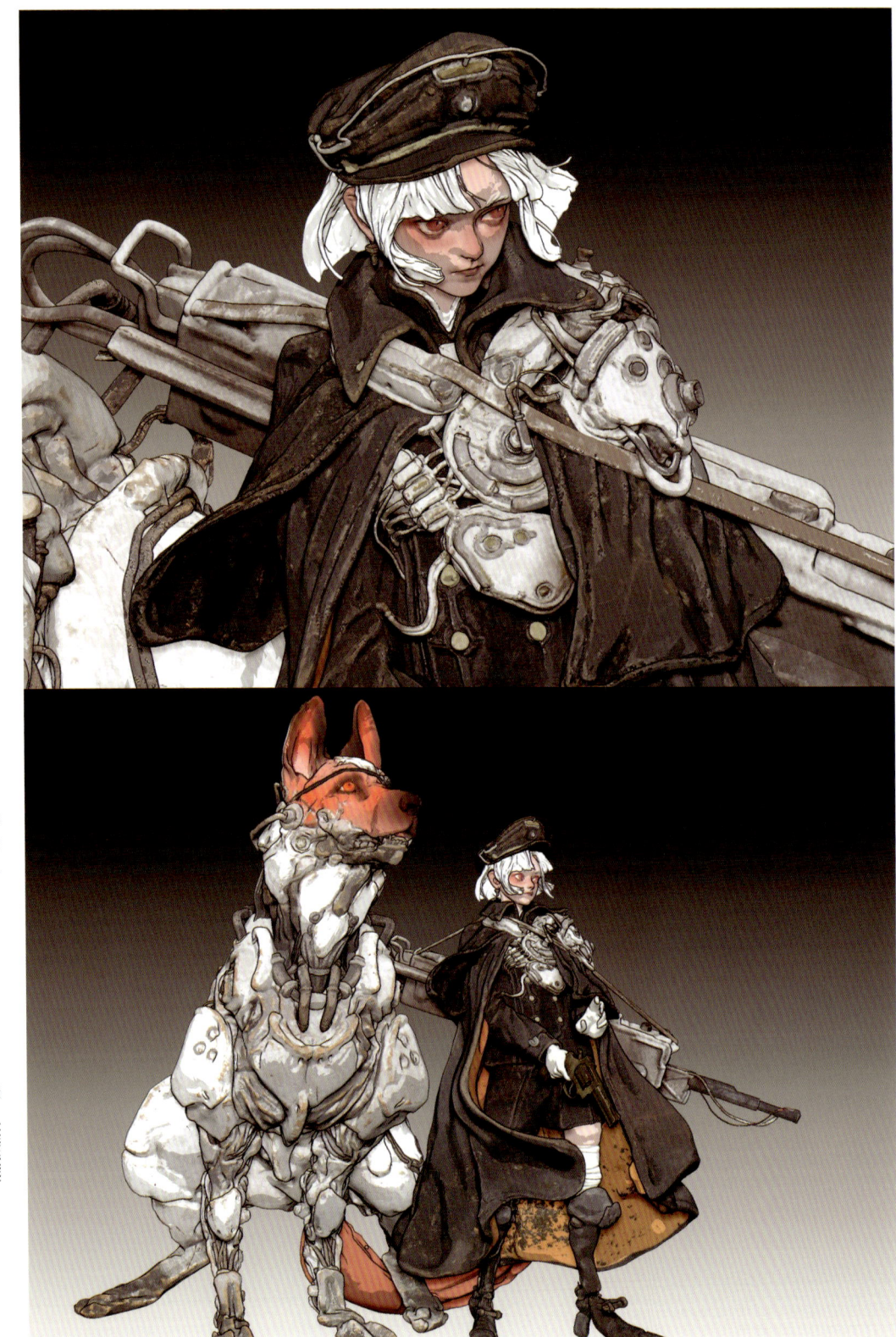

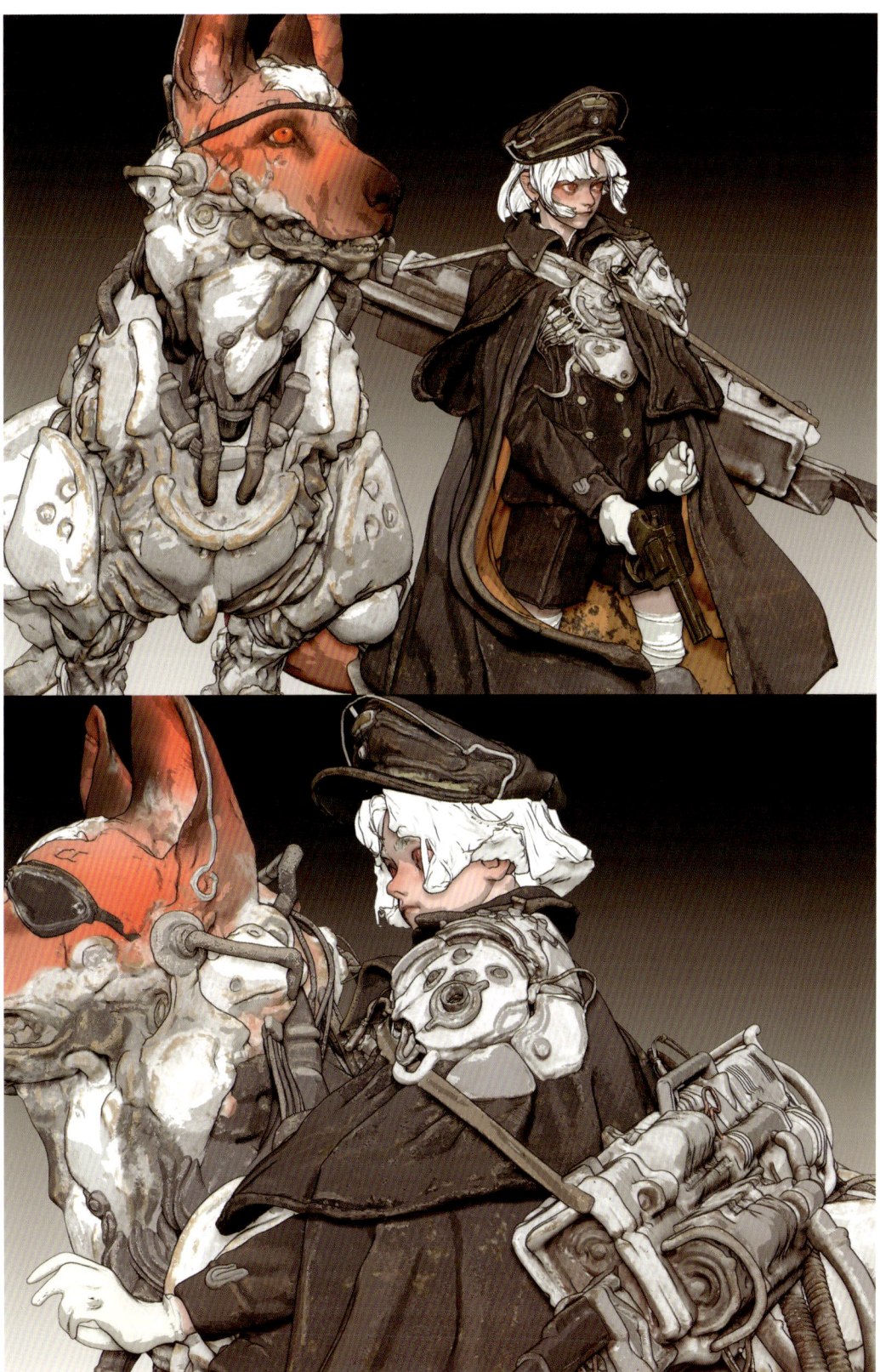

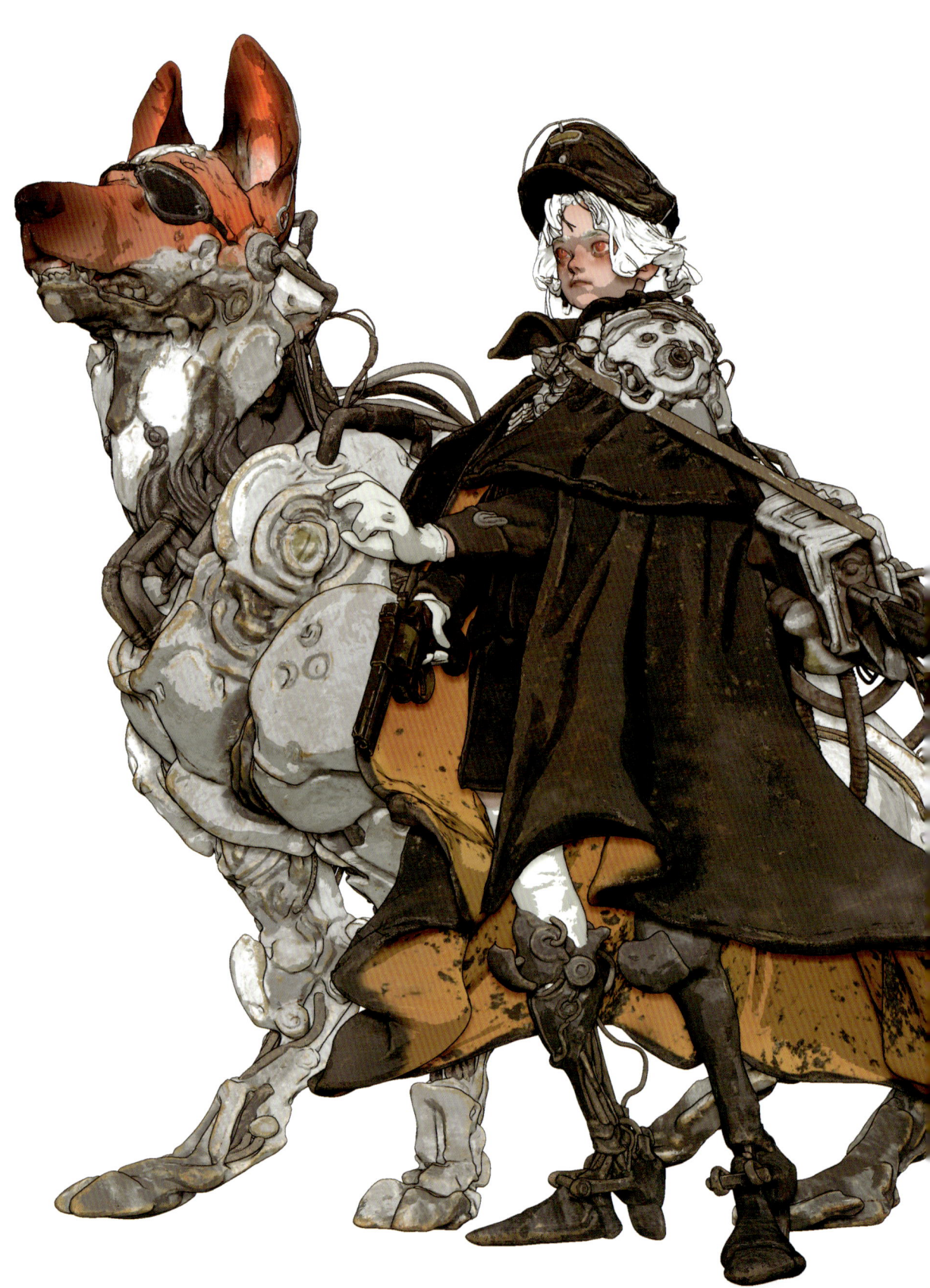

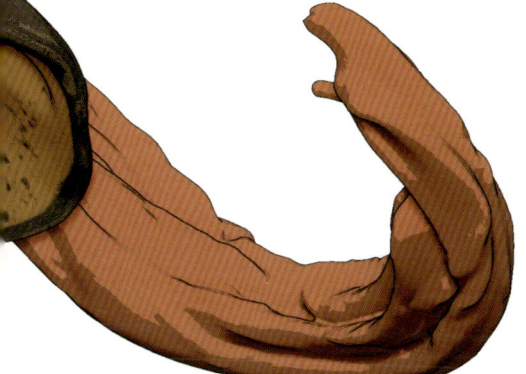

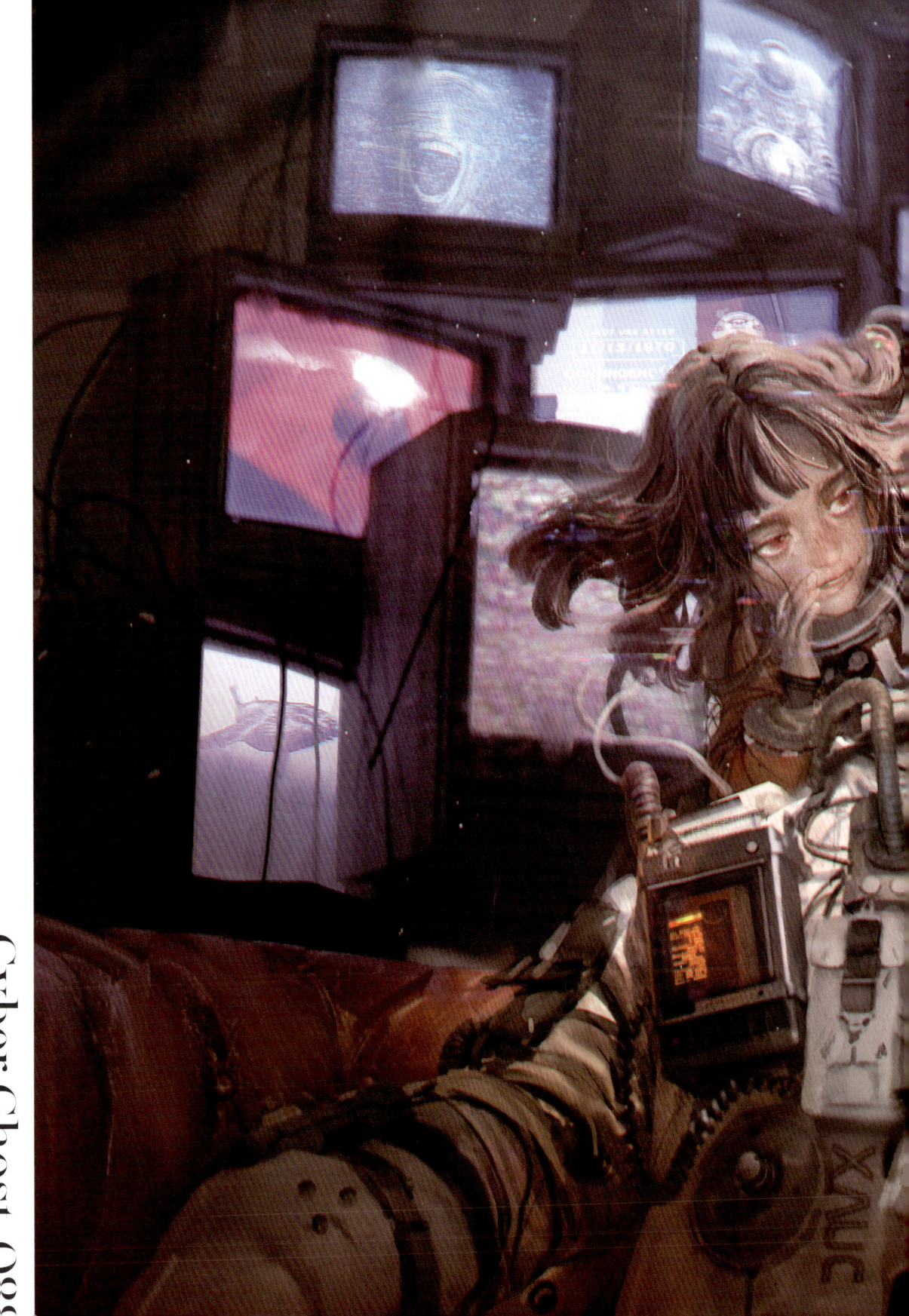

Cyber Ghost 088

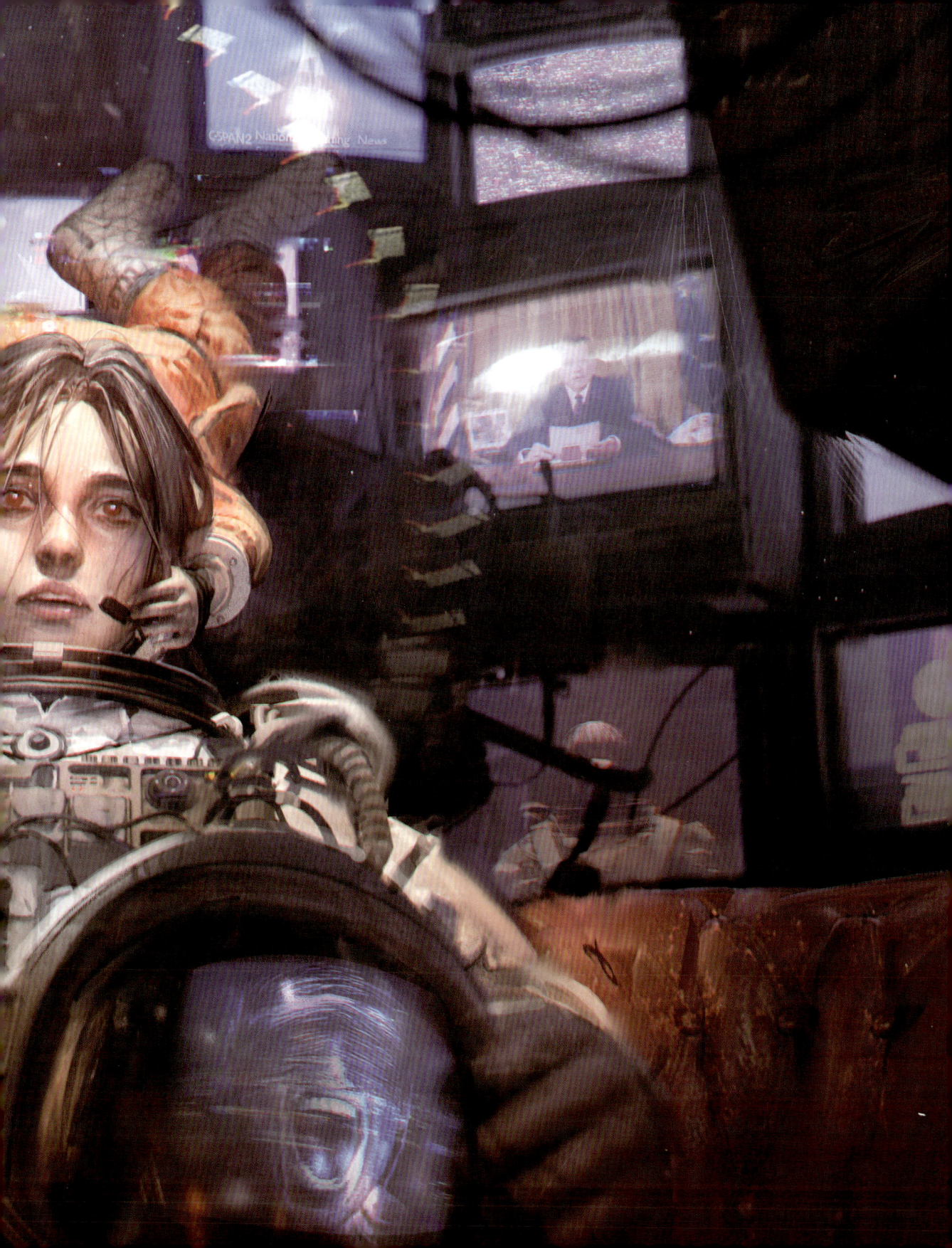

# Yakuza Medics 090

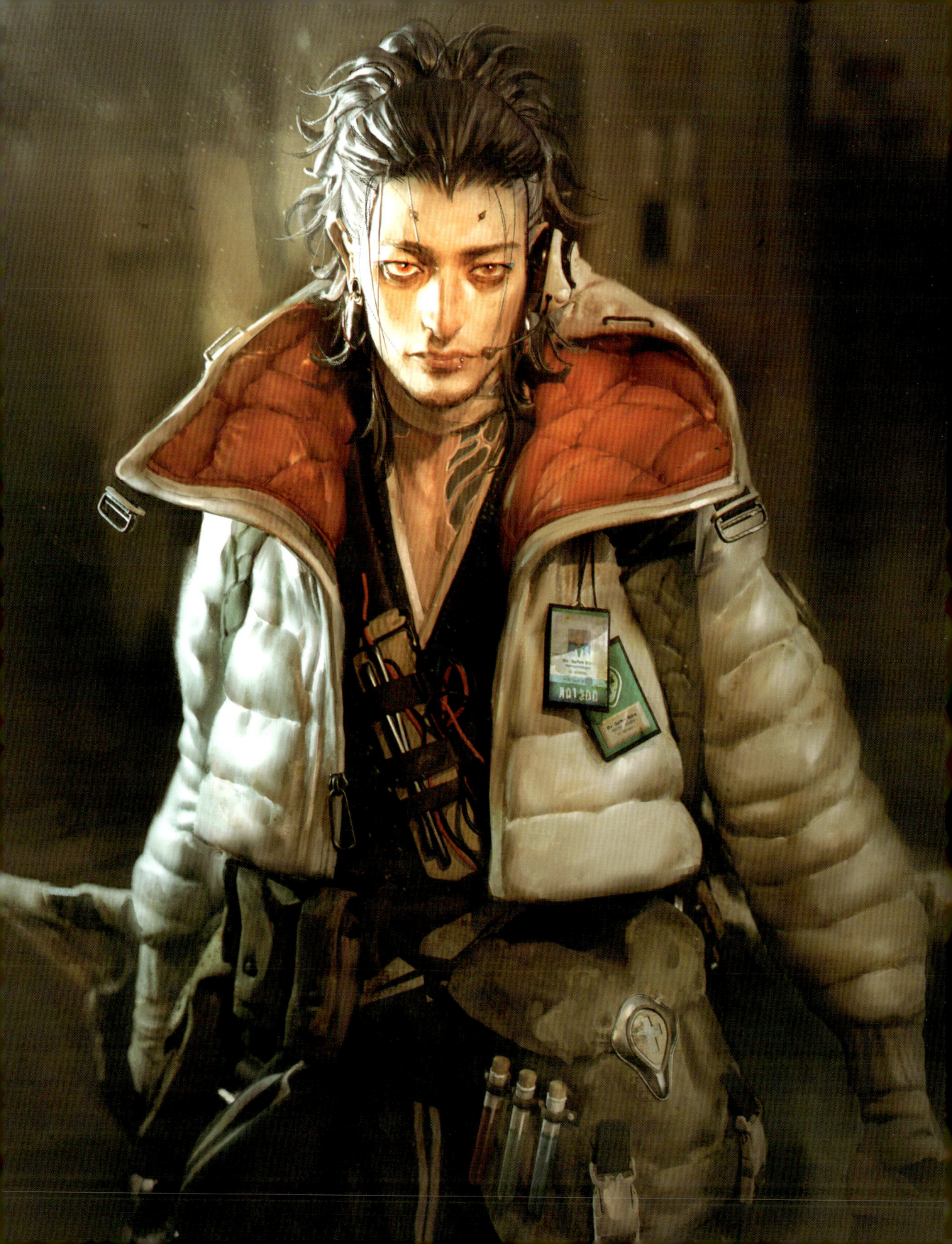

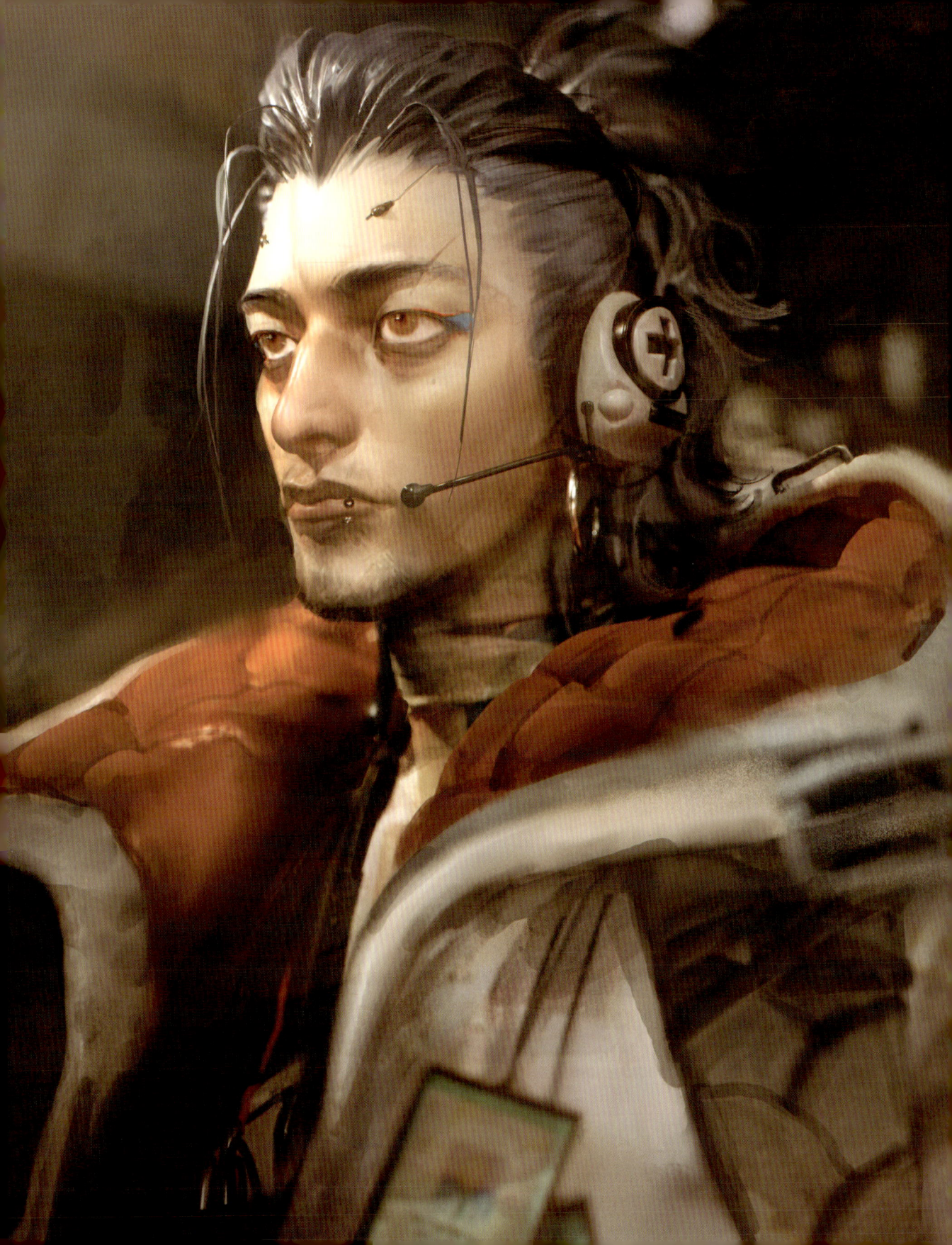

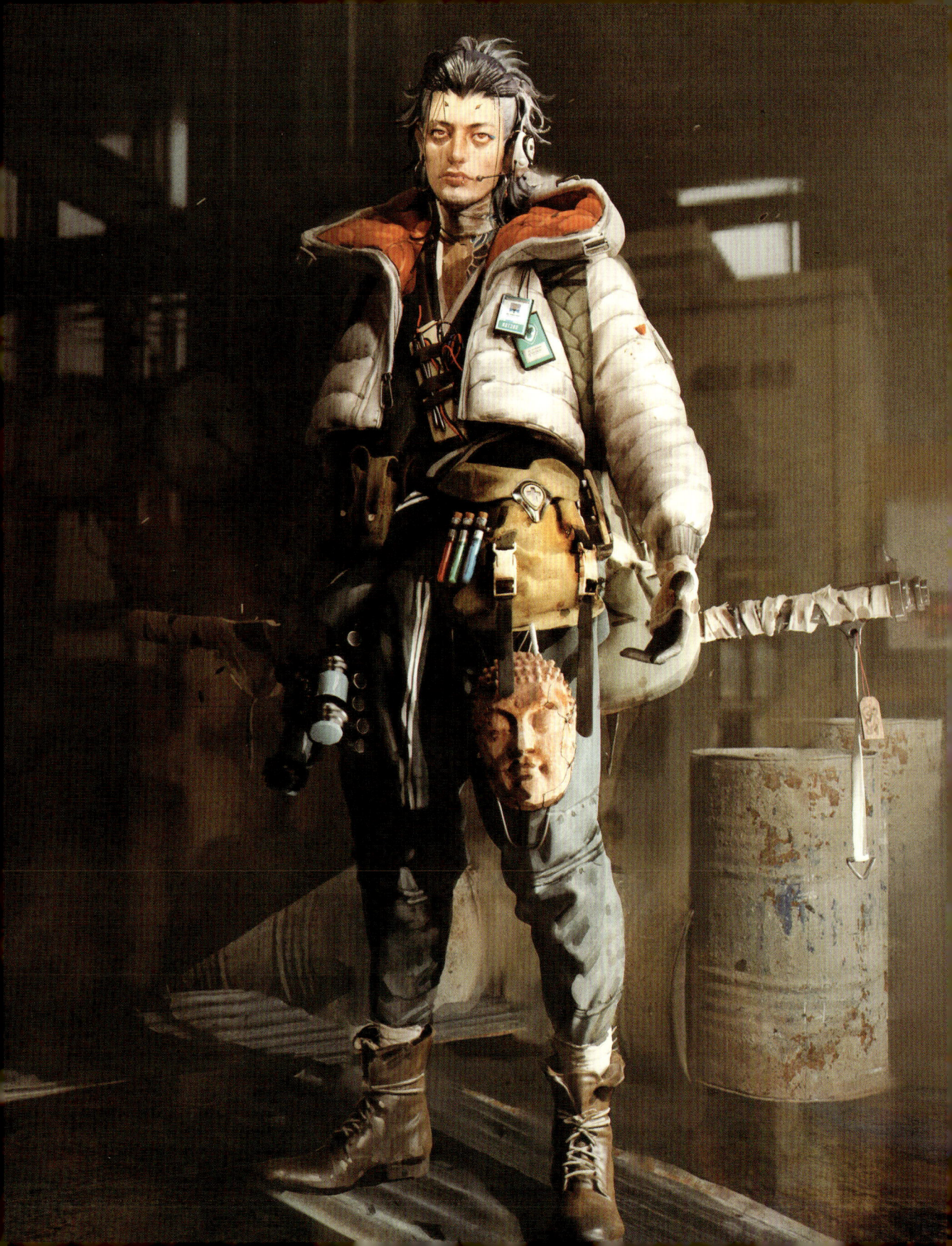

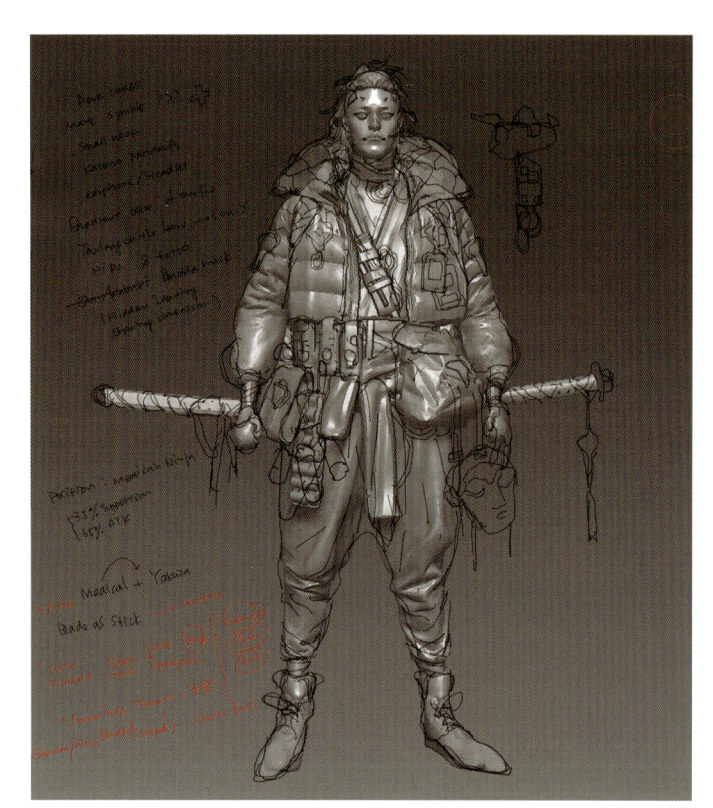

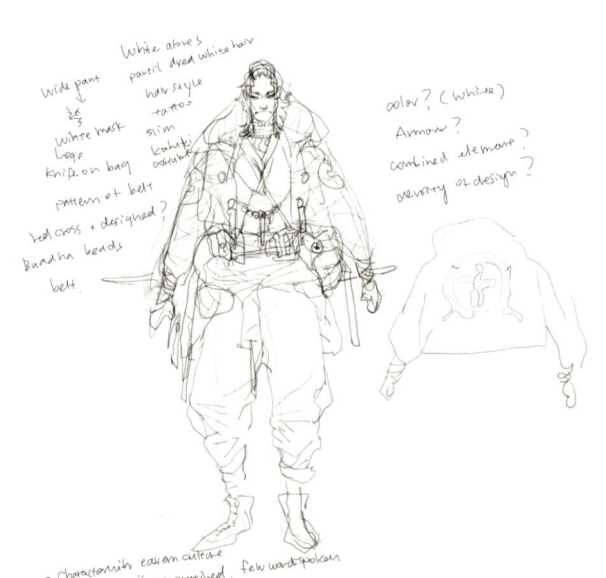

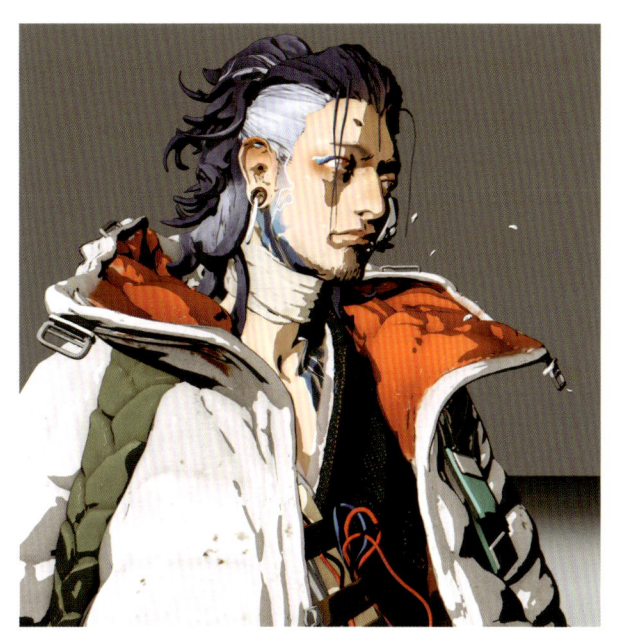
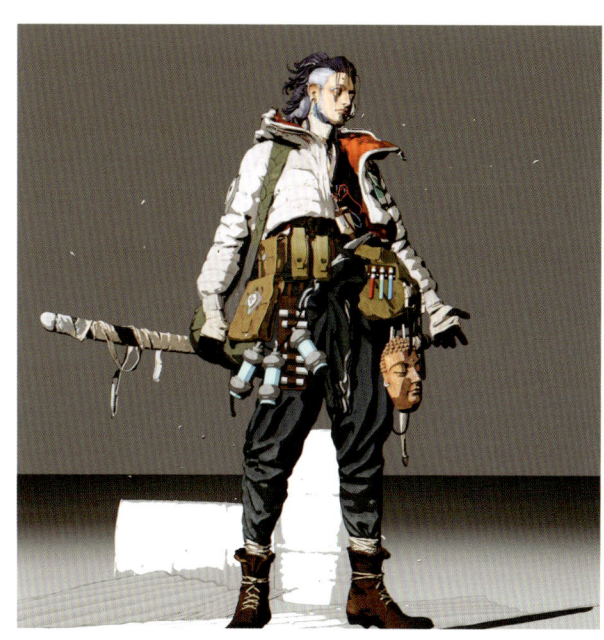
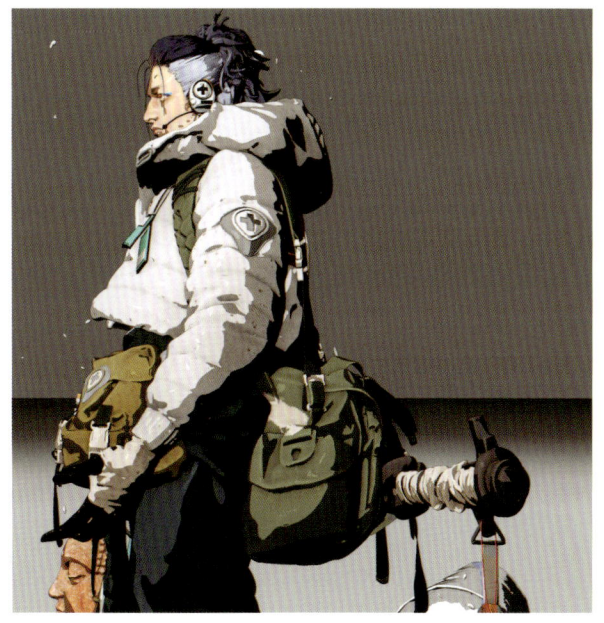
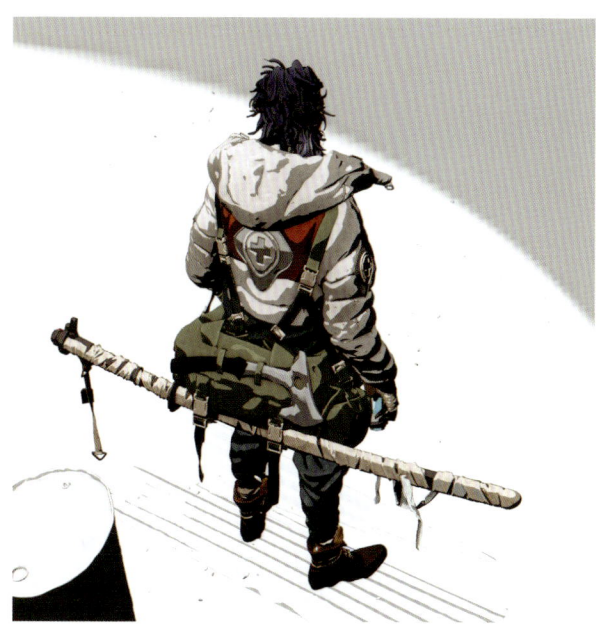

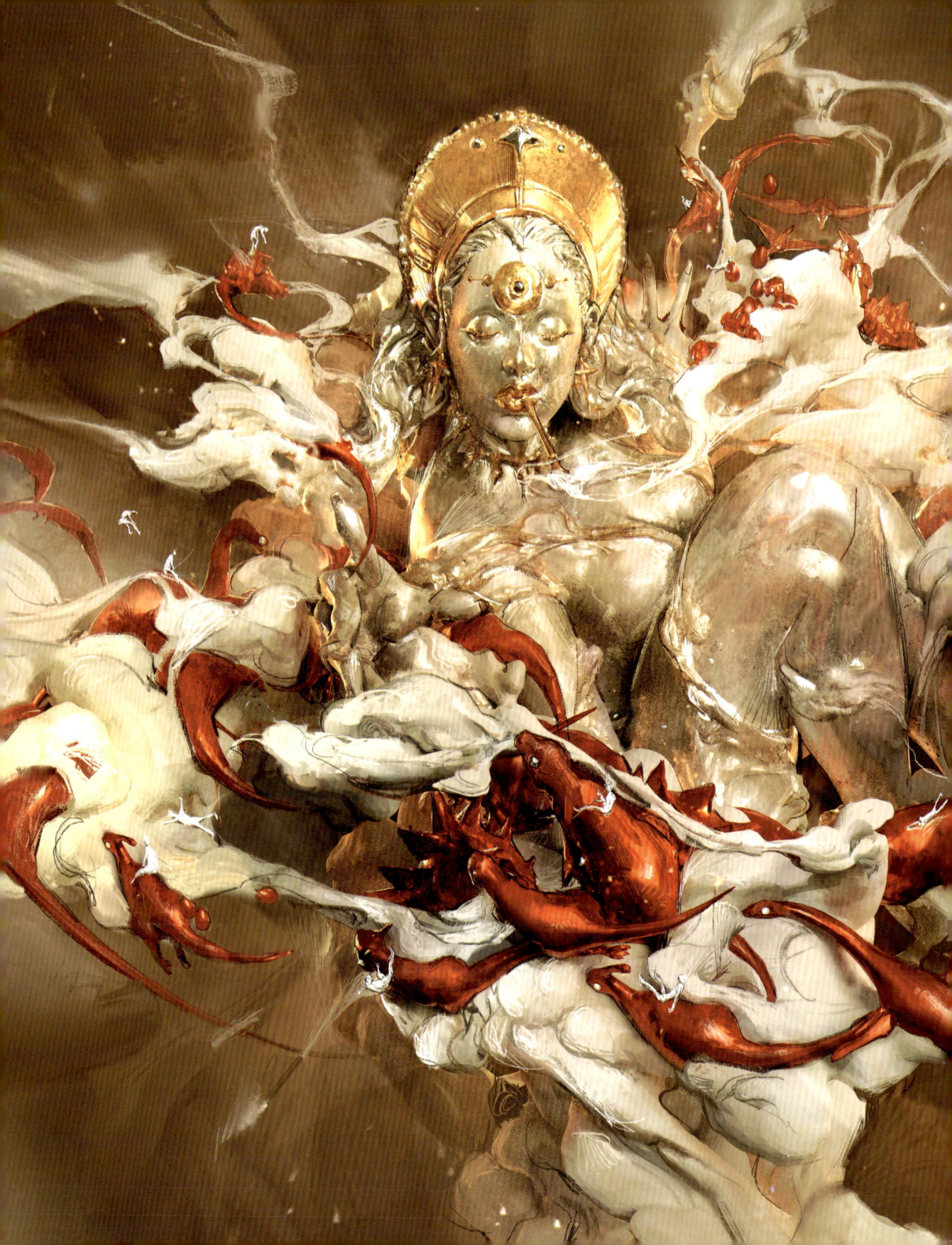

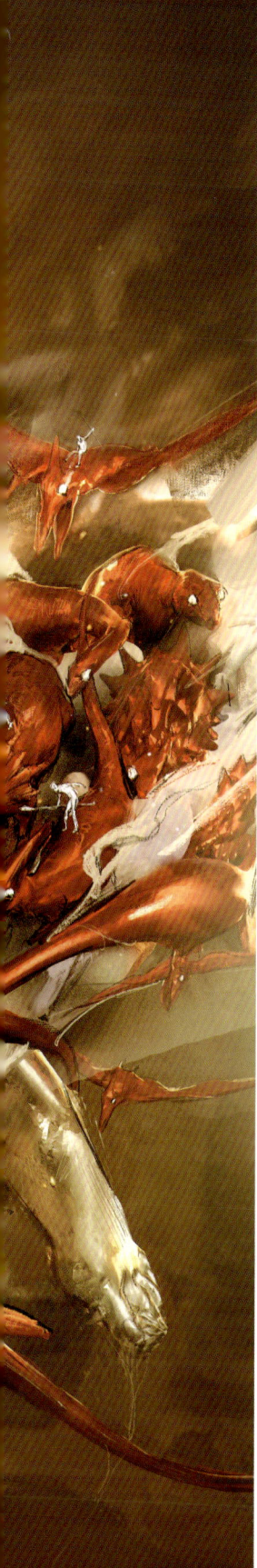

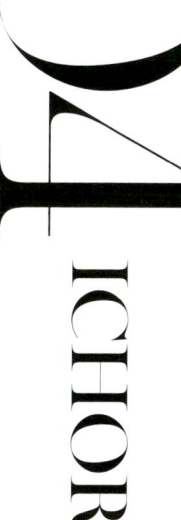
ICHOR(Divine Blood). What flows through an immortal is not blood, but the golden essence of eternity.

Some have described my style as akin to water—fluid, surging—yet also like metal—dense and shimmering. I've always felt that there is a deity upon the canvas, and I am but their most humble and devoted follower. We do not grasp the true essence of art merely by mastering techniques; facing creation is like facing a world filled with myriad uncertainties. I will forever follow my emotions and the beat of my heart as I wield my brush, always living within the process.

イコル（神の血）。不死の身体を流れるものは血ではなく、黄金に輝く永遠の本質です。

一部の人たちは私のスタイルを、流動的にうねる水のようでありながら、濃密でキラキラときらめく金属のようでもあると表現します。私はいつも、キャンバスの上には何らかの神性が存在していて、自分はその最も謙虚で献身的な信奉者に過ぎないと感じてきました。技術を極めるだけでは芸術の真髄は掴めません。創作と向き合うことは、無数の不確実性に満ちた世界と向き合うことに似ています。私はこれからもずっと自分の感情と心臓の鼓動に従って筆を振るい、常にその過程の中で生きてゆくでしょう。

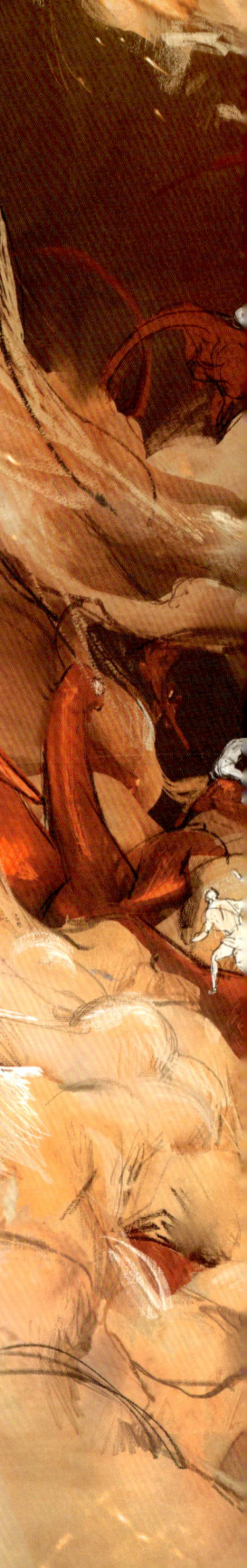

098

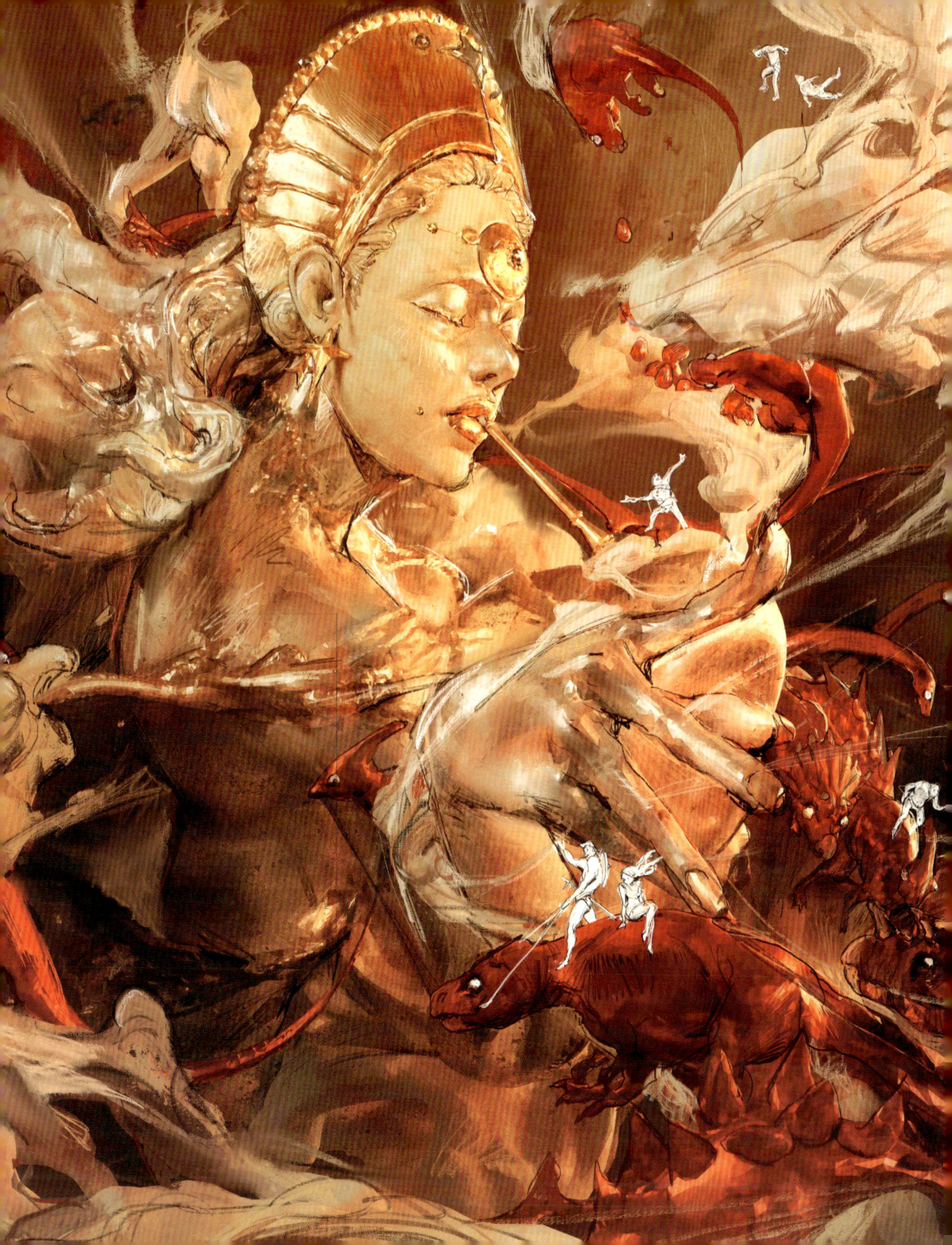

Lunaris 100

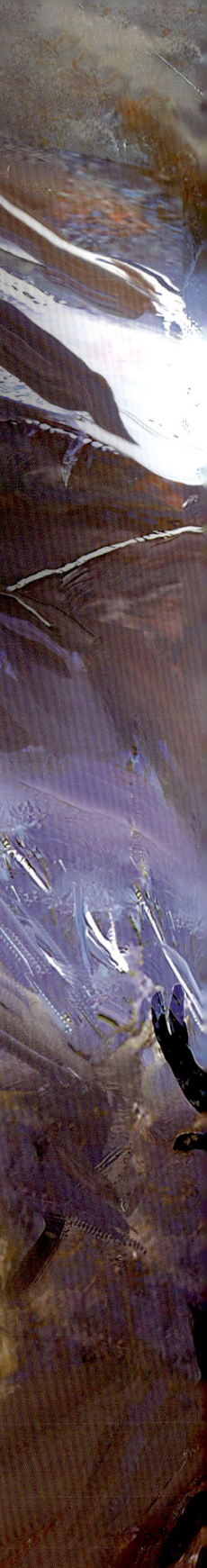

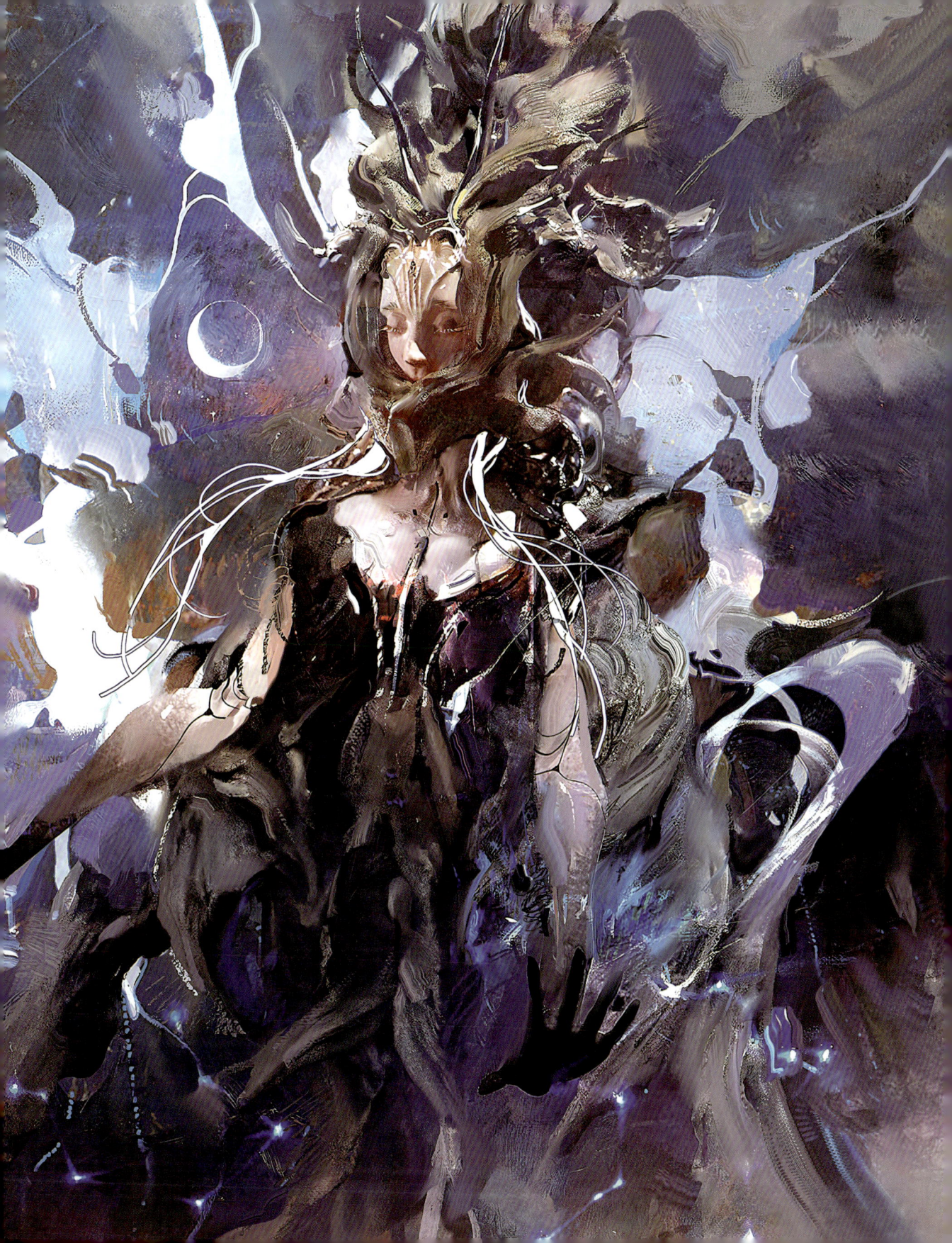

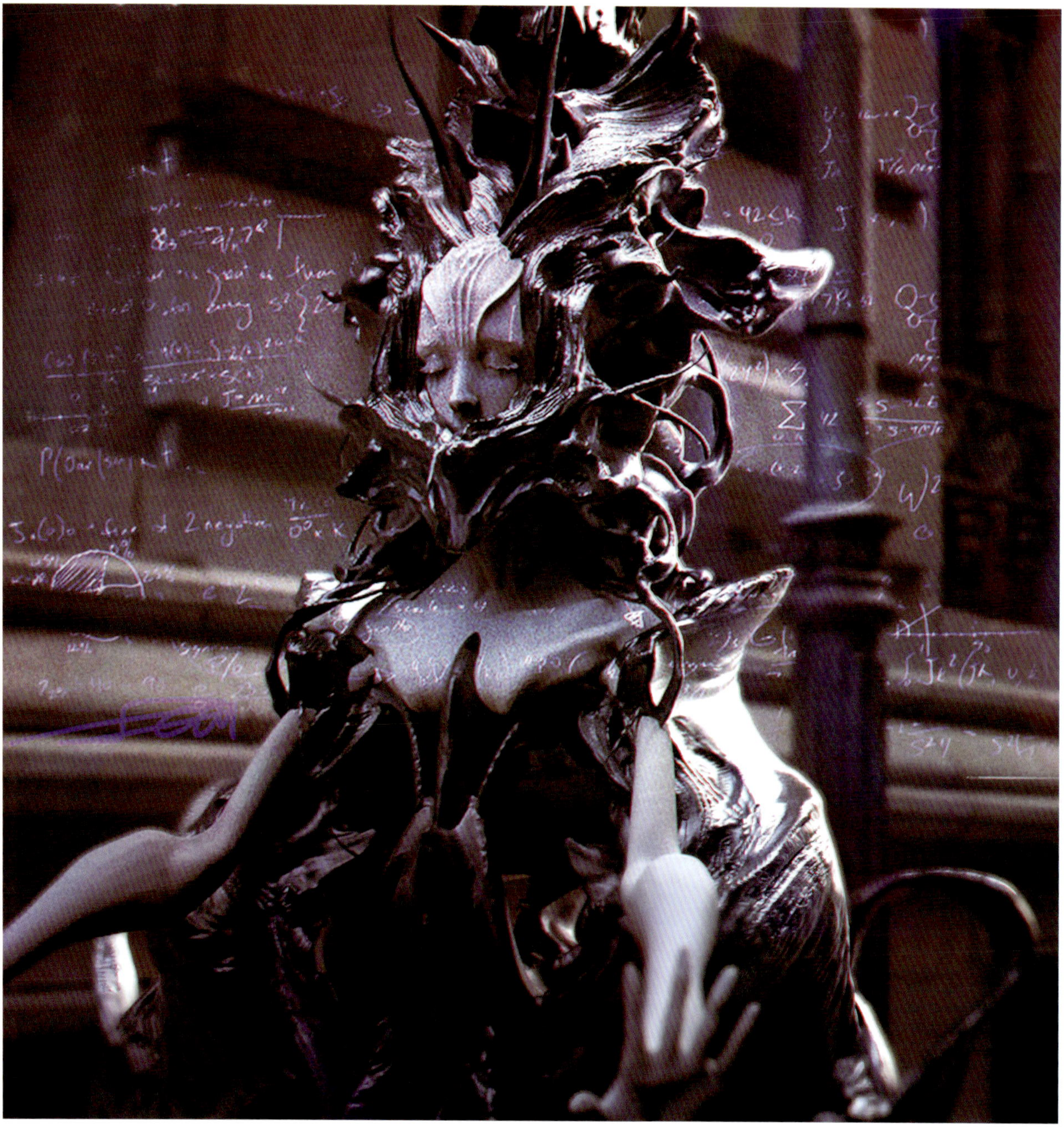

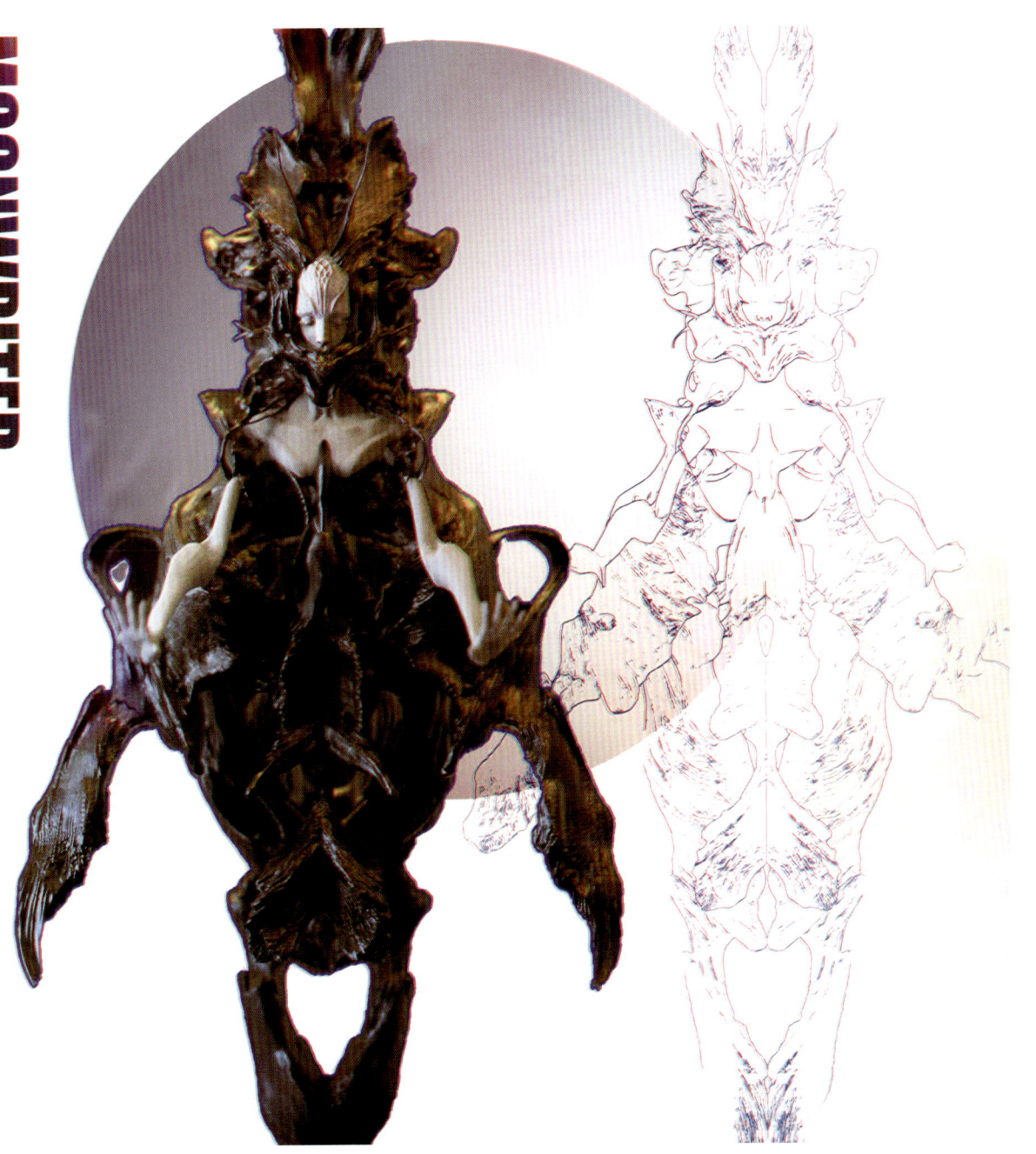

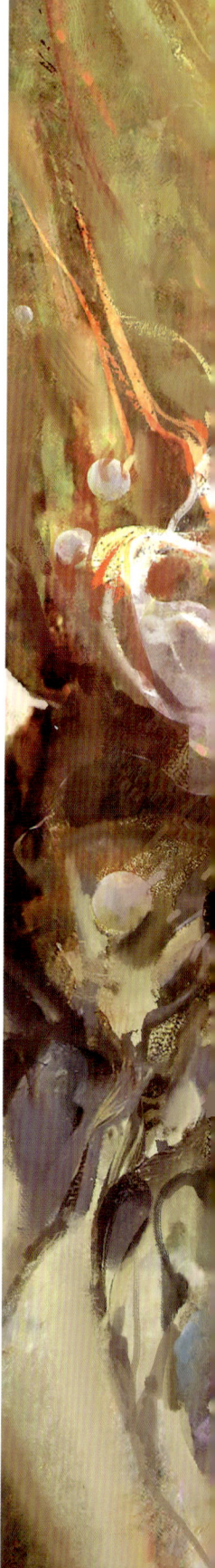

Solaris 10

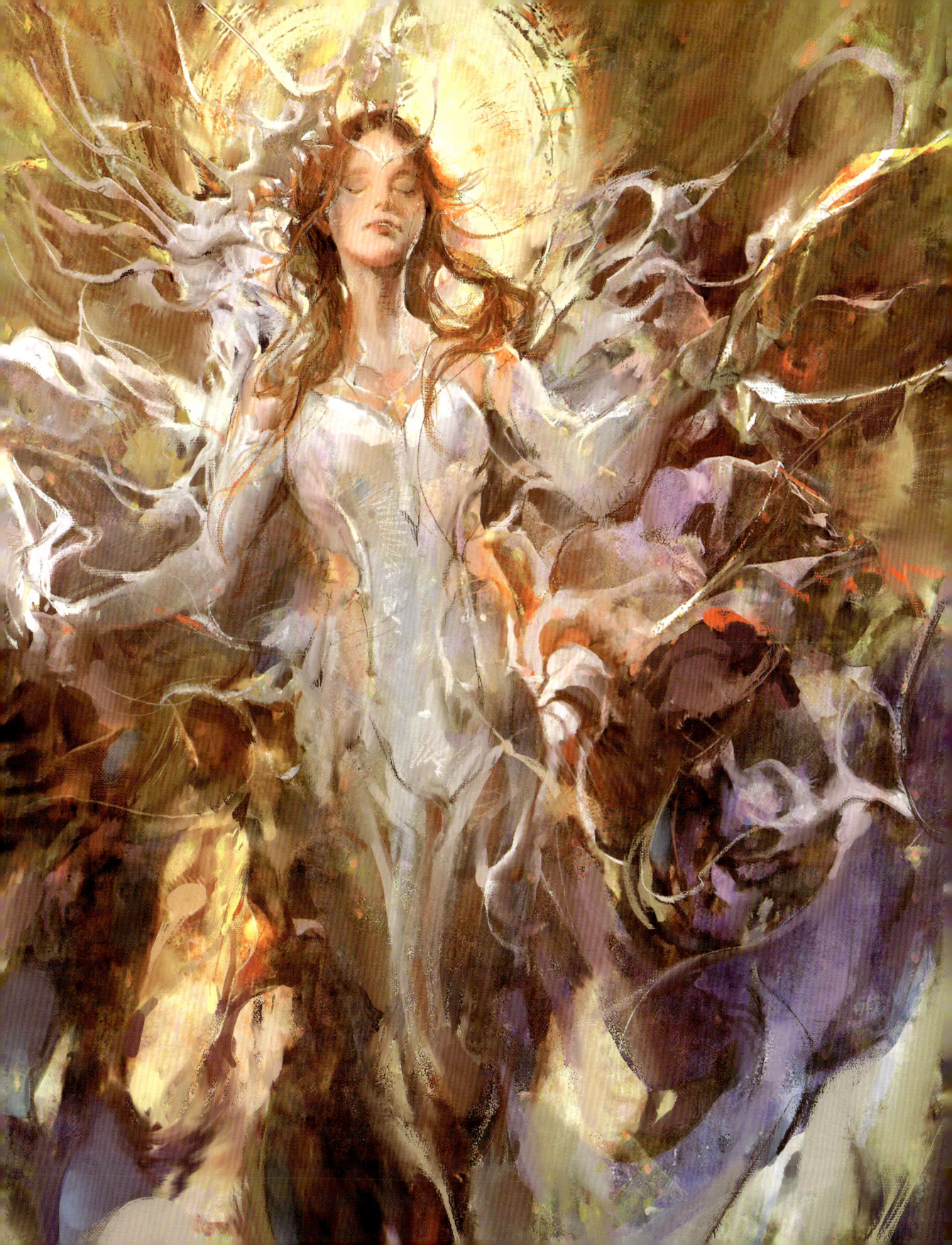

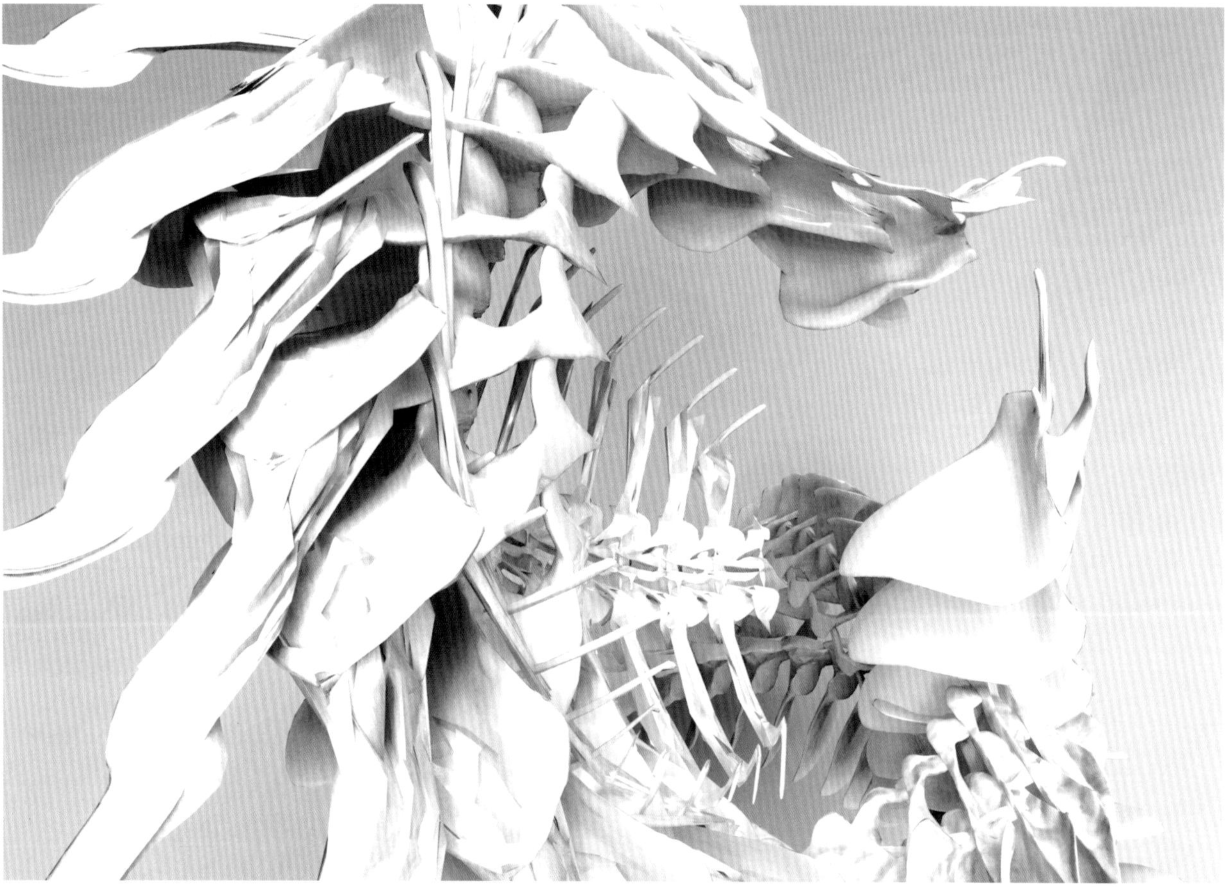

Part 01    The Connective Synthesis of Production

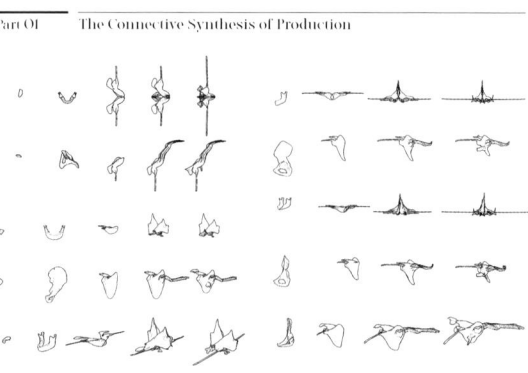

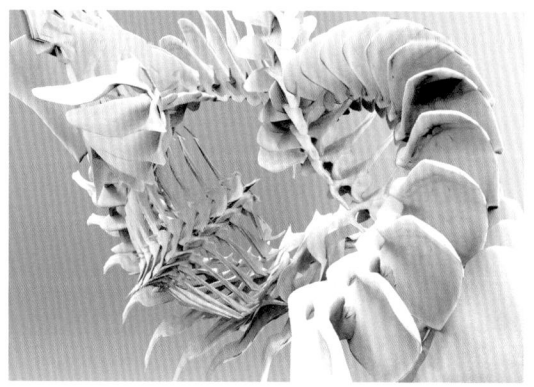

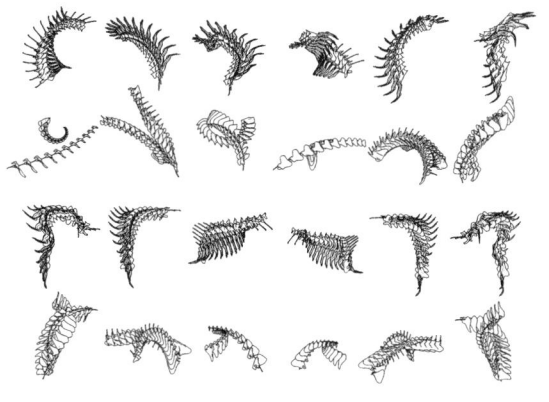

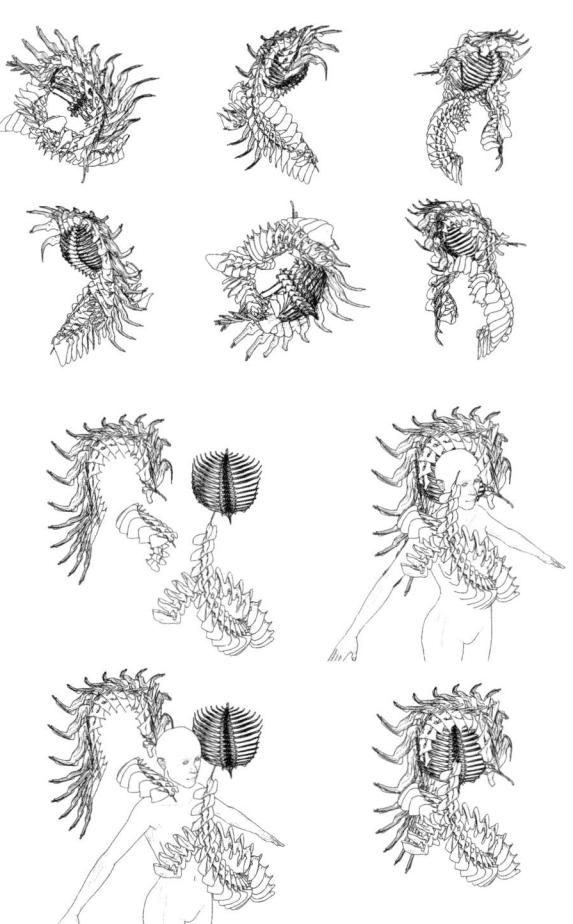

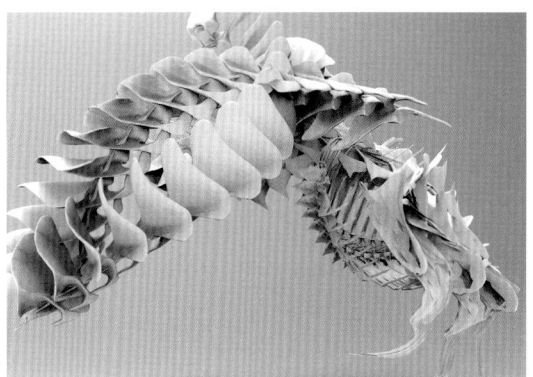

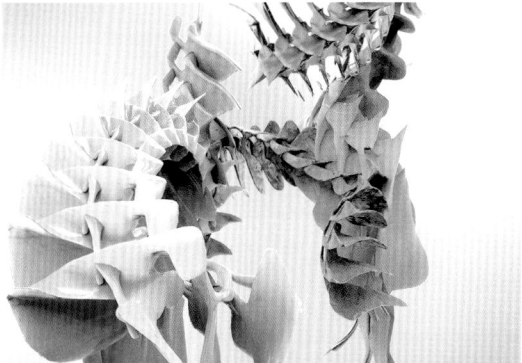

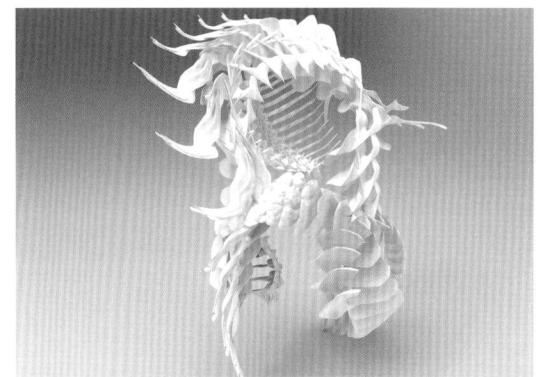

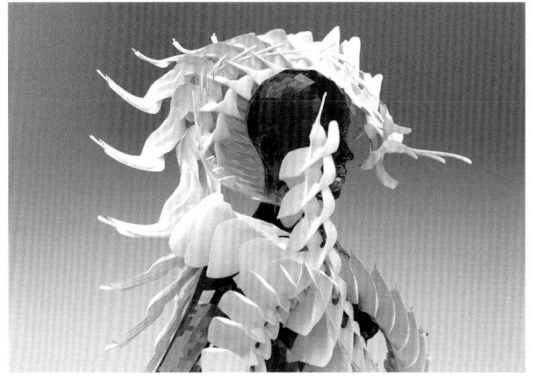

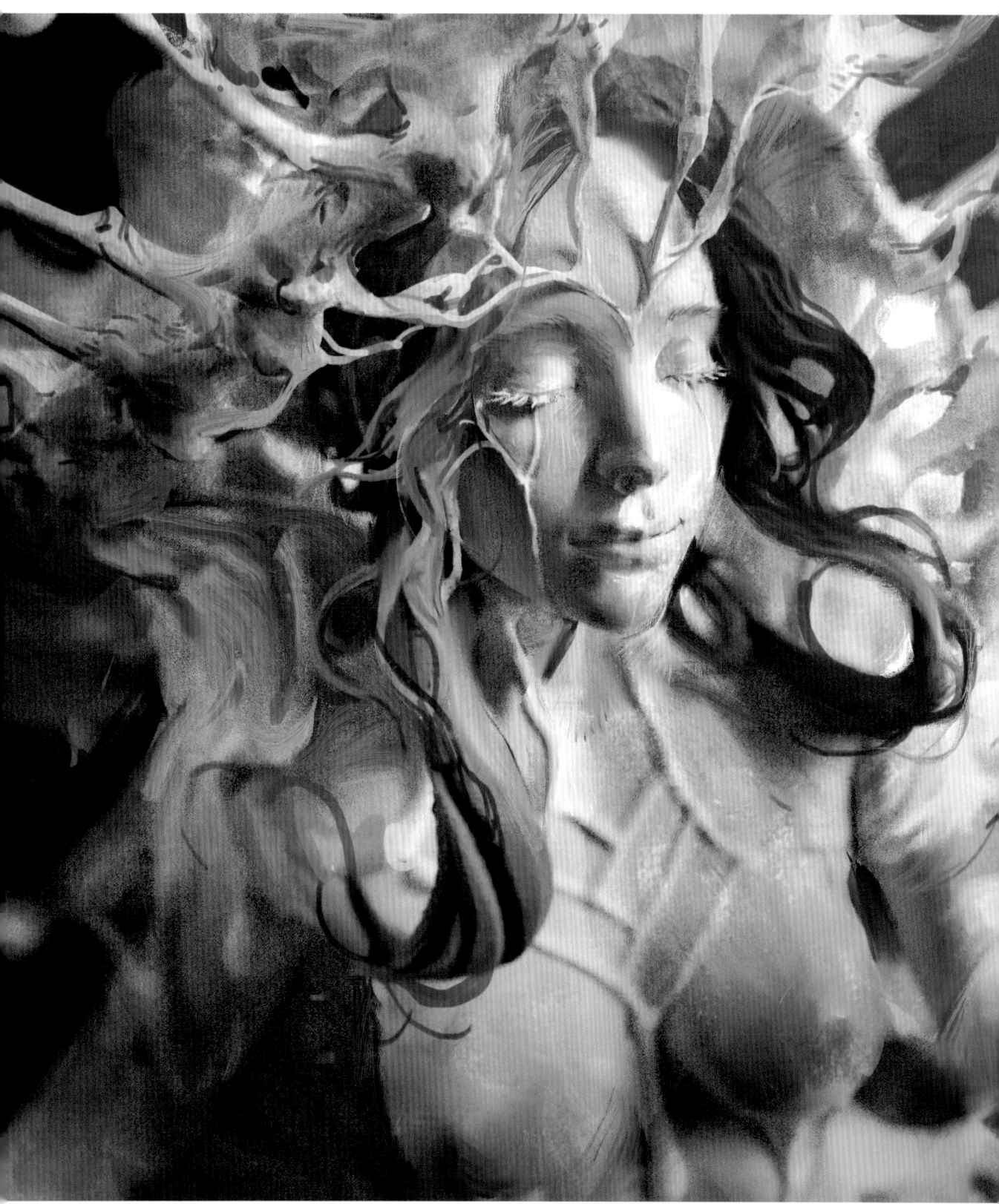

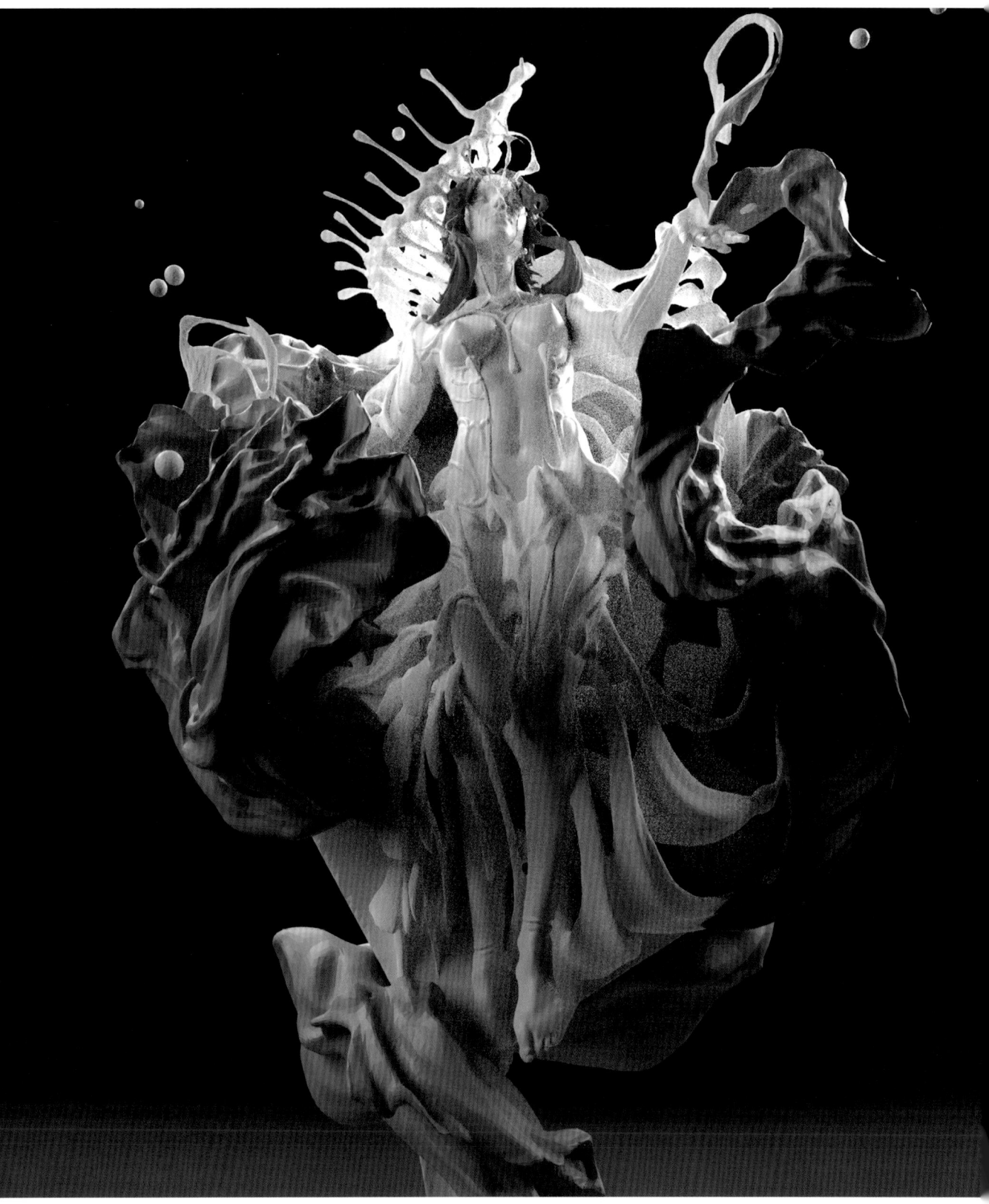

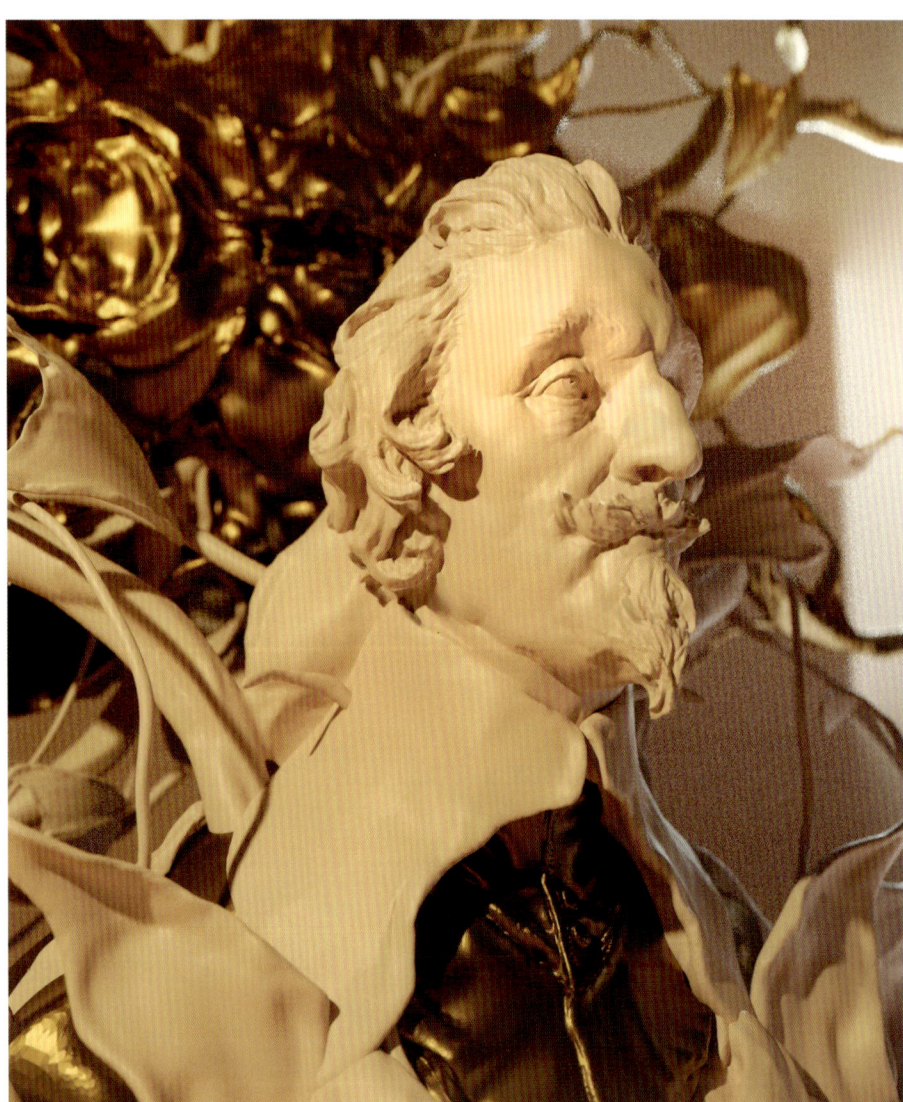

Richelieu 110

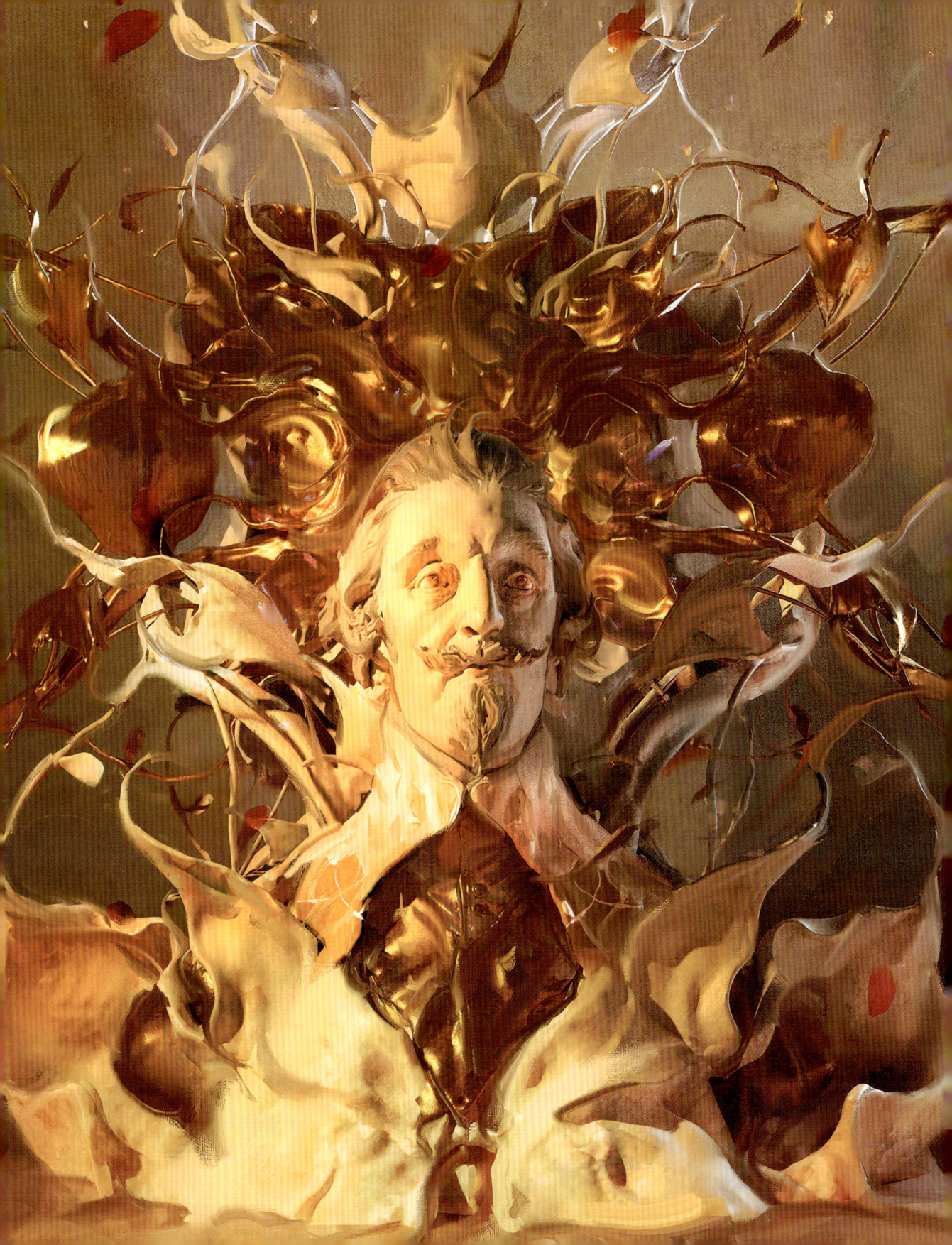

# Hanuman 112

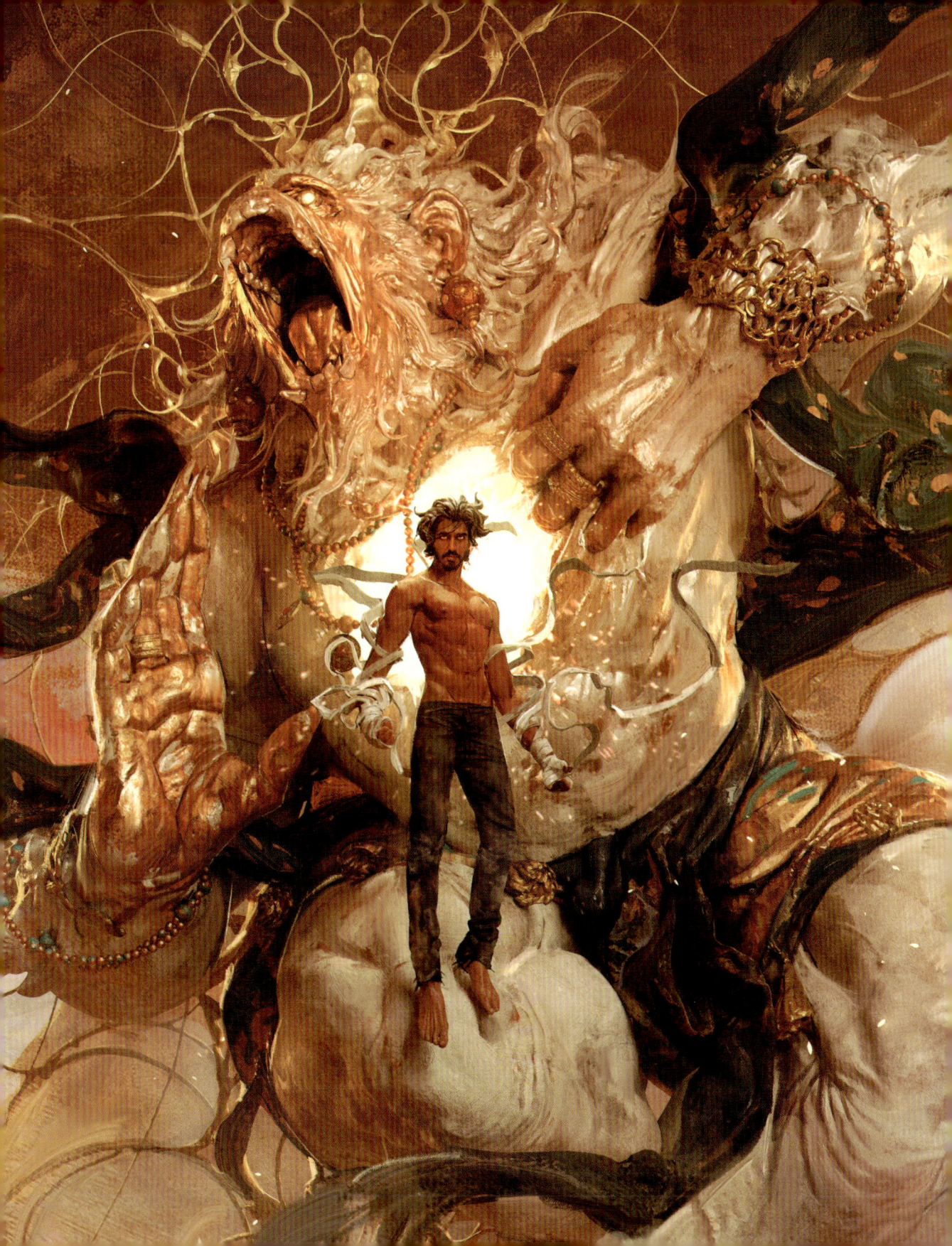

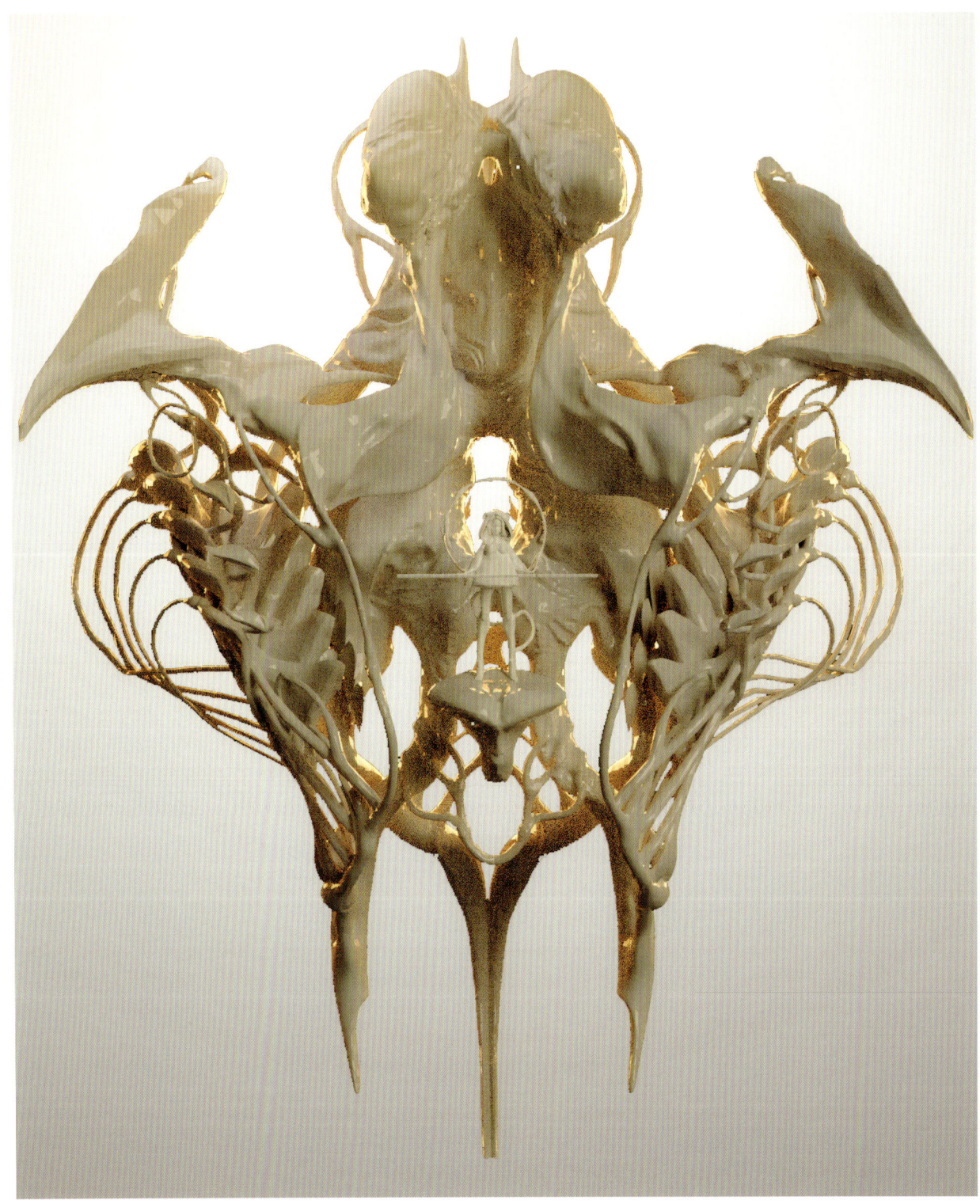

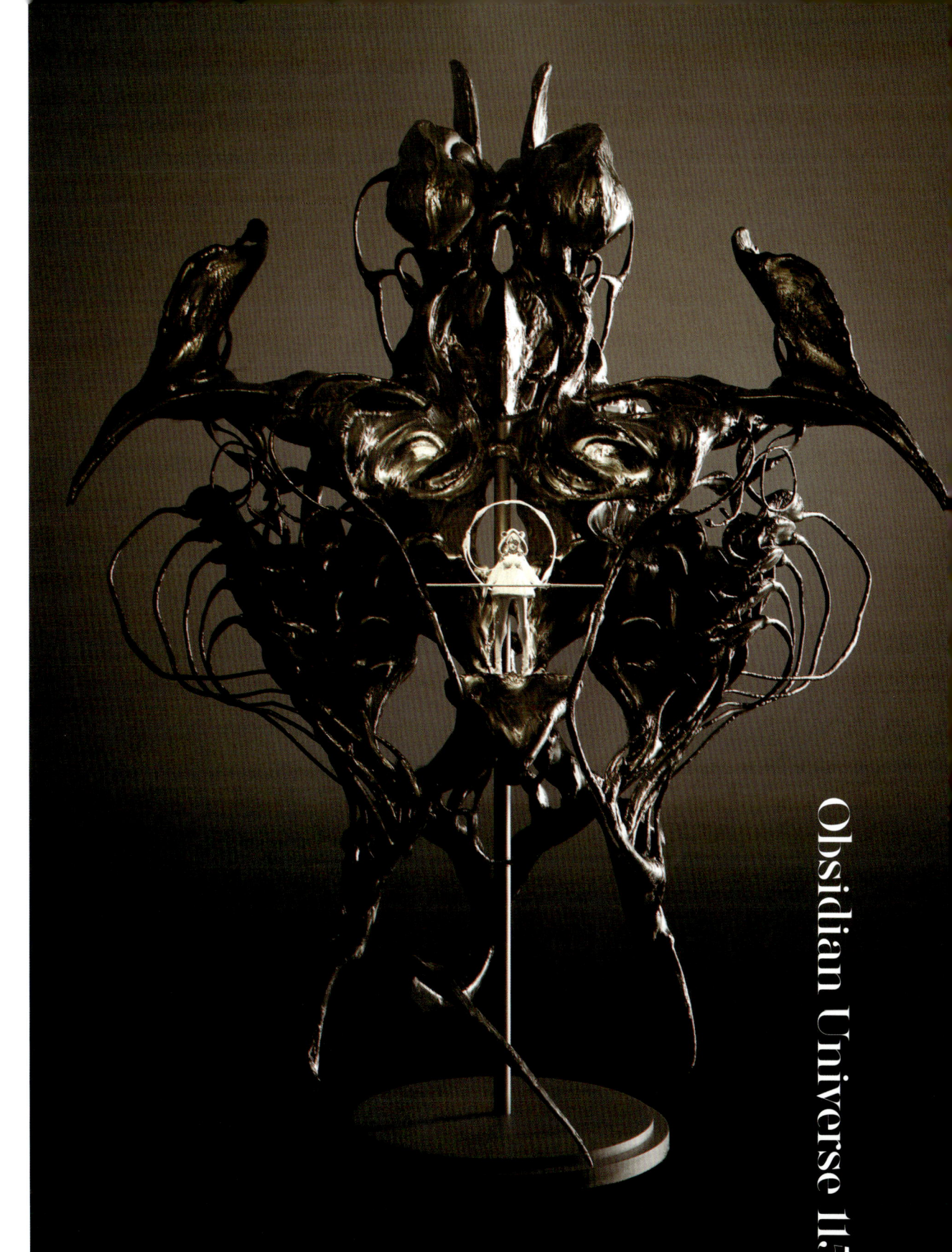

Obsidian Universe 115

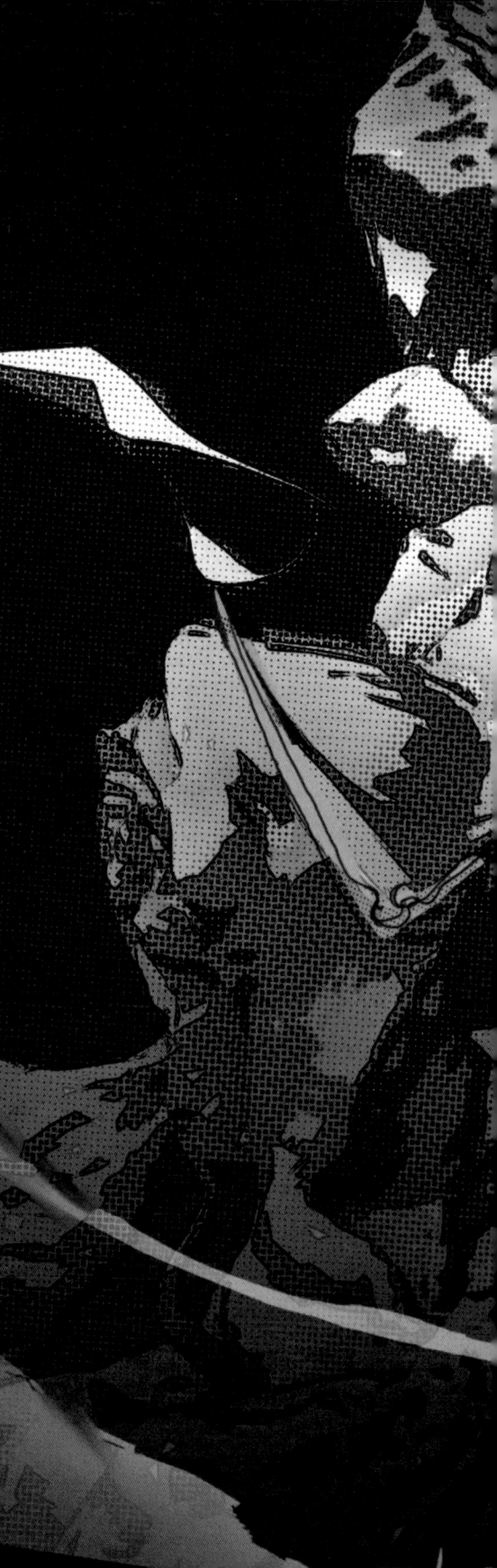

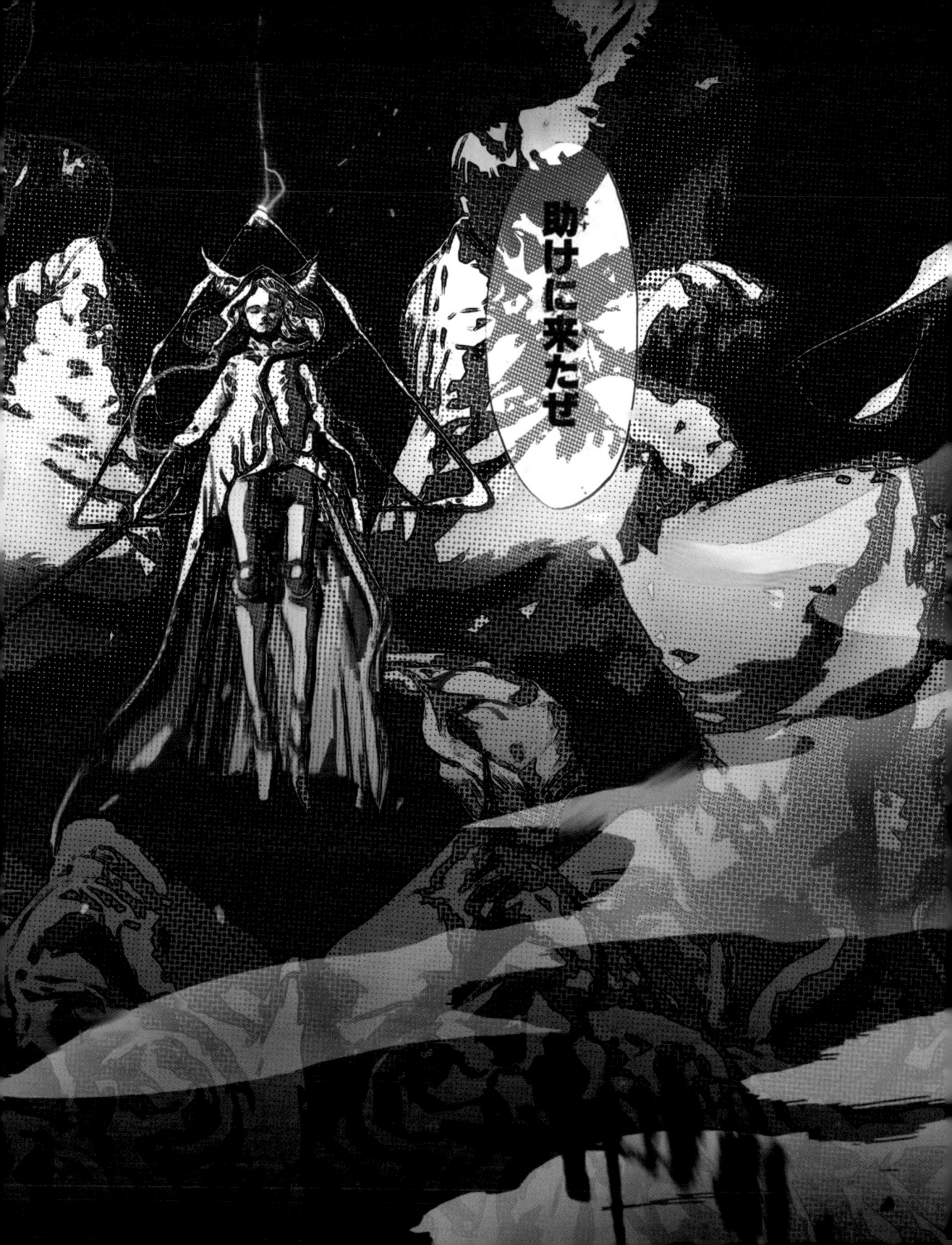

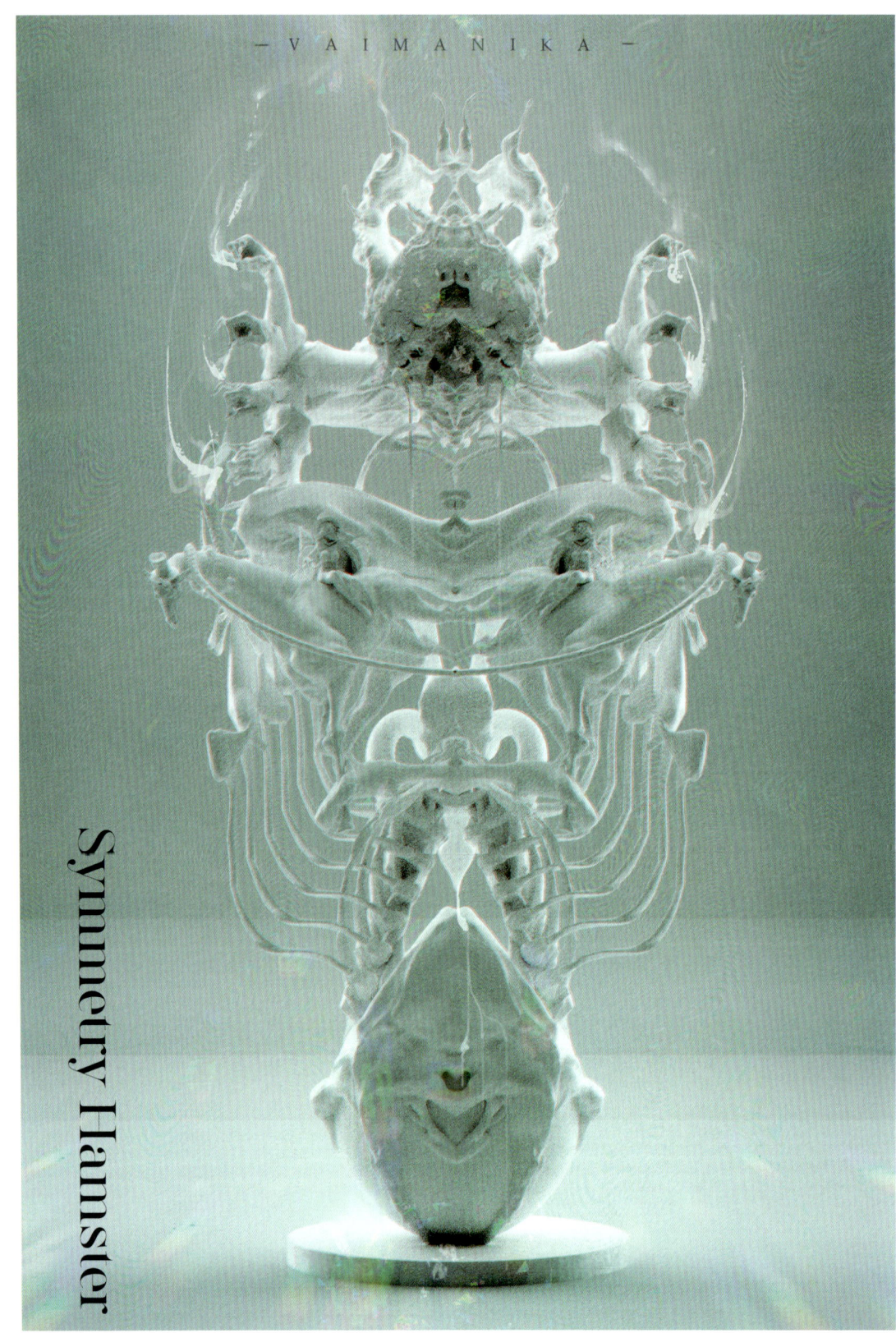

Symmetry Hamster

# Symmetry Hamster 118

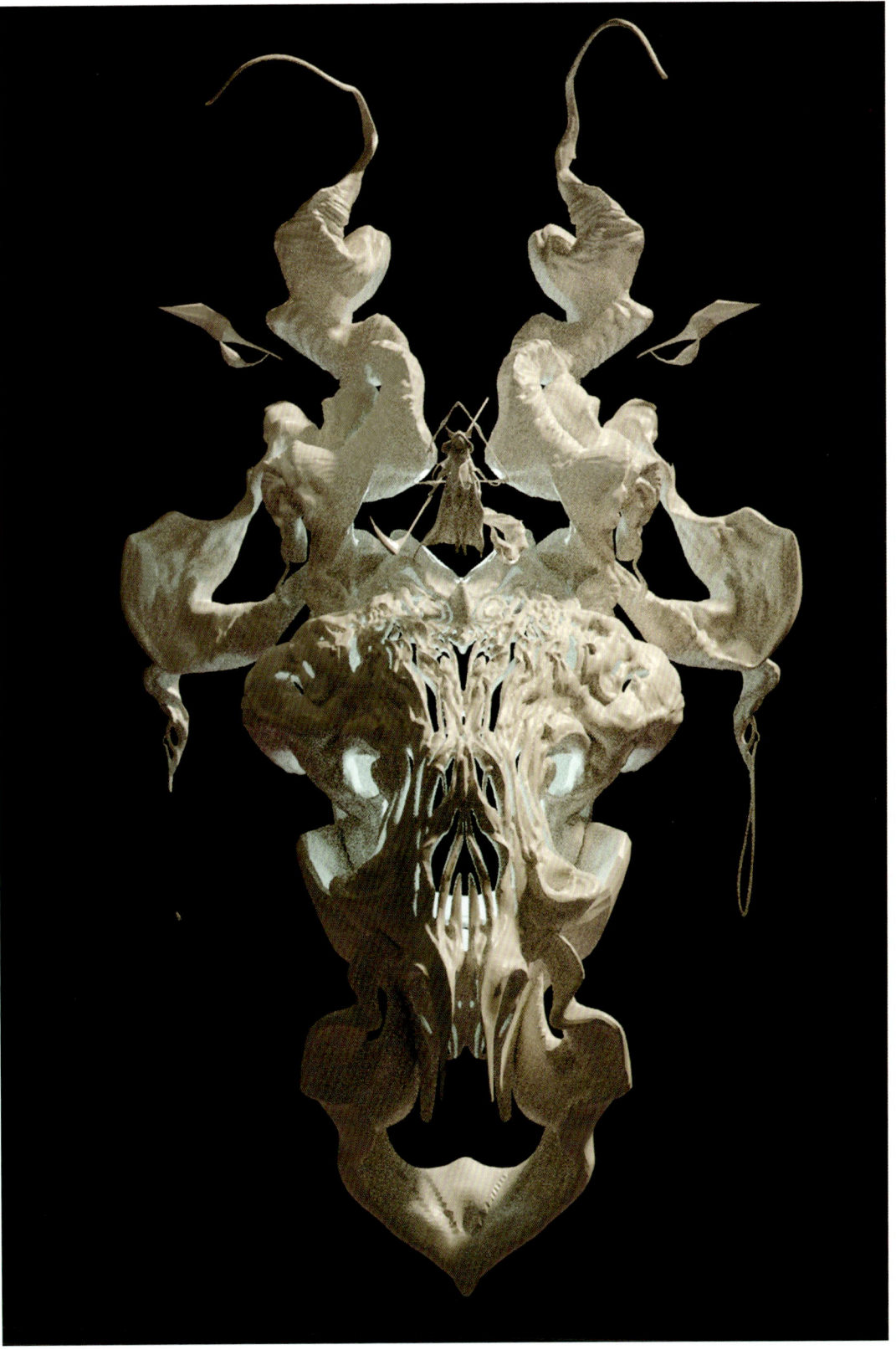

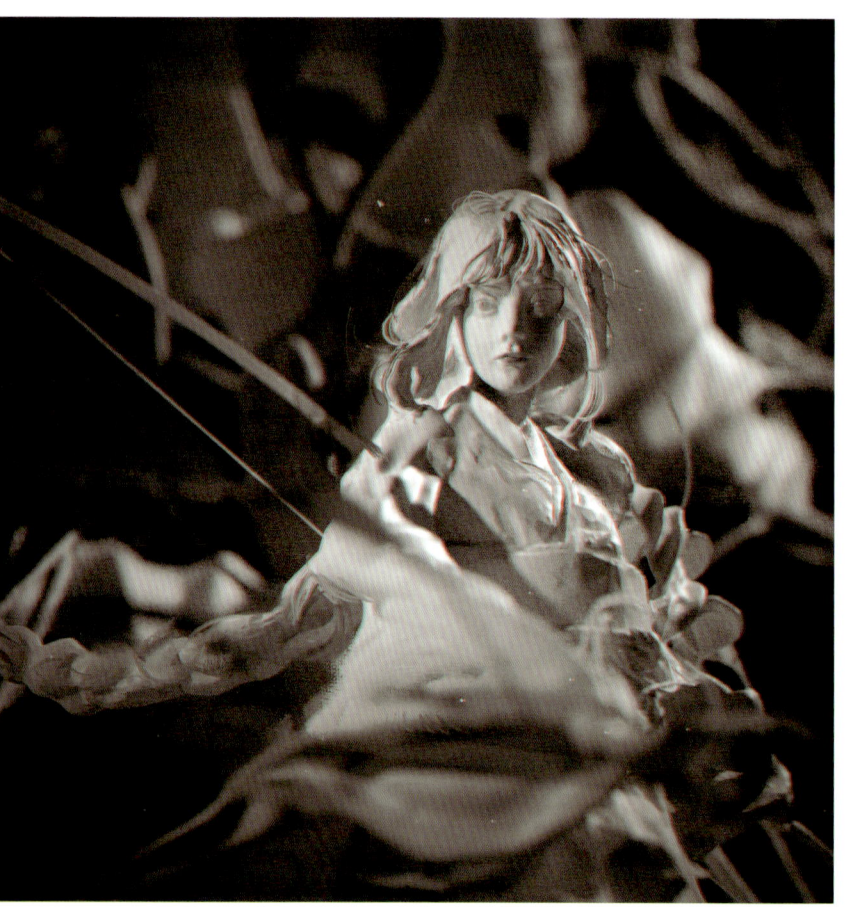

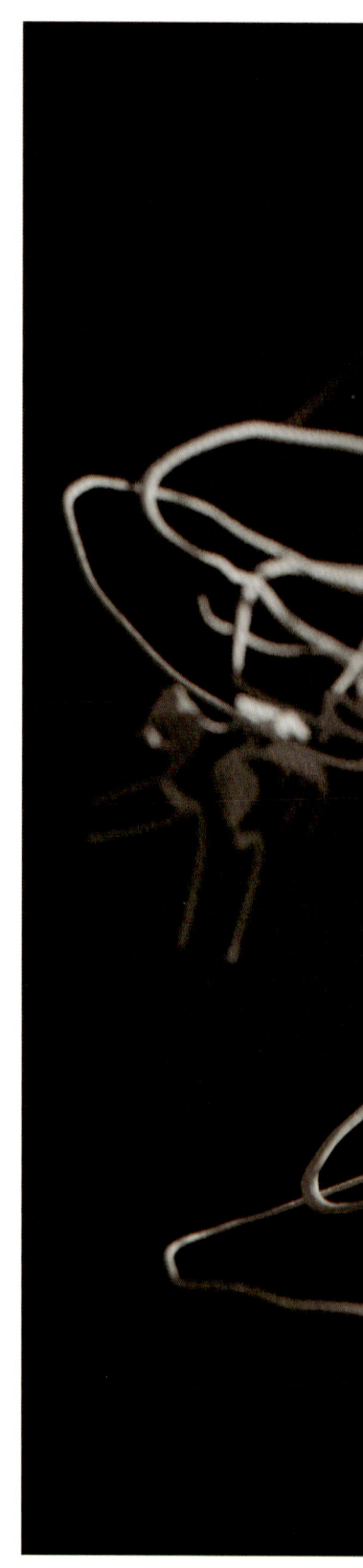

Rose Archer 120

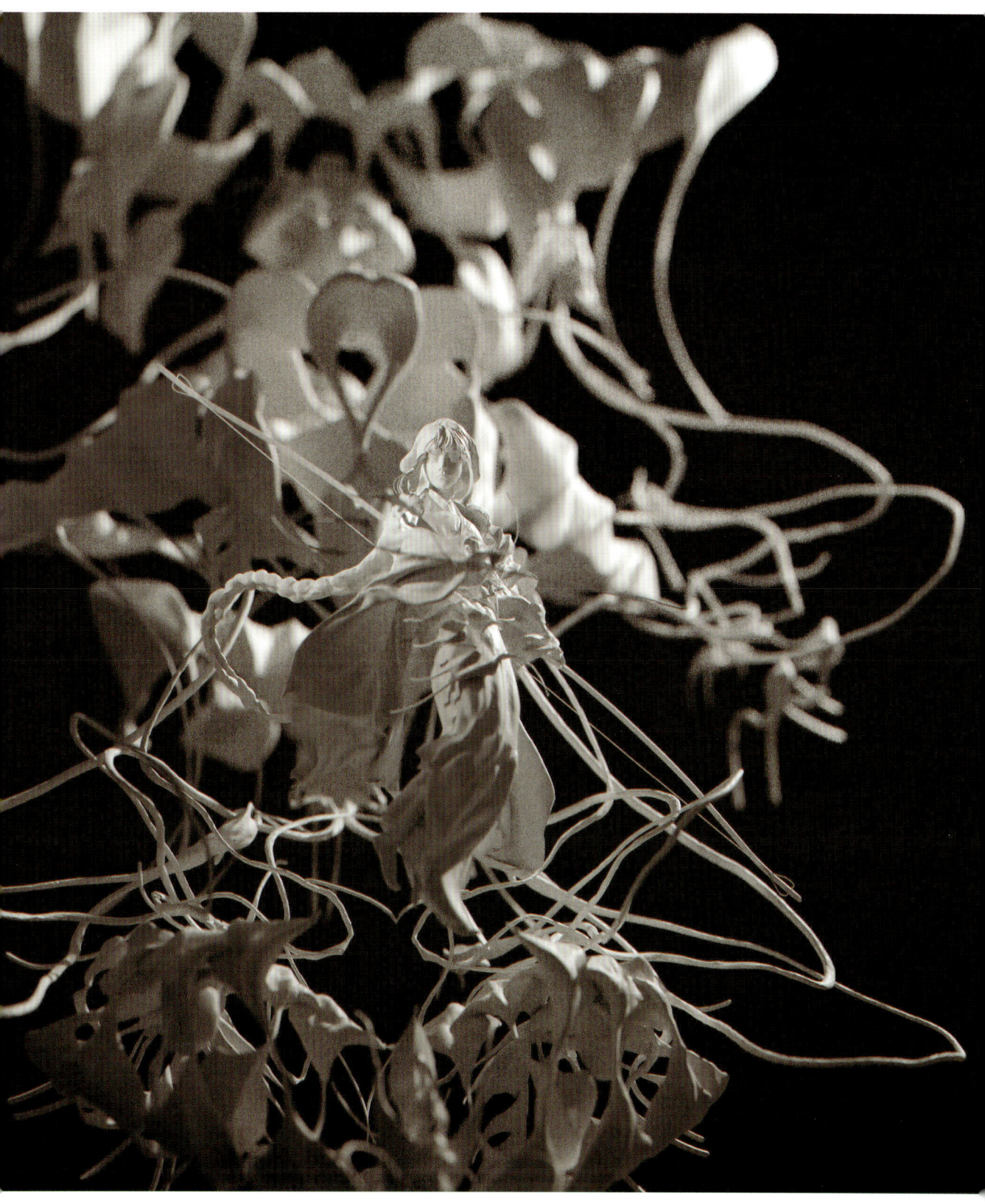

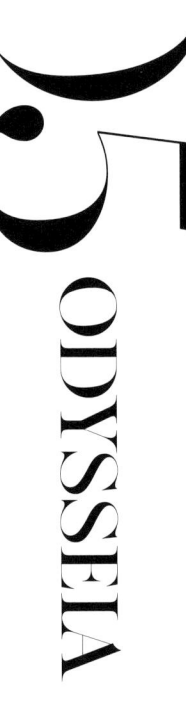

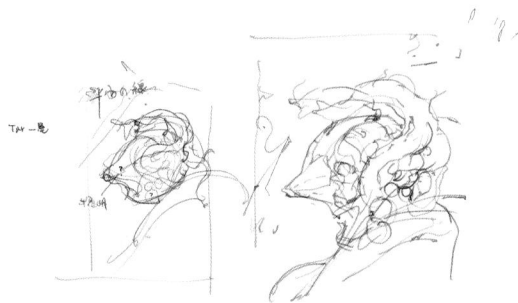

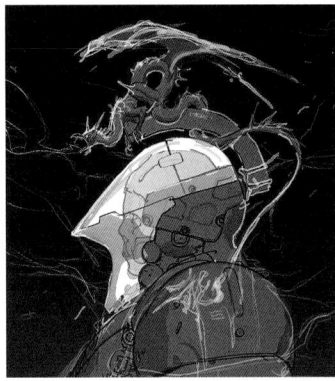

Just like its symbol character LUDENS, KOJIMA PRODUCTIONS pioneers the boundaries of game as an artistic medium with curiosity and a spirit of adventure. Here, a group of creative members gather to produce unique gaming experiences.

This chapter showcases some of the works I've created for KOJIMA PRODUCTIONS over the past few years. I am deeply grateful to Mr. Hideo Kojima and Mr. Youji Shinkawa for giving me the opportunity to interpret these captivating characters according to my own understanding. For instance, I designed a golden dragon sculpture for LUDENS' helmet shield to celebrate the Lunar New Year of the Dragon; a Halloween pumpkin rendition of the popular four-legged BT monster from DEATH STRANDING; and sculptures and illustrations of LUDENS GIRL.

based on Mr. Shinkawa's original designs, but incorporatinged my own style and trying with new dynamics and color textures. The depiction captures the moment she leaps up to plant a flag during her conquest of untamed Lands.

The works I can publicly disclose as a concept artist at KJP are just the tip of the iceberg. I hope the art created with the team can be showcased in the form of games in the future.

コジマプロダクションは、そのシンボルキャラクターであるルーデンスのように、好奇心と冒険心を胸に芸術メディアとしてのゲームの限界を開拓します。ここはクリエイティブなメンバーが集まり、唯一無二のゲーム体験を生み出す場です。

この章では、過去数年間にコジマプロダクションのために制作した作品の一部を紹介します。魅力的なキャラクターたちを私なりに解釈する機会を与えてくださった、小島秀夫氏と新川洋司氏に深く感謝します。例えば、辰年の旧正月を祝うためにルーデンスのヘルメットに黄金の龍の造形をデザインしたり、「DEATH STRANDING」に登場する4本足の人気キャラクター、BTキャッチャーをハロウィンのカボチャで表現したりしました。ルーデンス・ガールの造形とイラストでは、新川氏のオリジナルデザインをベースに、私自身のスタイルを取り入れて、新しいダイナミクスと色彩のテクスチャーに挑戦しました。

この肖像は、未開の地を探検する際に跳び上がり、旗を立てる瞬間を捉えたものです。

コジマプロダクションでコンセプトアーティストとして手掛けた作品のうち、私が公開できるものはほんの一部です。チームとともに作り上げた芸術が、ゲームという形で将来披露されることを願っています。

Dragon New Year Art 12:

From "LUDENS" by KOJIMA PRODUCTIONS

©KOJIMA PRODUCTIONS Co.,Ltd.

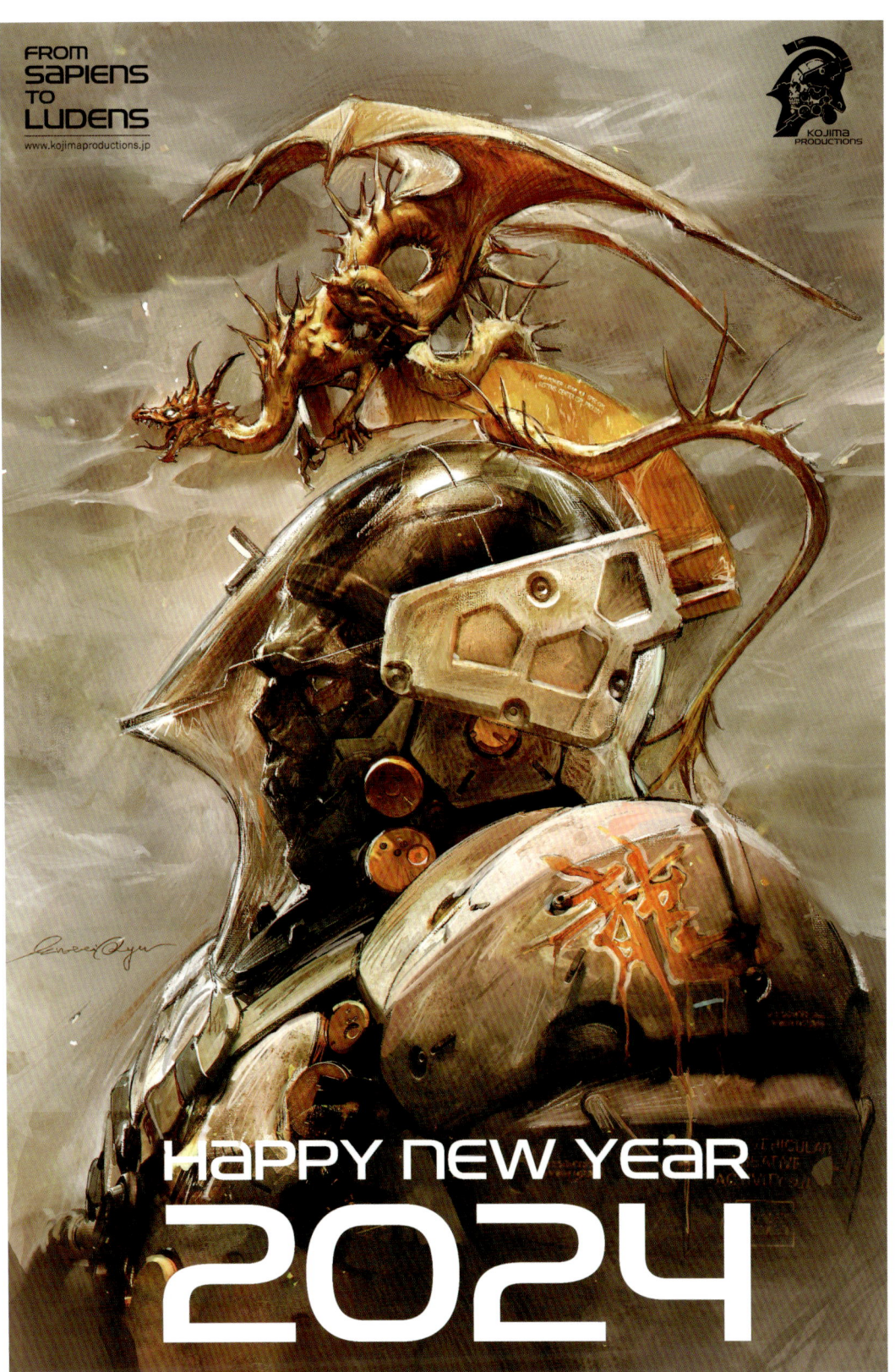

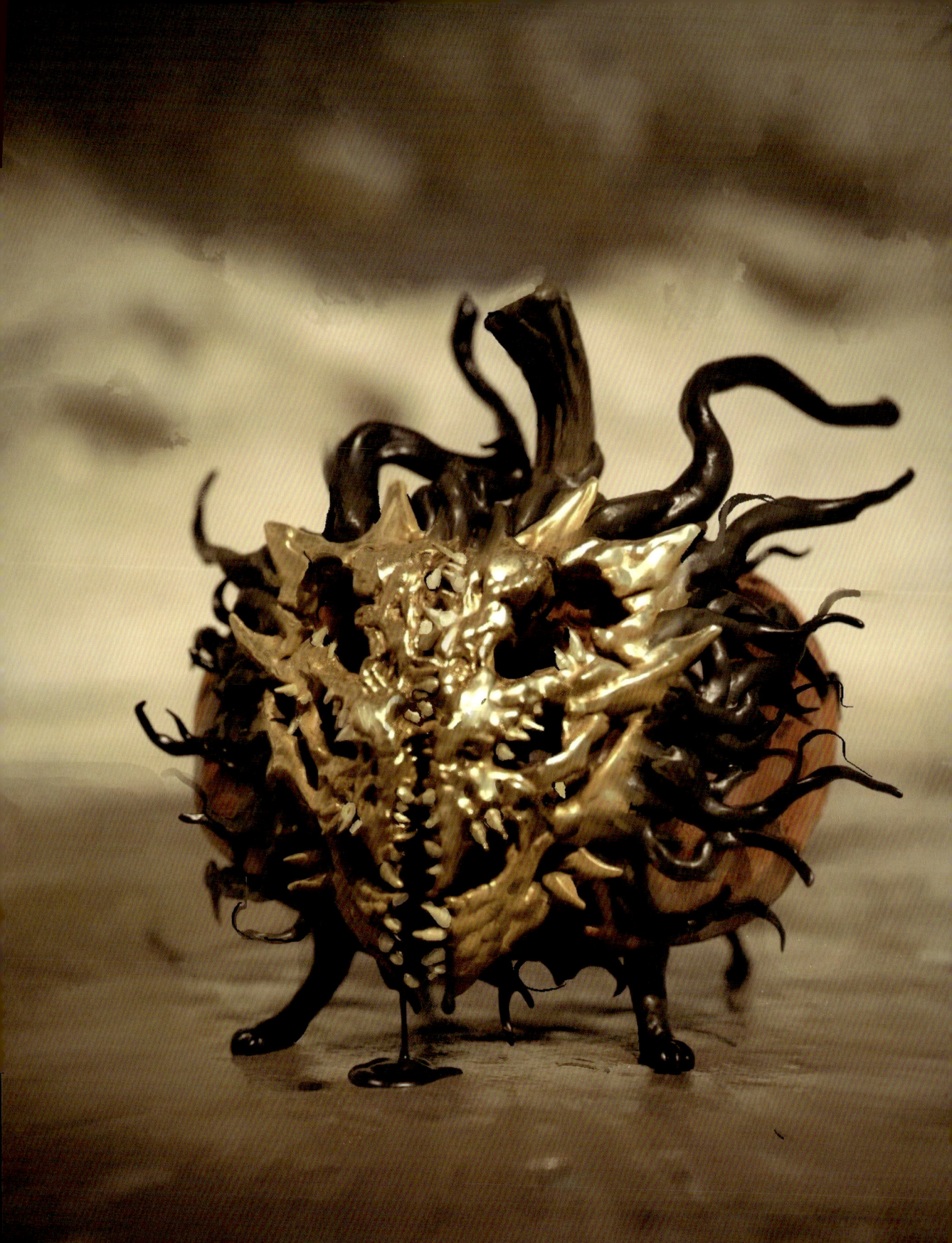

キャッチャー
＋
カボチャー
↓
キャッボチャー

Halloween BT Mascot 125

From DEATH STRANDING by KOJIMA PRODUCTIONS

© Sony Interactive Entertainment Inc. / KOJIMA PRODUCTIONS Co., Ltd. / HIDEO KOJIMA

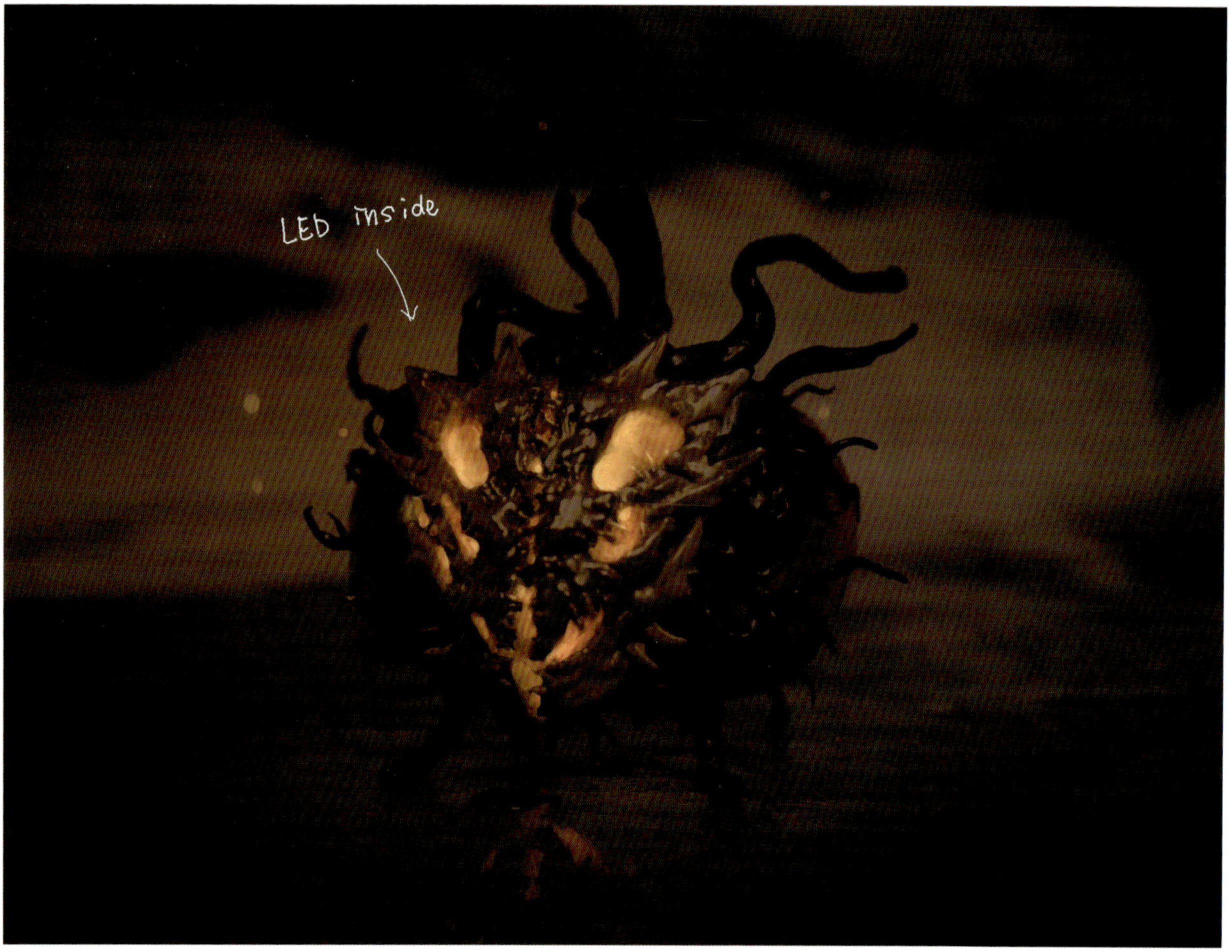

LED inside

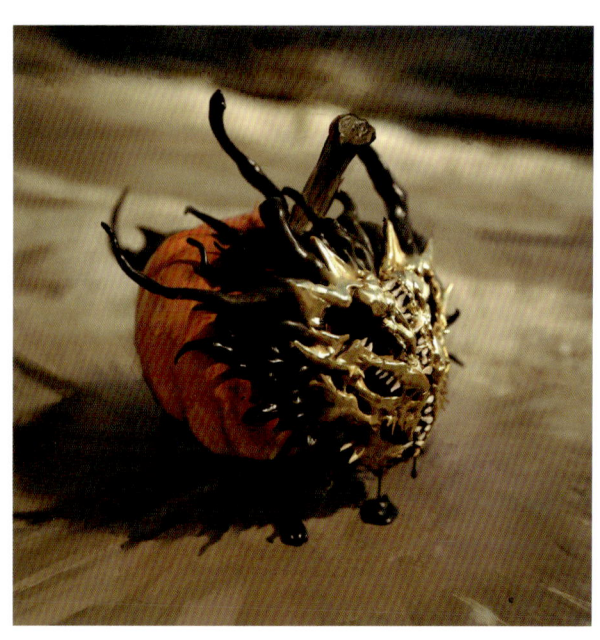

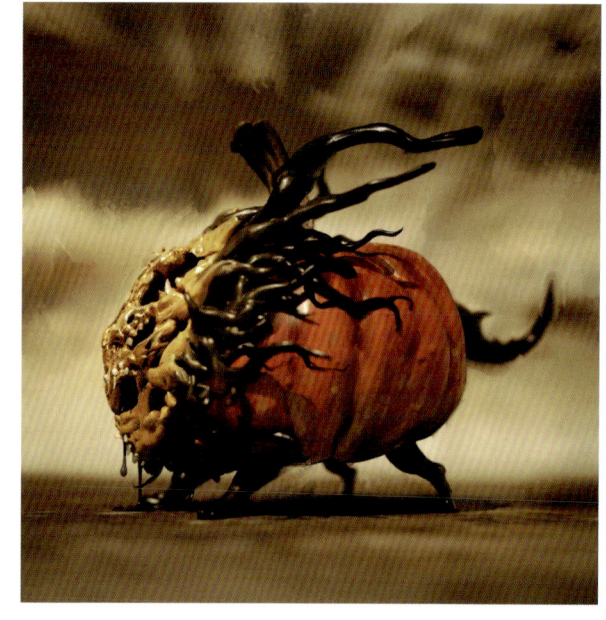

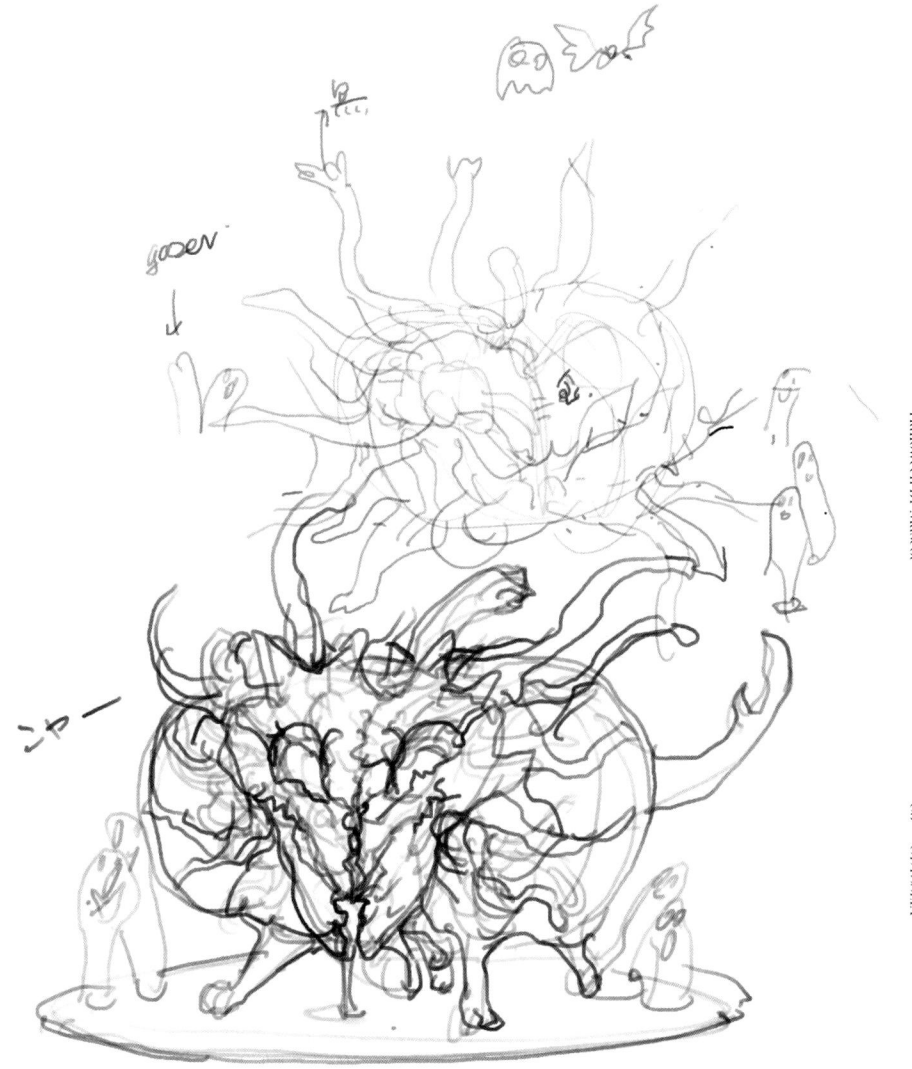

ニャー

gooen

# Ludens (X) 128

From "LUDENS" by KOJIMA PRODUCTIONS

© KOJIMA PRODUCTIONS Co.,Ltd.

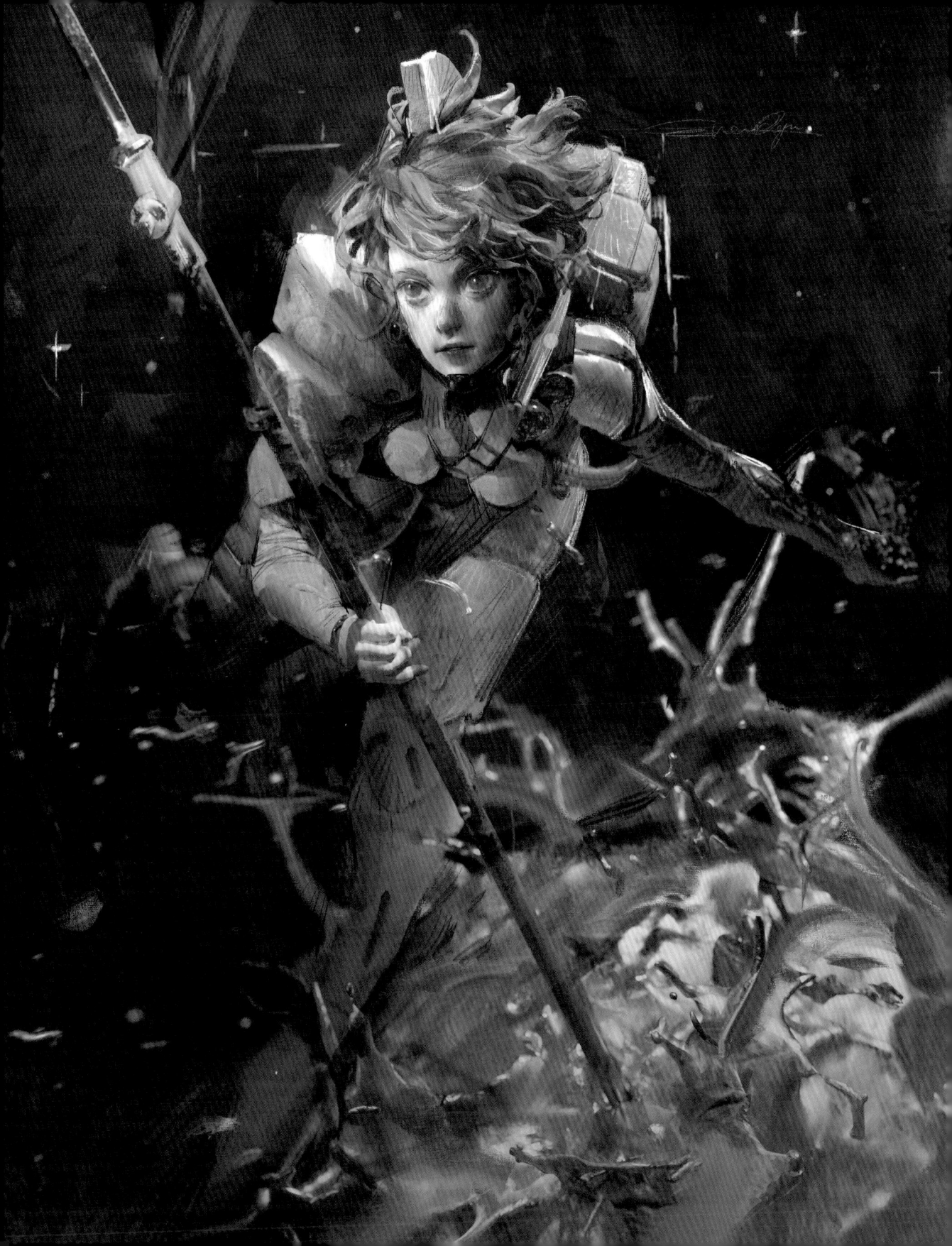

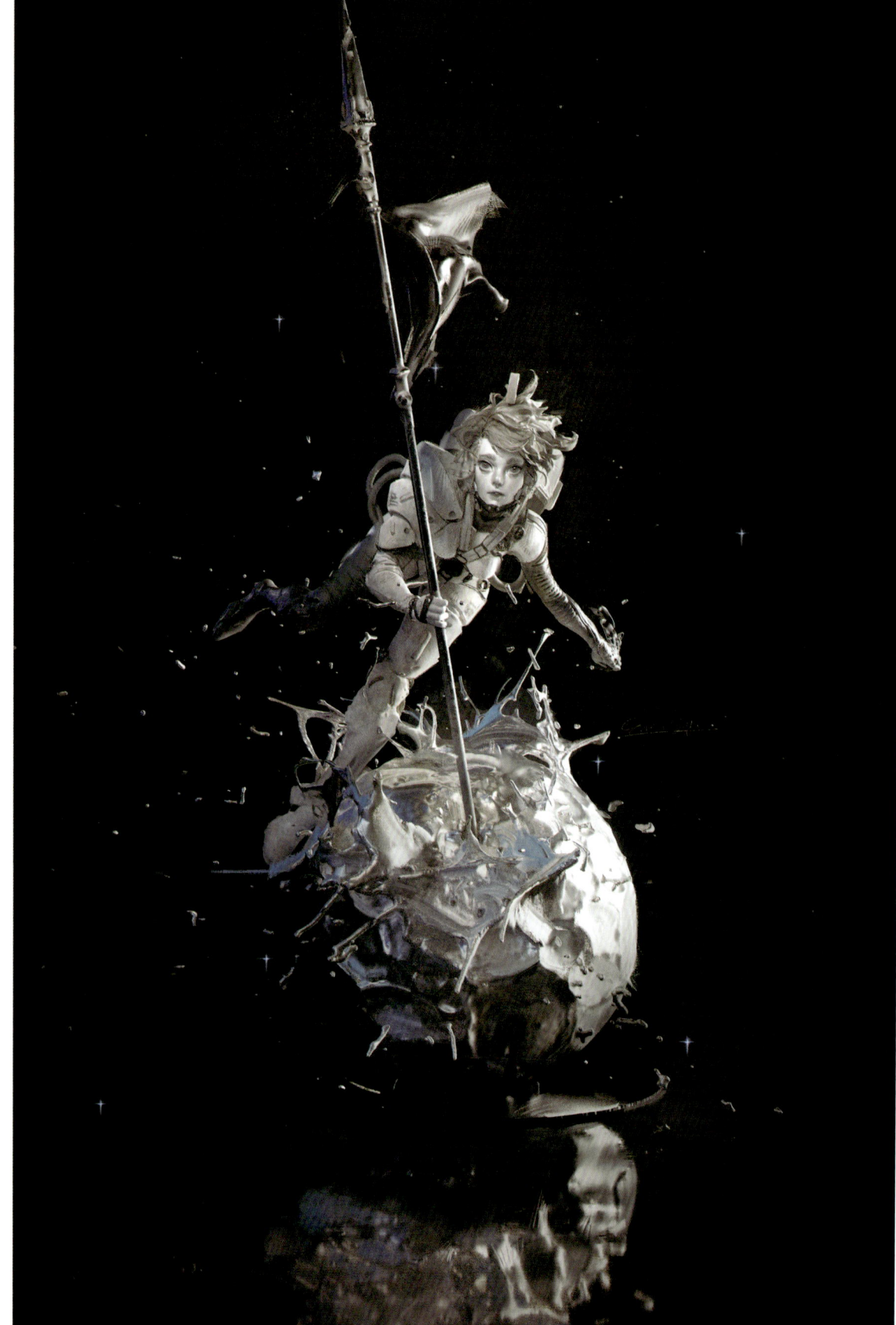

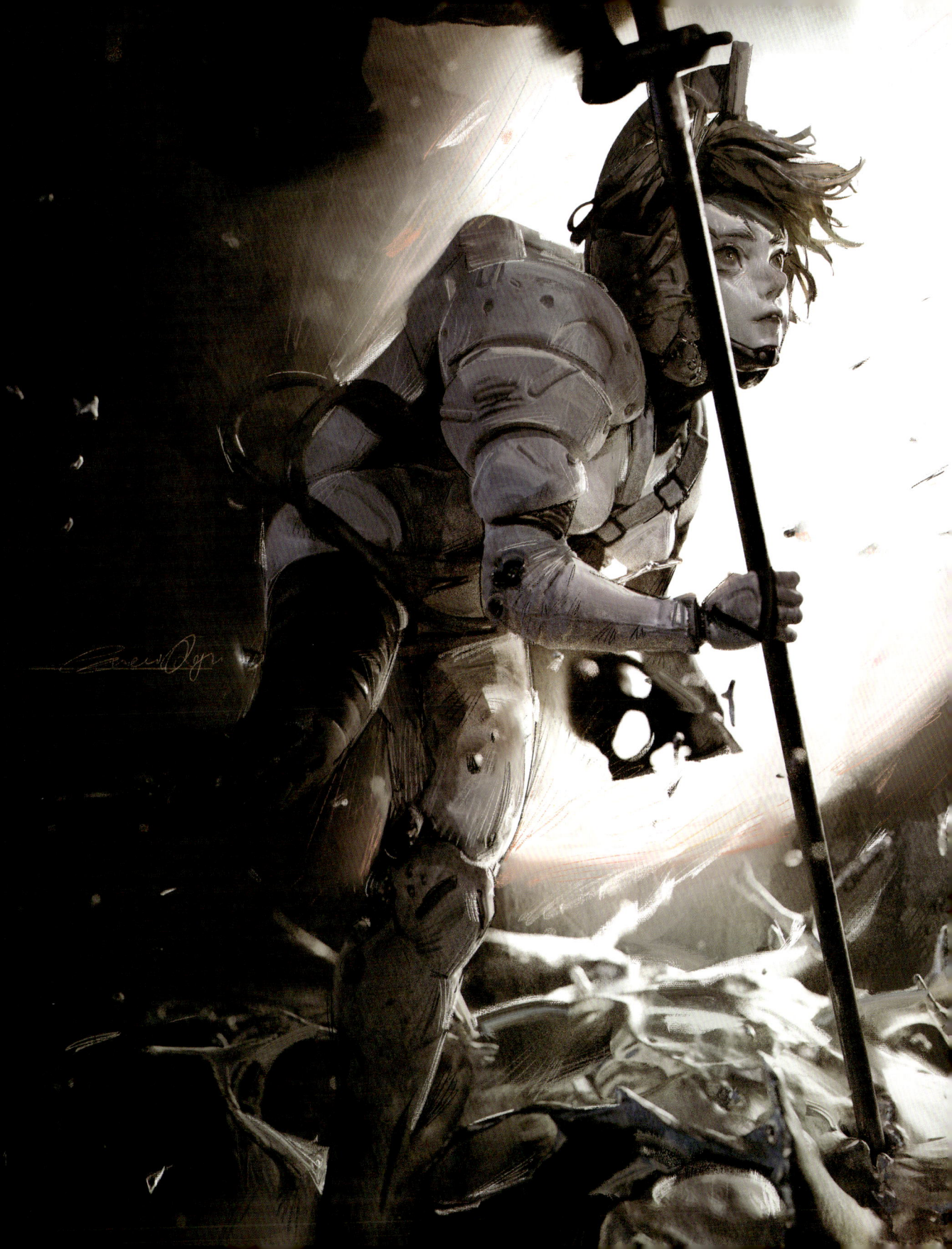

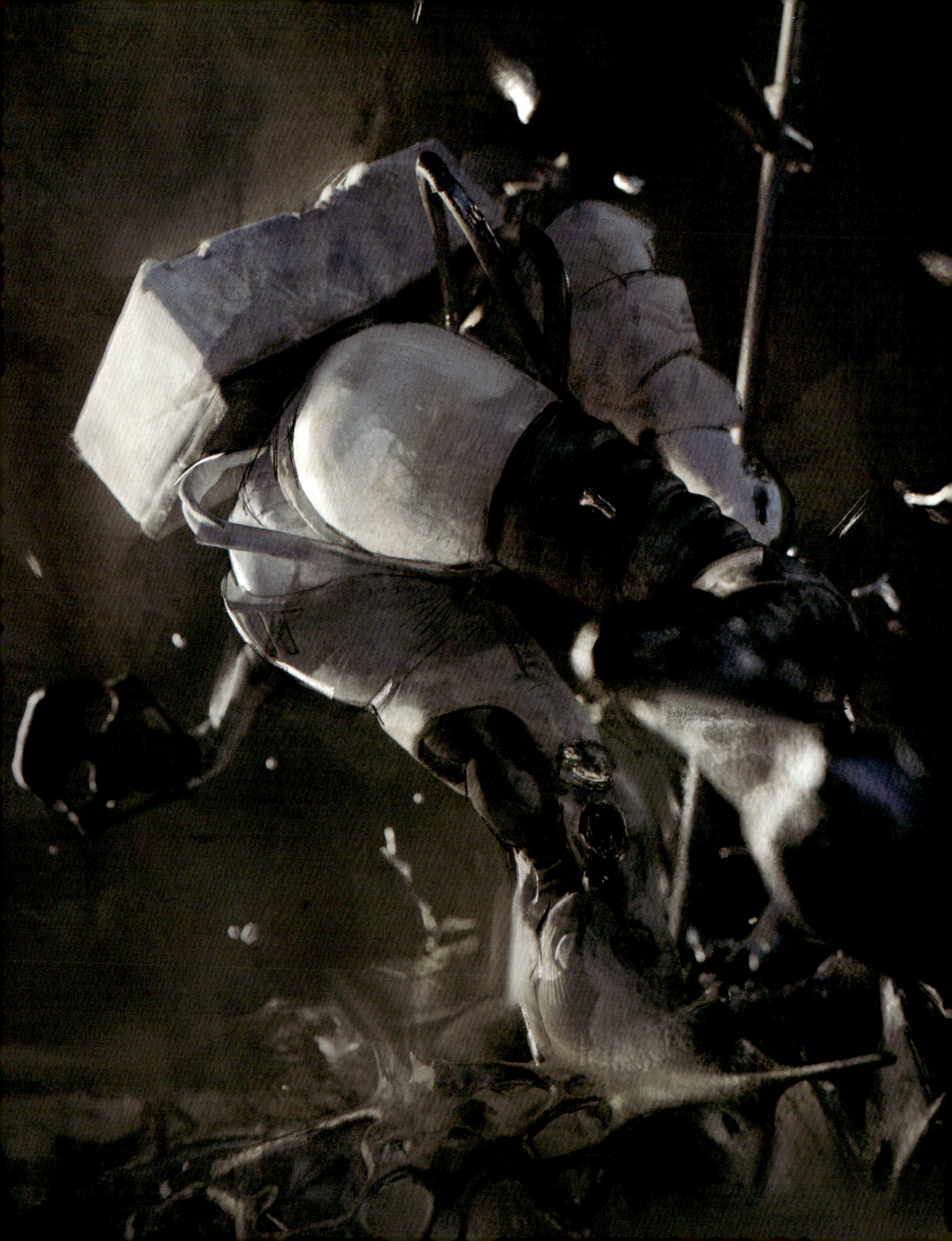

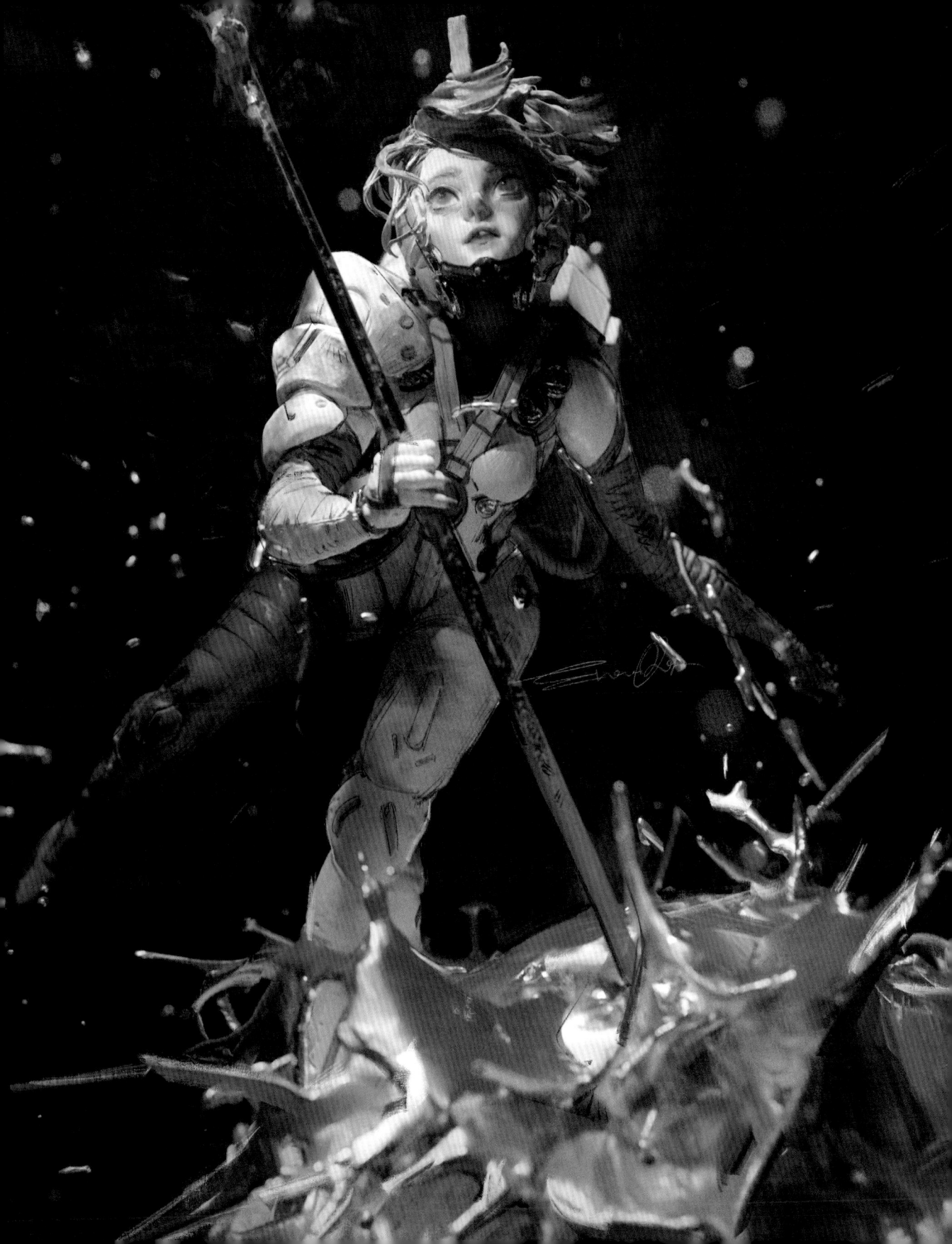

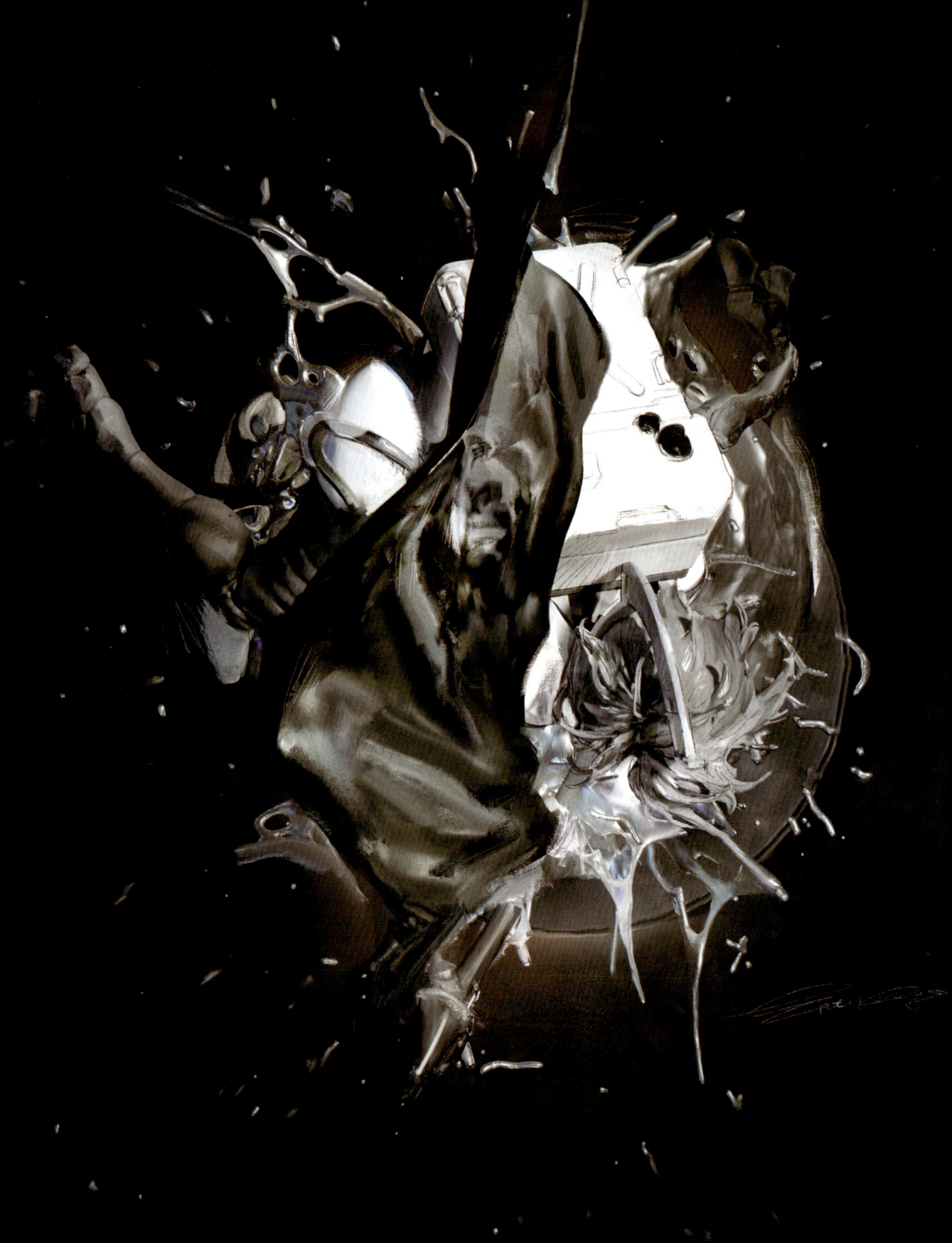

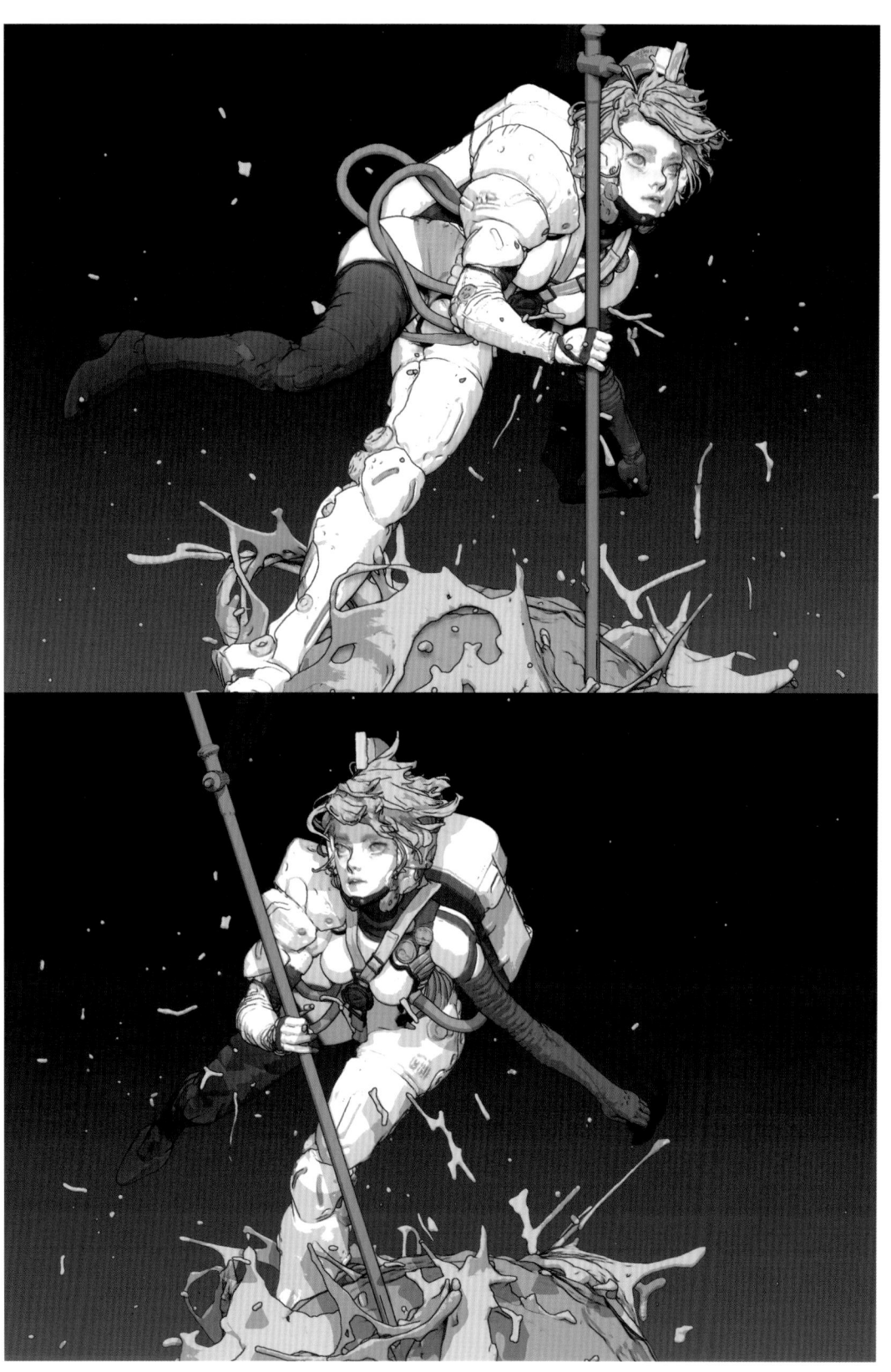

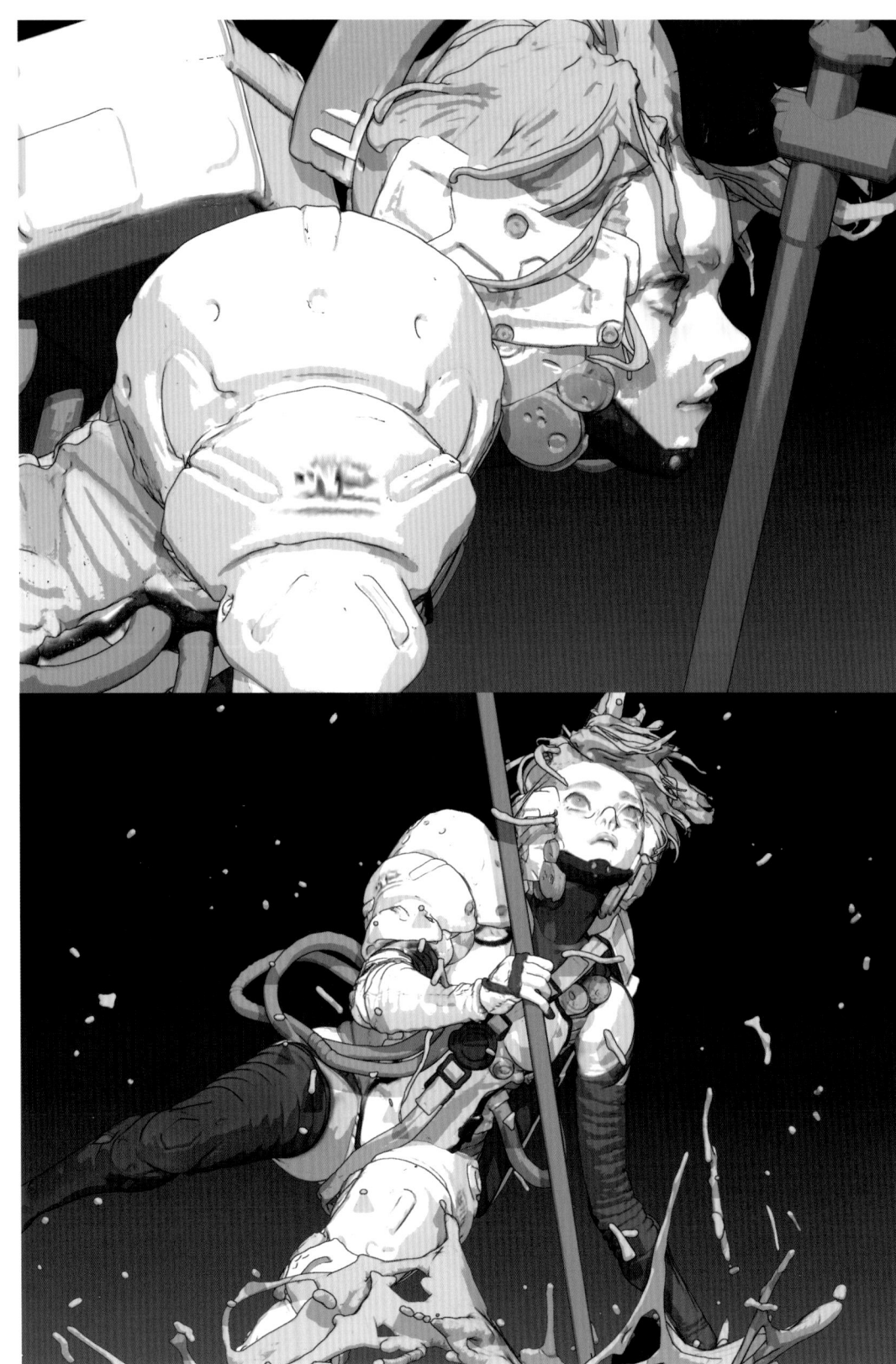

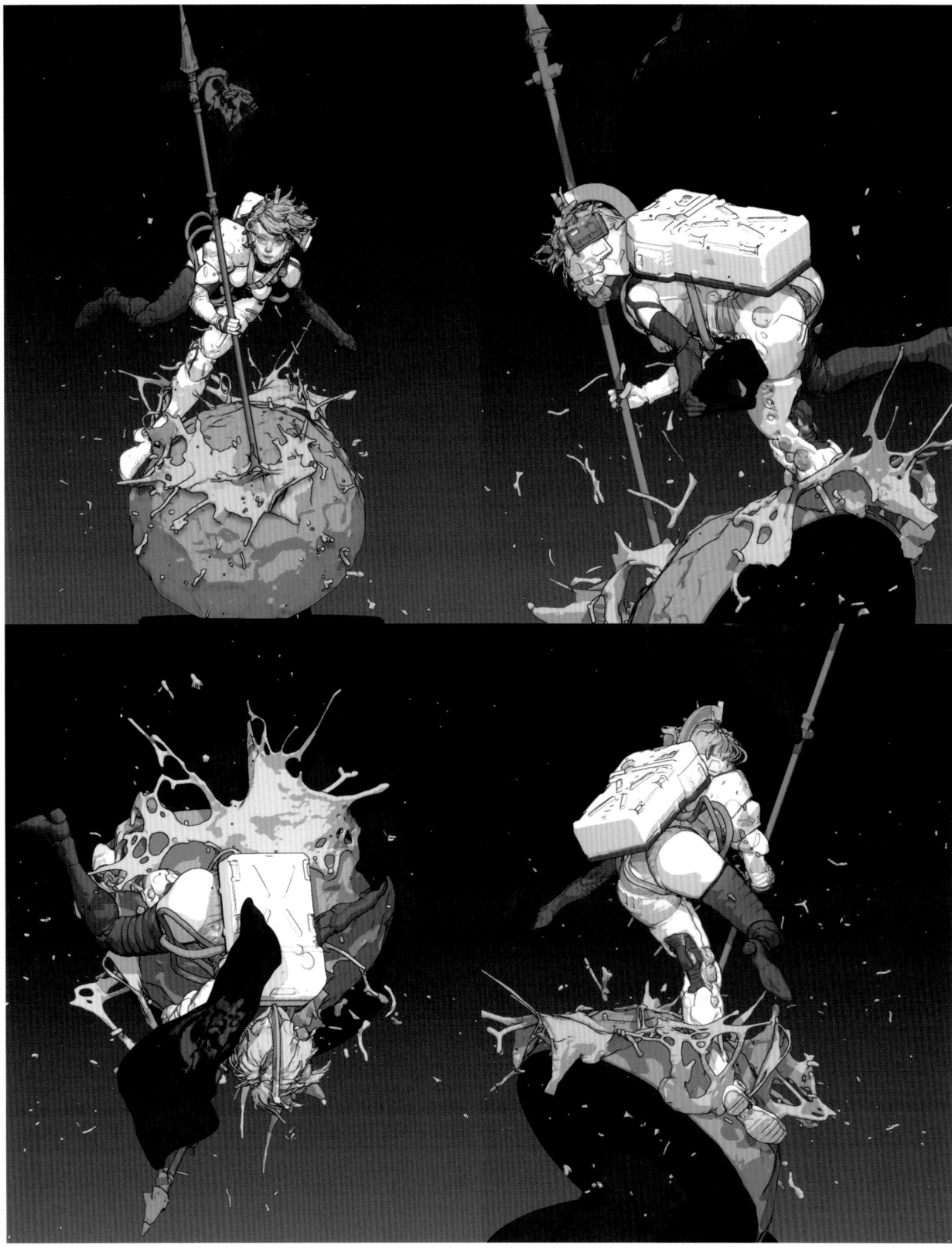

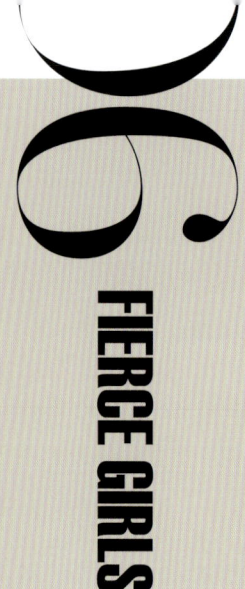

# 06

## FIERCE GIRLS

"Every girl has a fierce beast living inside her heart."
From the moment they are born, society labels girls with various stereotypes—liking pink, liking dolls, being emotional, and delicate. But in truth, everyone should beallowed to be uniquely different, even bizarre. Under societal norms, from schools to corporations, people are categorized and labeled as if they were commercial products, polished to have no edges, living similar lives. As a character designer, I prefer to depict those who are off the beaten path. I enjoy portraying girls with distinct personalities and traits, even if they are not perfect enough, or even a bit rebellious. FIERCE GIRLS is a product of this mindset.
I am fond of sculpting animals, capturing realistic and detailed specifics. The fur, scales, and claws of wild beasts make it impossible to distinguish whether they are merely animal coats worn by the girls or if they are inherently part of them.

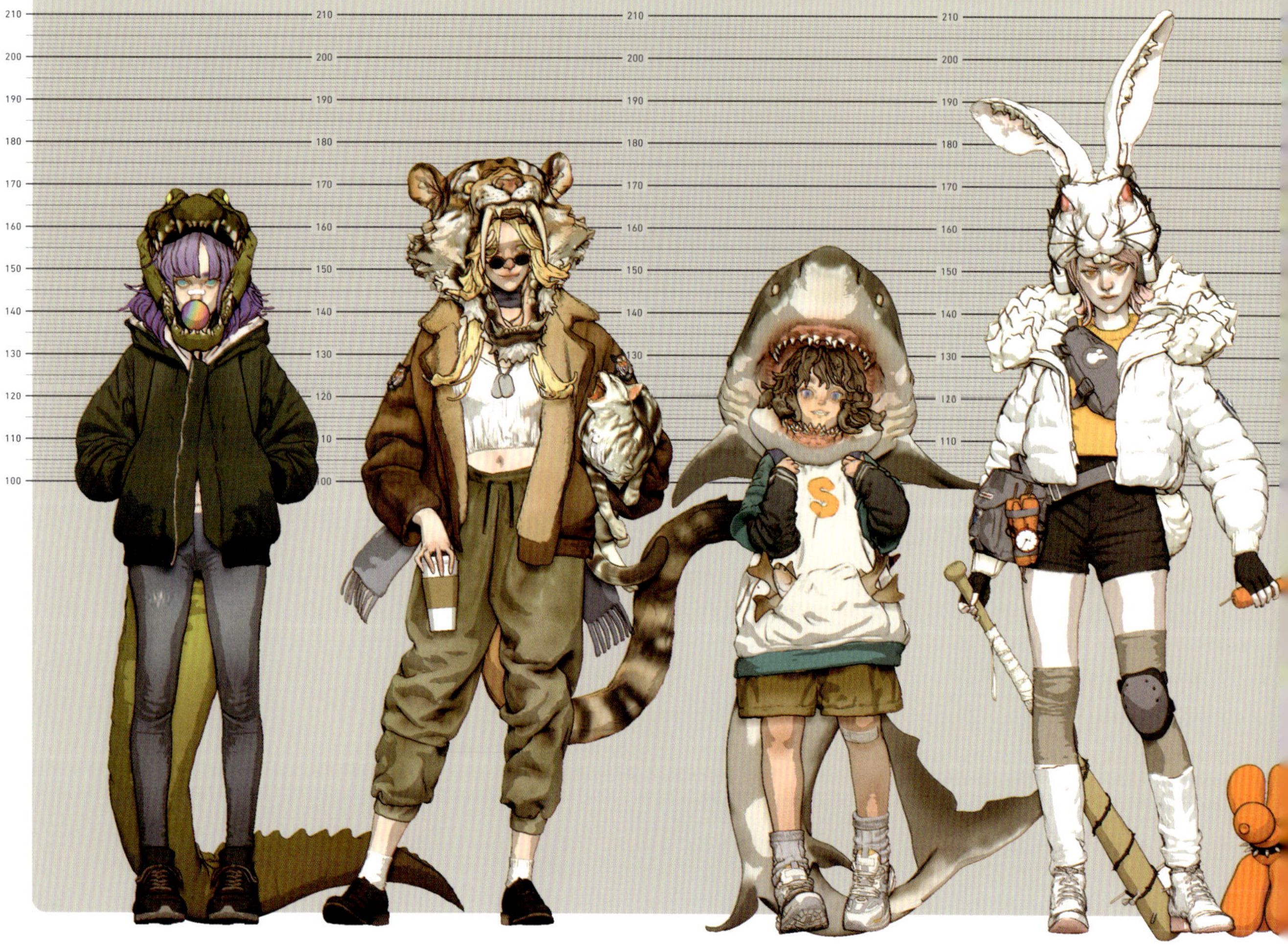

「すべての少女の心の中には猛獣が棲んでいる」

少女たちは生まれた瞬間から、社会によって固定化されたさまざまなレッテルを貼られます——ピンクが好きだとか、人形が好きだとか、感情的だとか、繊細だとか。しかし本当は、誰もが他の人とは違っていて、それどころか奇妙でさえあるはずなのです。学校から会社まで、社会規範の下で人々はまるで製品であるかのように分類され、レッテルを貼られ、似たような生活を送るように角が取れるまで磨かれます。私はキャラクターデザイナーとして、他人とは違う我が道を行く人たちを描くのが好きです。たとえ完璧でなくても、少し反抗的であっても、個性や特徴のはっきりした女の子を描くのが好きです。「FIERCE GIRLS」は、こうした考えから生まれました。

細部までリアルに捉えた動物の造形は、私のお気に入りです。猛獣の毛皮、鱗、爪によって境界が曖昧になります。それは少女たちがまとう動物のコートに過ぎないのでしょうか、それとも生まれつき備わった彼女たちの一部なのでしょうか。

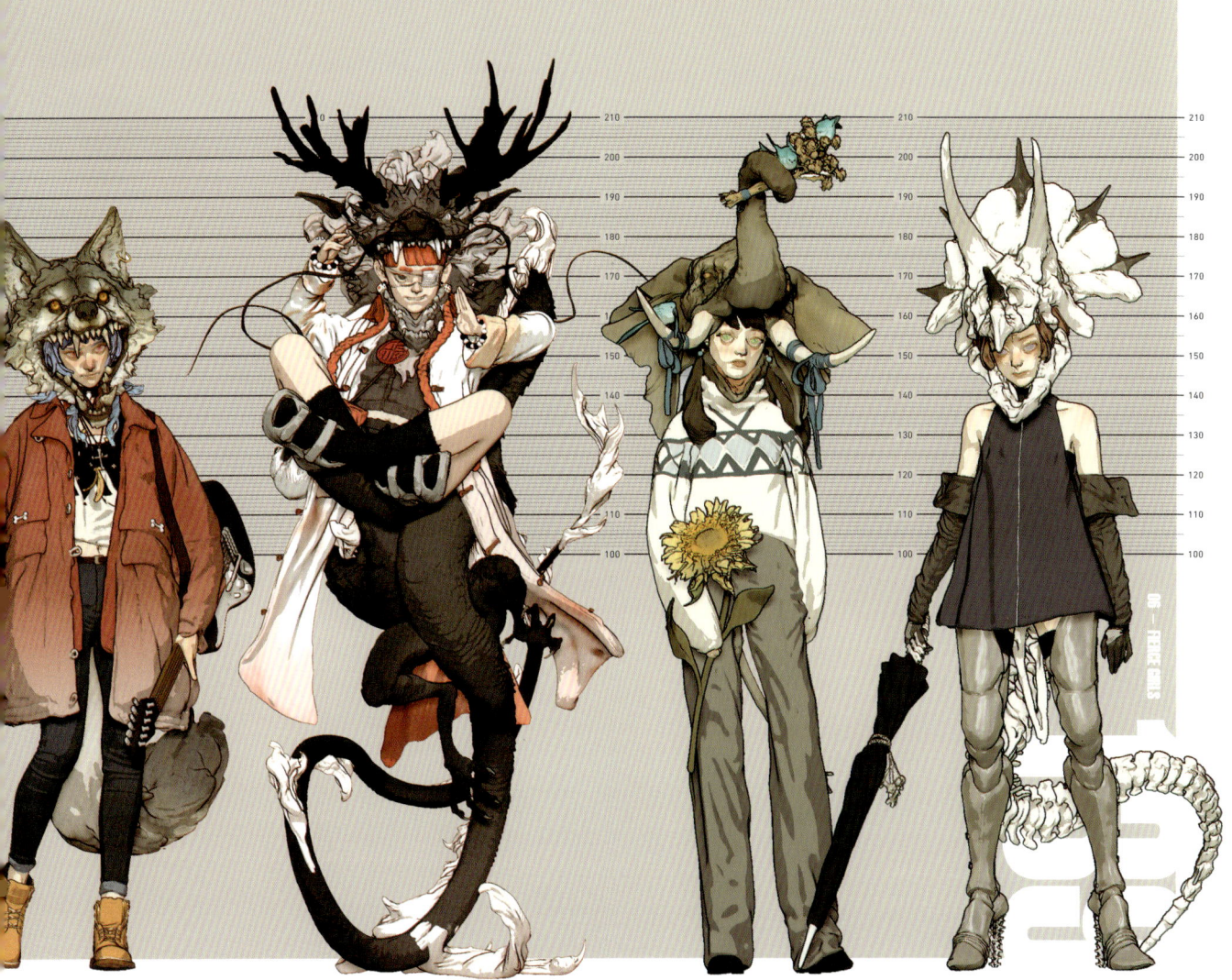

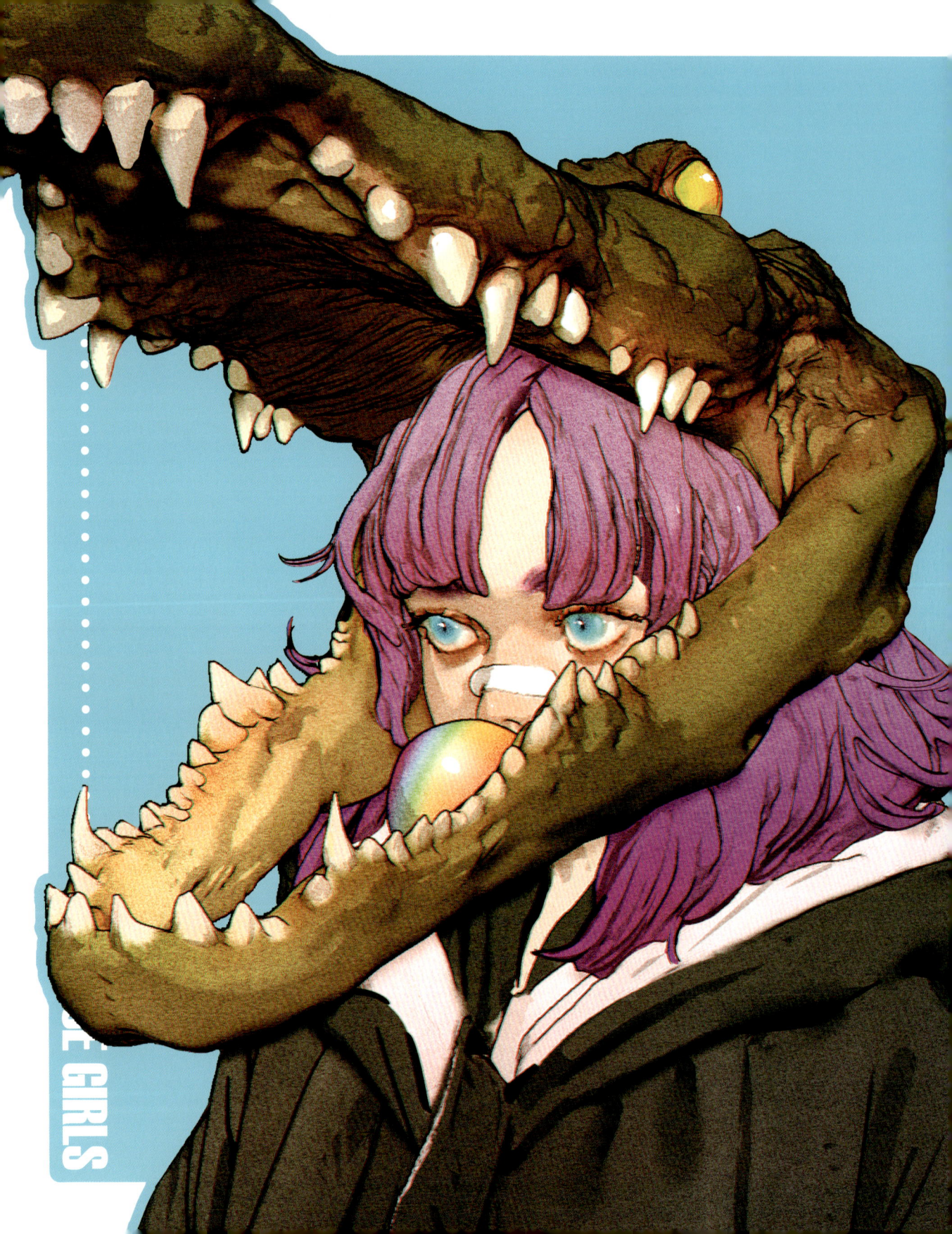

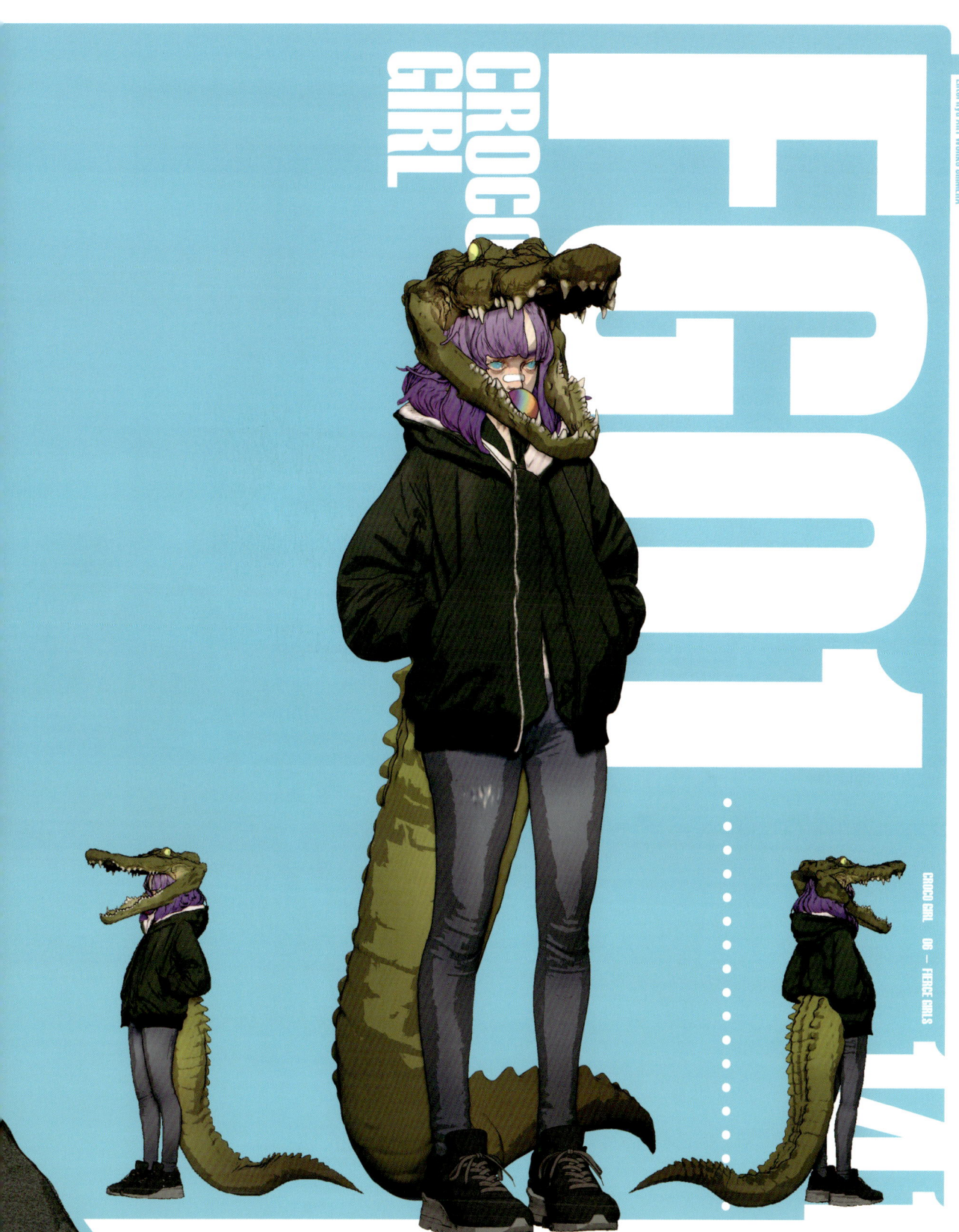

CROCO GIRL

CROCO GIRL

FG01

FG 01

CROCO GIRL  06 — FIERCE GIRLS

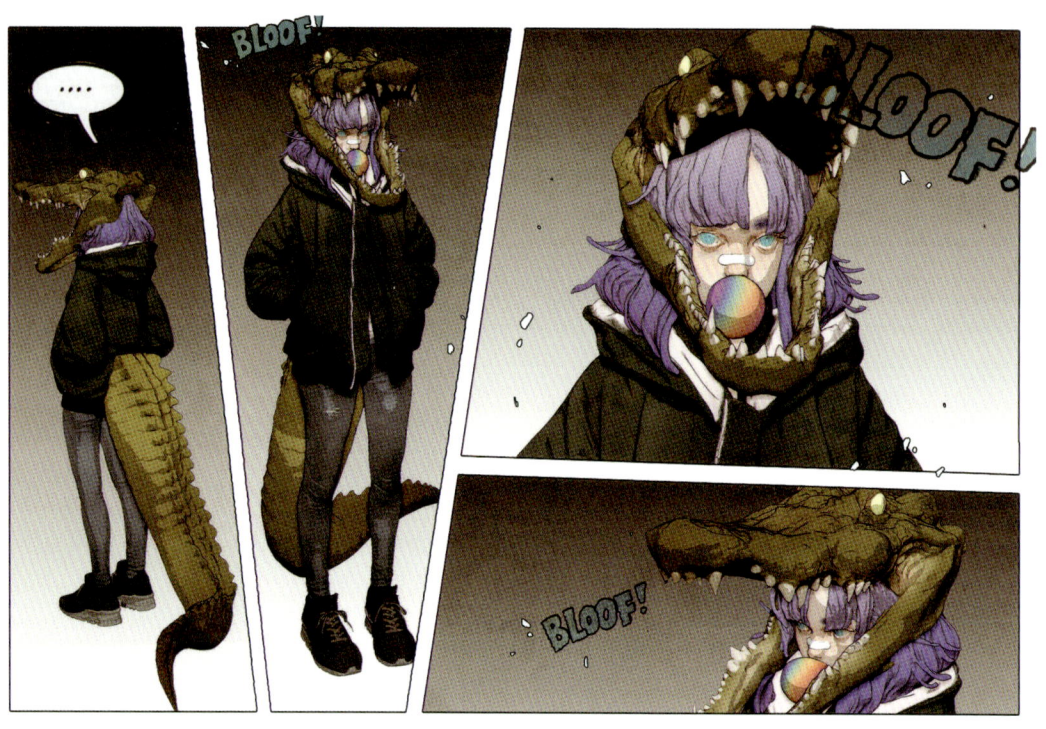
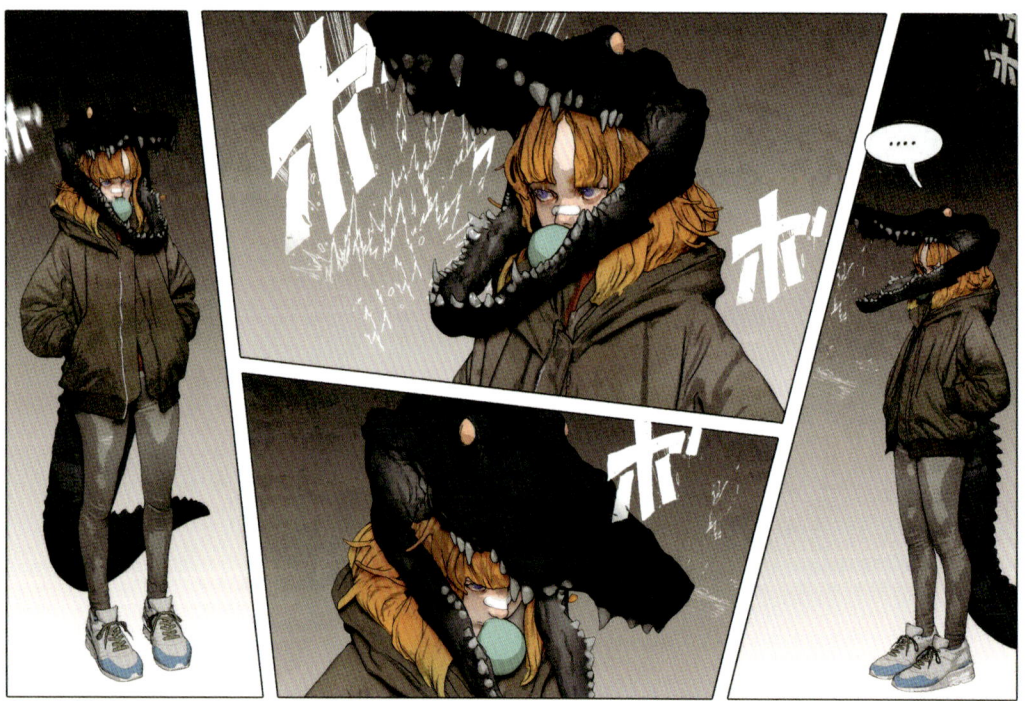

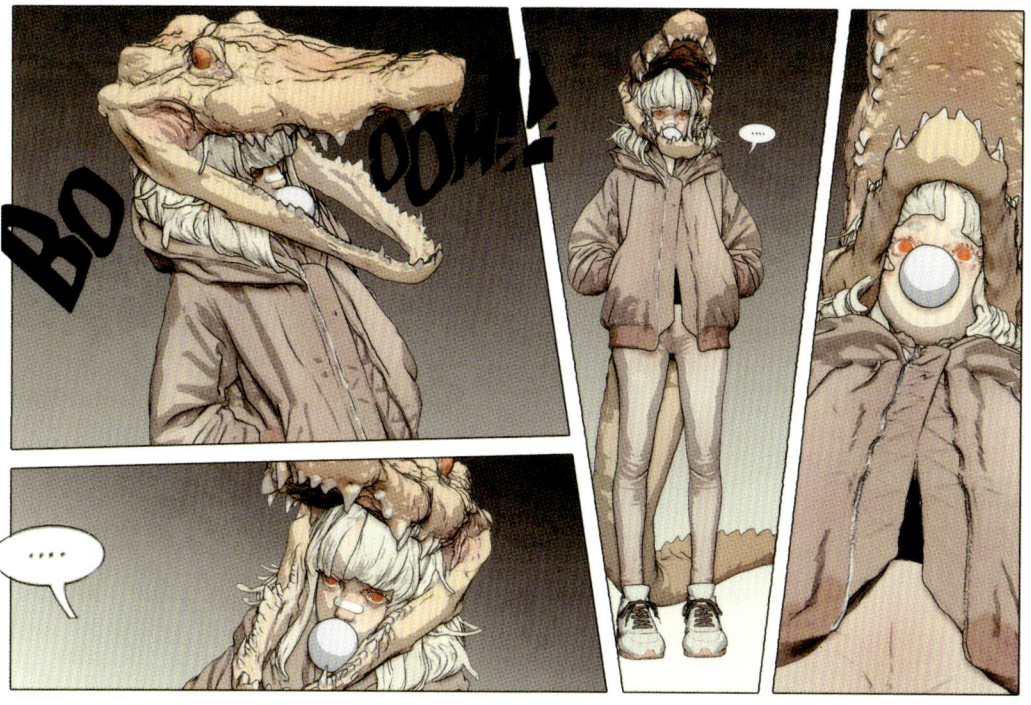

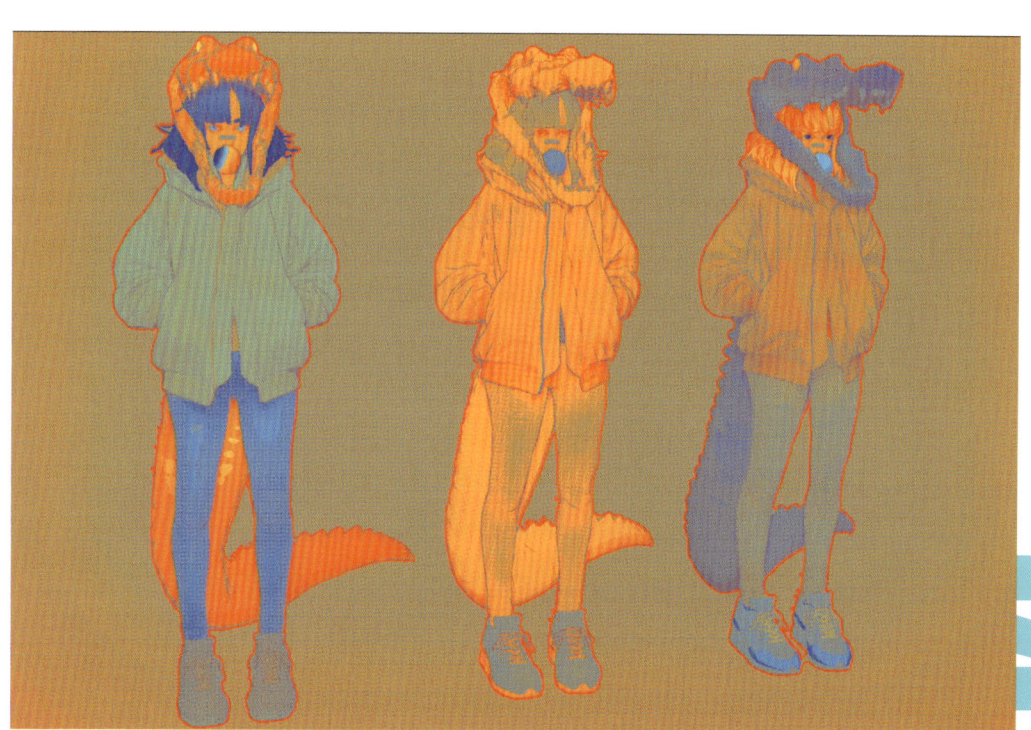

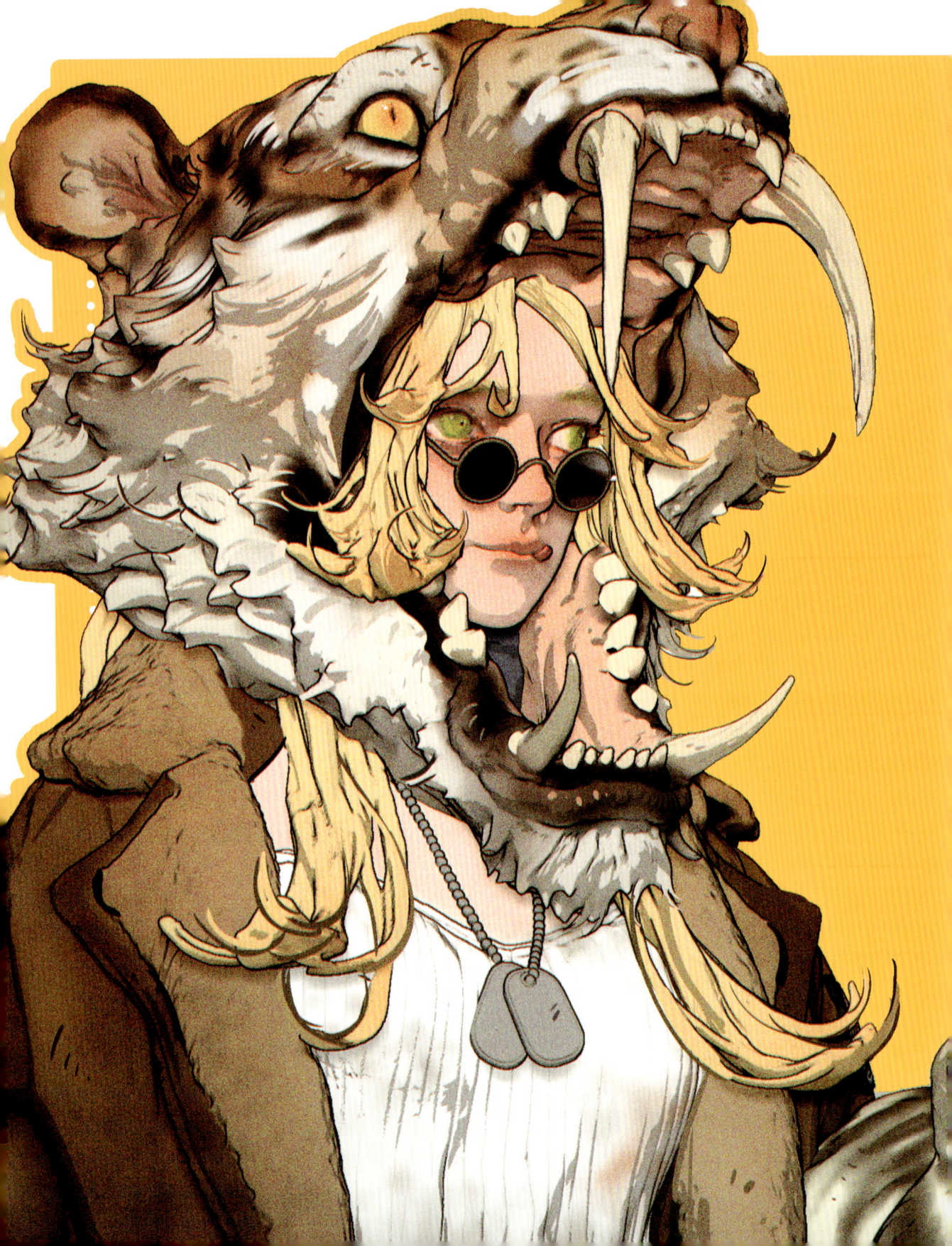

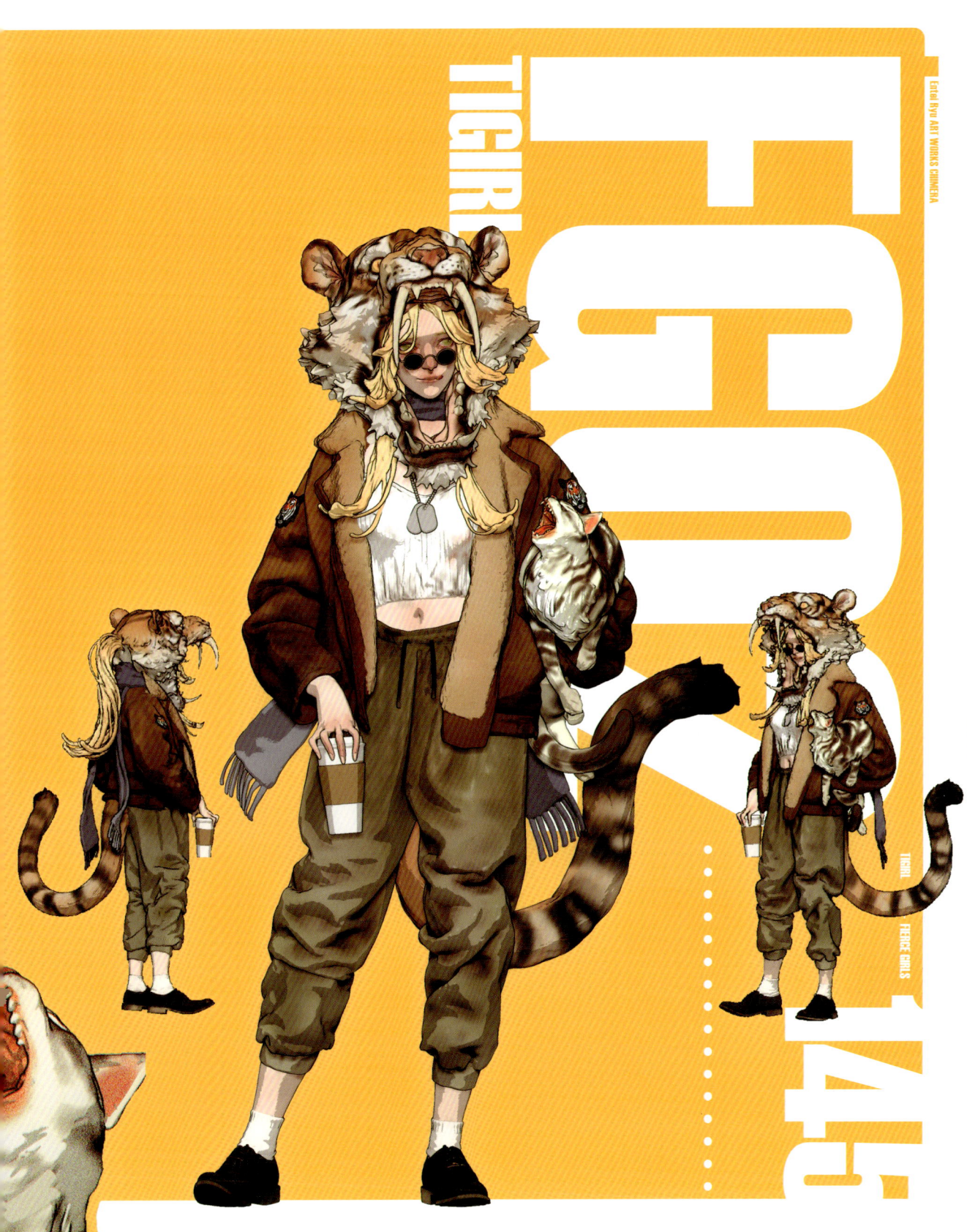

Entei Ryu ART WORKS CHIMERA

FGO[2] 145

TIGIRL

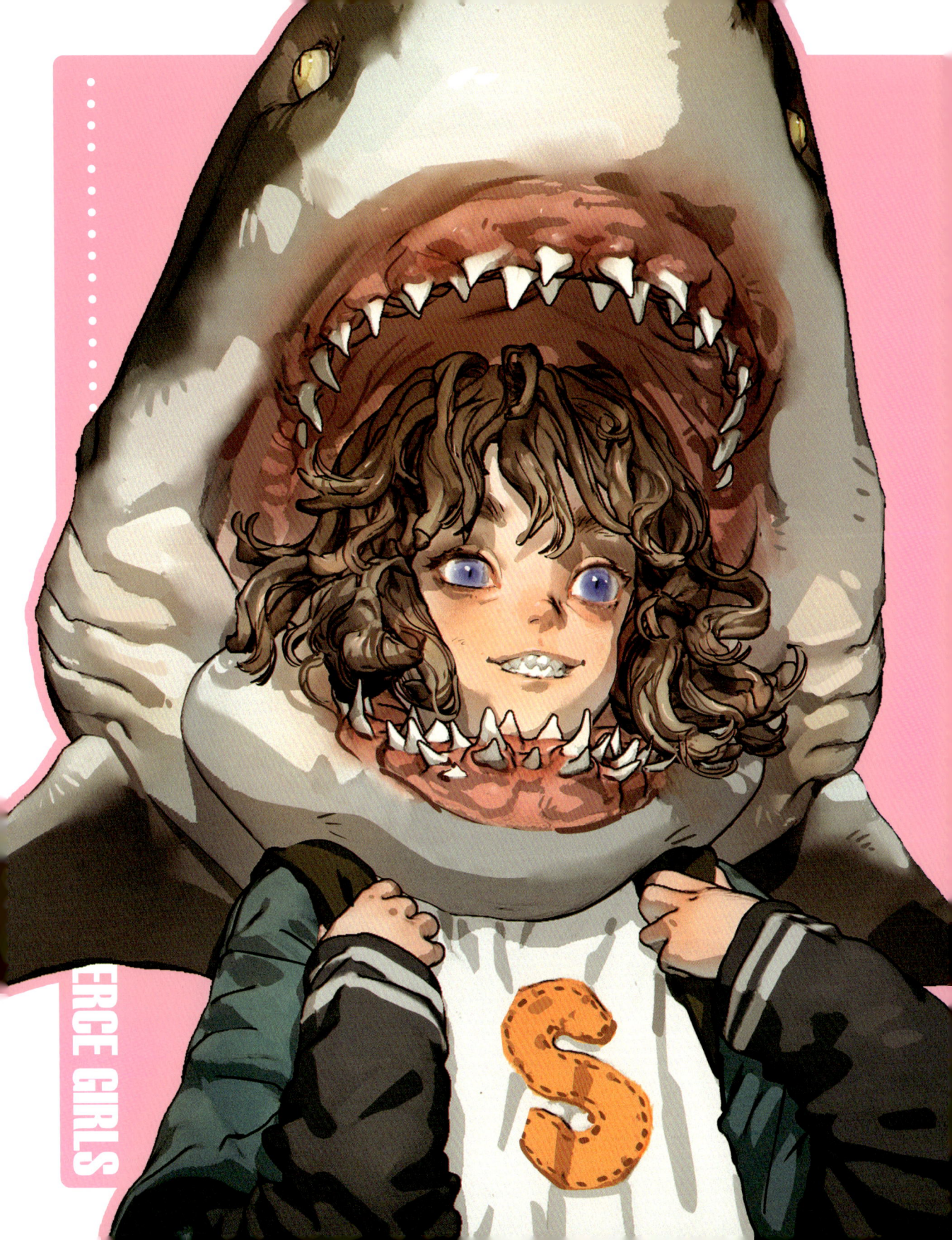

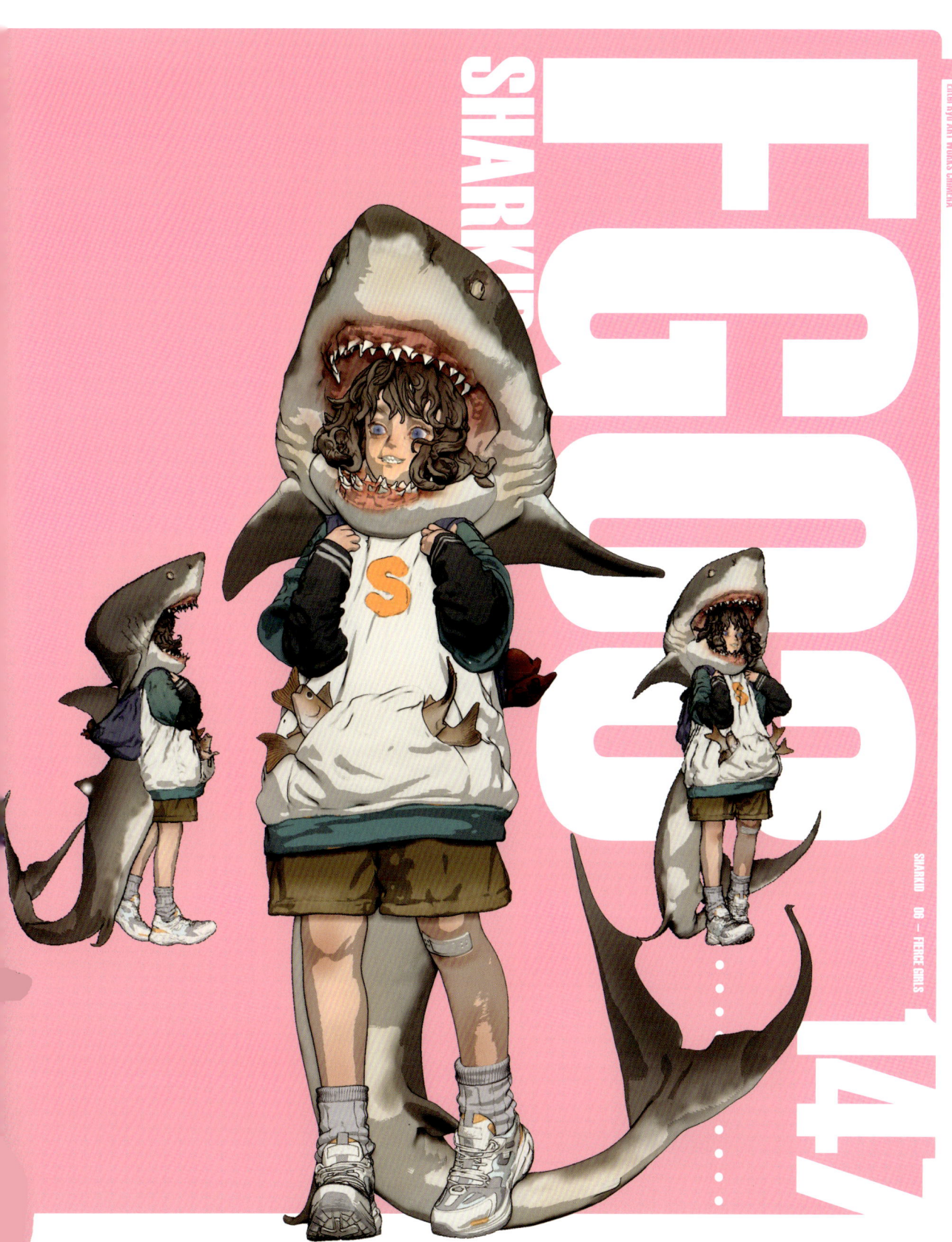

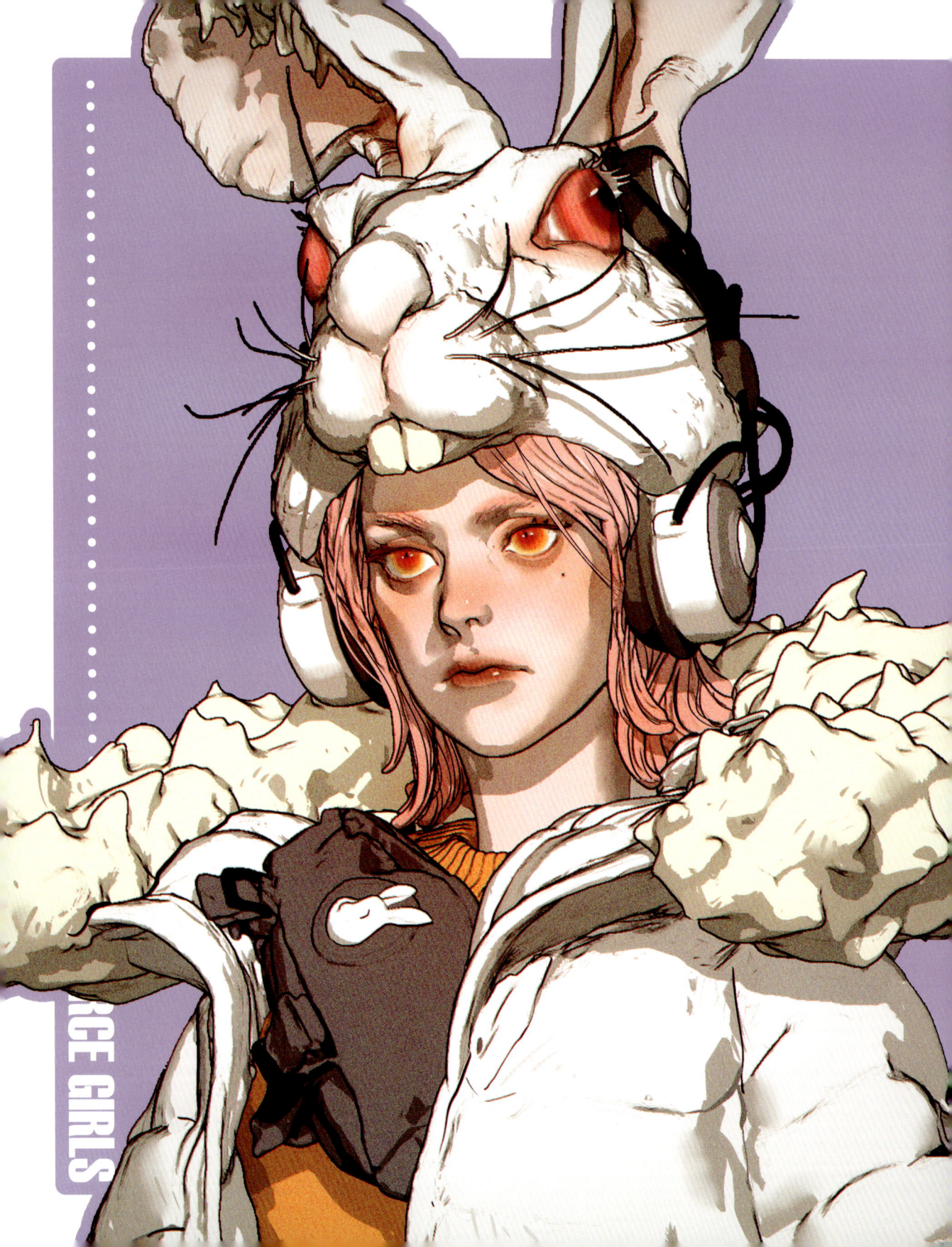

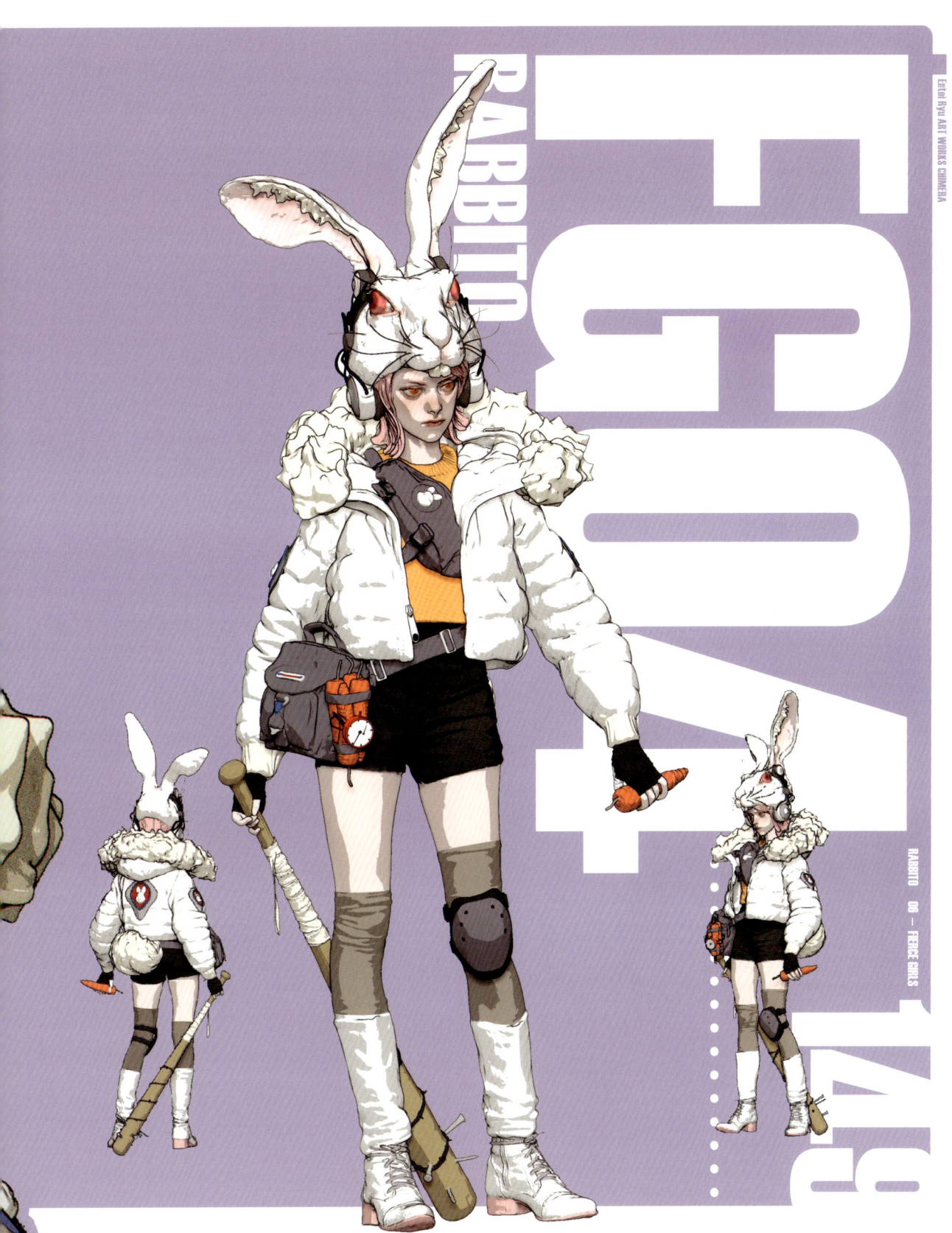

Entei Ryu ART WORKS CHIMERA

RABBITO

EGO4

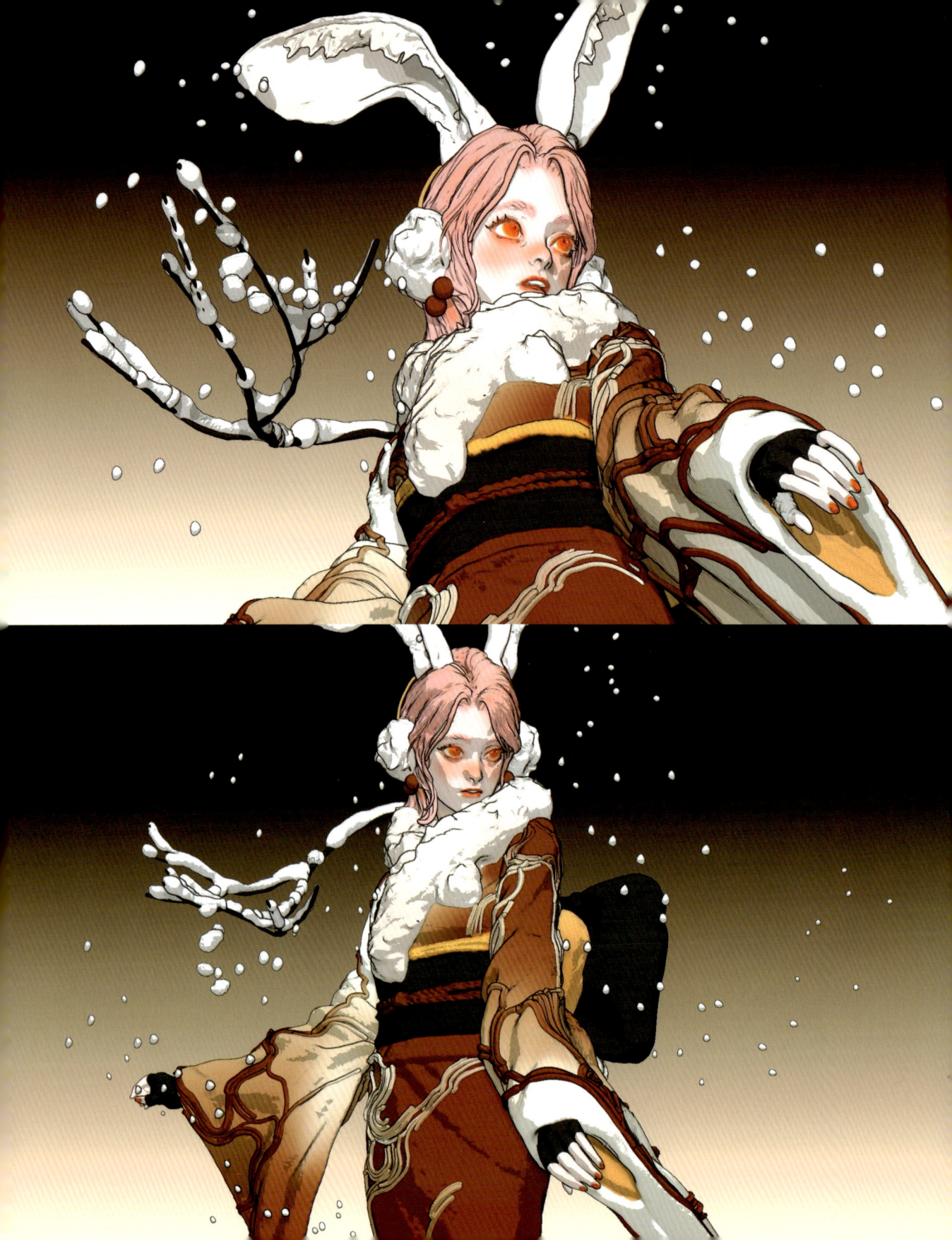

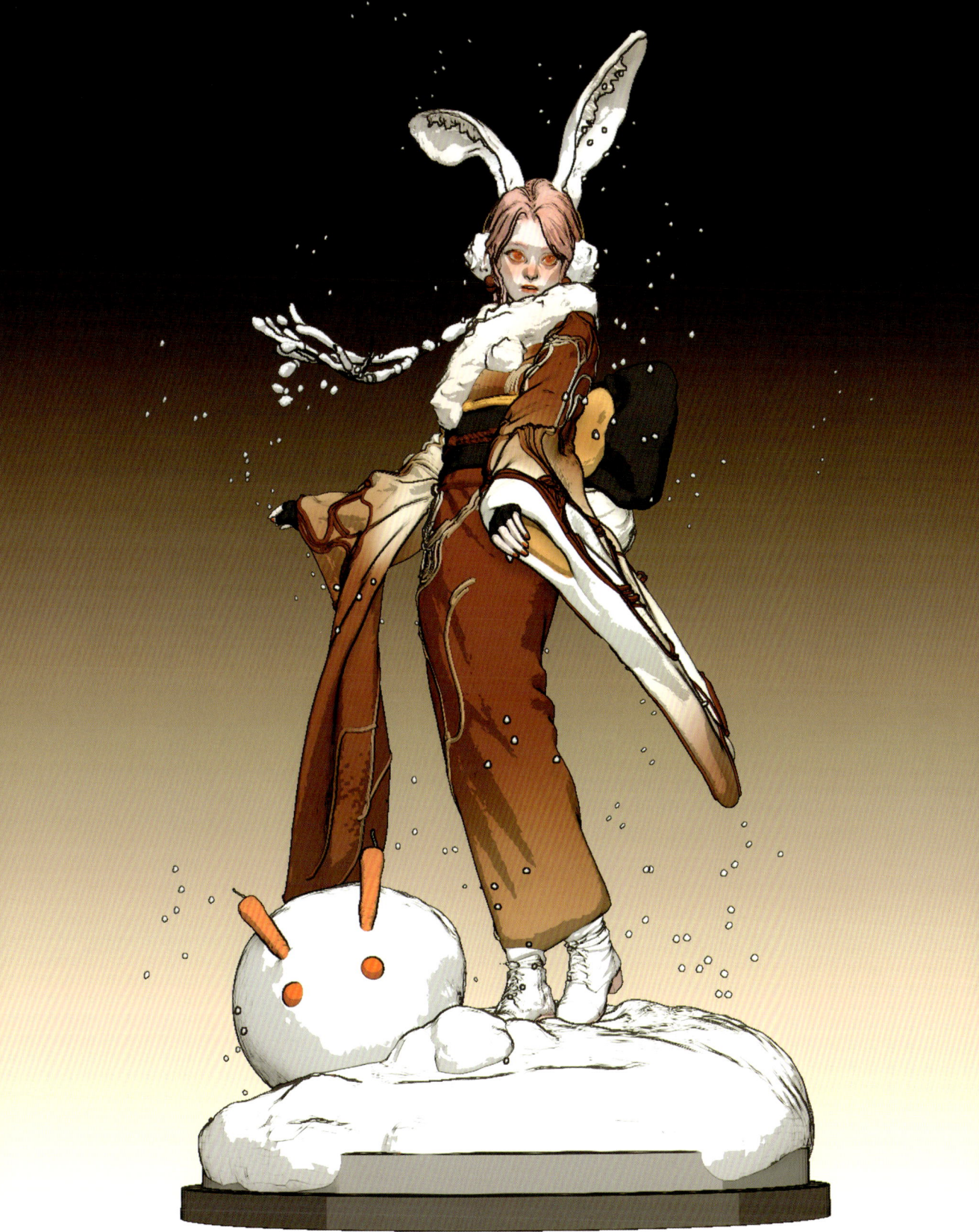

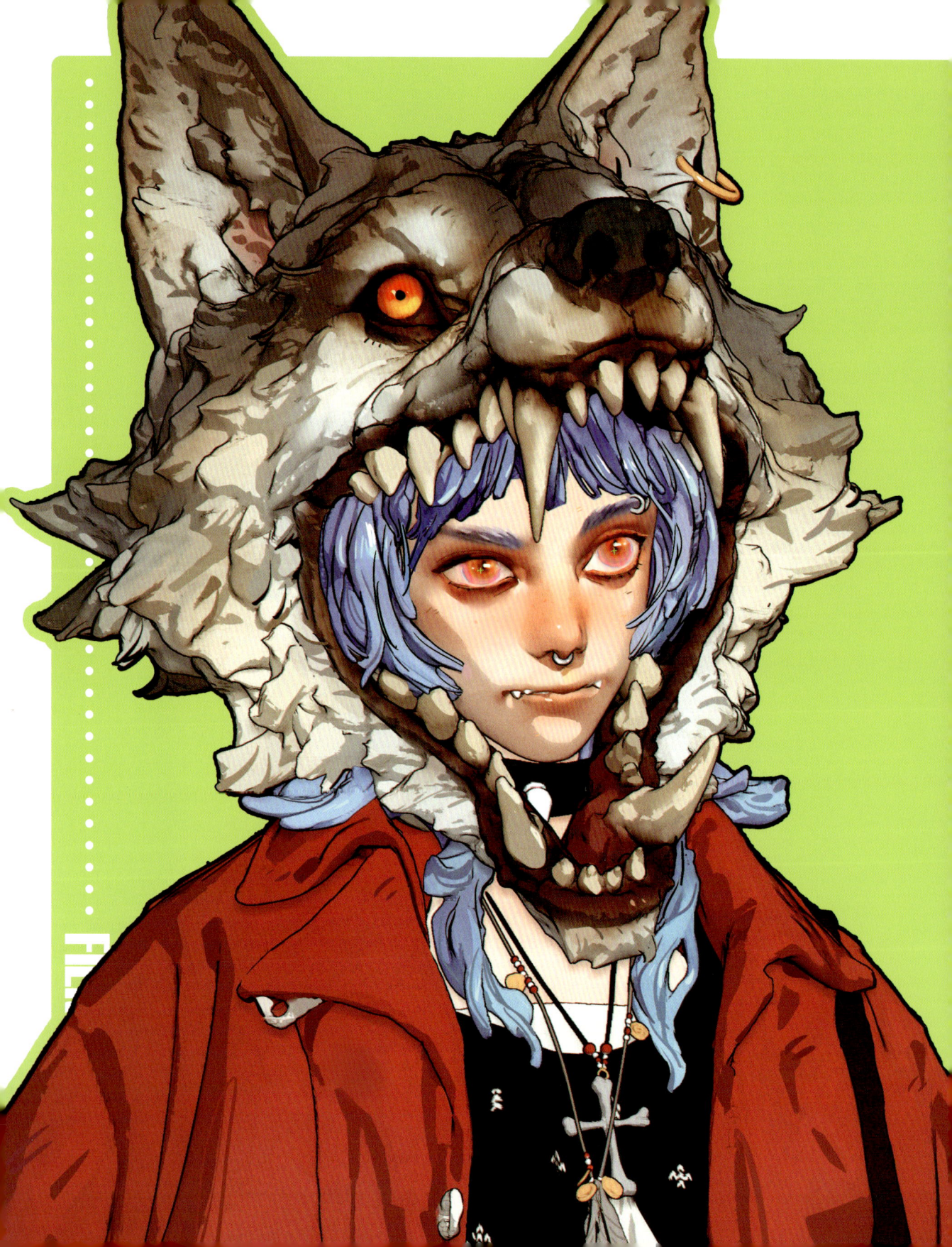

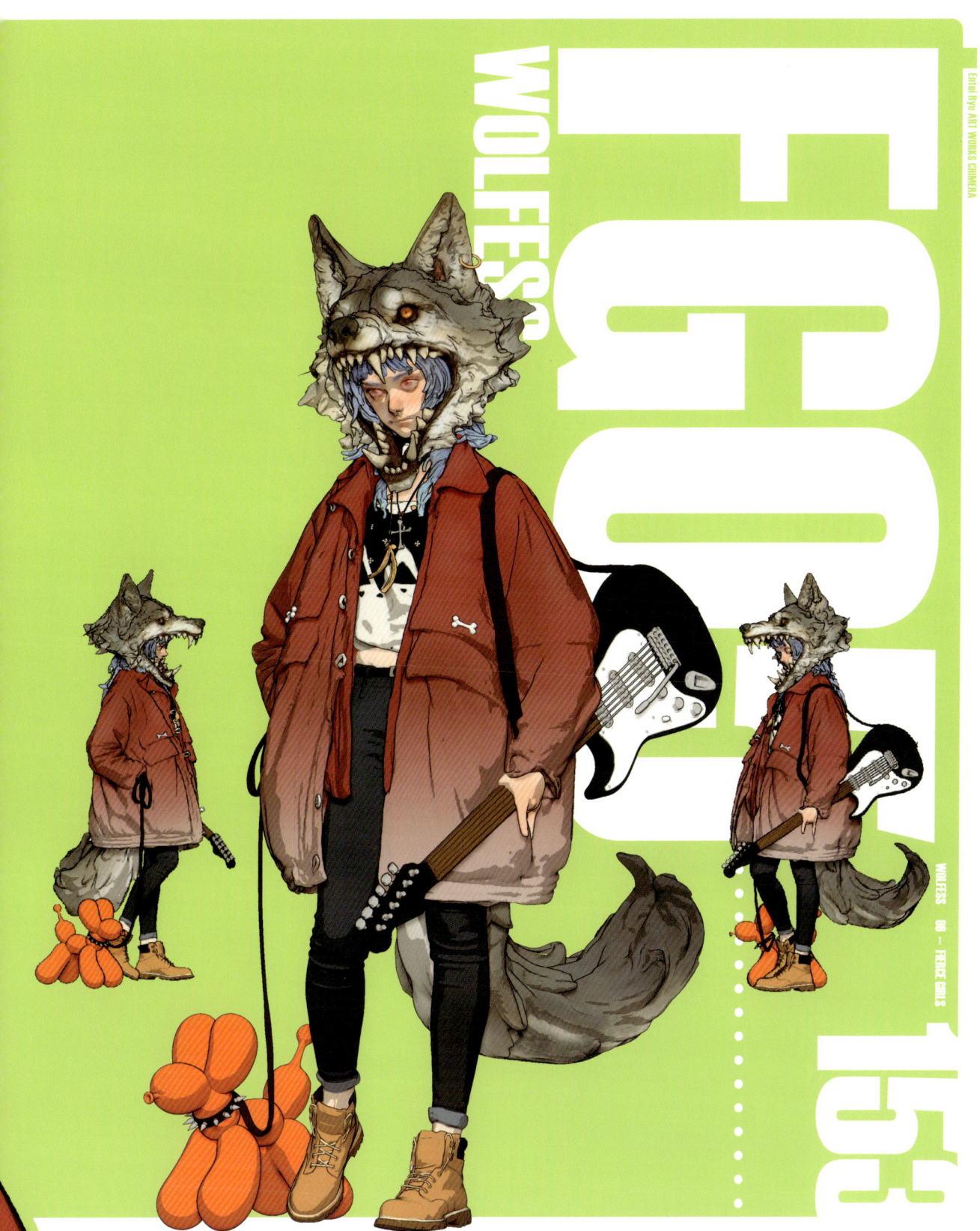

Enfui Ryu ART WORKS CHIMERA

FOOL

WOLFESS

WOLFESS

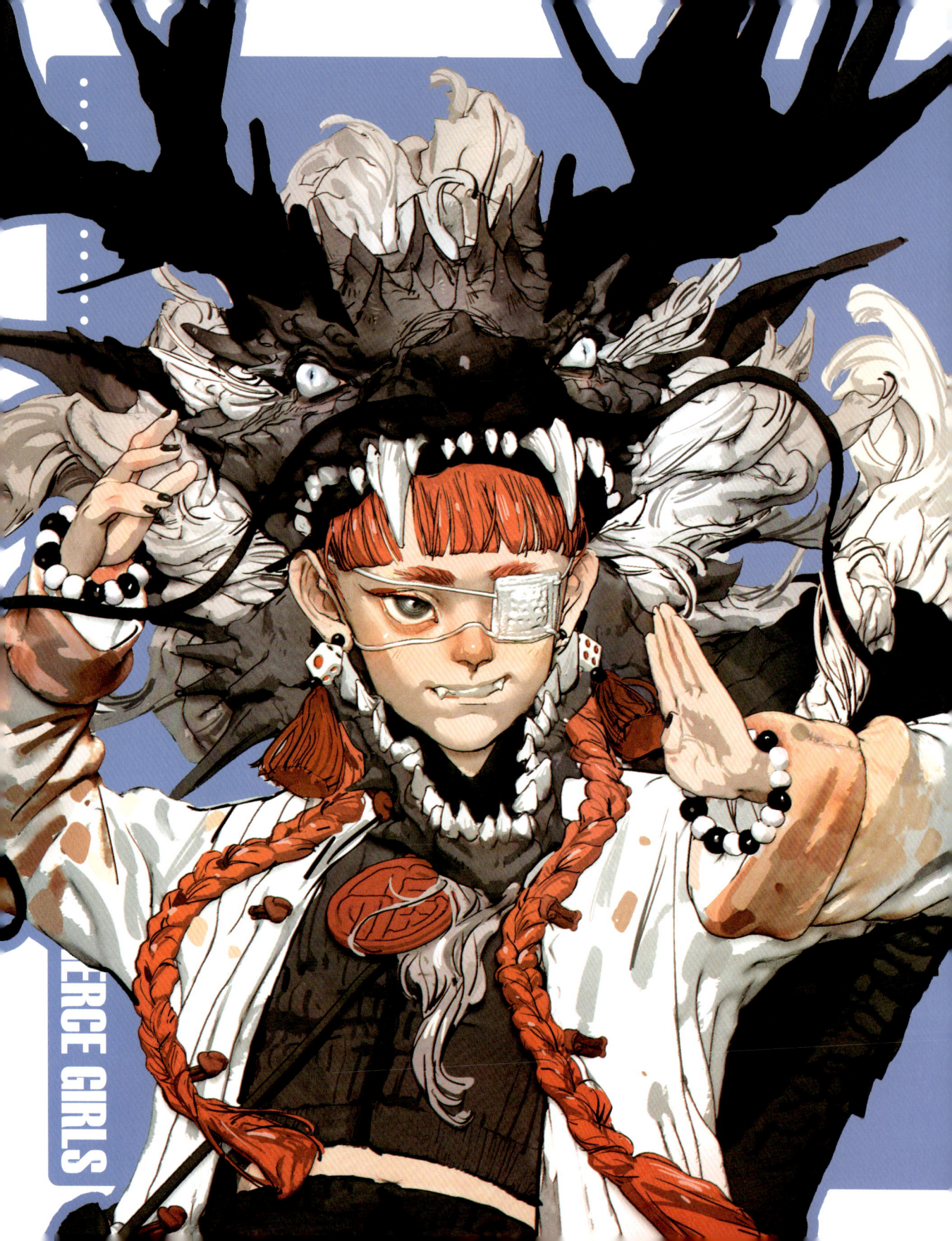

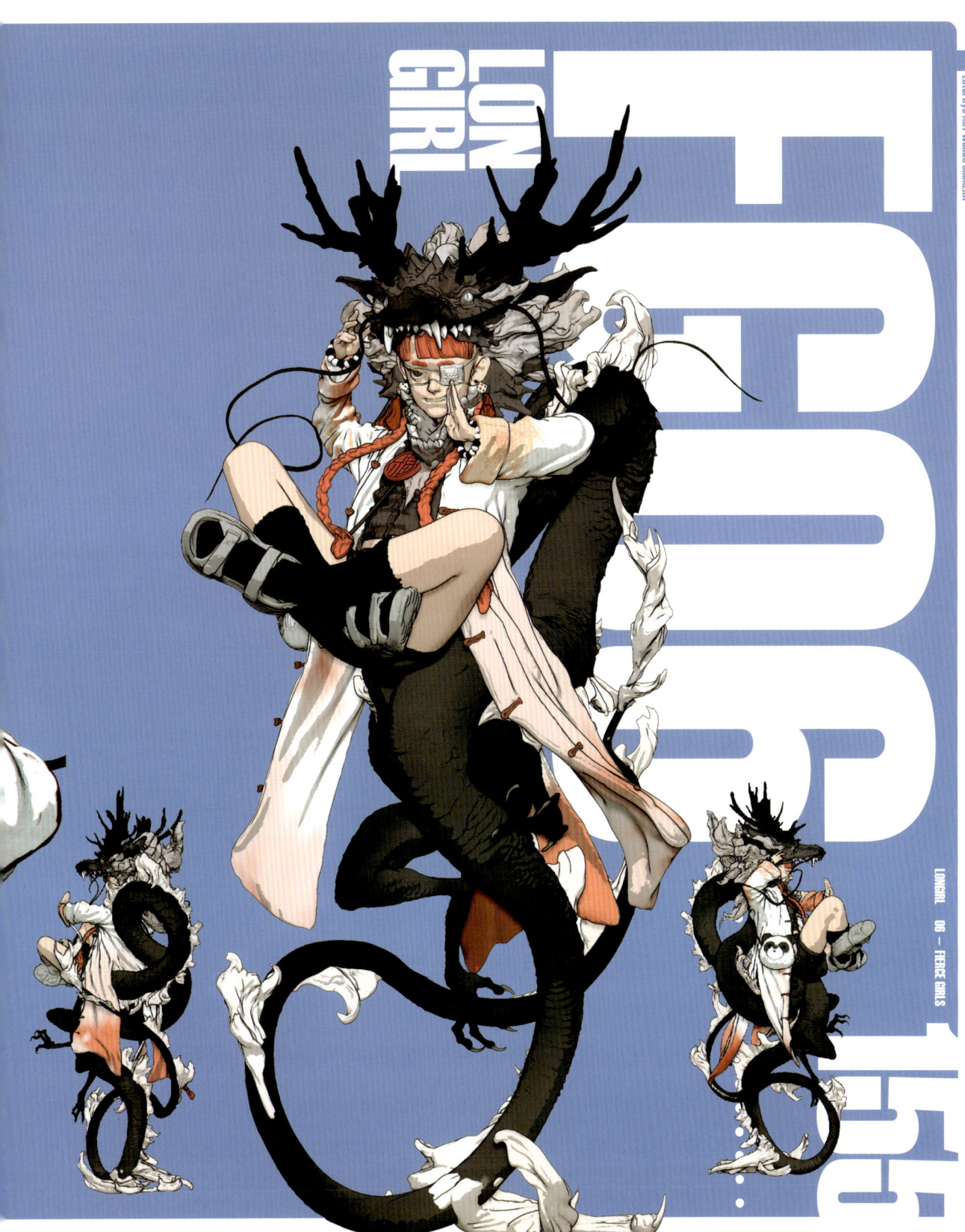

Entei Ryu ART WORKS CHIMERA

LON GIRL

E06

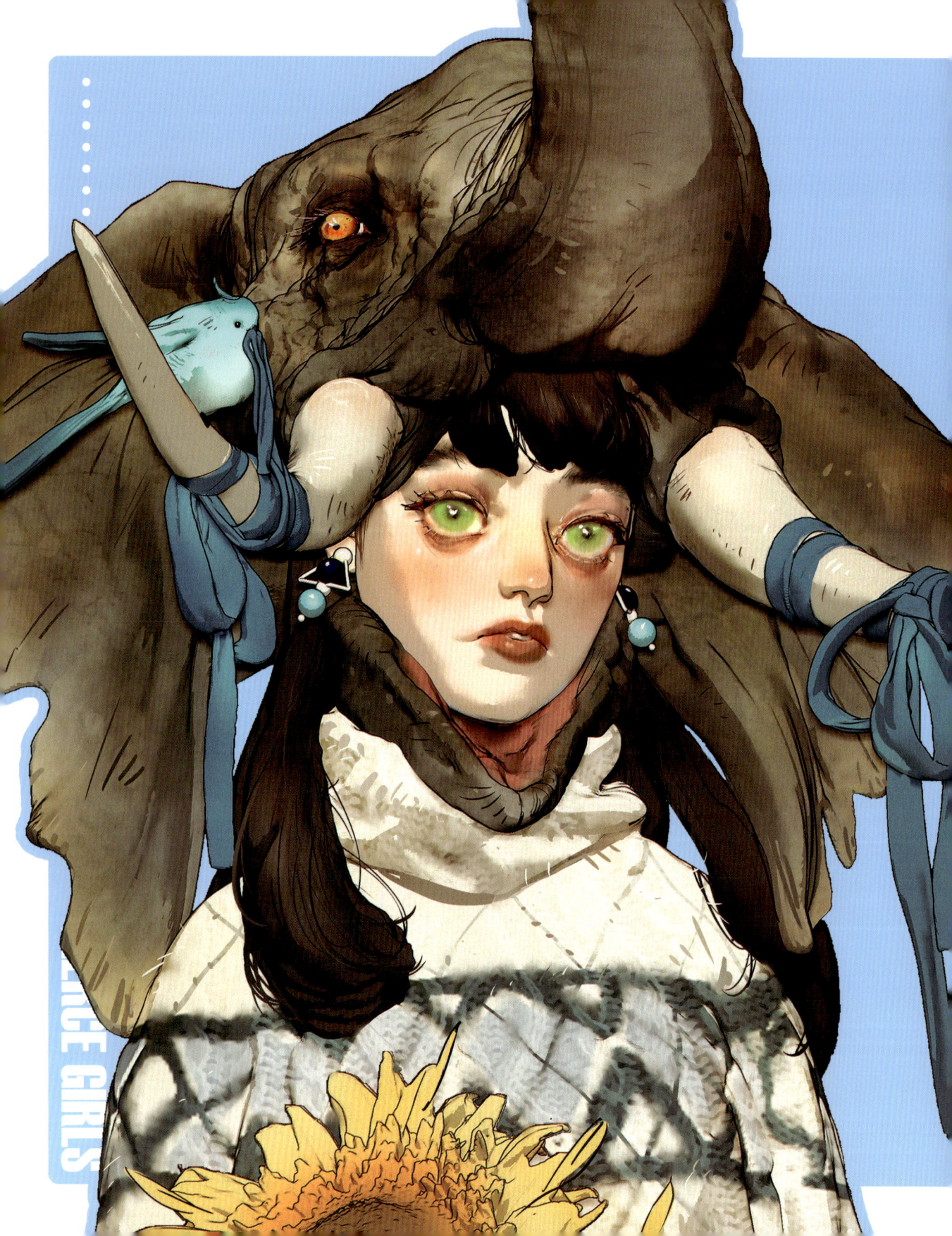

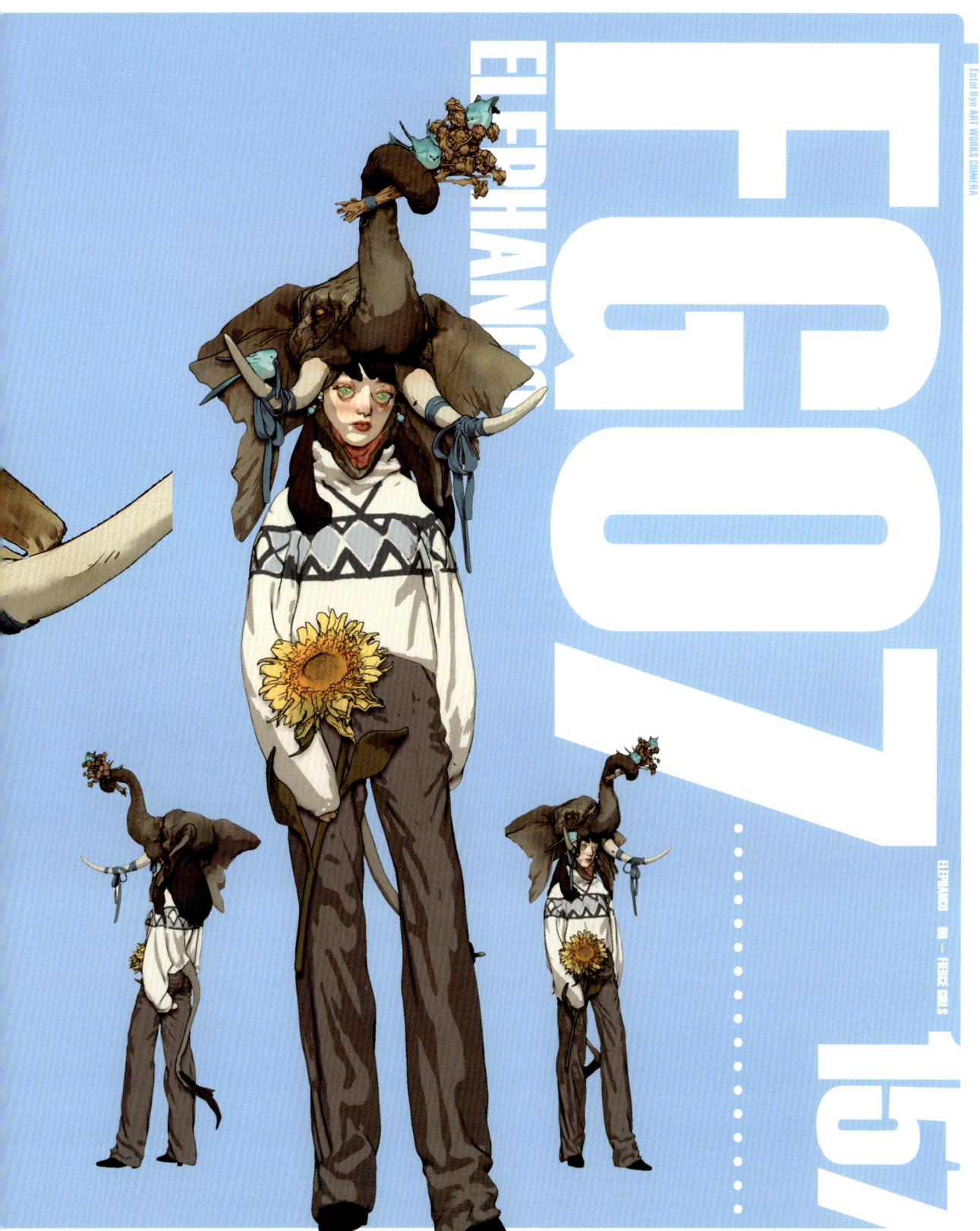

Edited Ryo ART WORKS CHIMERA

ELEPHANGI

ELEPHANGI 06 — FIERCE GIRLS

157

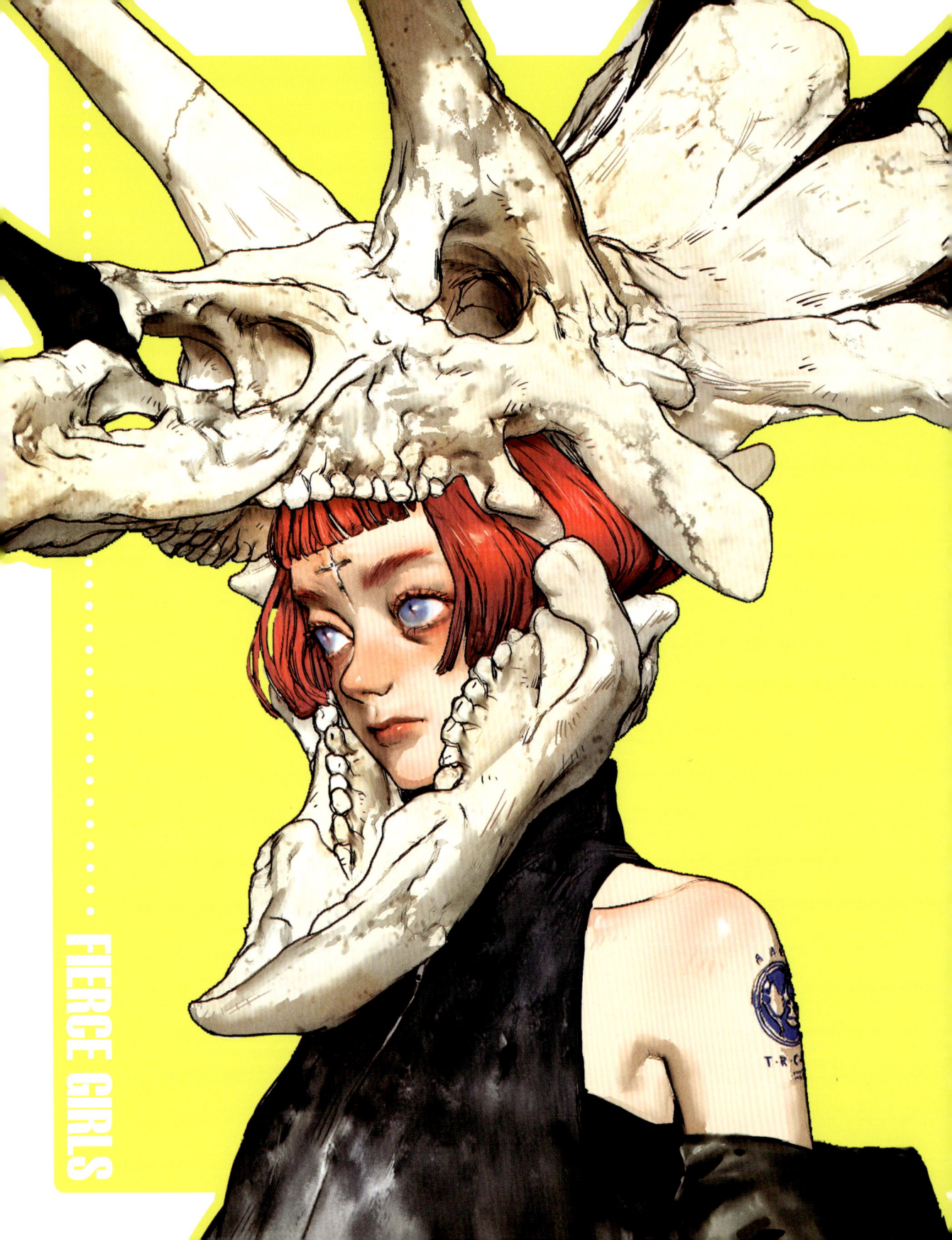

FIERCE GIRLS

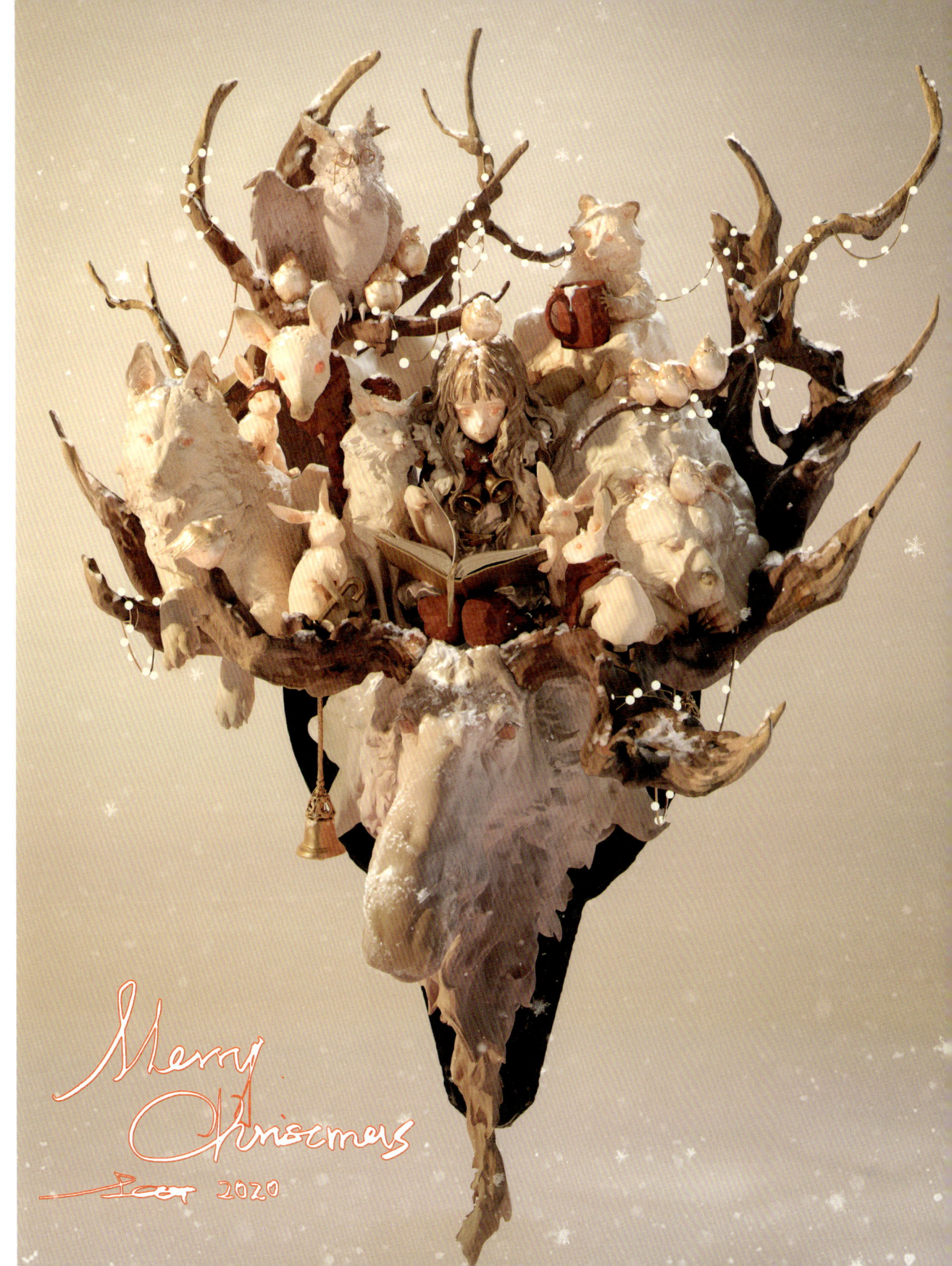

Merry
Christmas
2020

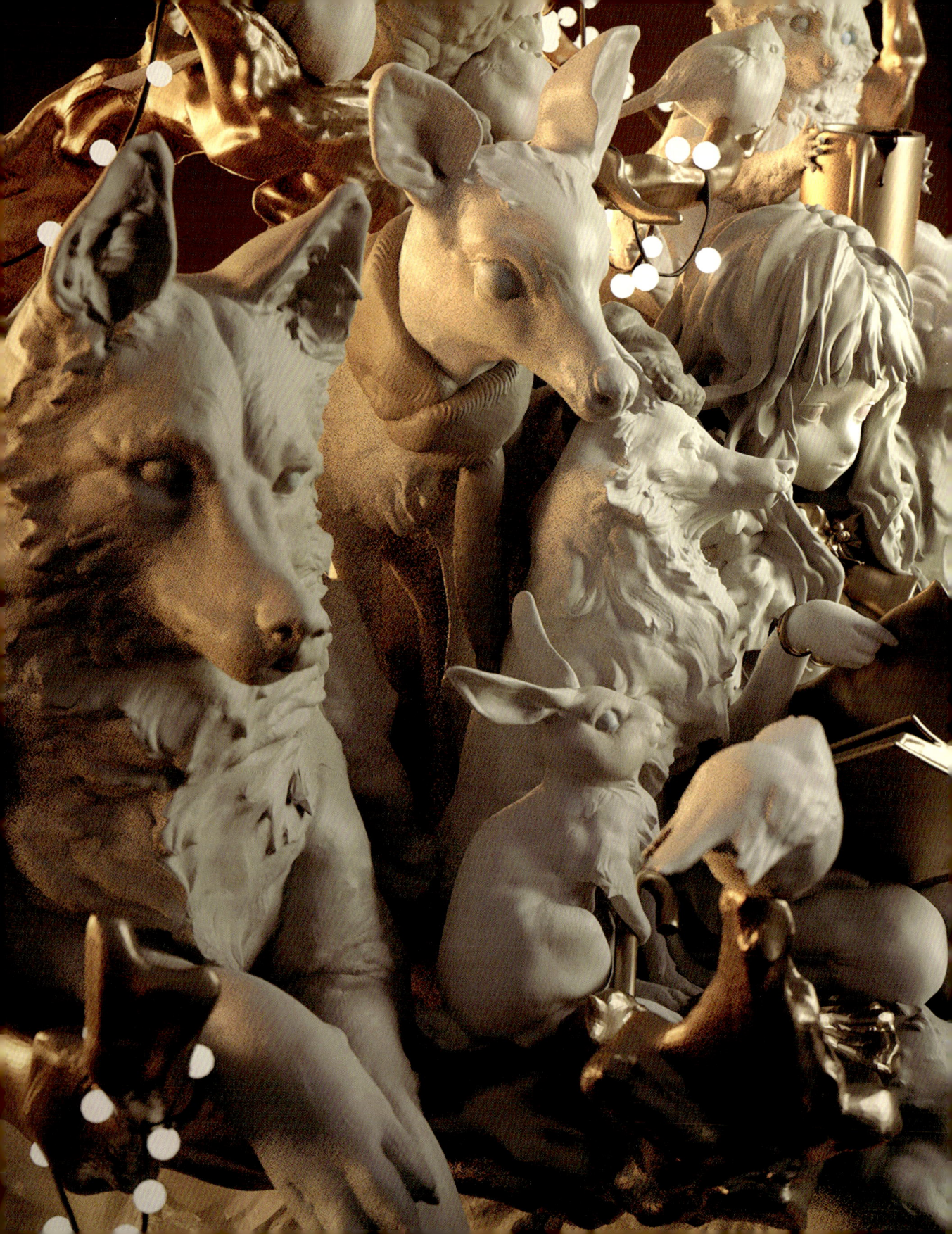

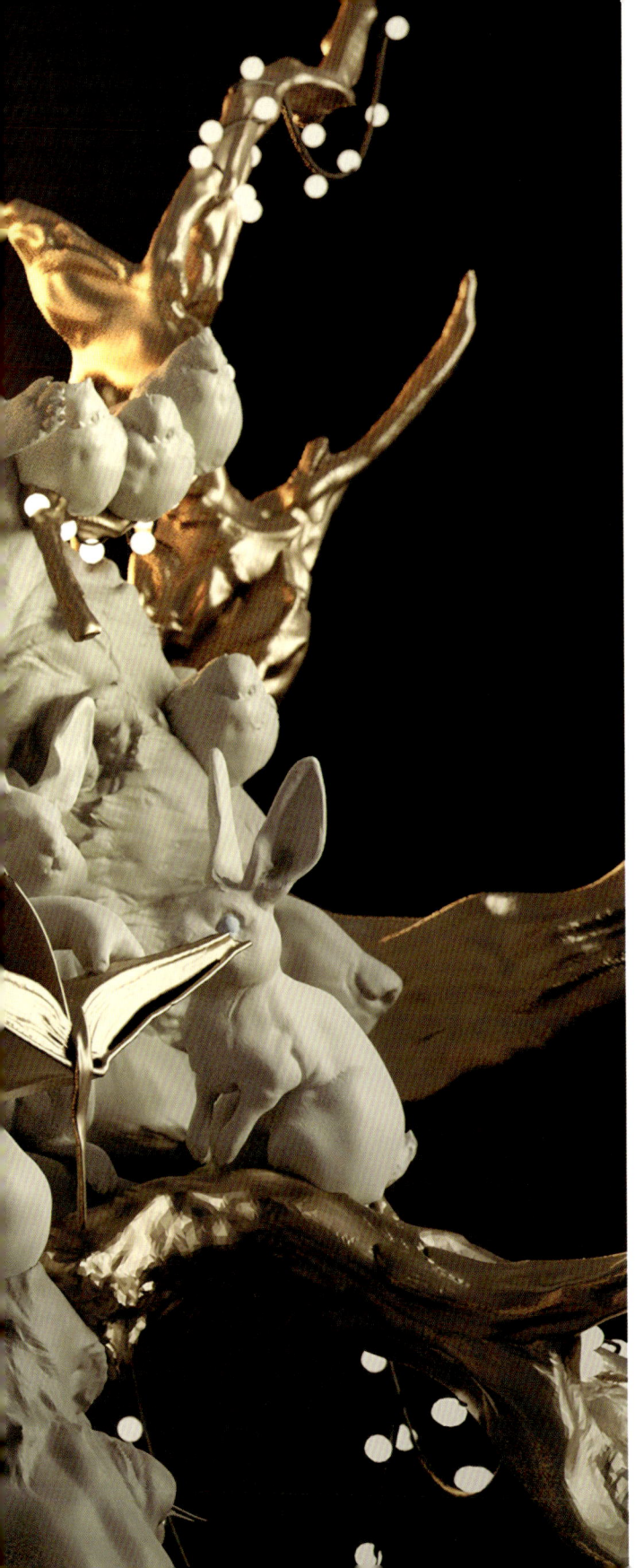

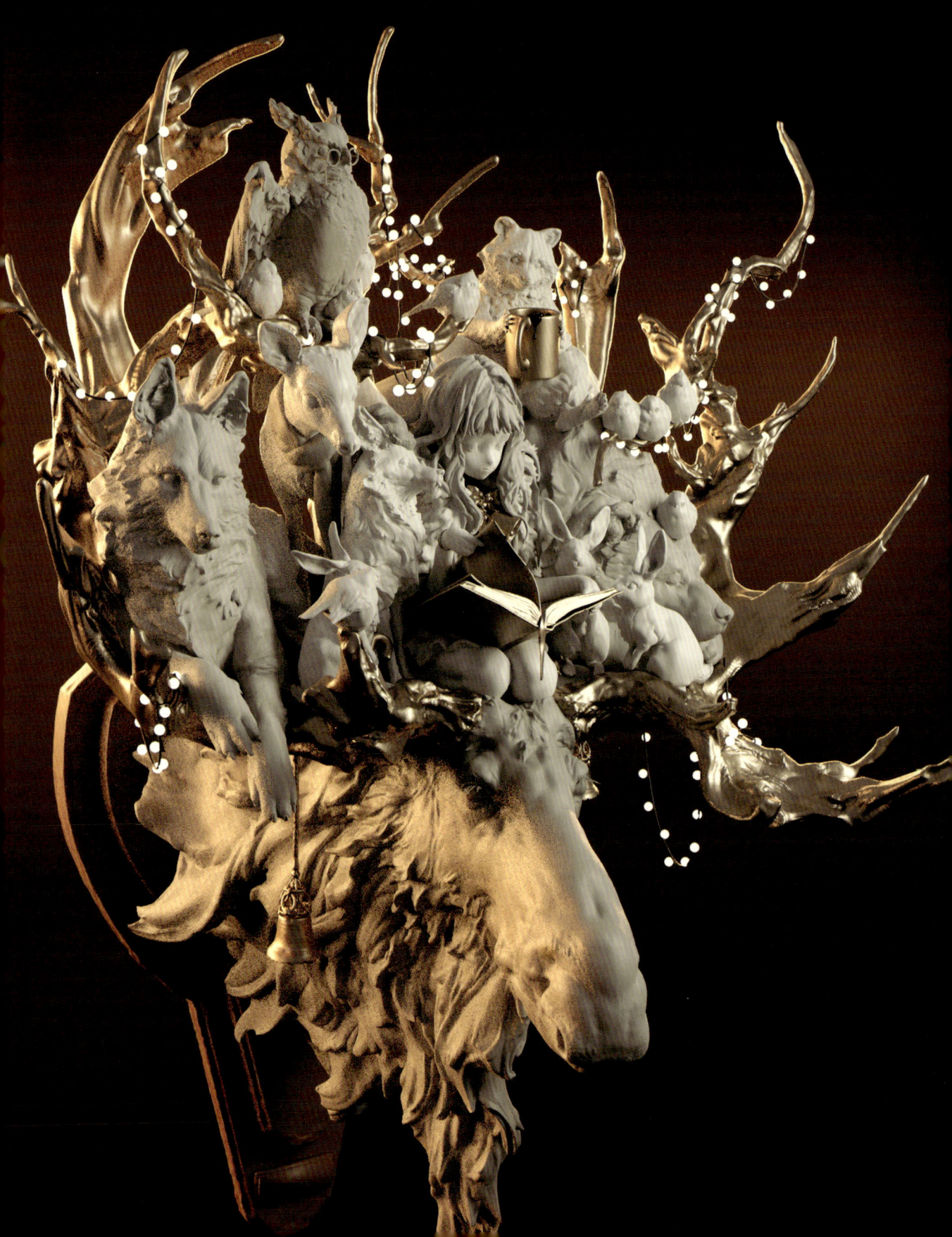

**FOREST FRIENDS**

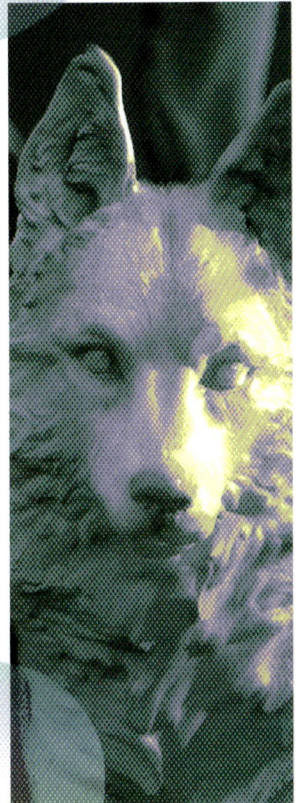

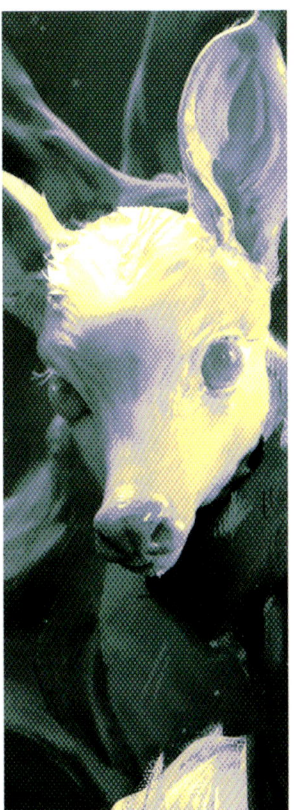

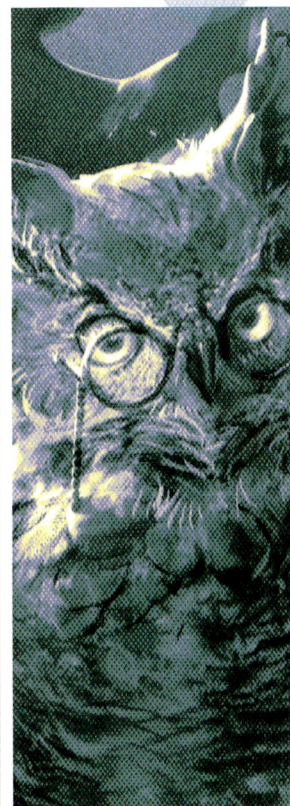

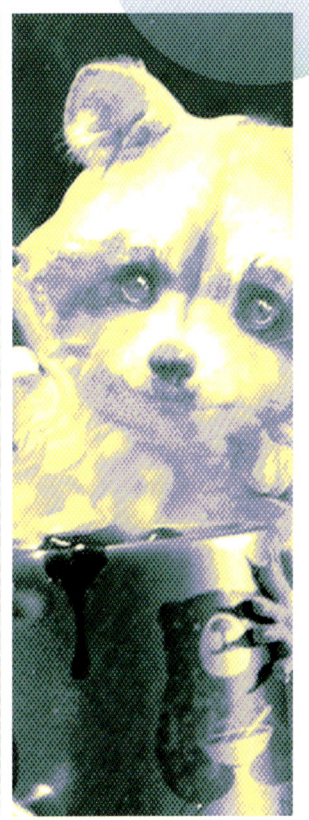

**Volume A**
The north wind, the cedar,
and the friends of the forest
are surrounding the
mysterious winter chanter...

**Volume B**
The Sound of Bell, the Frost
Flower, the friends of the
forest are guarding the
mysterious winter chanter...

**Volume C**
The Scroll, The Moon, the
friends of the forest are
listening to the mysterious
winter chanter...

**Volume D**
Gold foil candy wrapppers,
Cocoa gingerbread, the
friends of the forest said
goodbye to the winter
chanter.

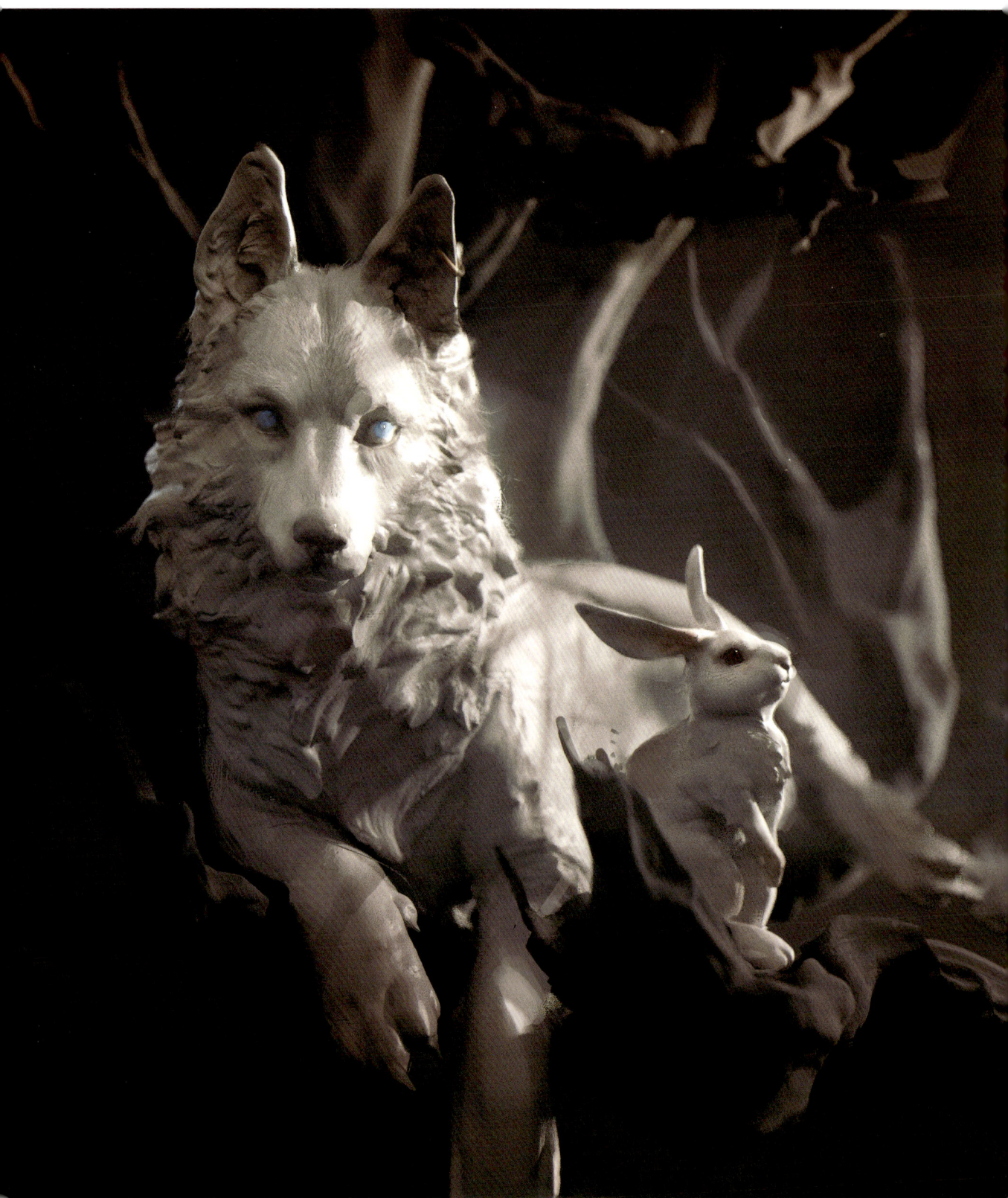

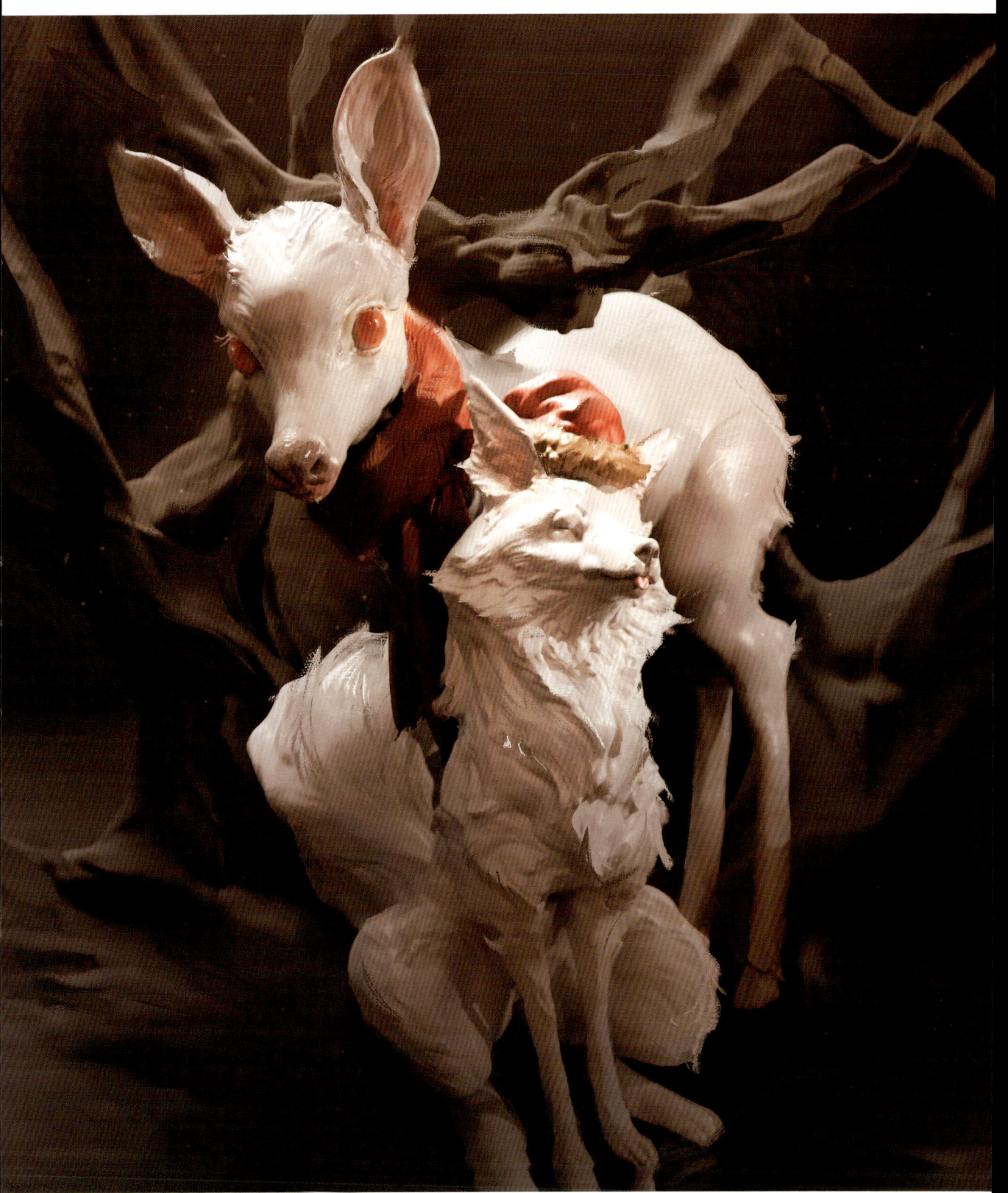

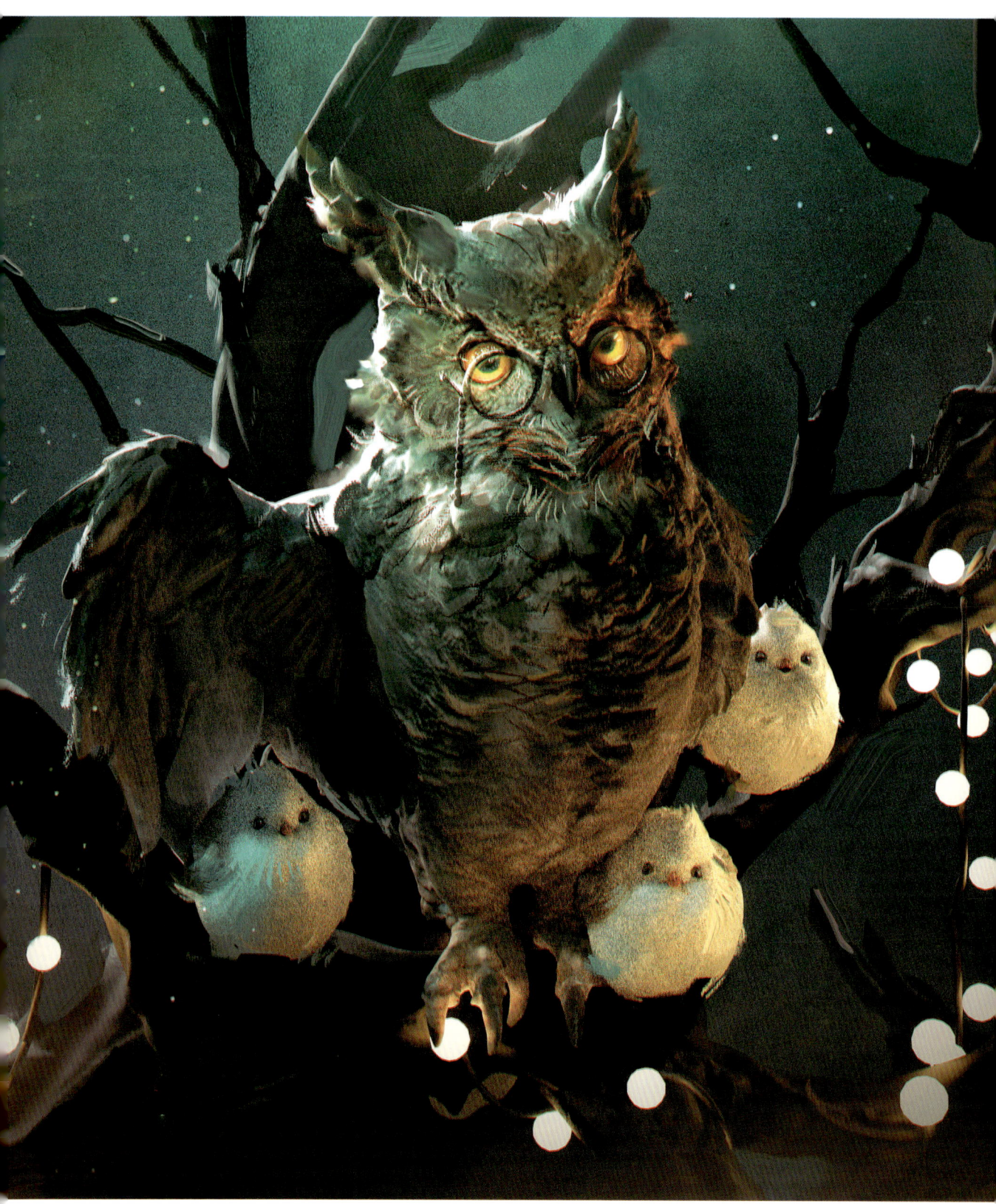

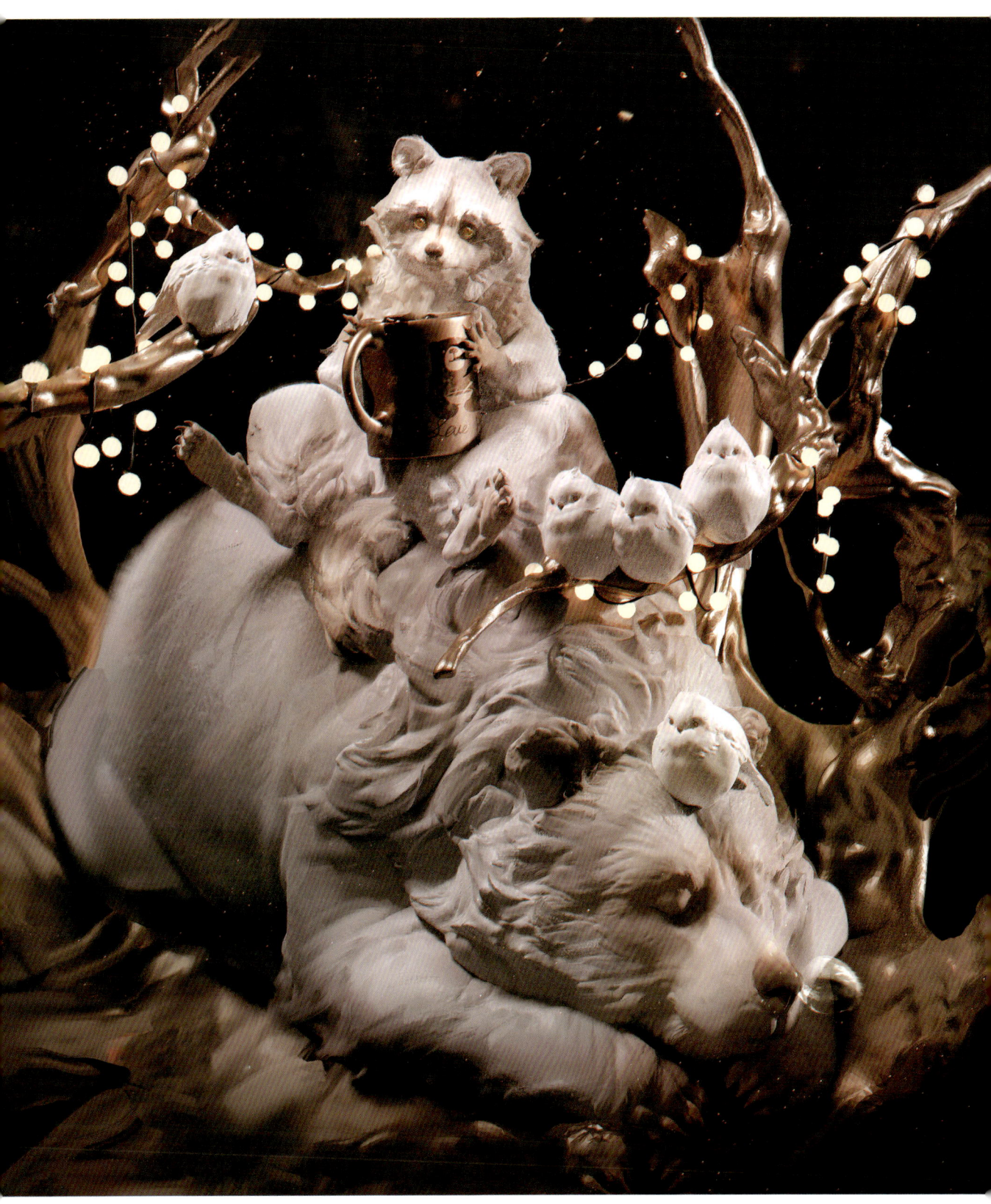

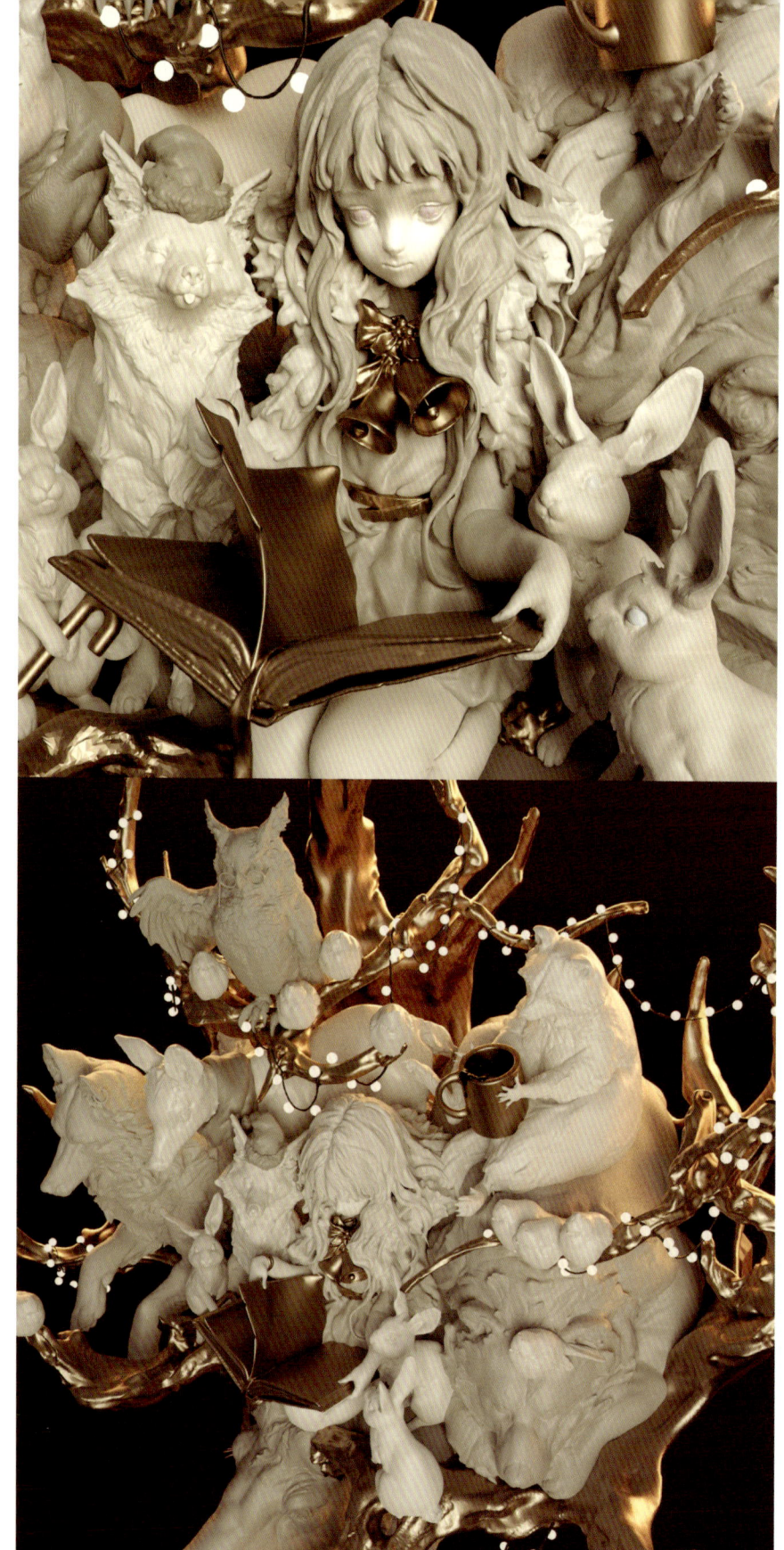

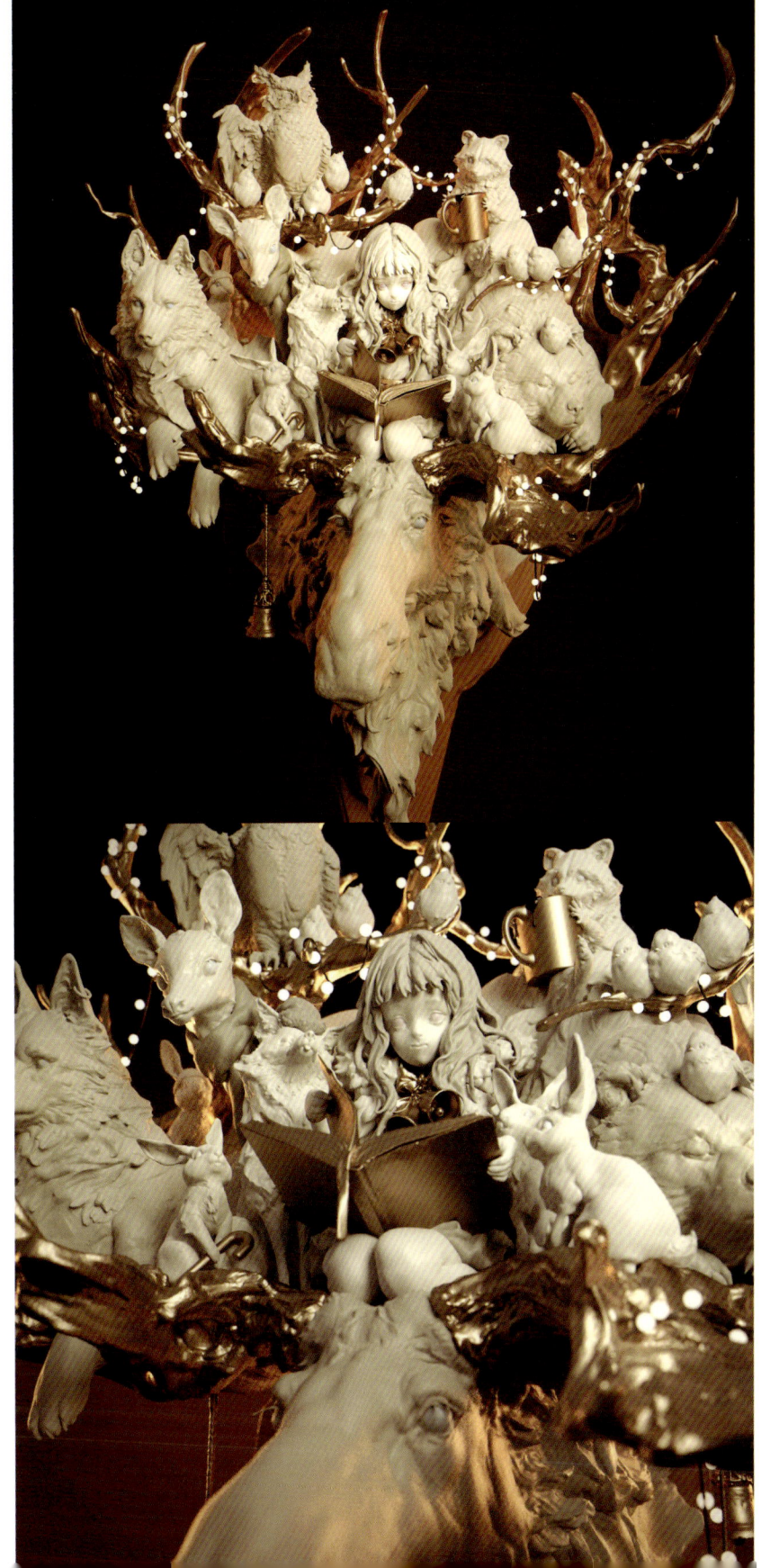

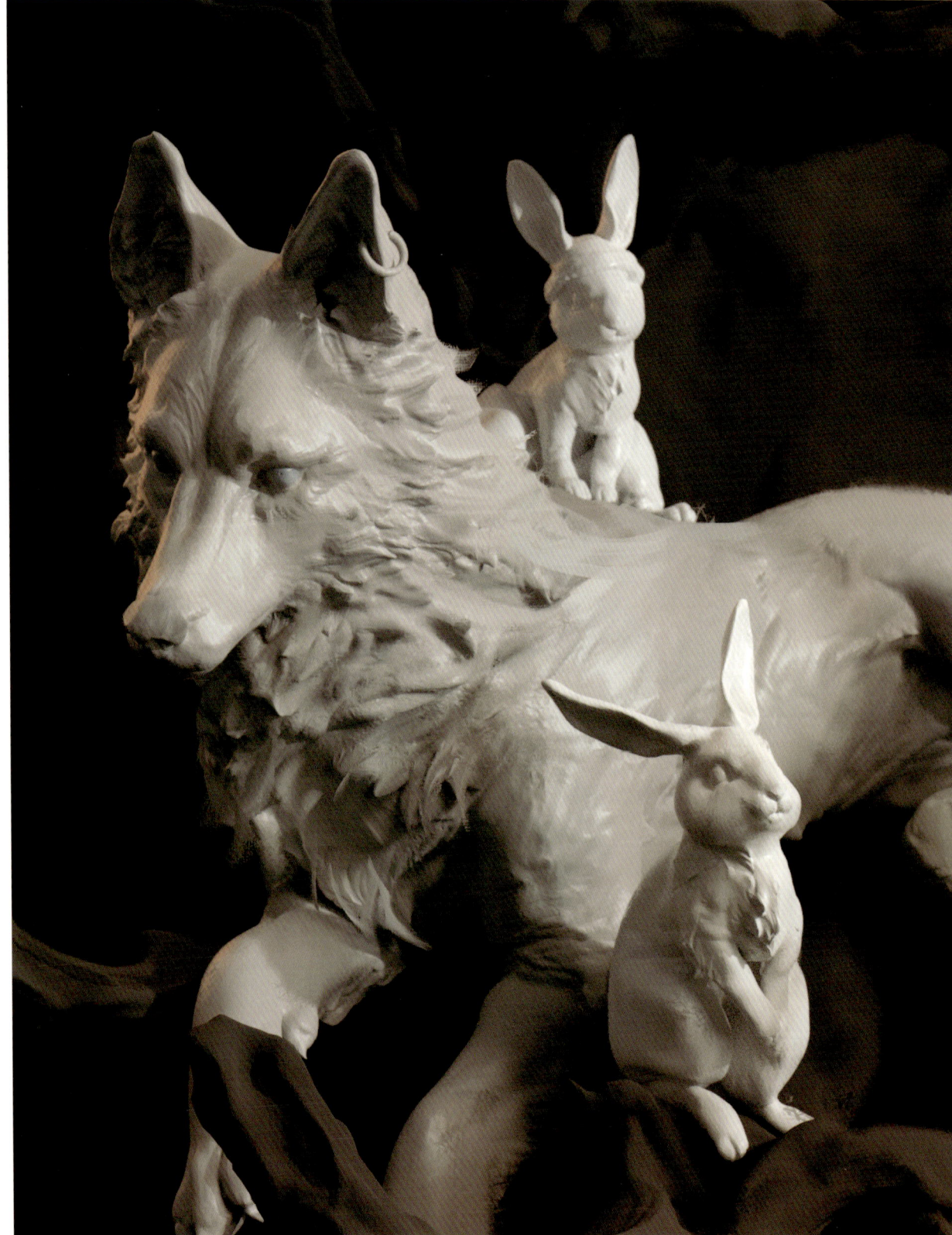

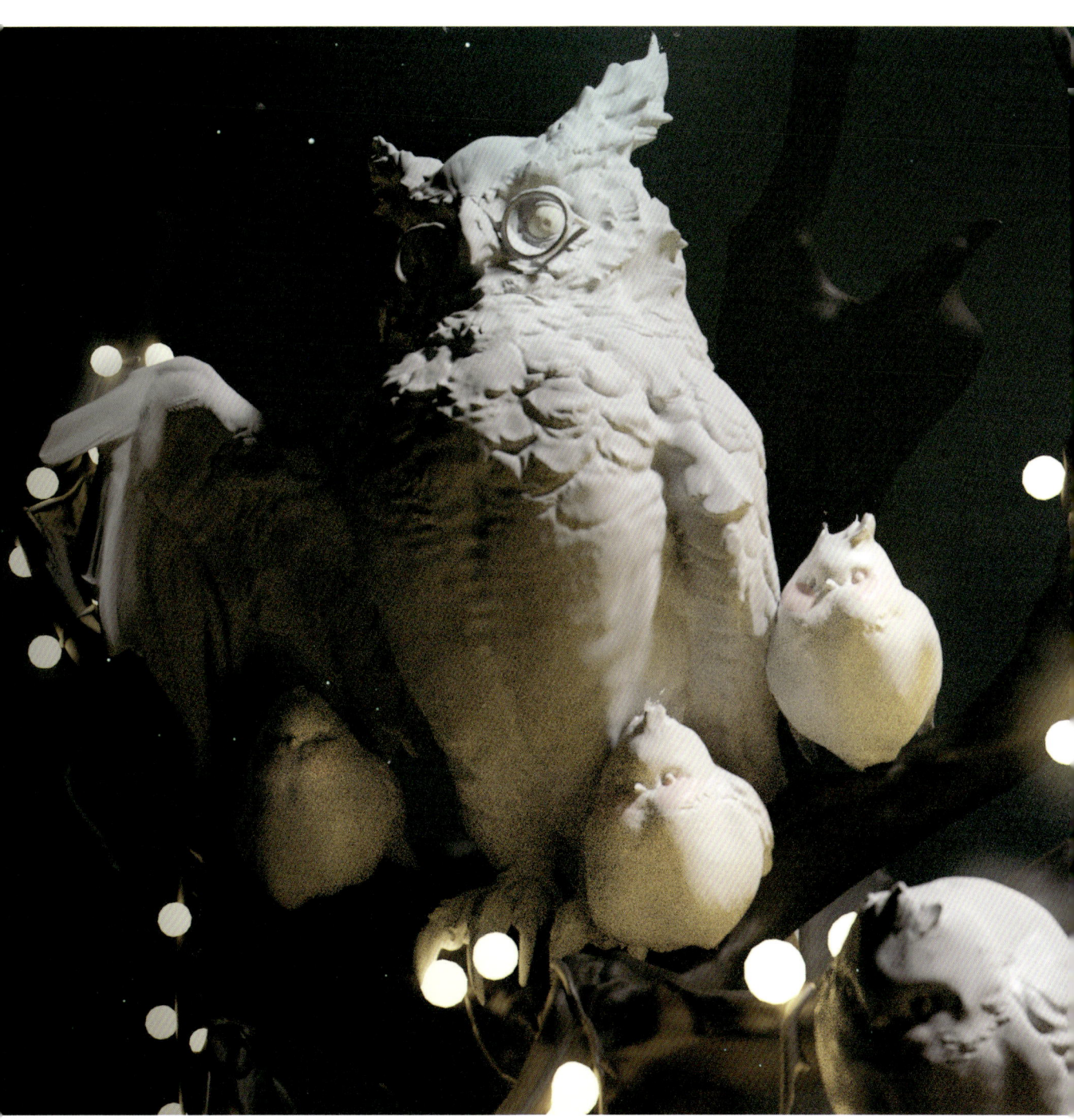

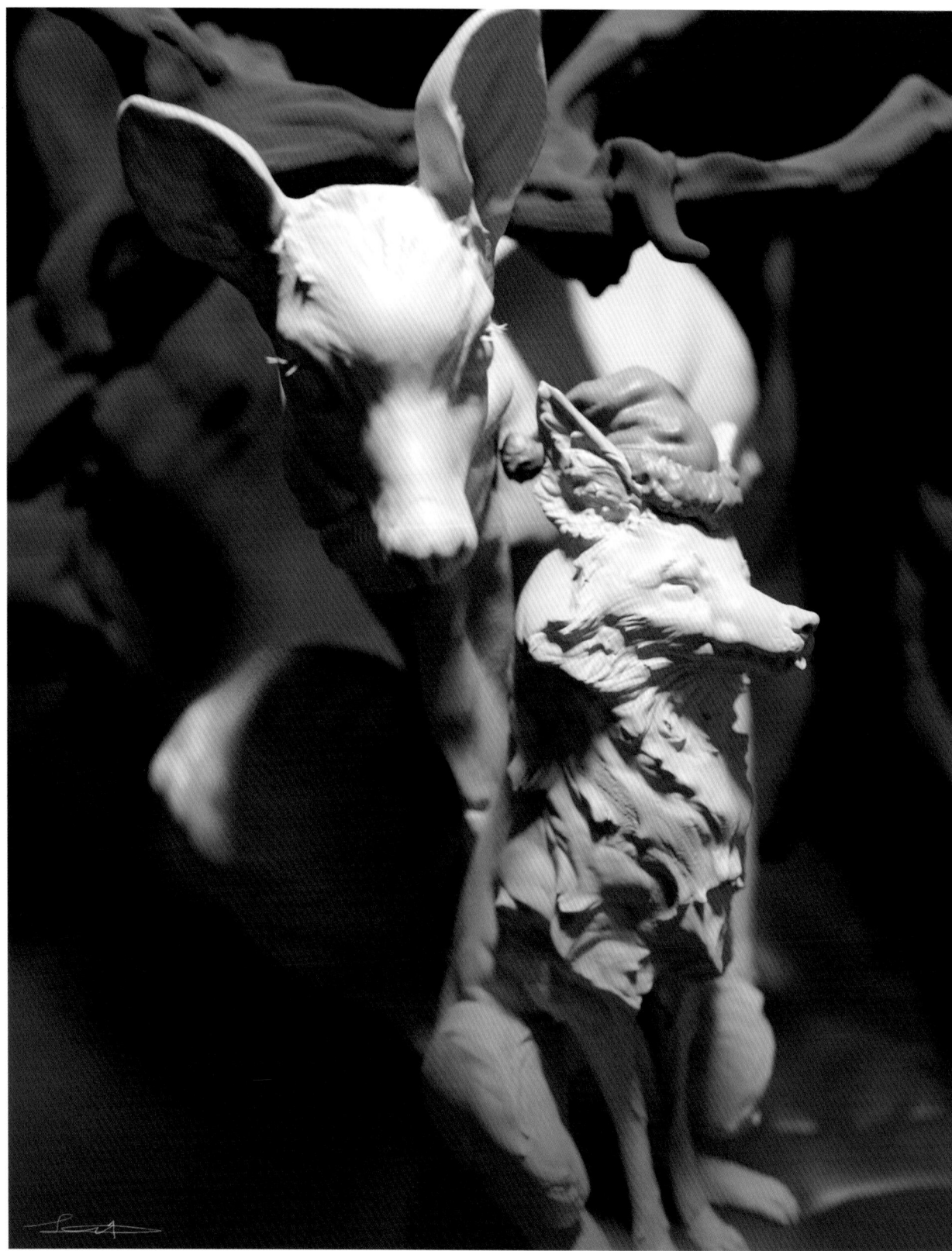

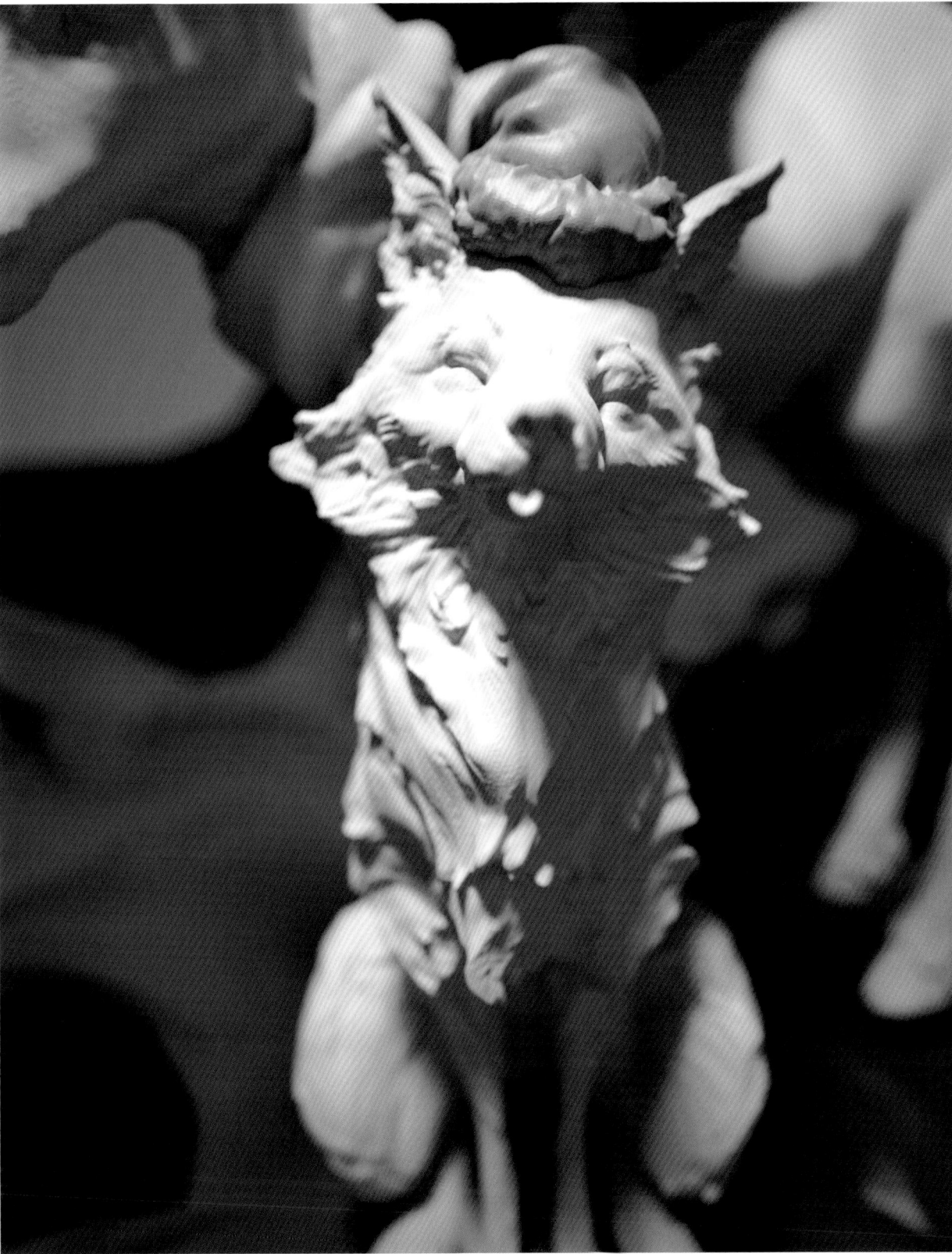

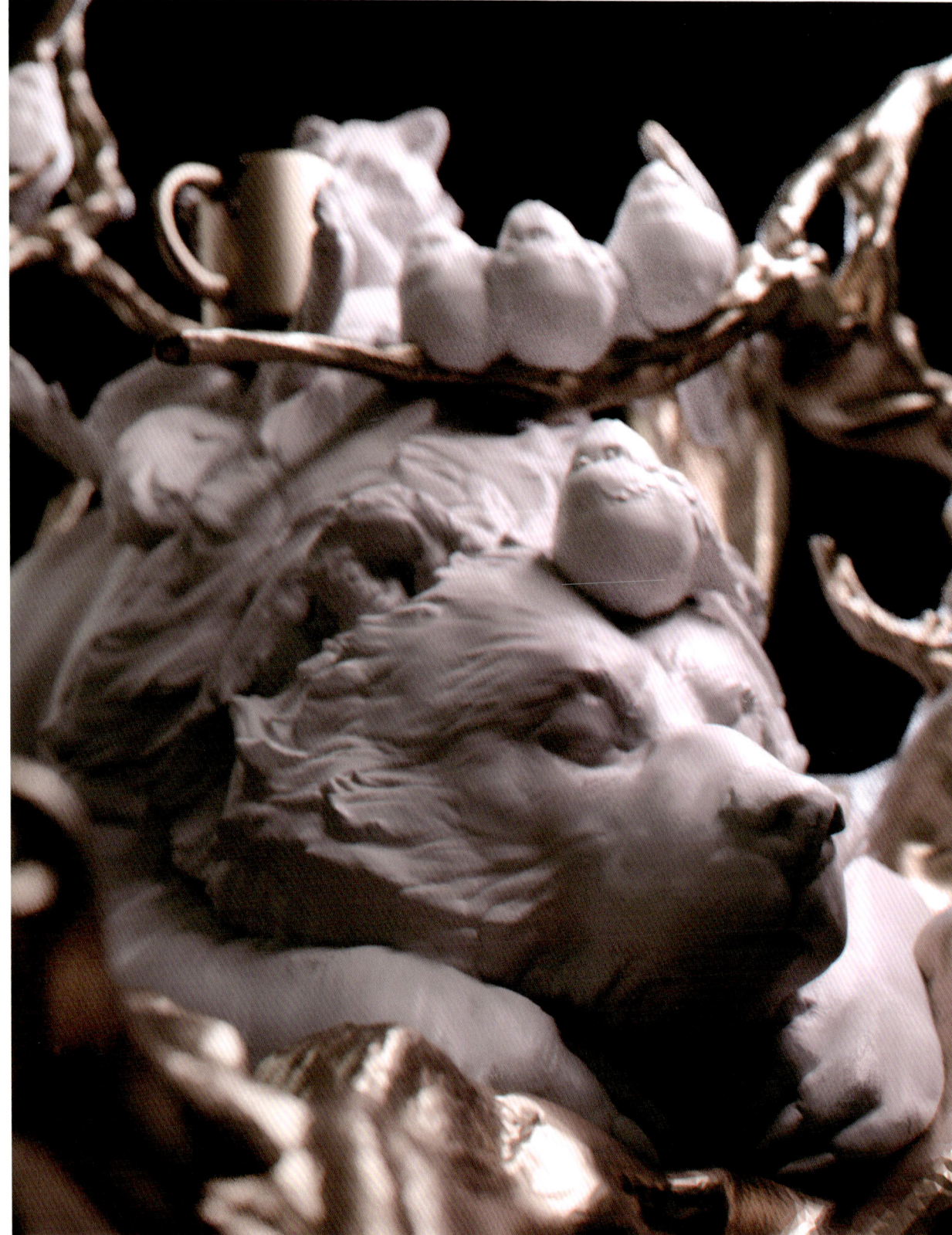

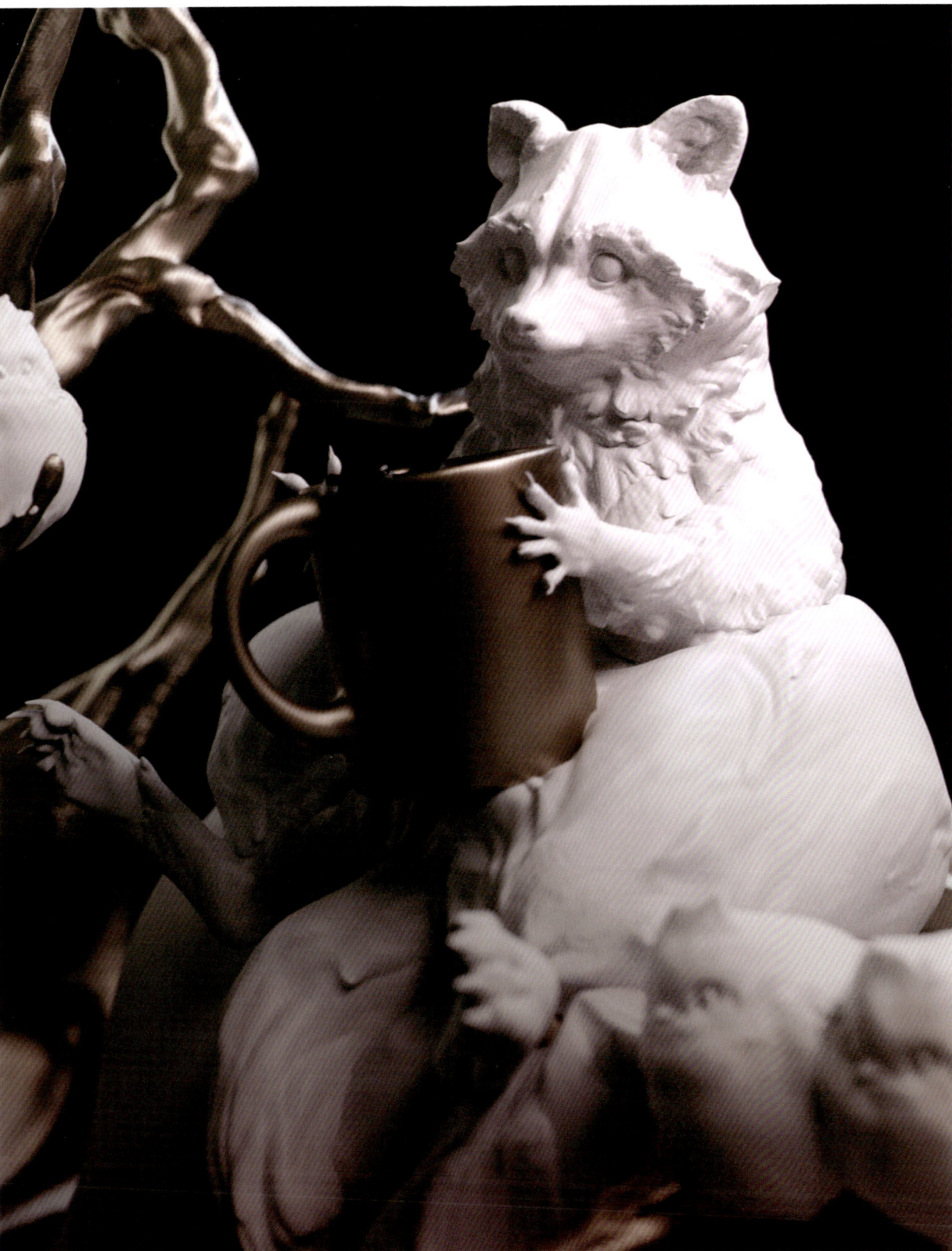

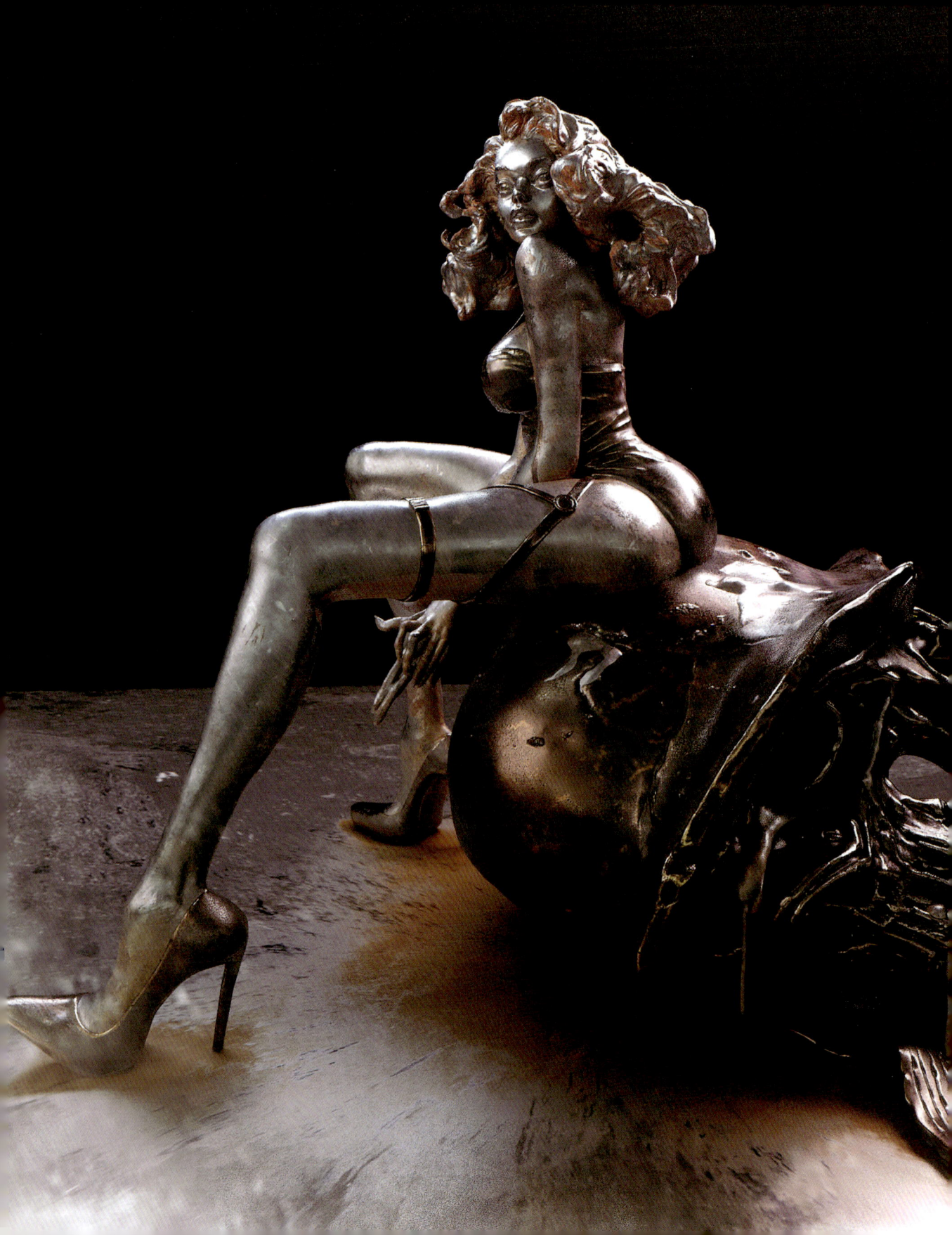

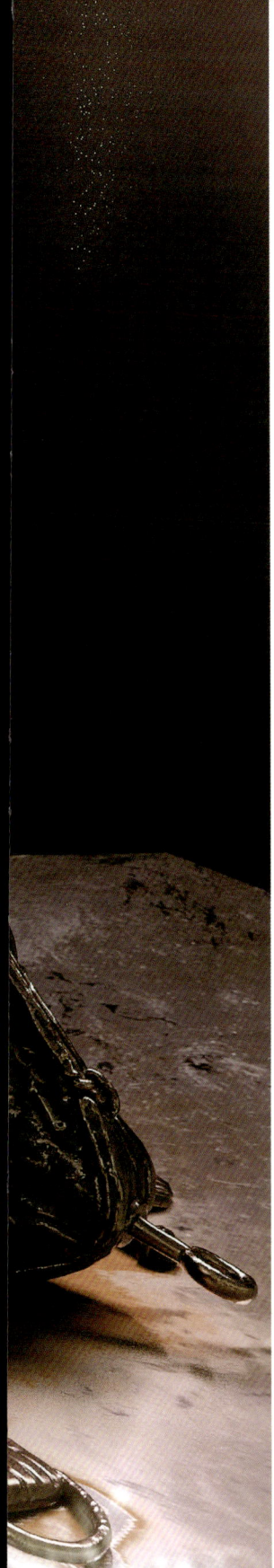

# Majestic Perch 18:

Artwork by Ashley Wood
Majestic Perch by 2021 7474 pay ltd

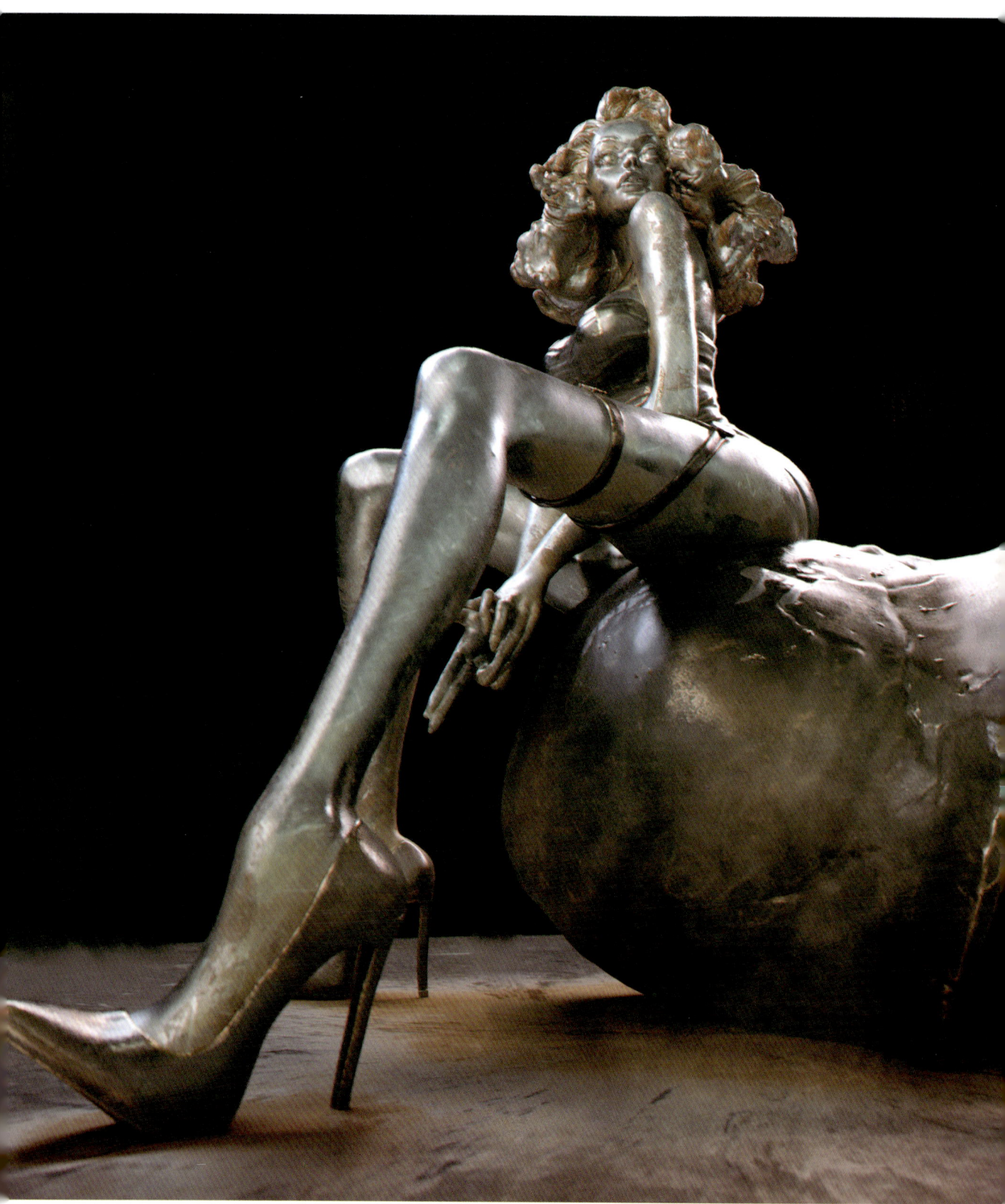

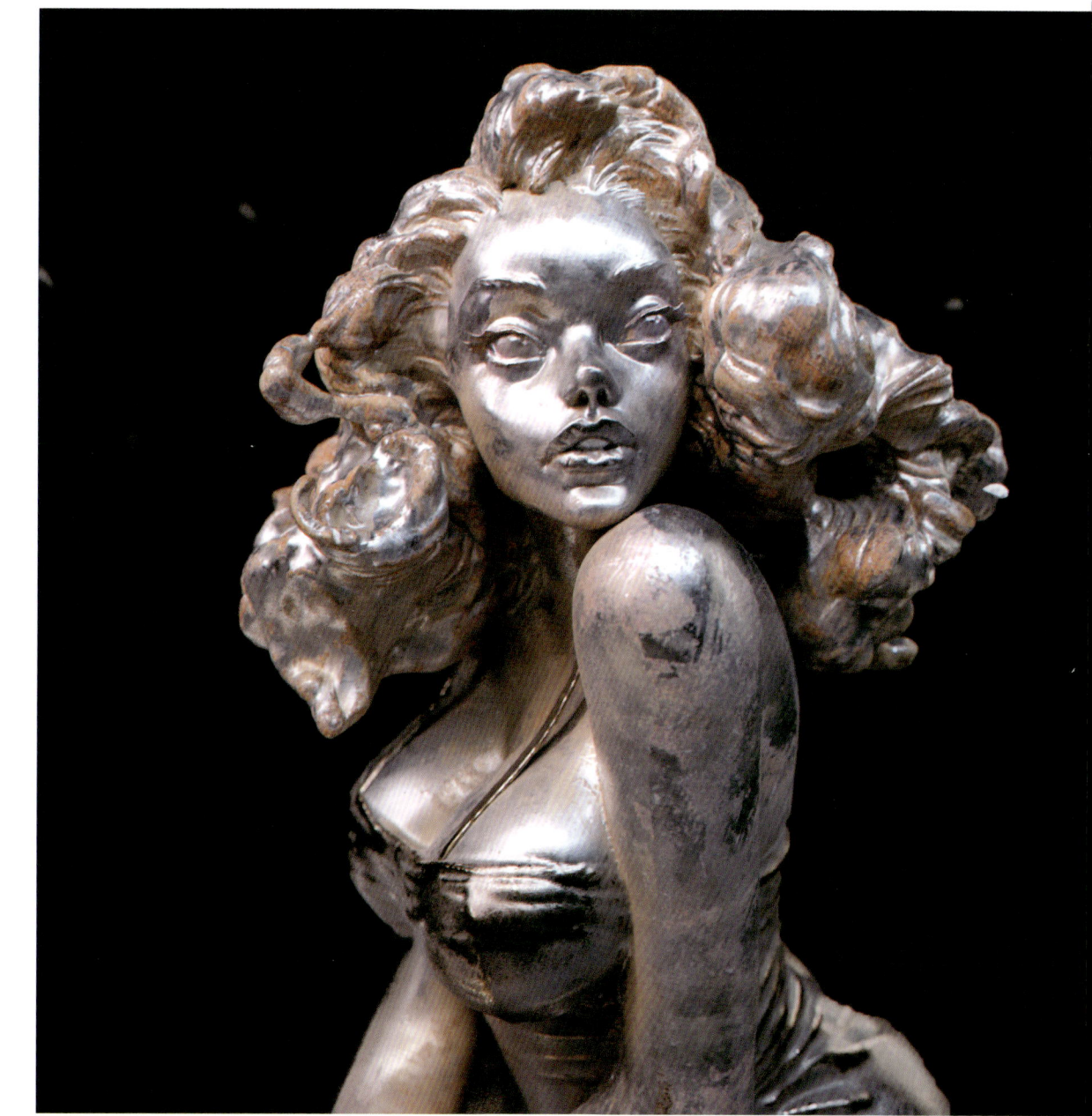

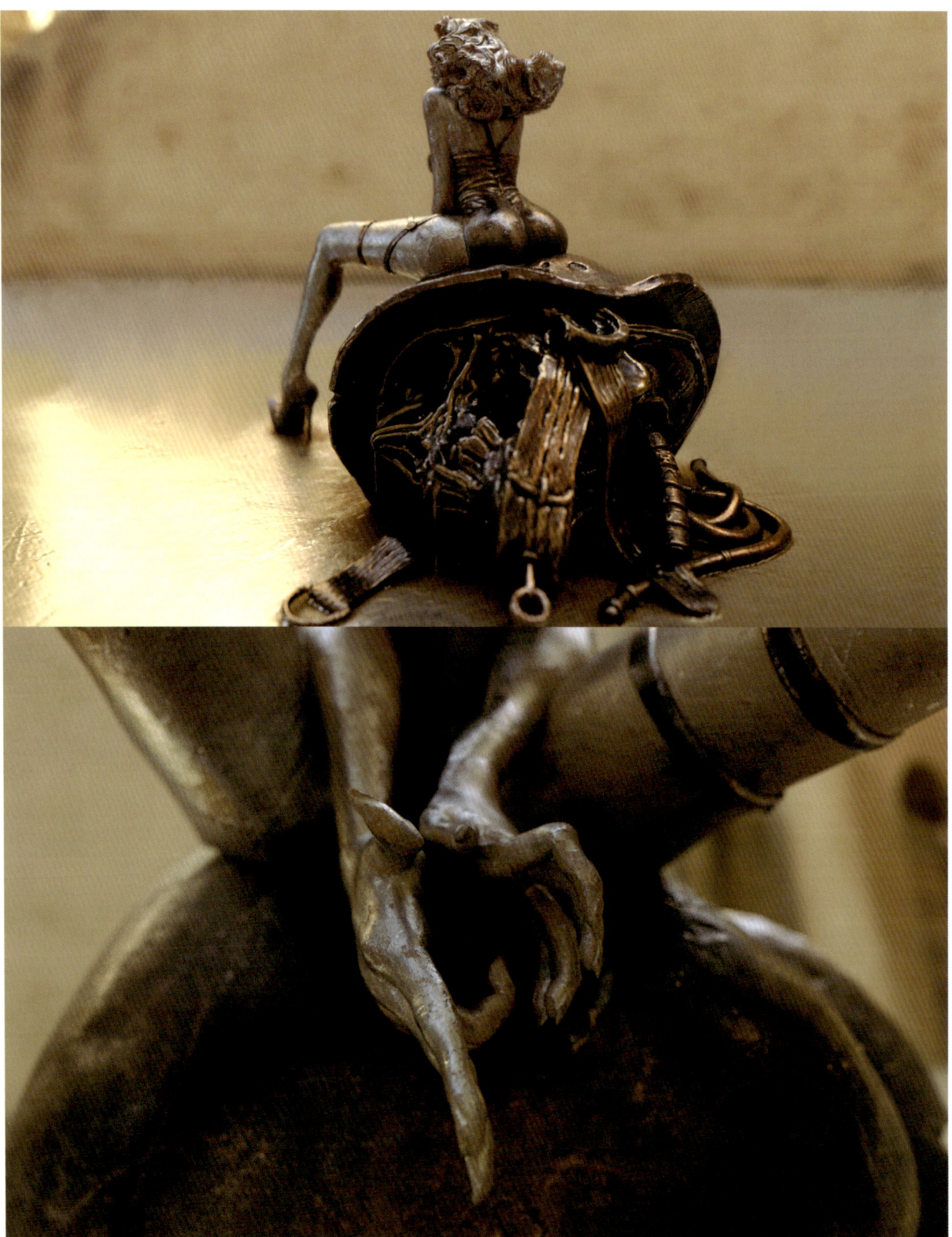

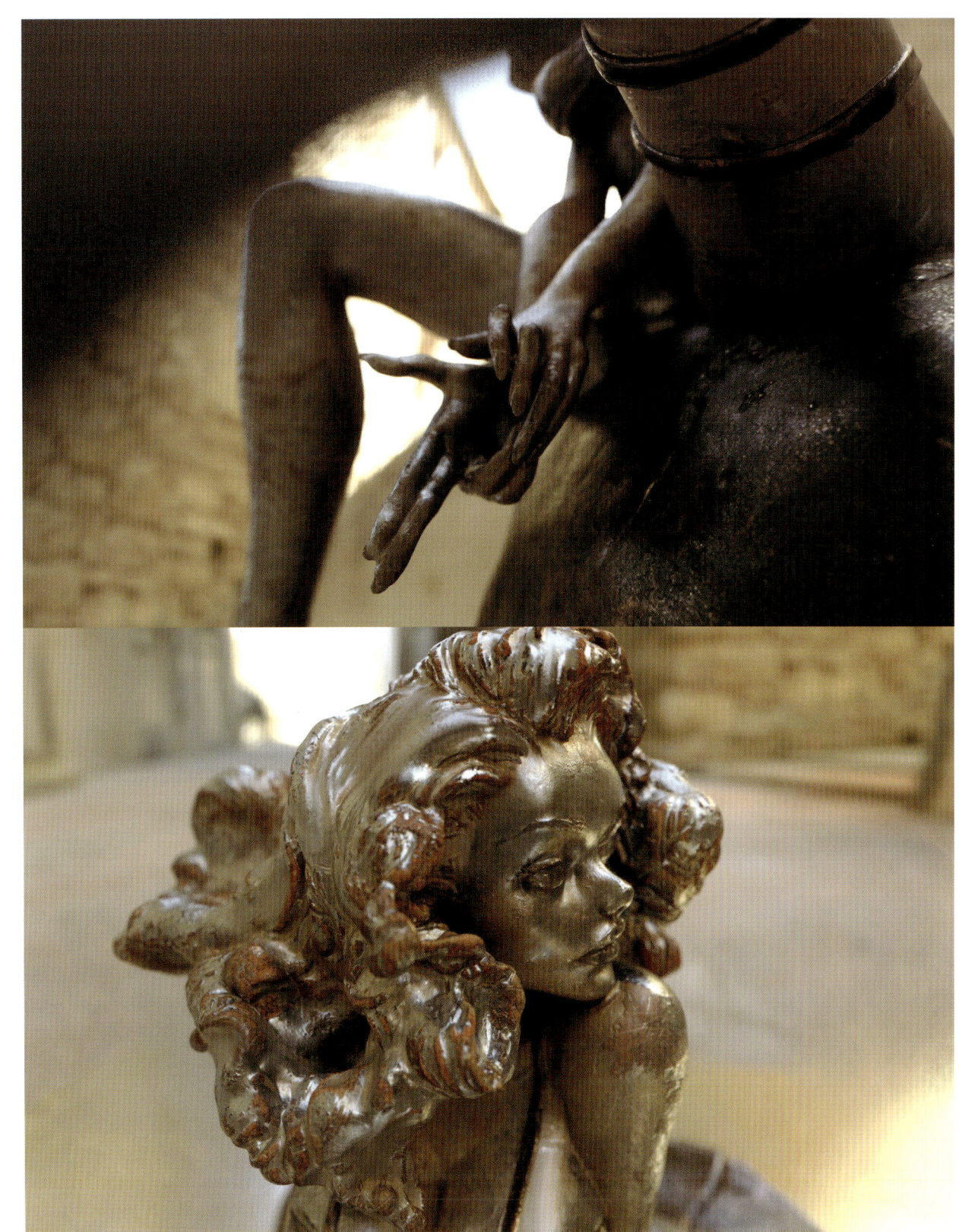

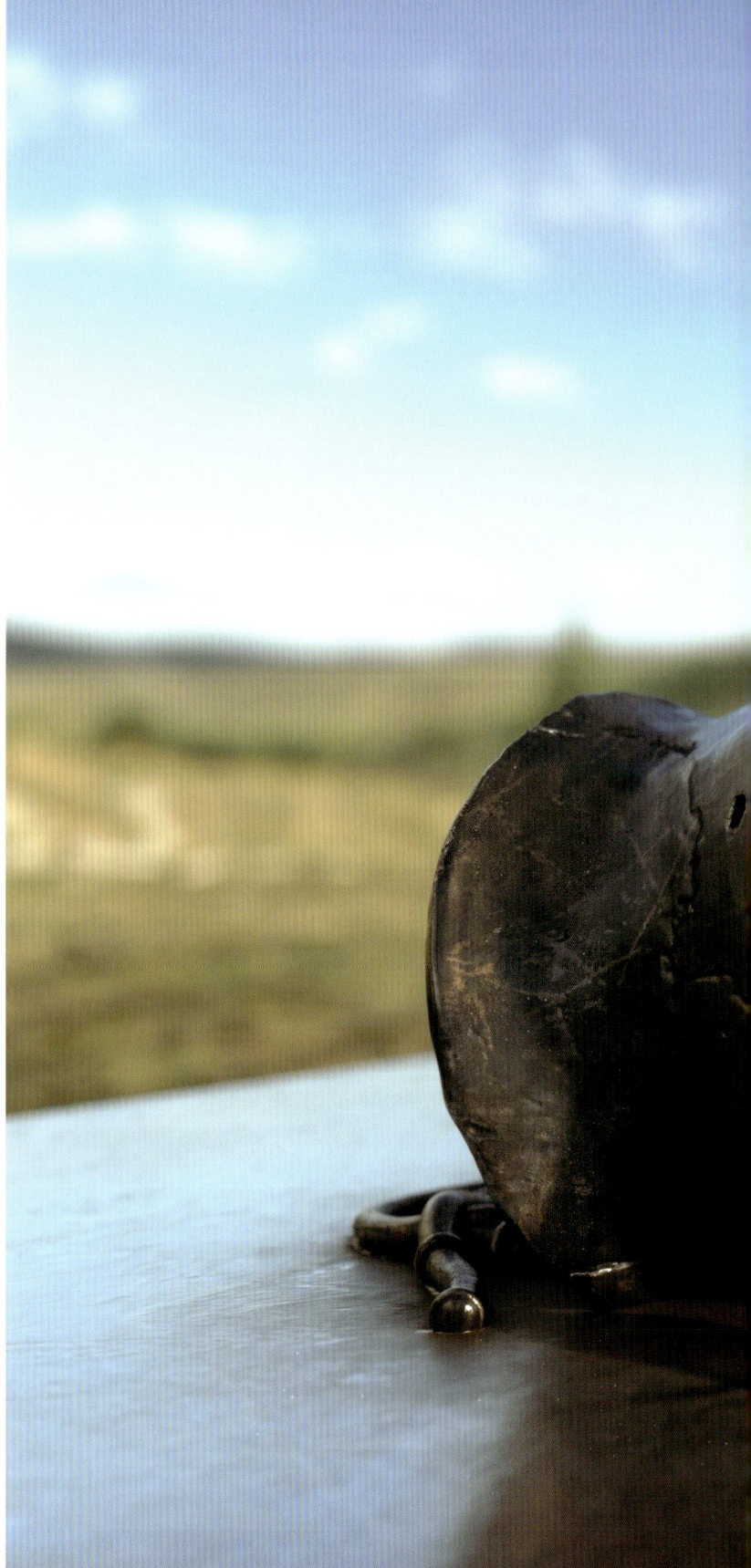

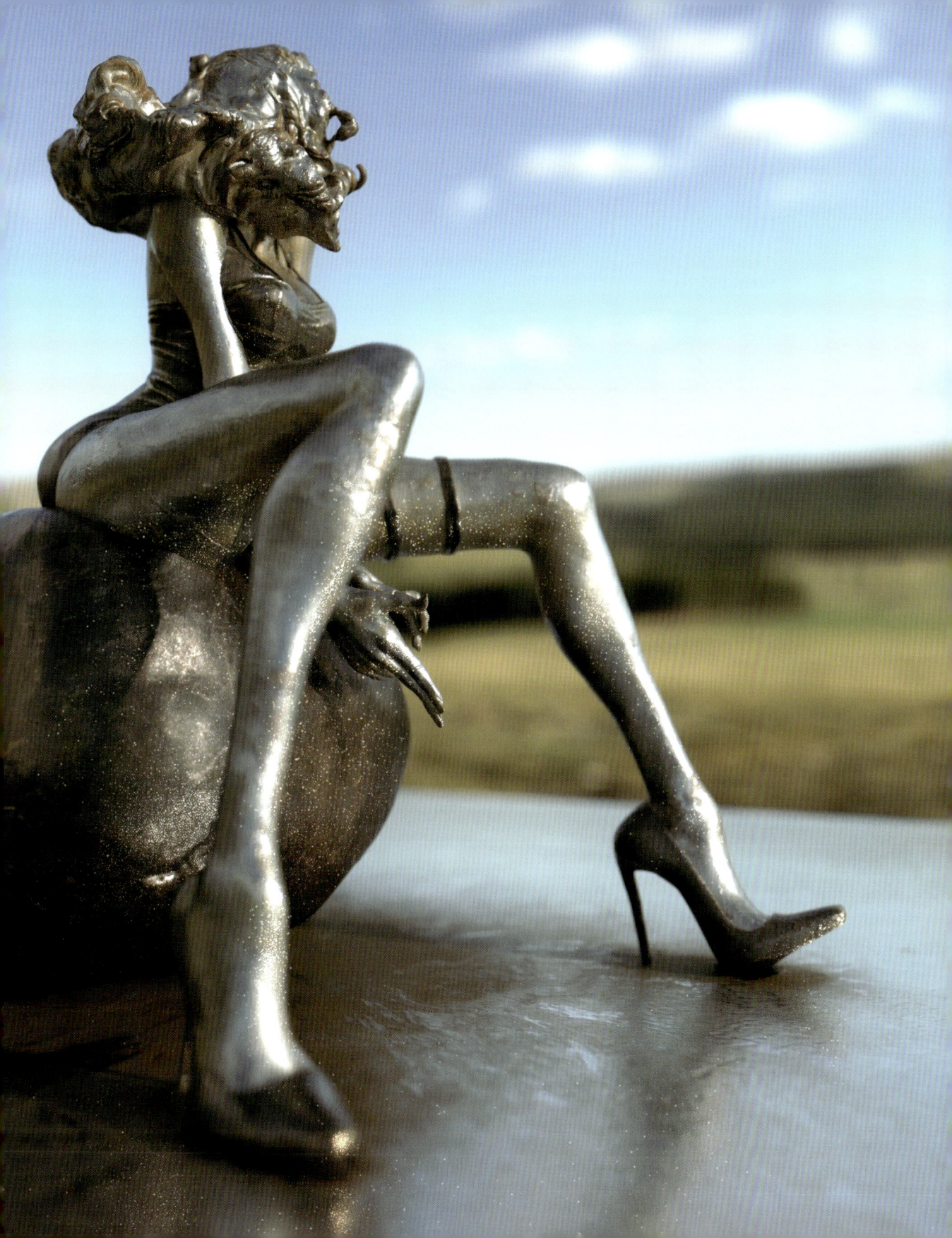

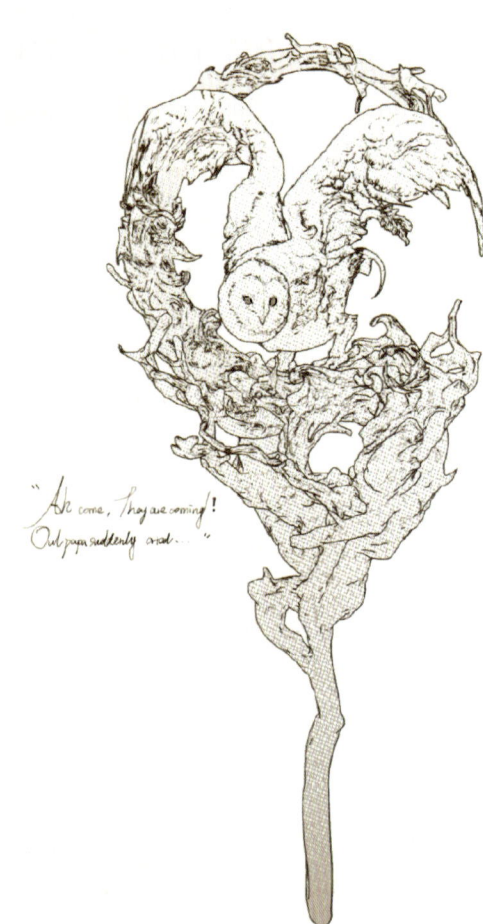

"Ah come, they are coming!"
Owl pages suddenly aread…"

Owl Brooch 19(

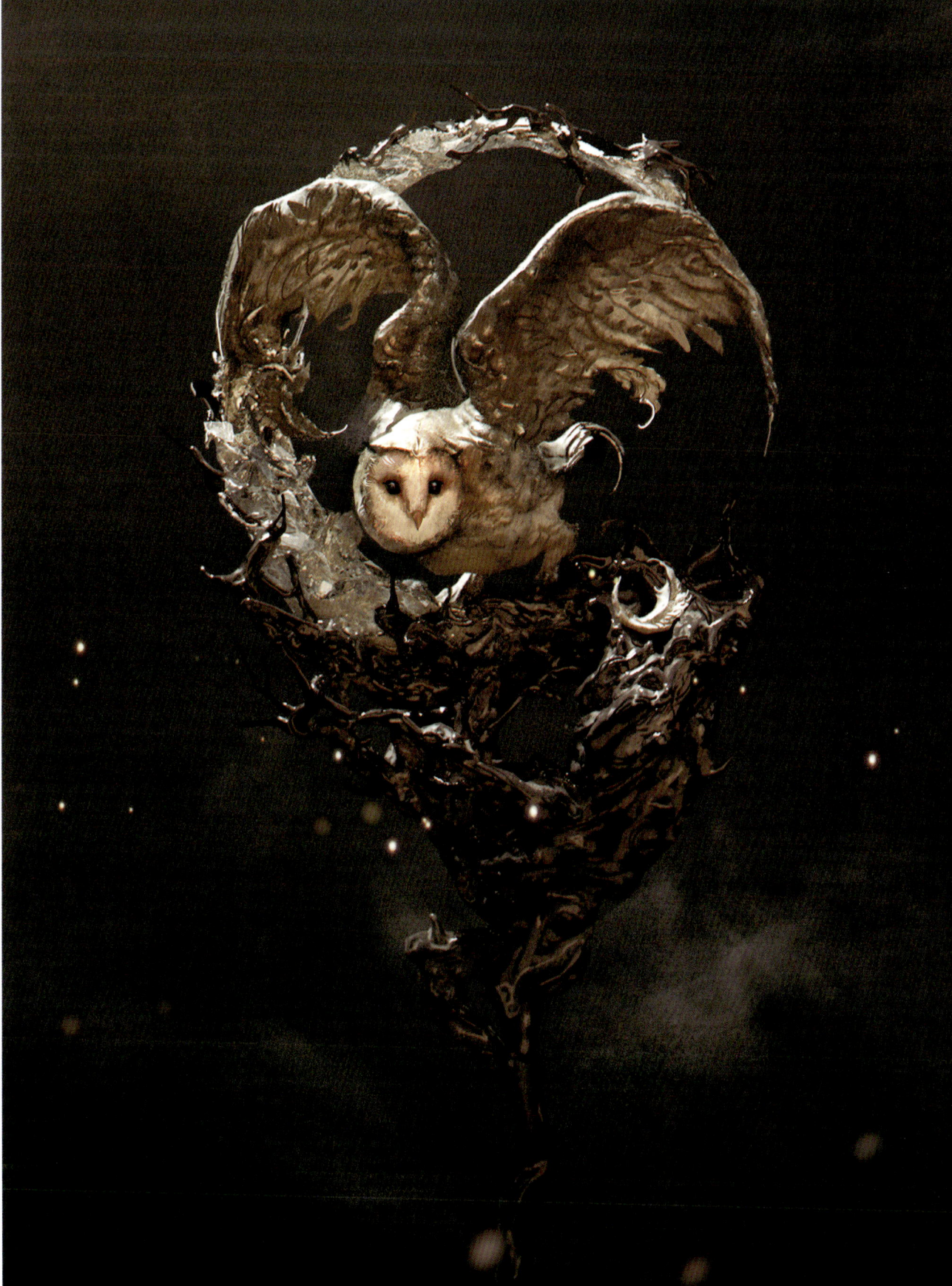

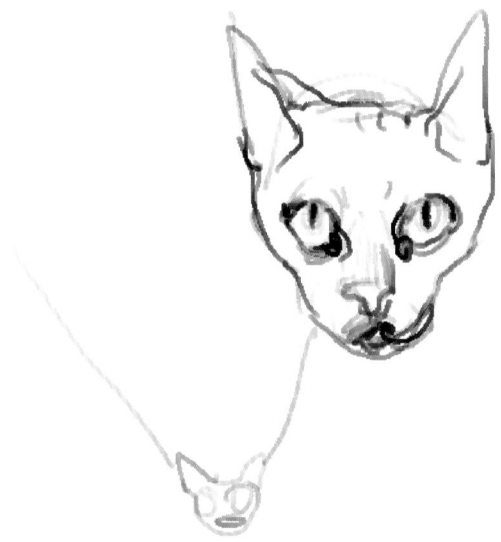

Necolace 192

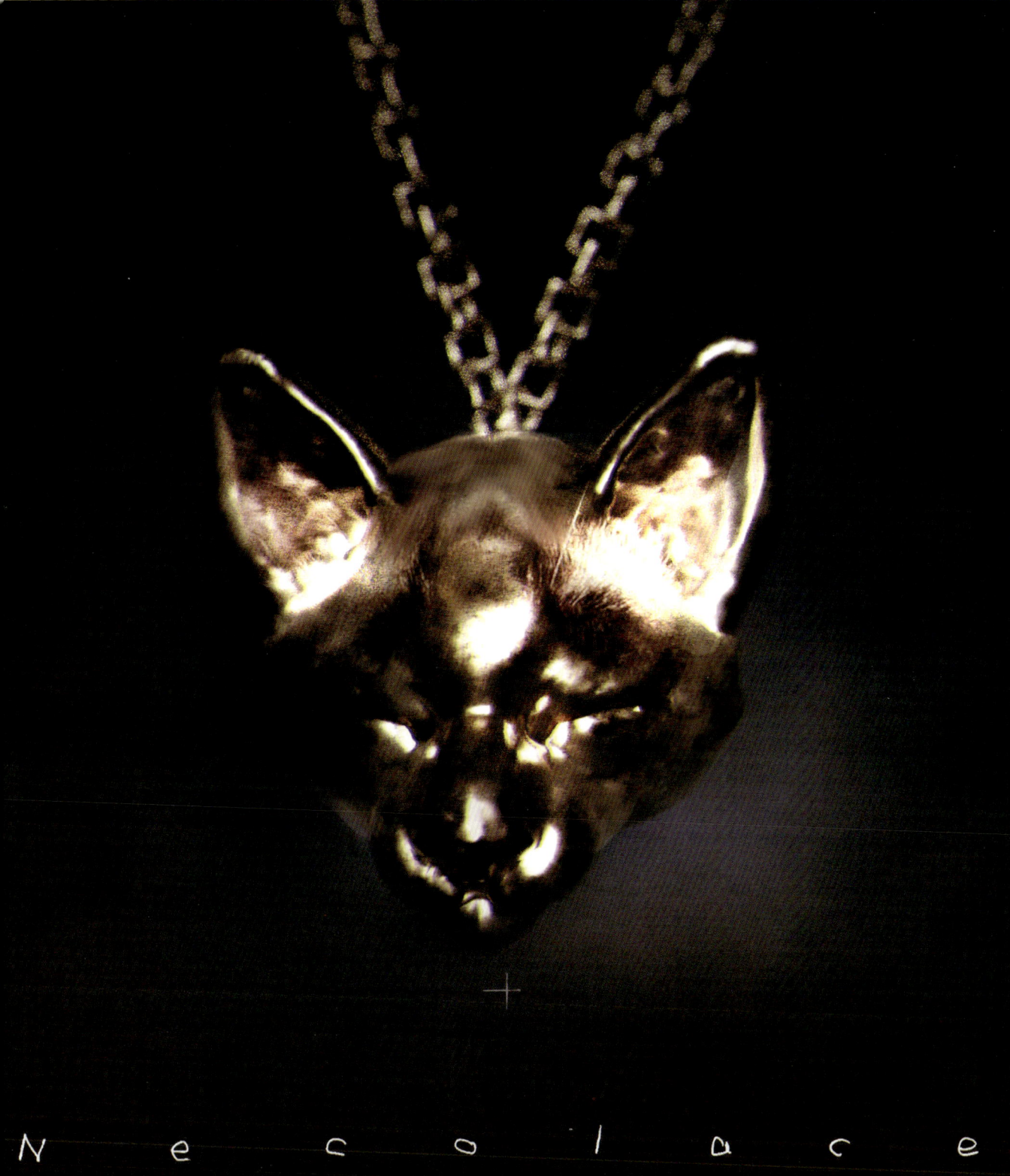

Necklace

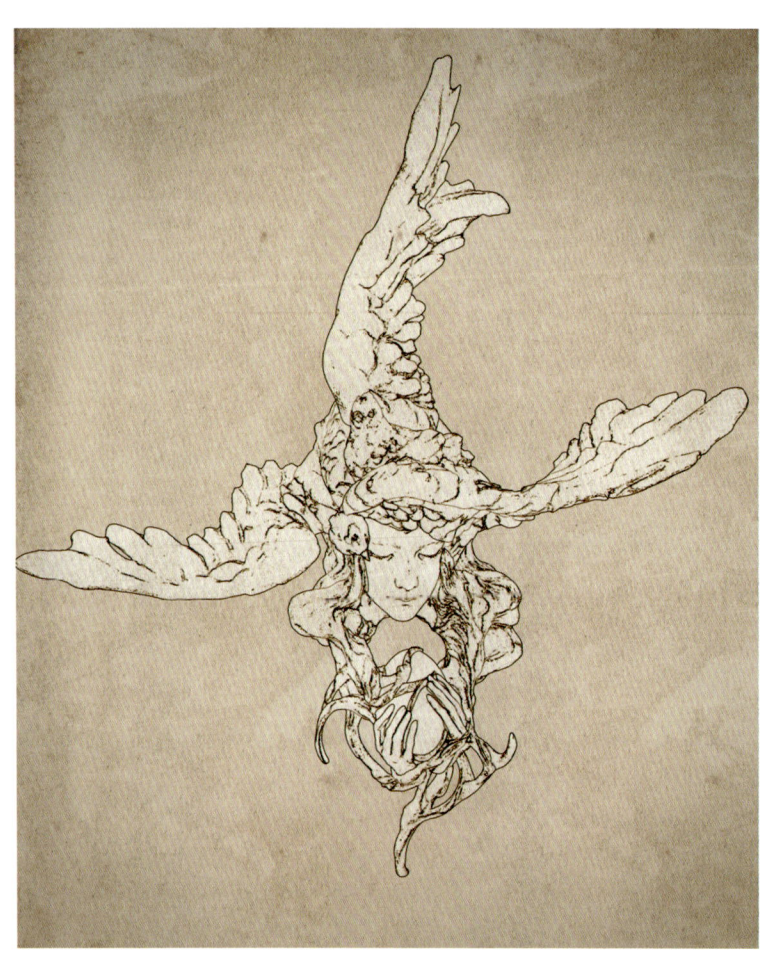

Crow's Witch 194

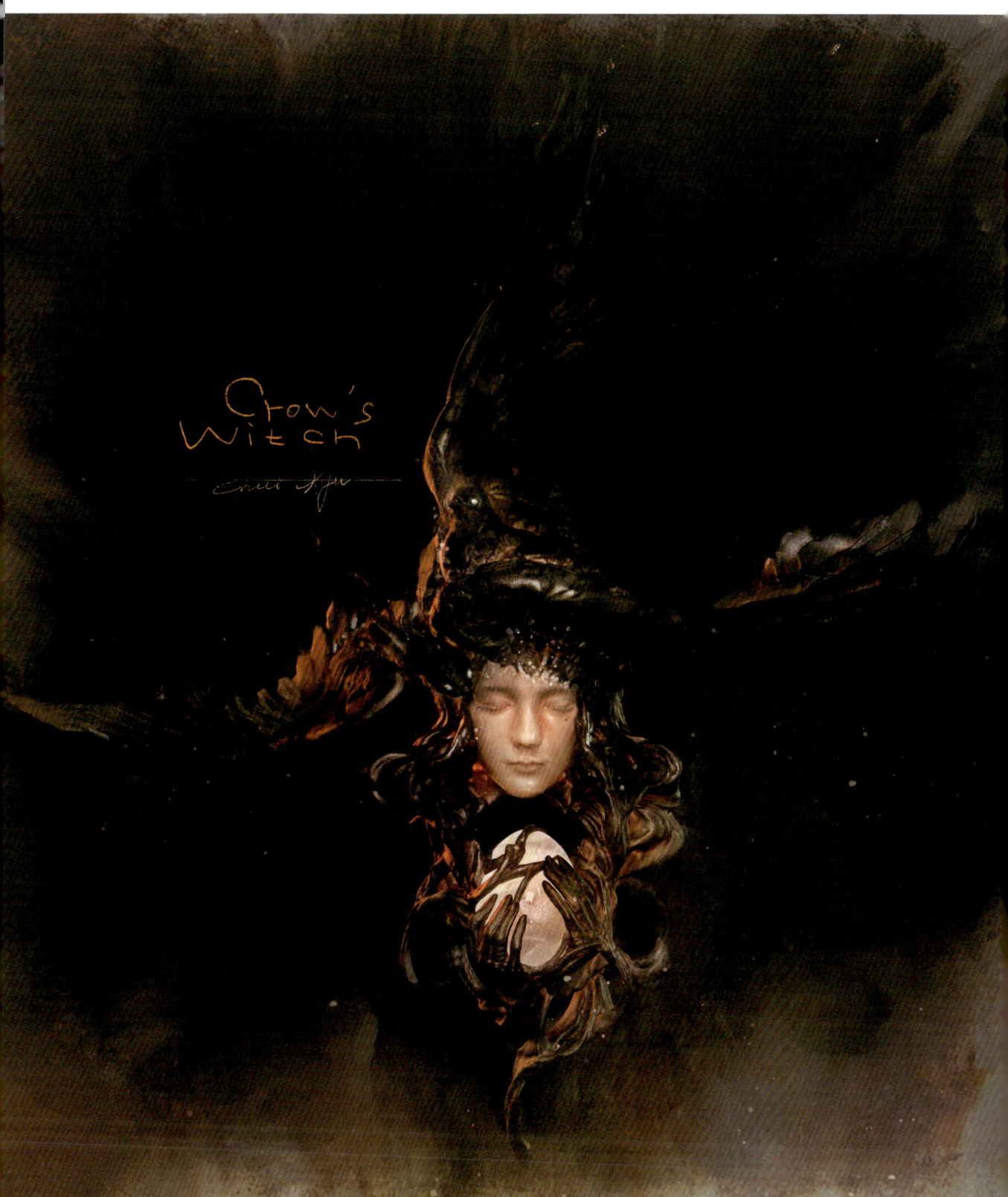

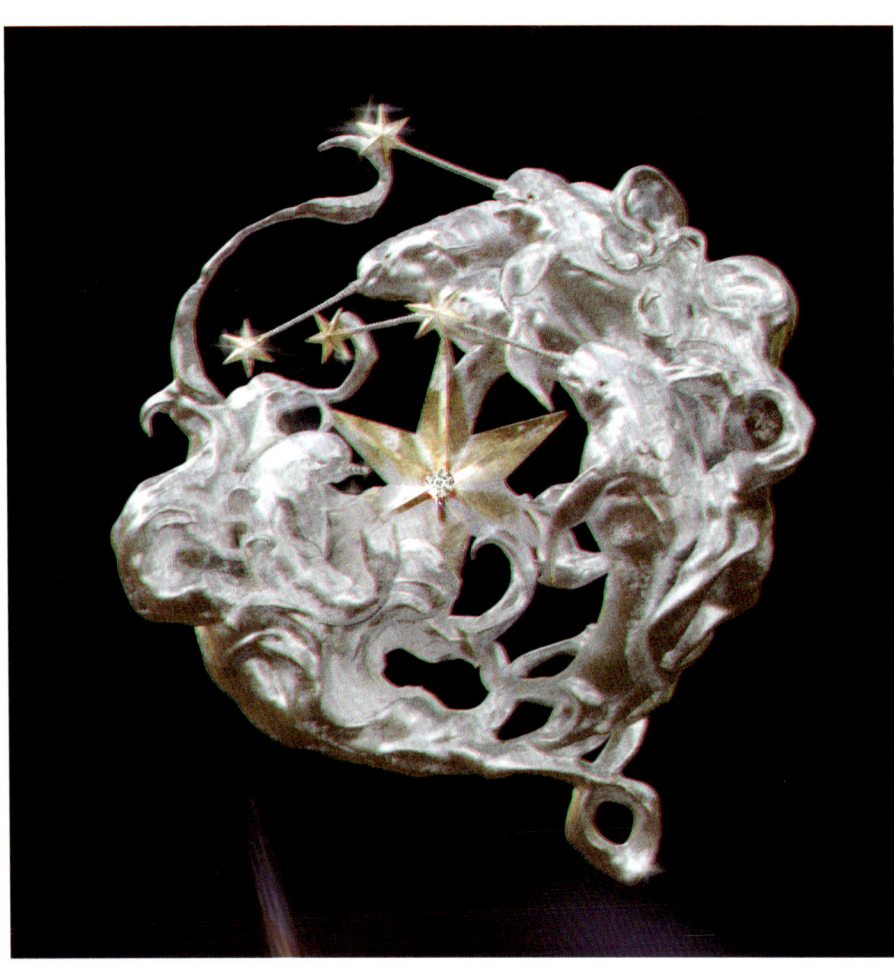

Narwhale Brooch 190

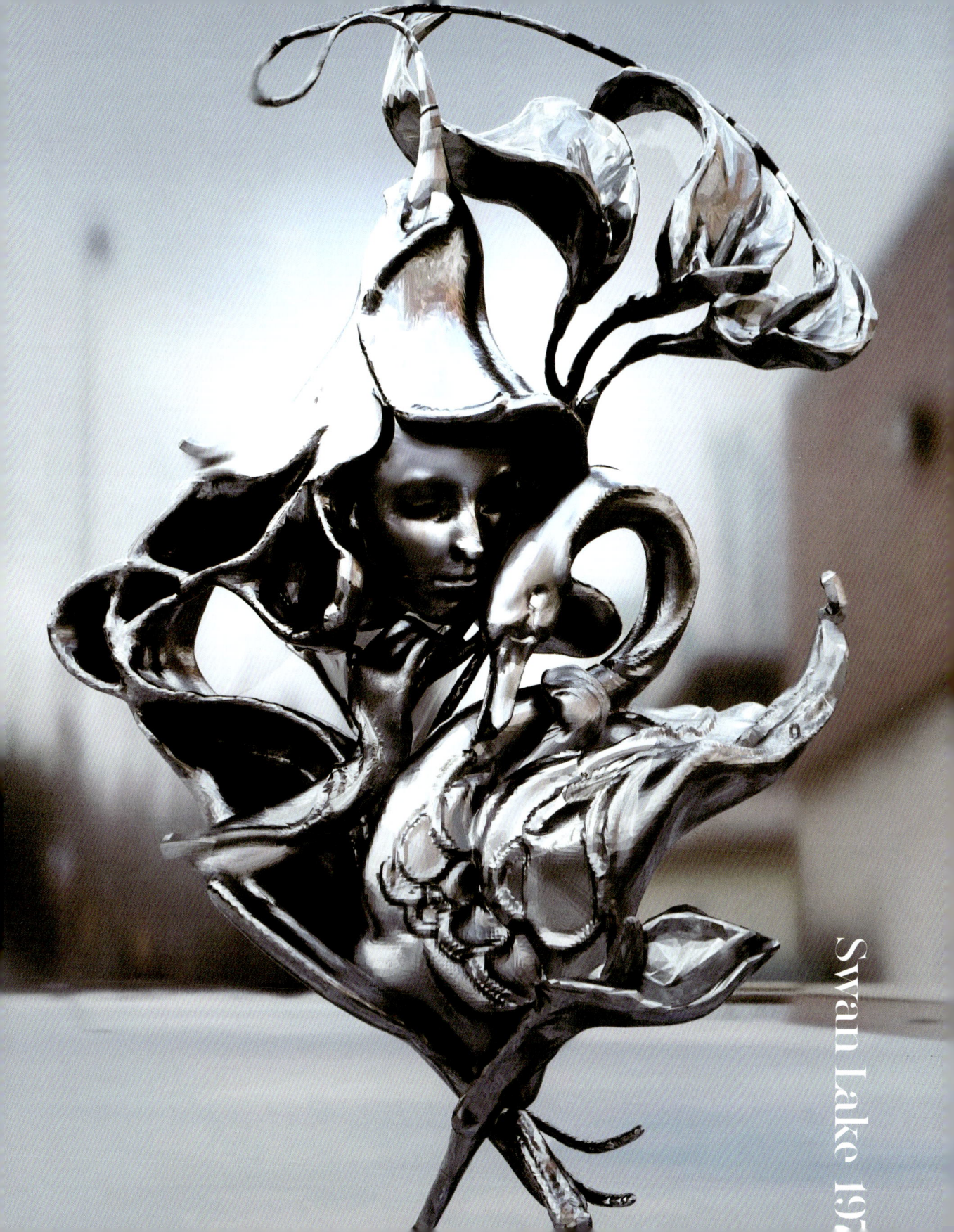

Swan Lake 197

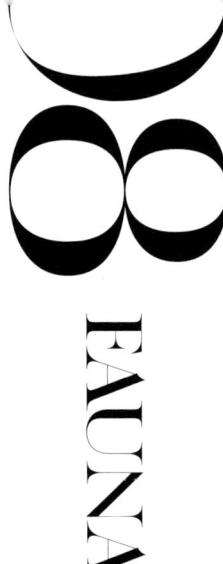

08

FAUNA

Animals, are the Creator's masterpieces. With their robust bones and leaping postures, their bodies showcase millions of years of evolution and the miraculous craftsmanship of the Creator. When I designing and sculpting mythical creatures, my art mentor is the observation of nature. Drawing from the anatomical essence of natural creatures, their unique adaptations to their environments, and the complex ecological networks among species, these insights can be transformed into the language of design. This enables the creation of both realistic and enchanting mythical creatures and fantasy worlds.

動物とは、創造主が創り上げた最高傑作です。頑丈な骨と跳躍の姿勢を備えるその身体は、何百万年にも及ぶ進化と創造主による奇跡的な匠の技を示します。神話上の生物をデザインして造形するとき、私の芸術の師となるのは自然観察です。自然界の生物が備える身体構造上の本質、環境に対する独自の適応、そして種の間の複雑な生態学的ネットワークを足場として、こうした洞察はデザインという言語に変換できます。このようにして、リアルであると同時に魅惑的な神話上の生物とファンタジーの世界を創り出すことができるのです。

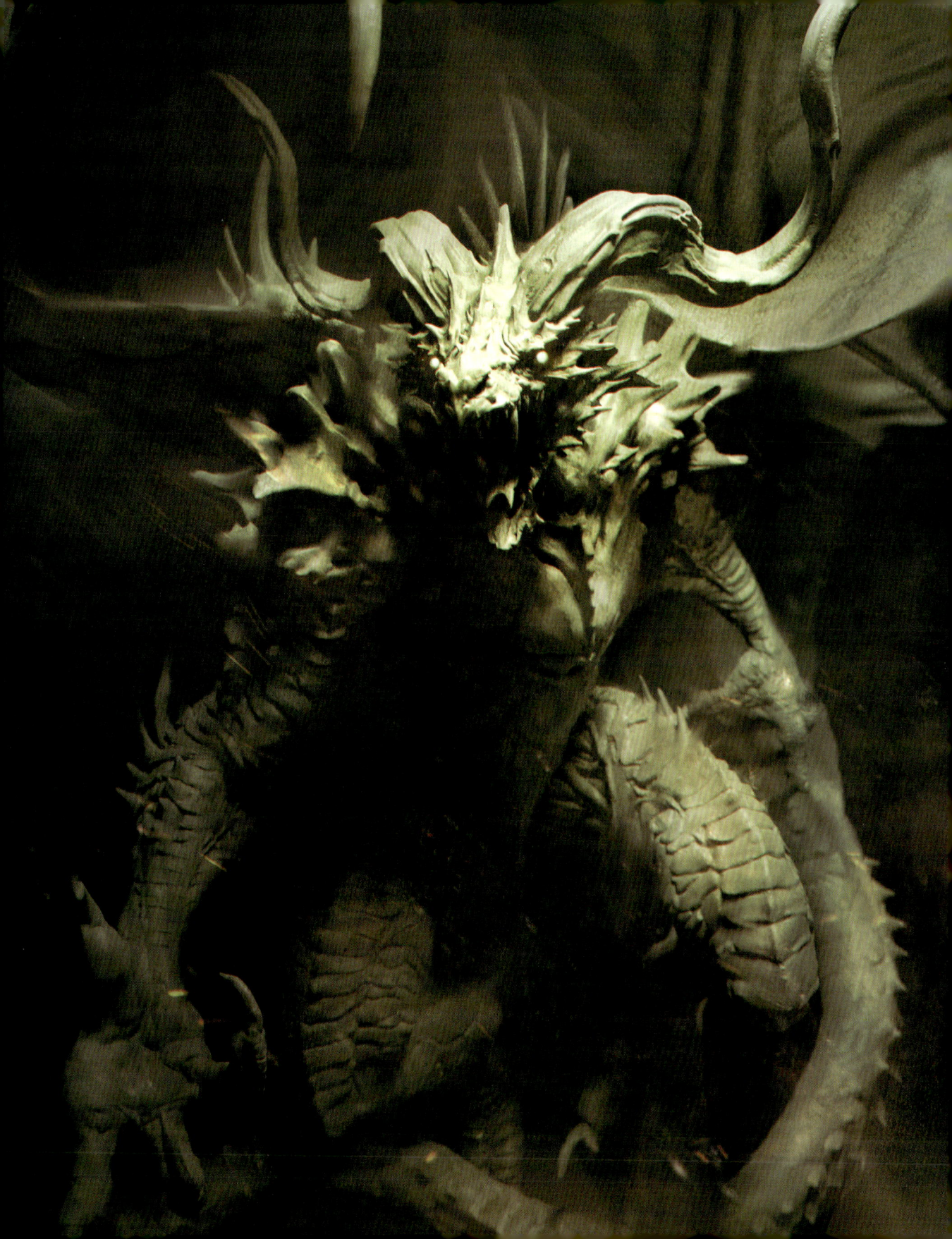

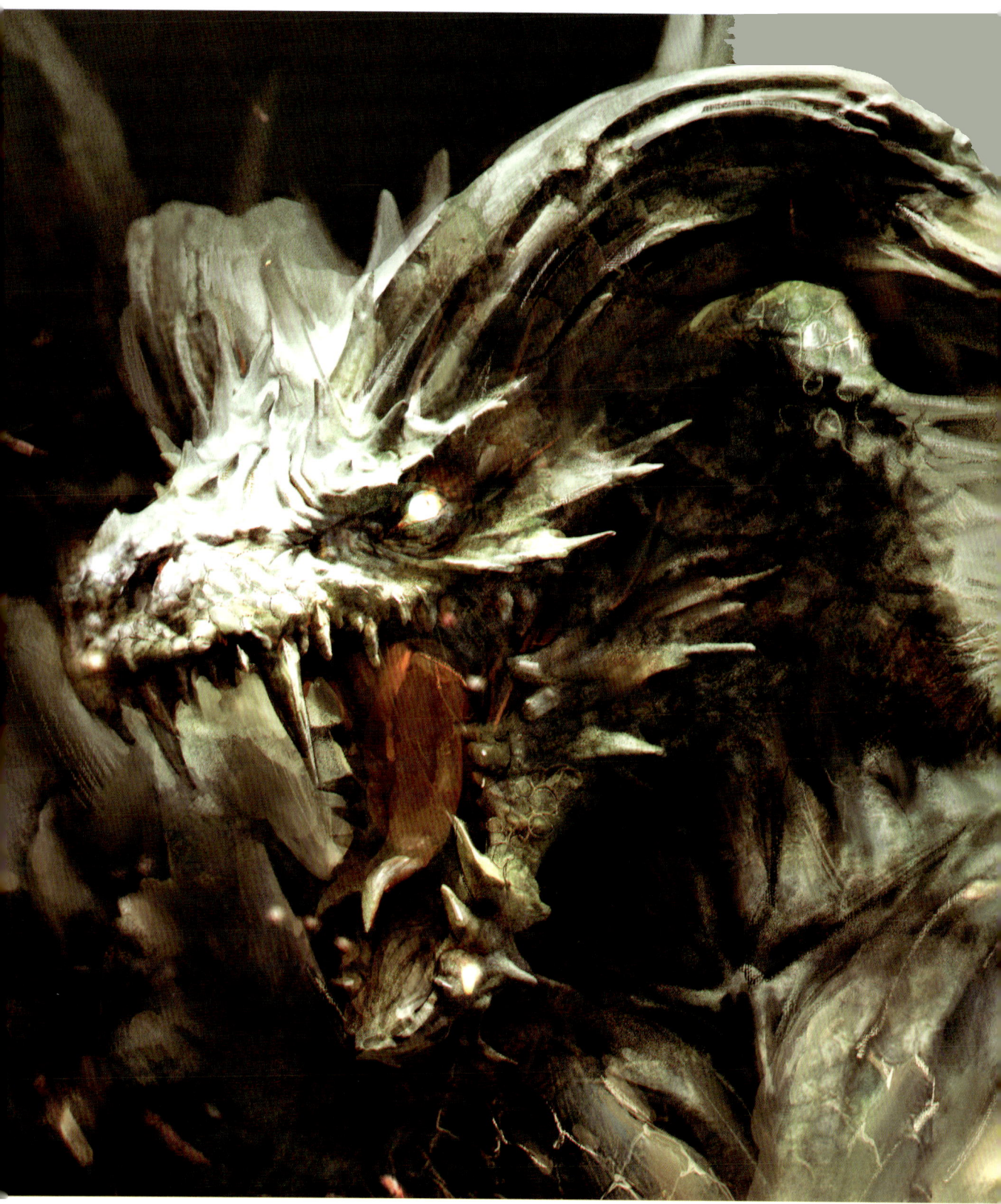

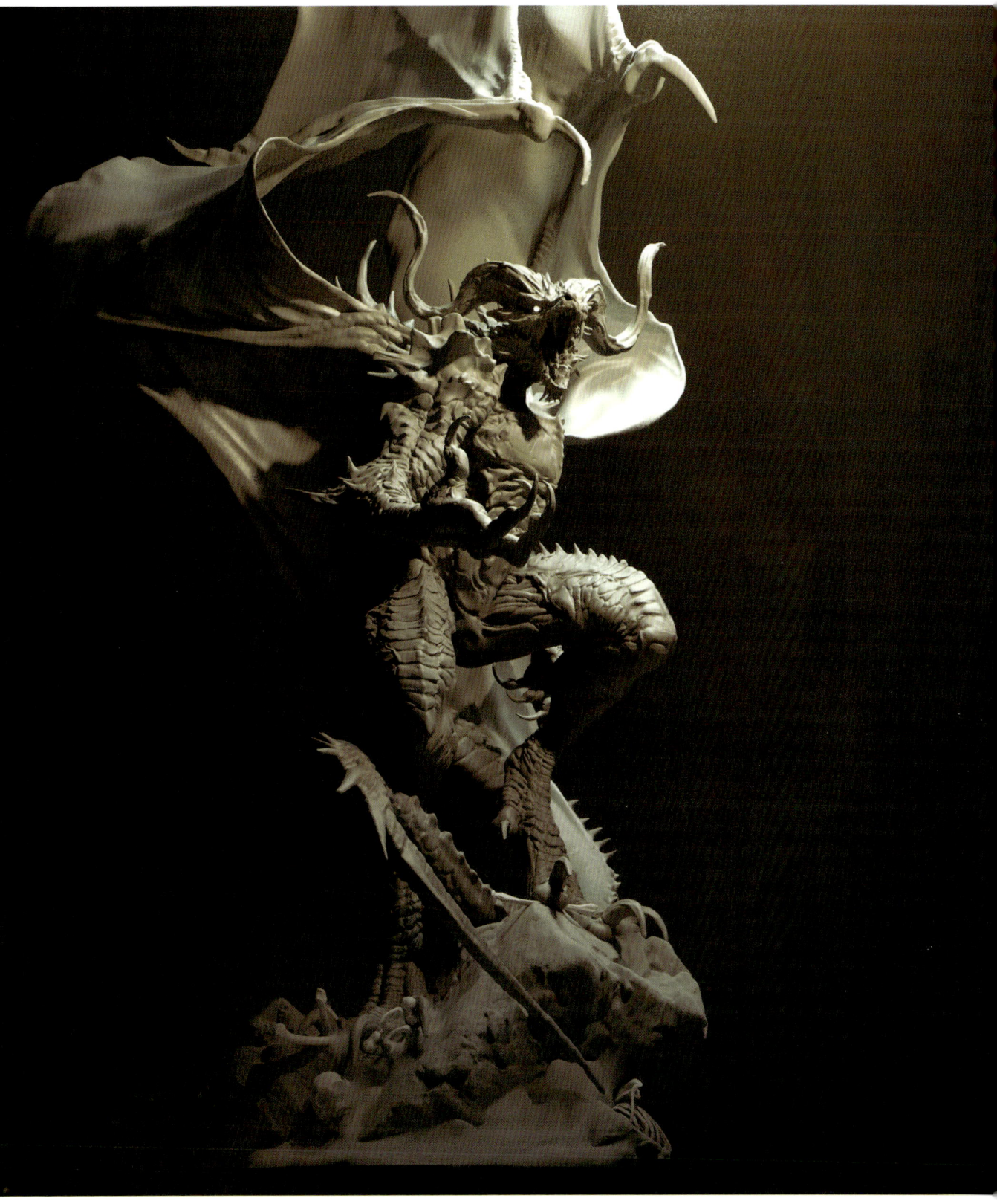

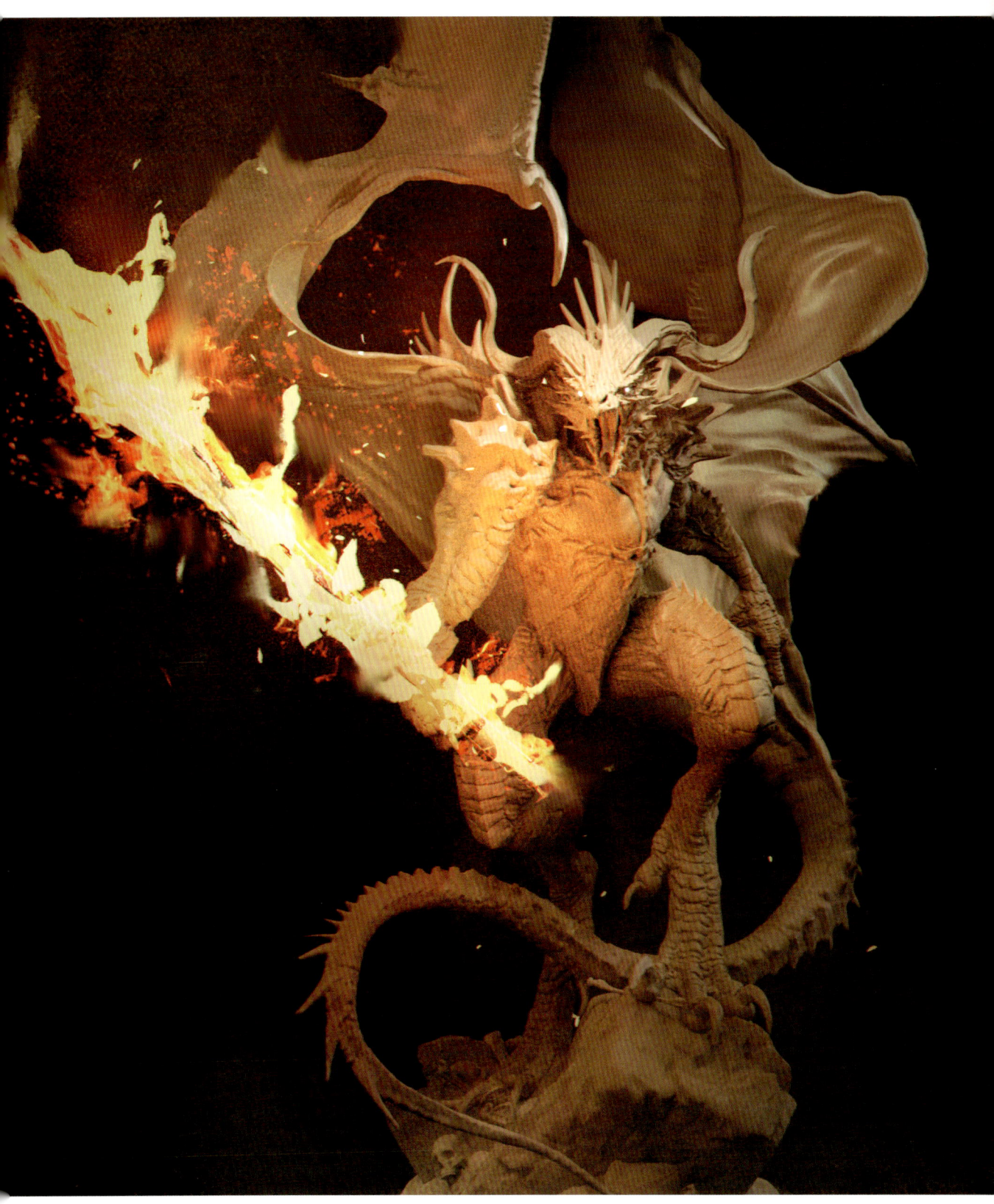

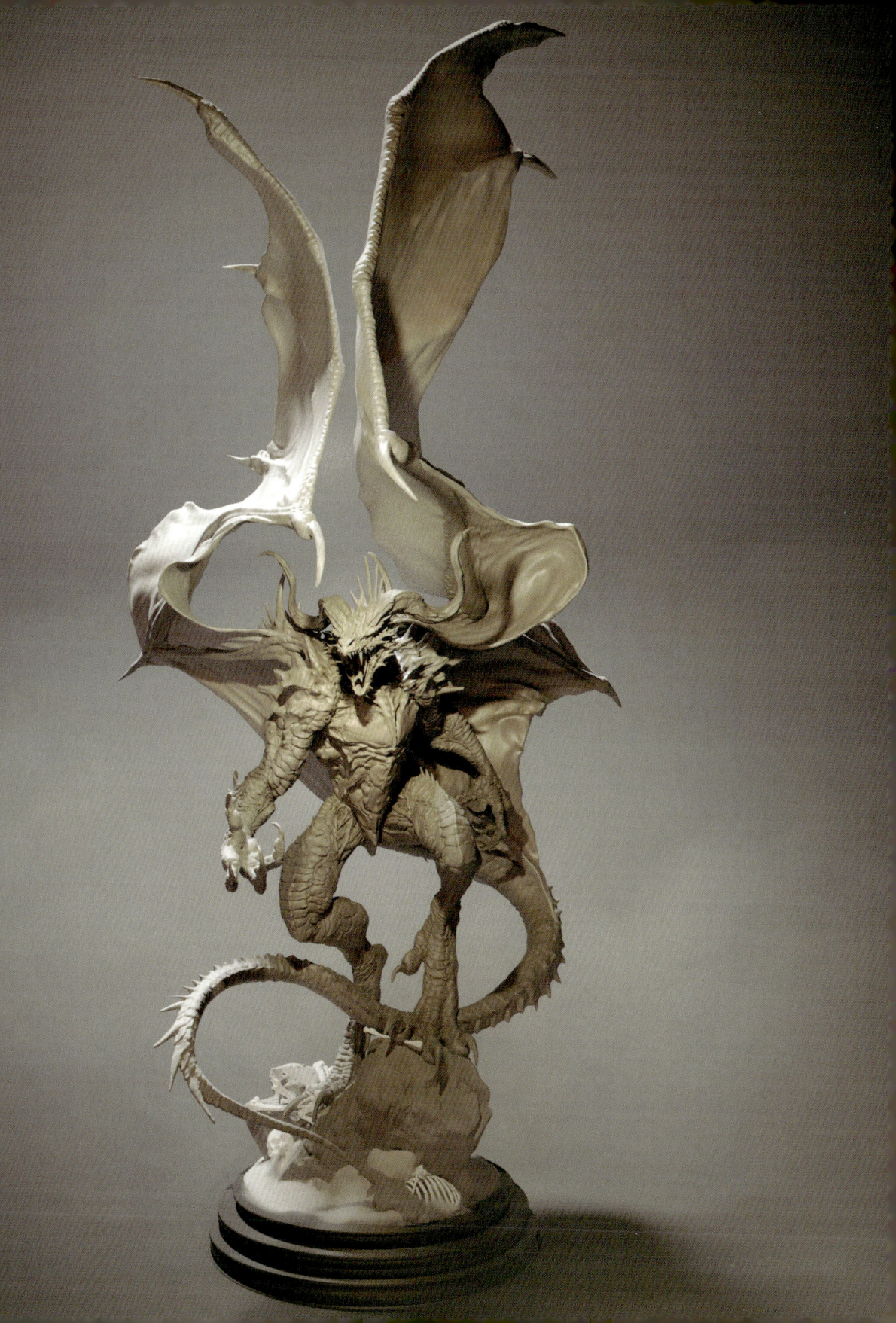

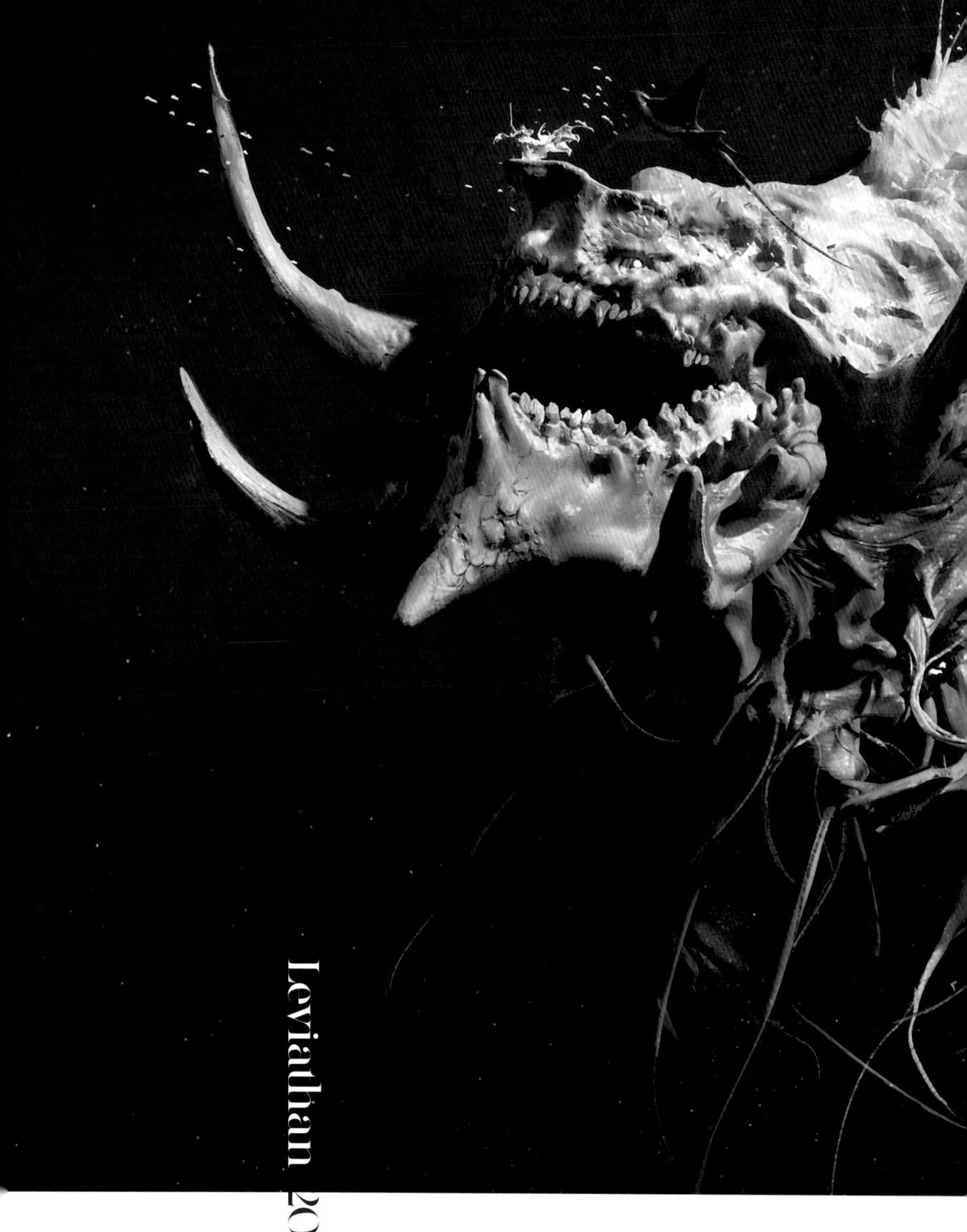

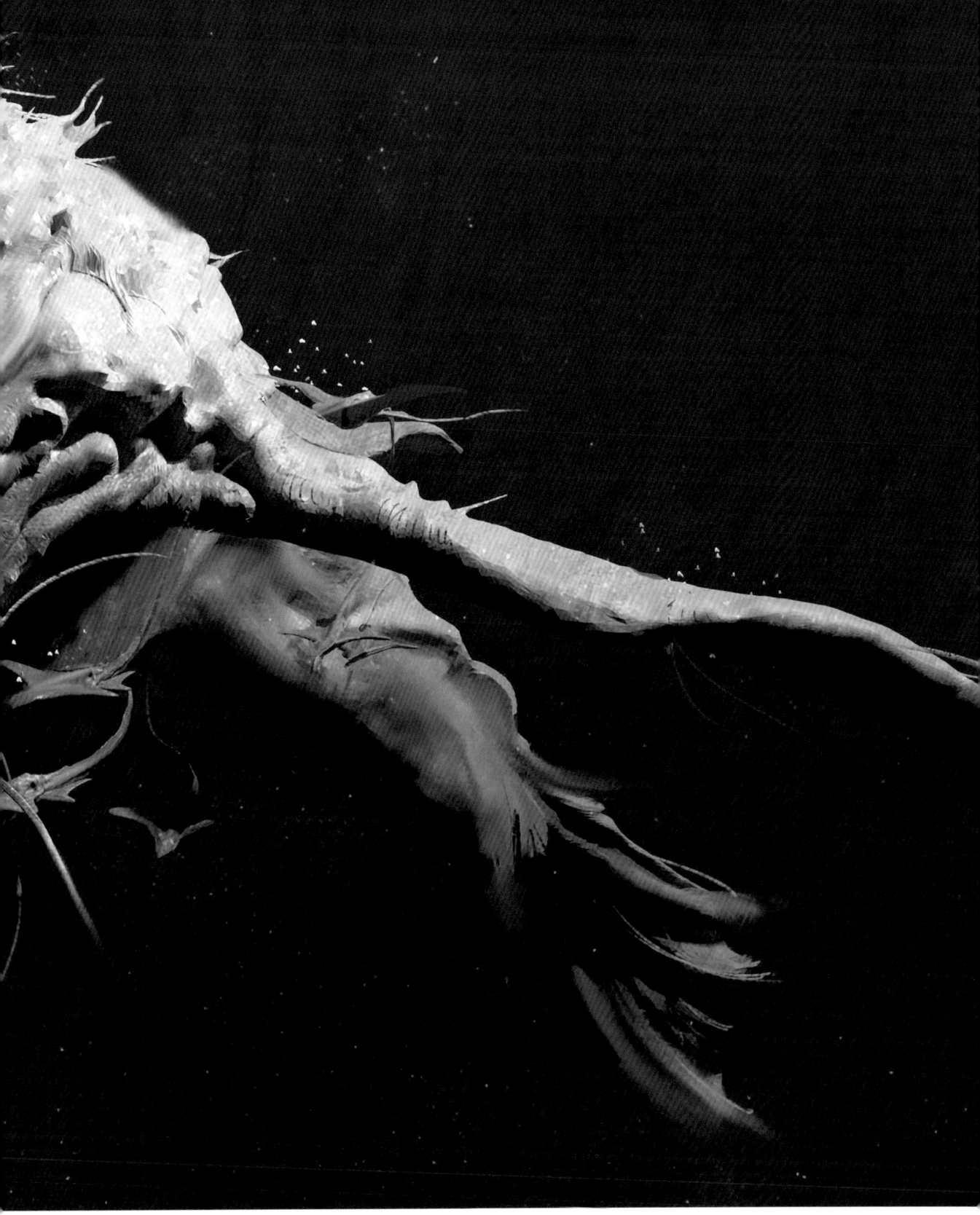

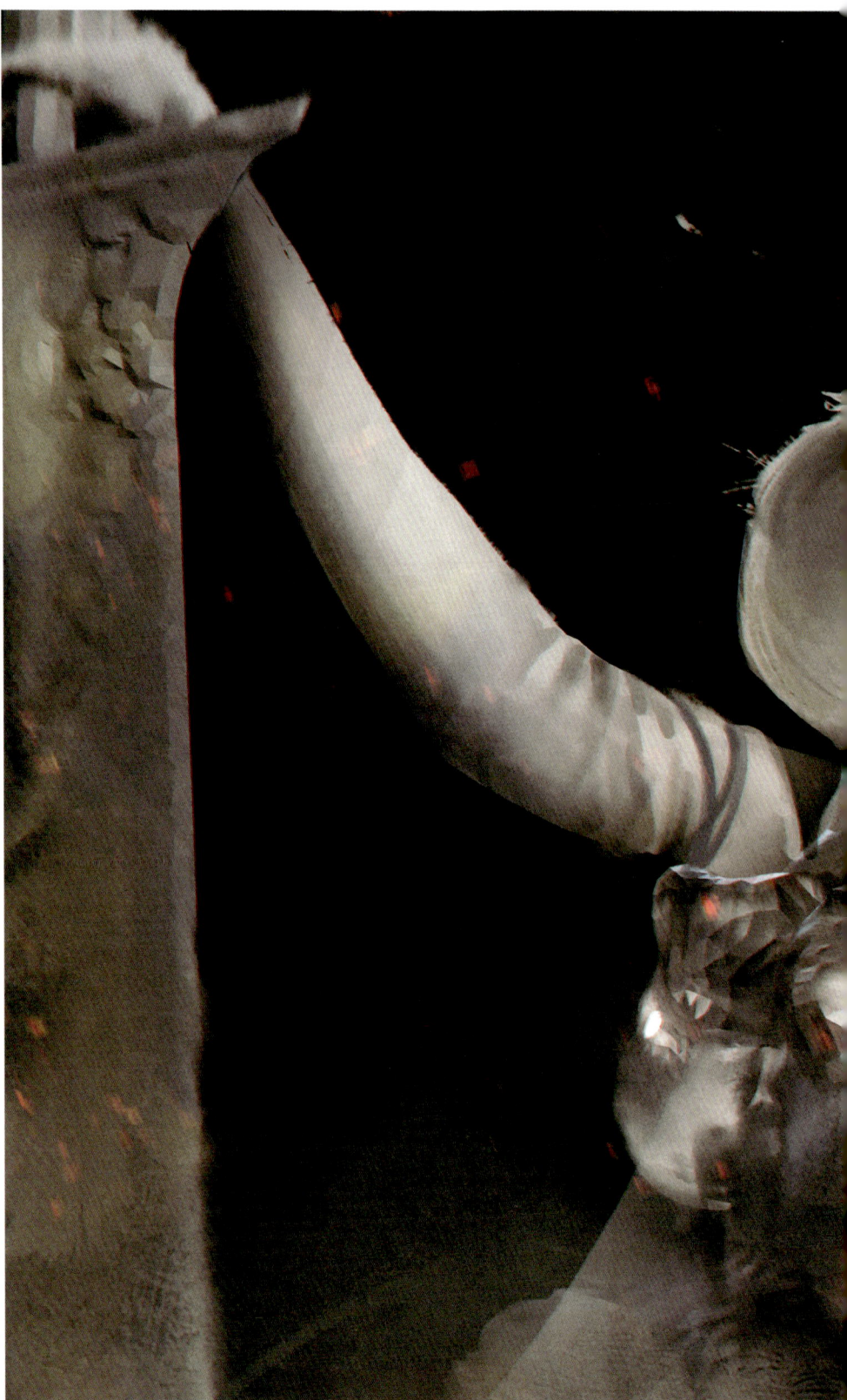

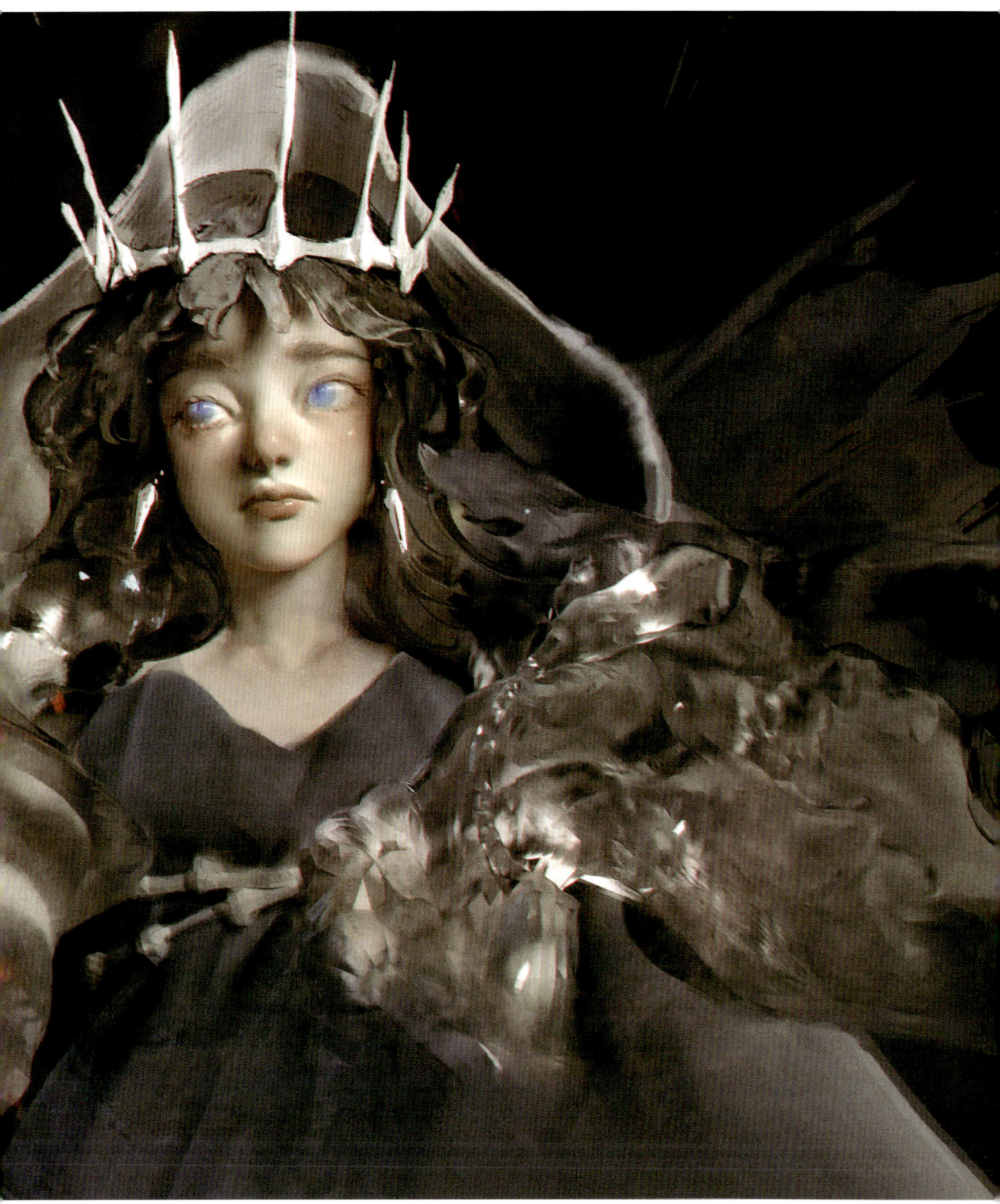

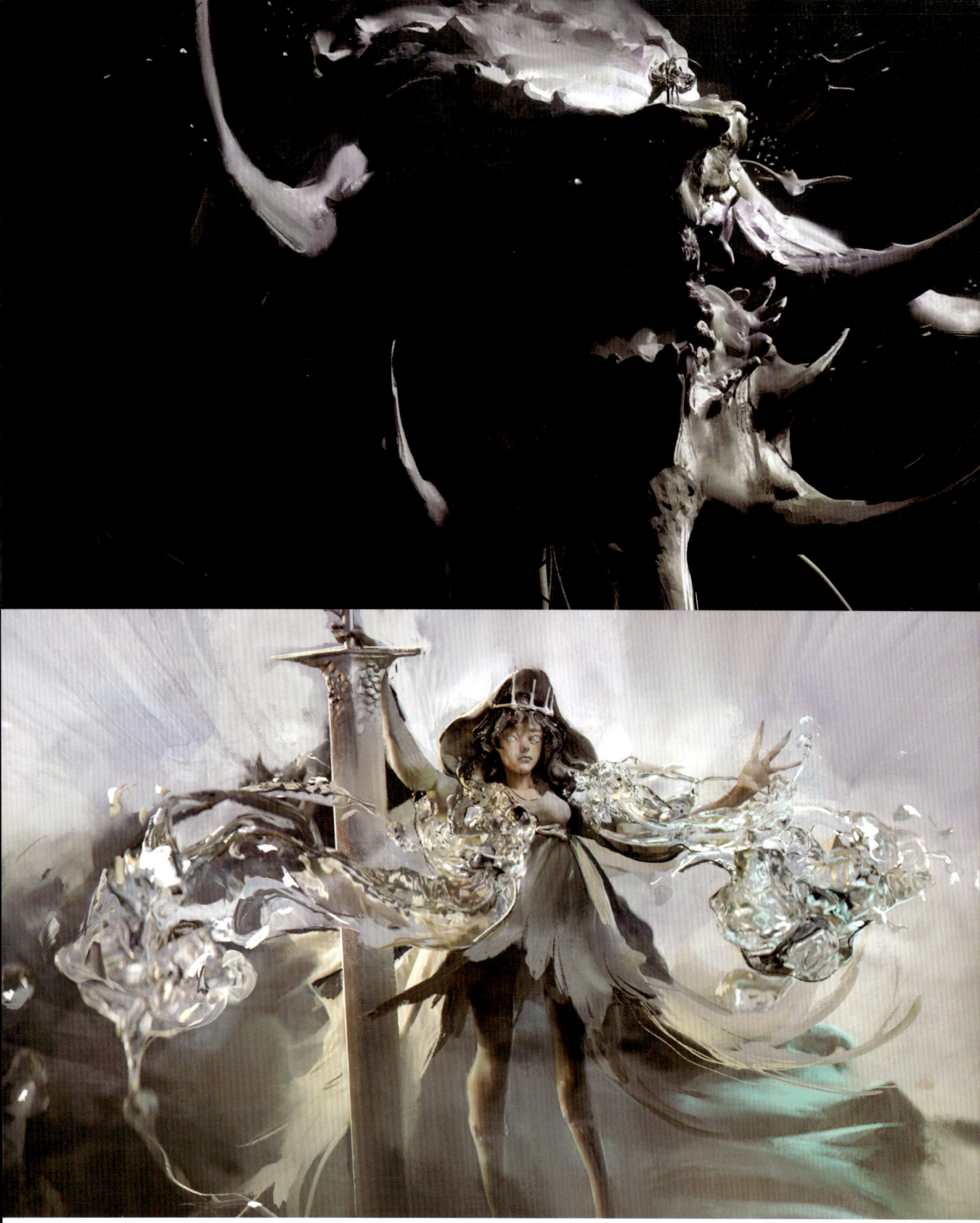

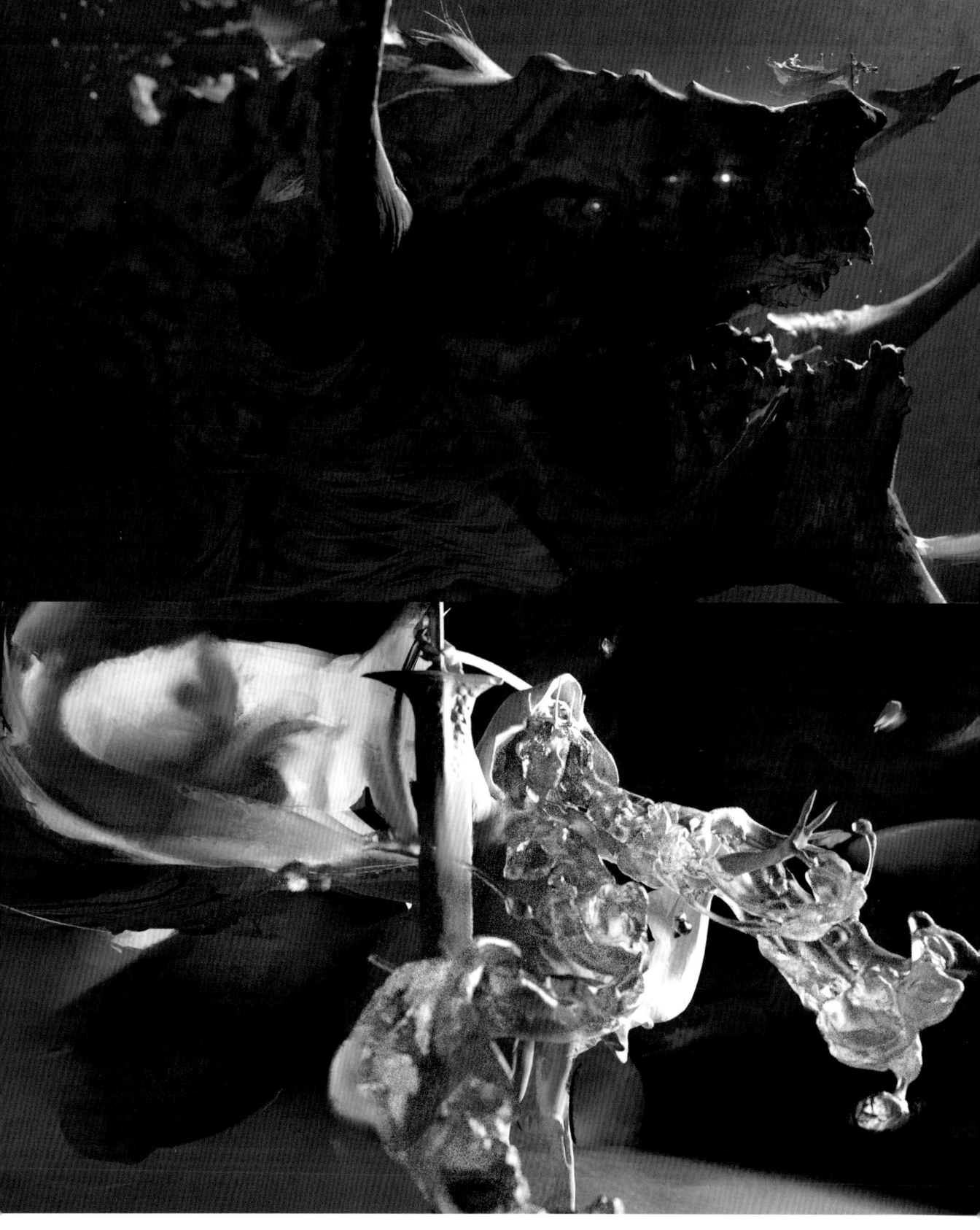

Jurassic Vegetarian 20

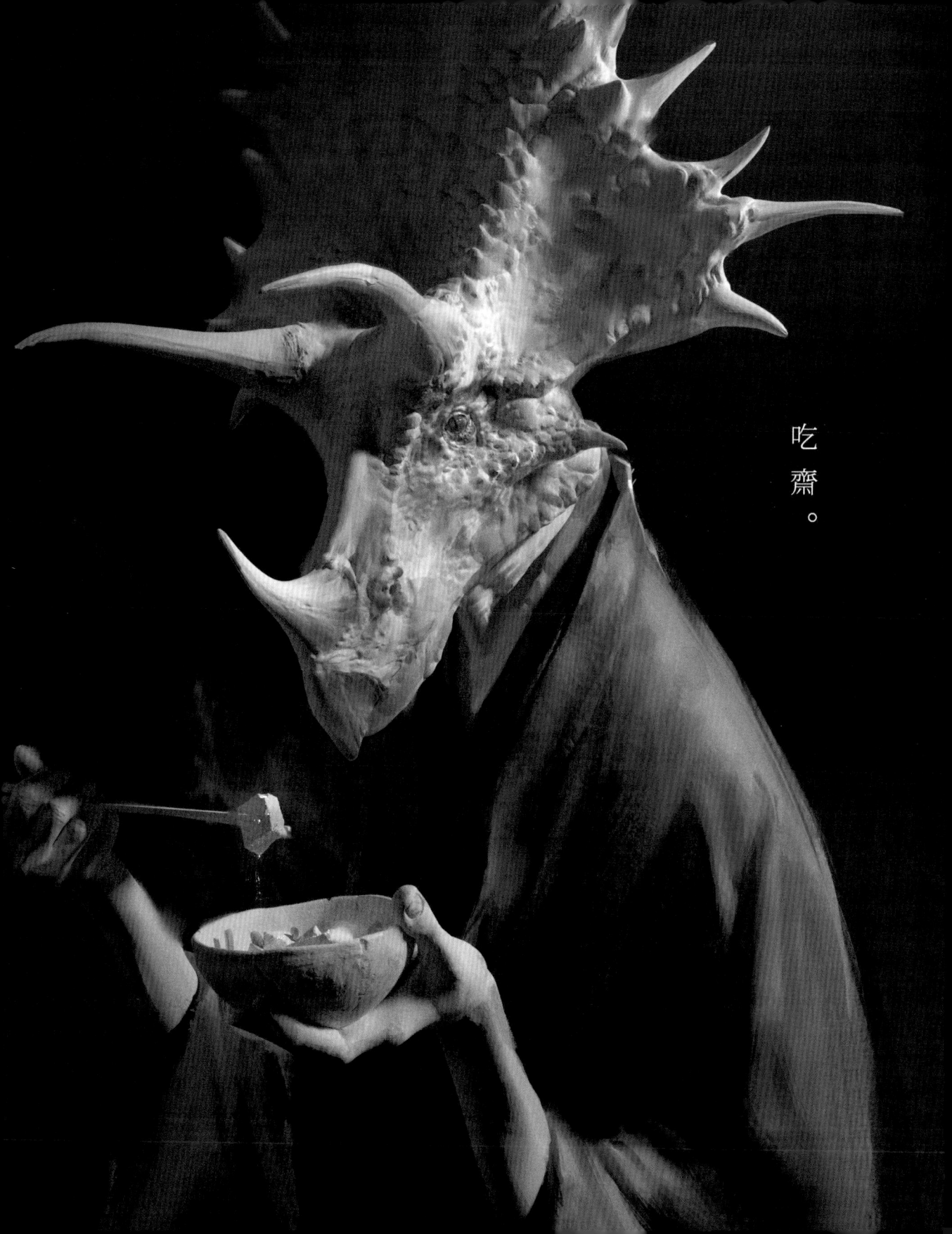

吃
齋
。

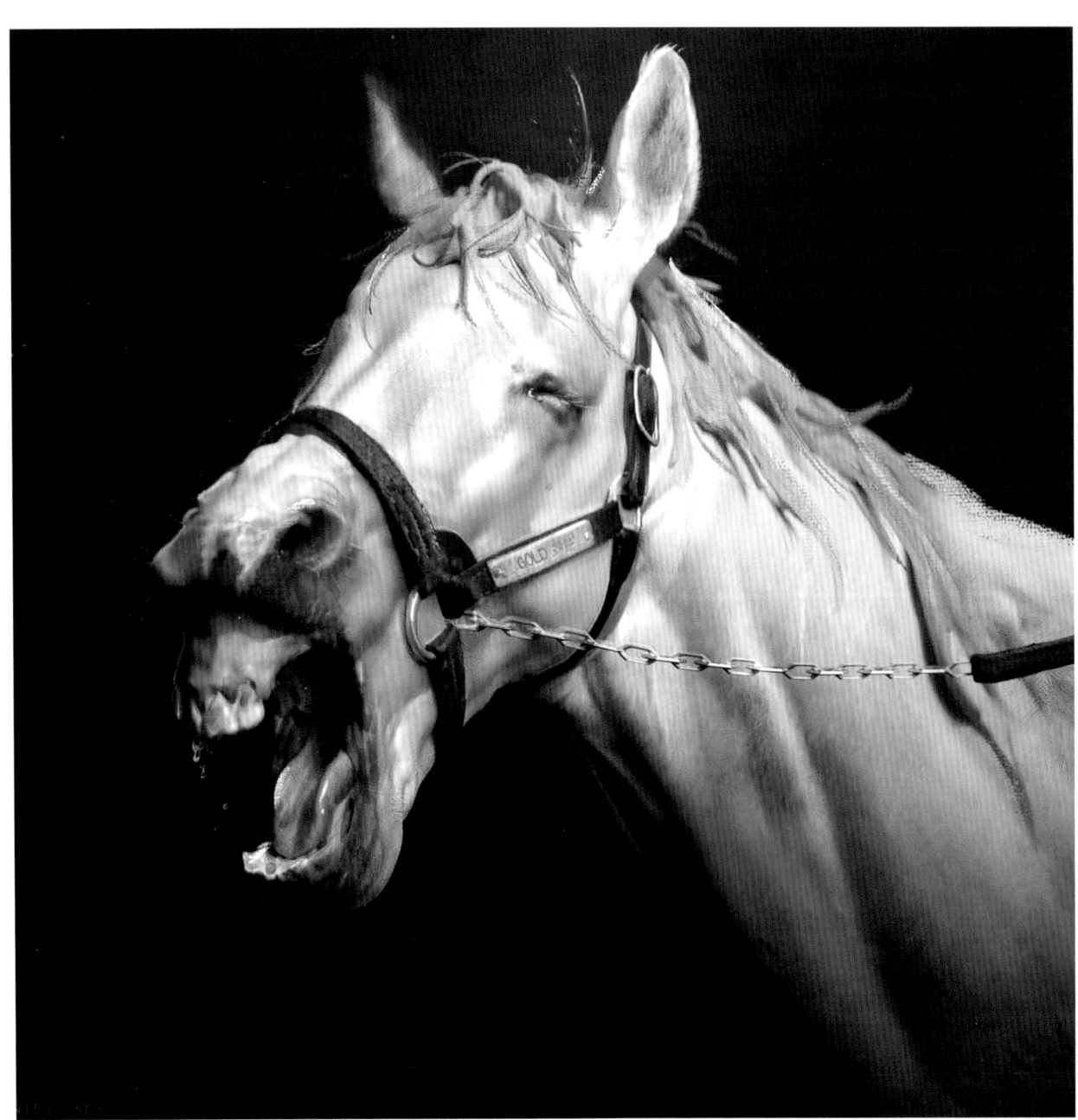

Smile 212

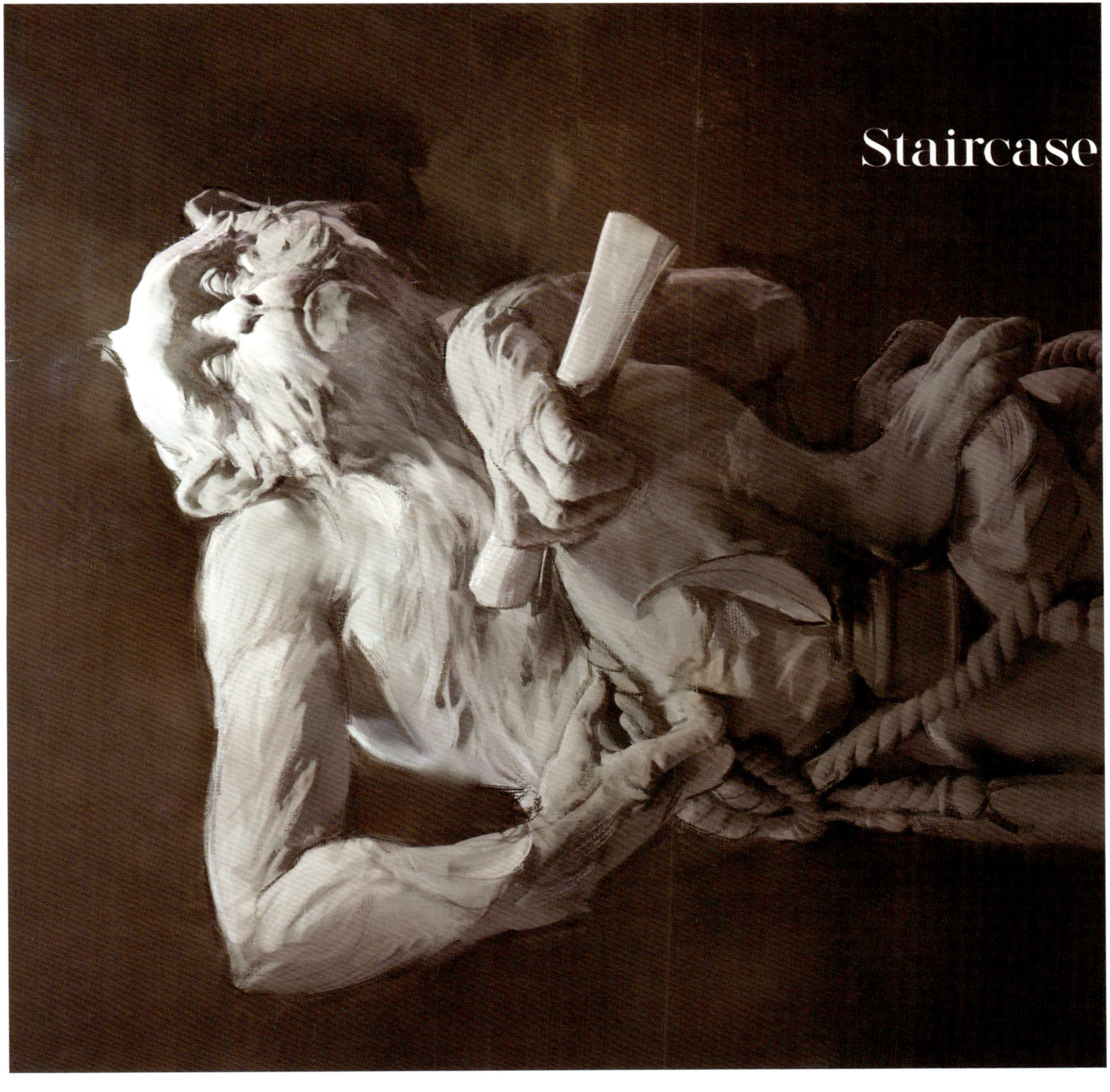

Staircase Monkey    21:

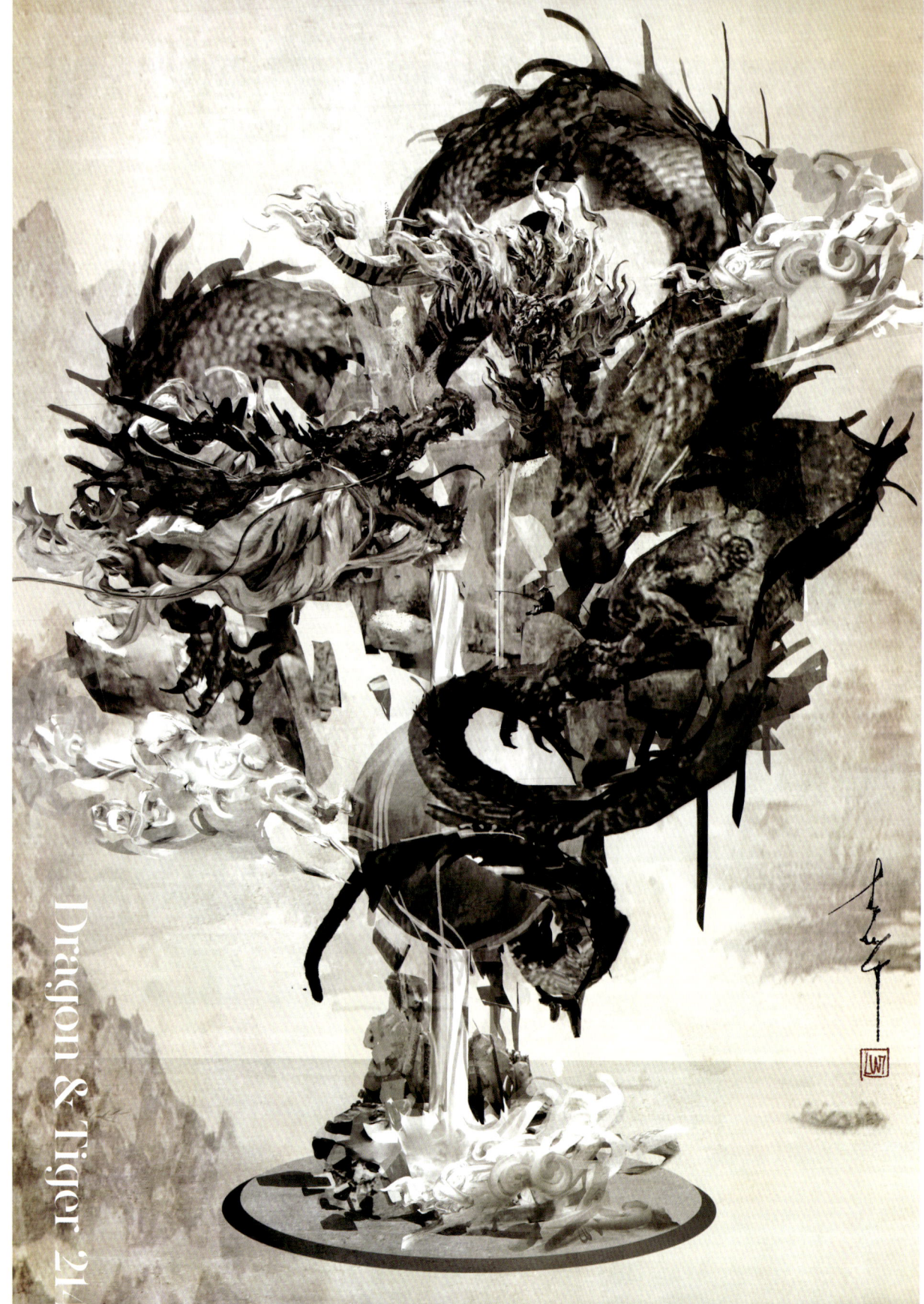

Dragon & Tiger 21

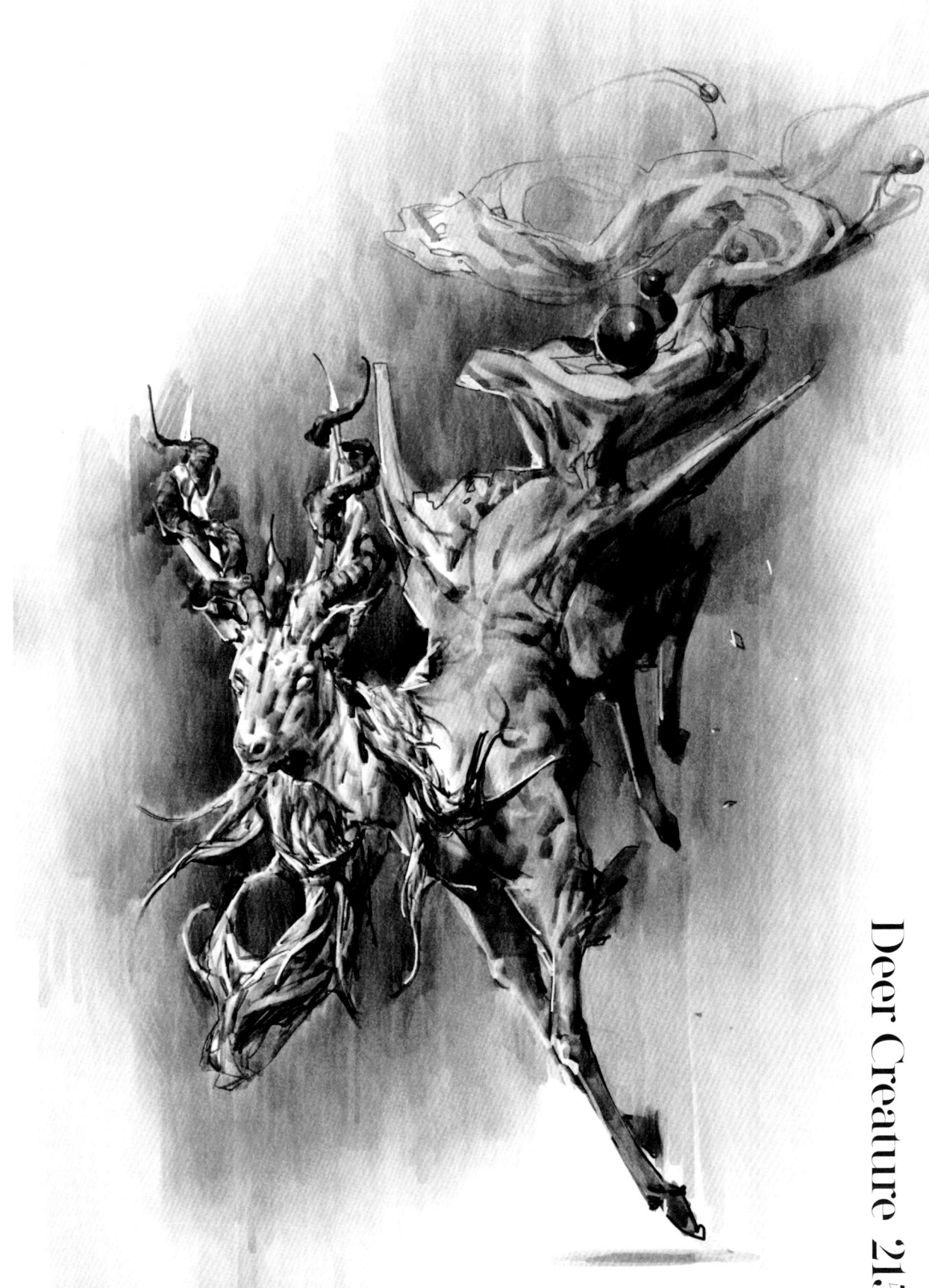

Deer Creature 21E

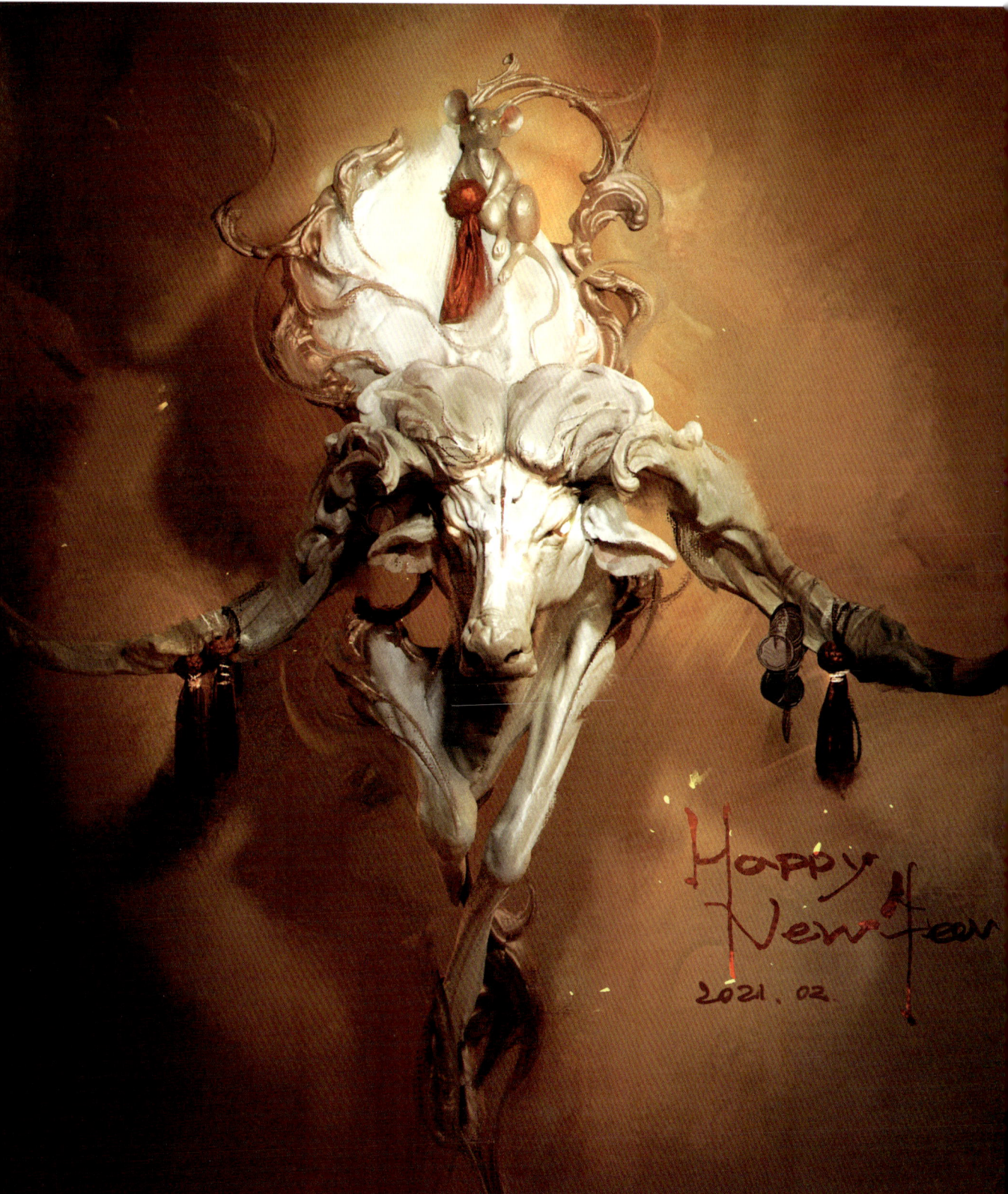

Happy
New Year
2021. 02.

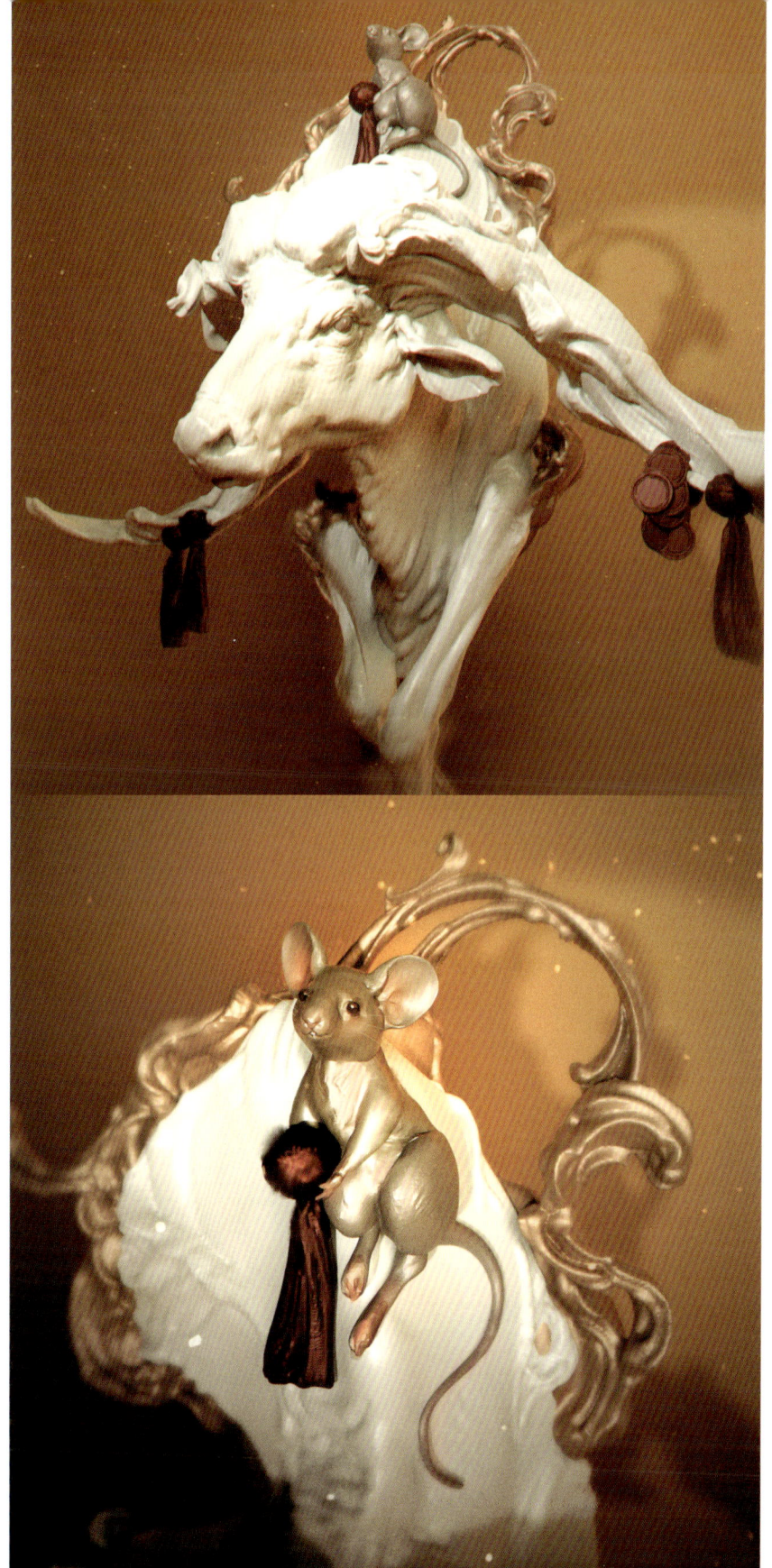

Year of The ox  21

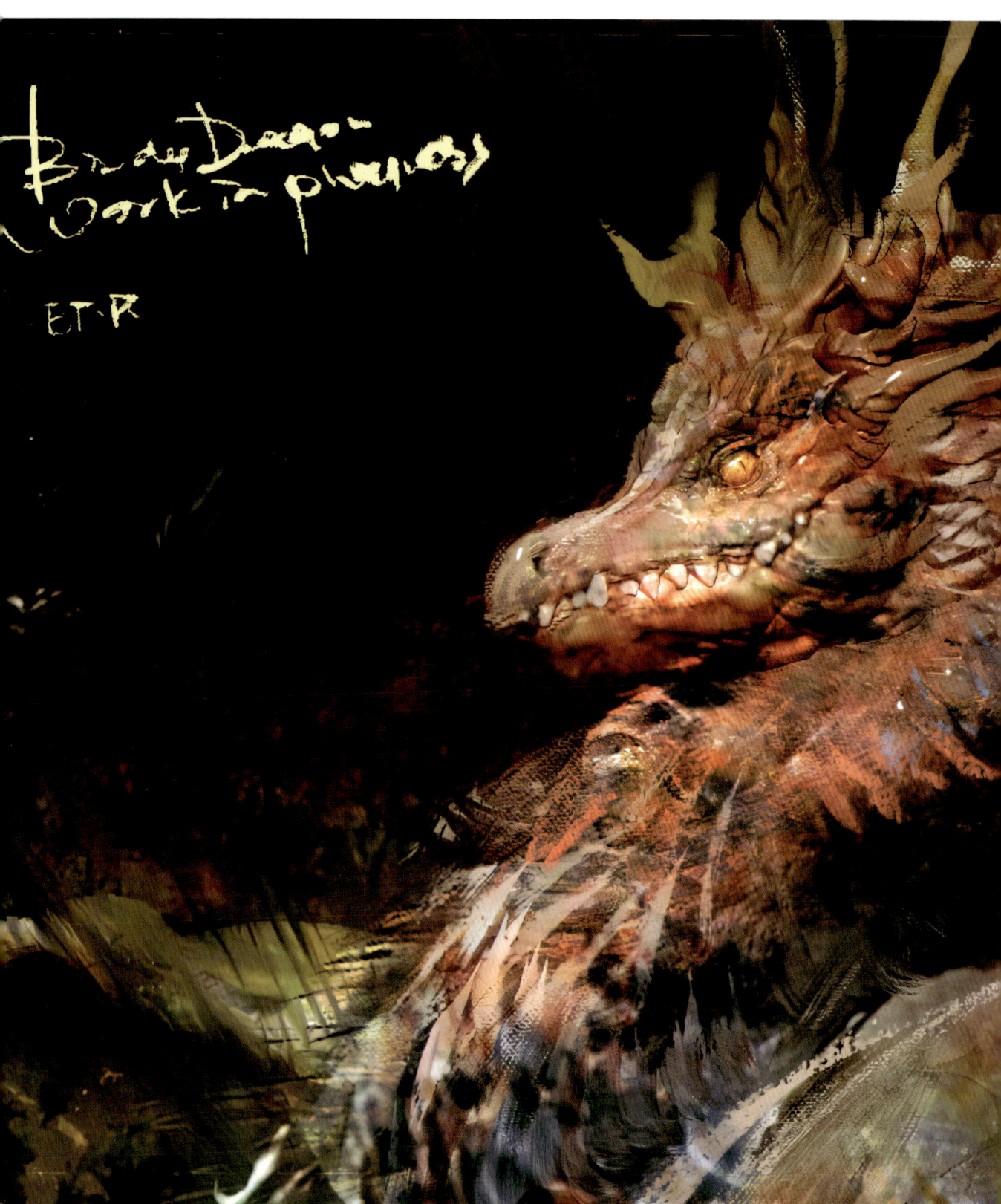

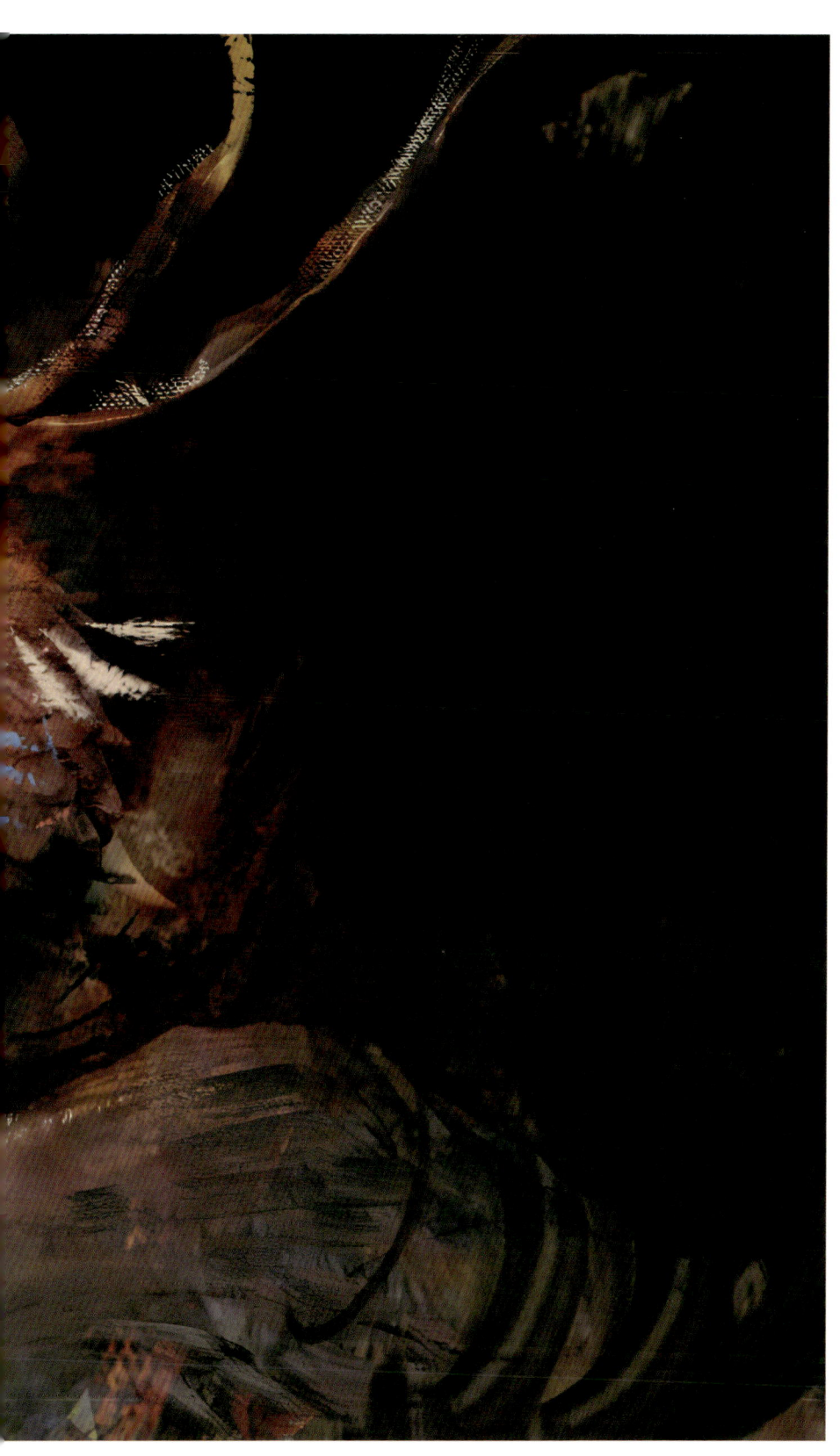

Ostrich Dragon 219

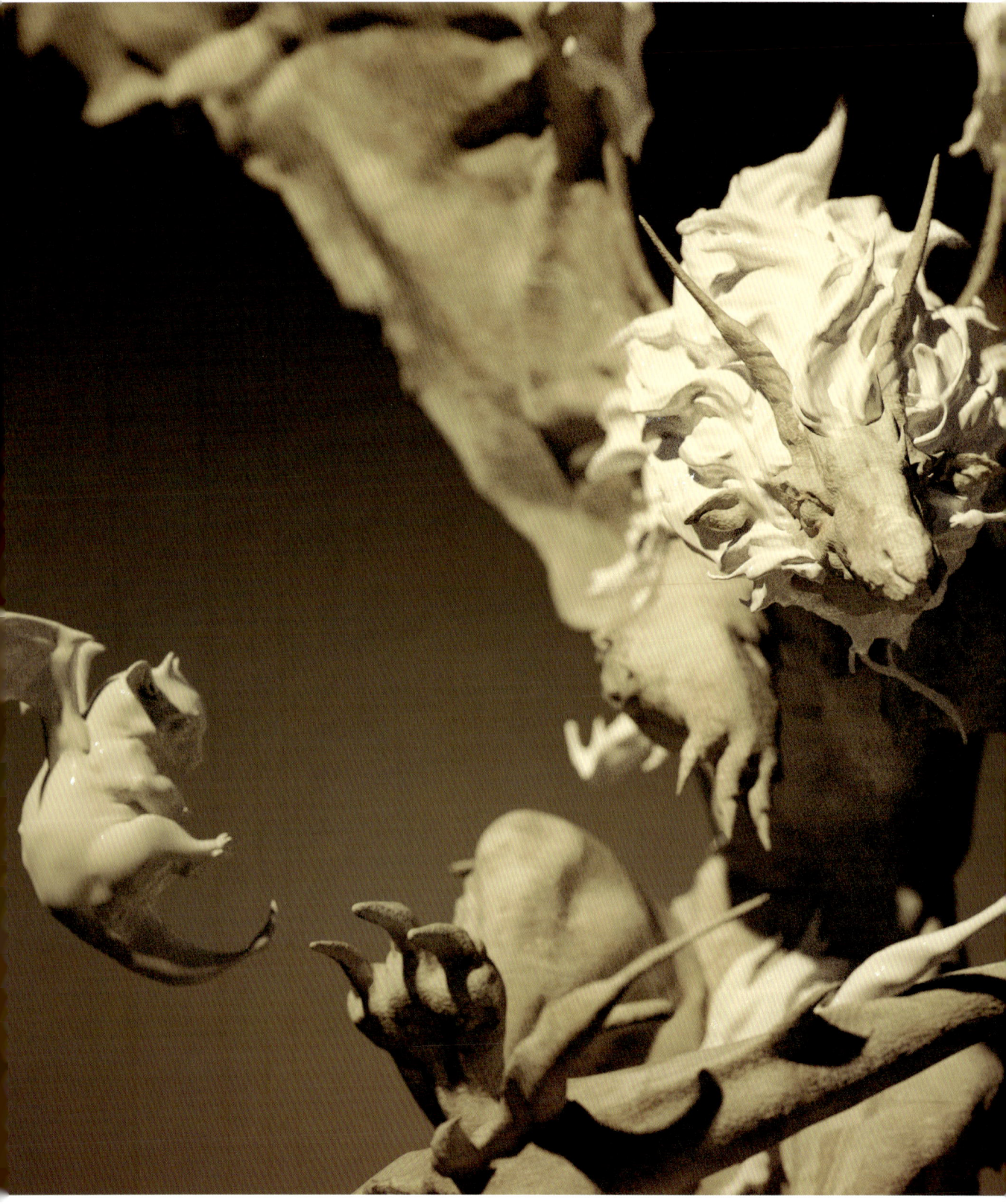

Moth Dragon '22

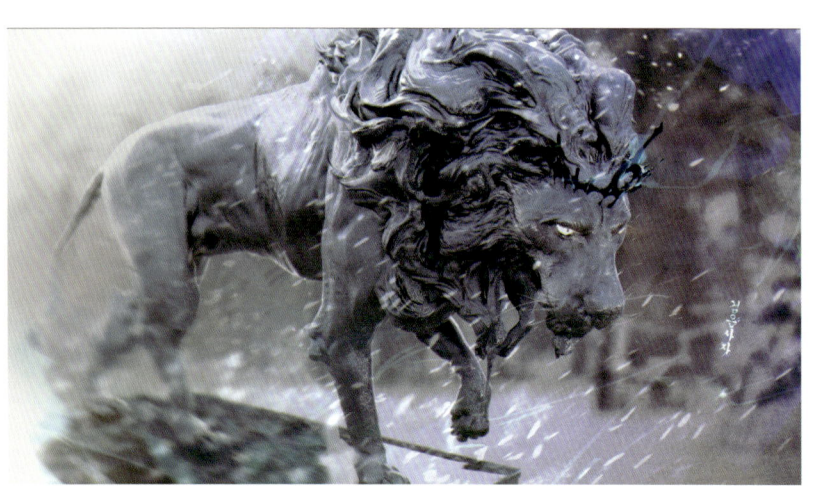

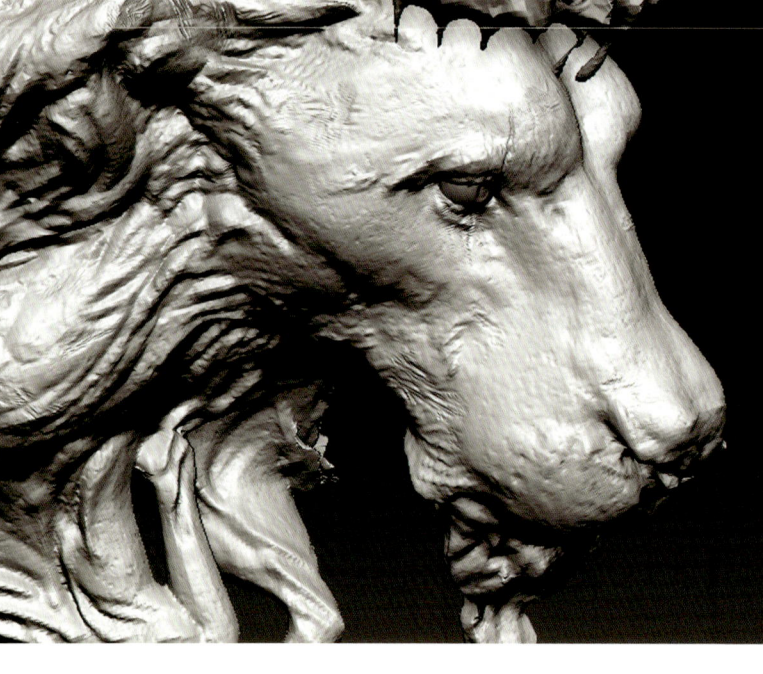

Black kitty 222

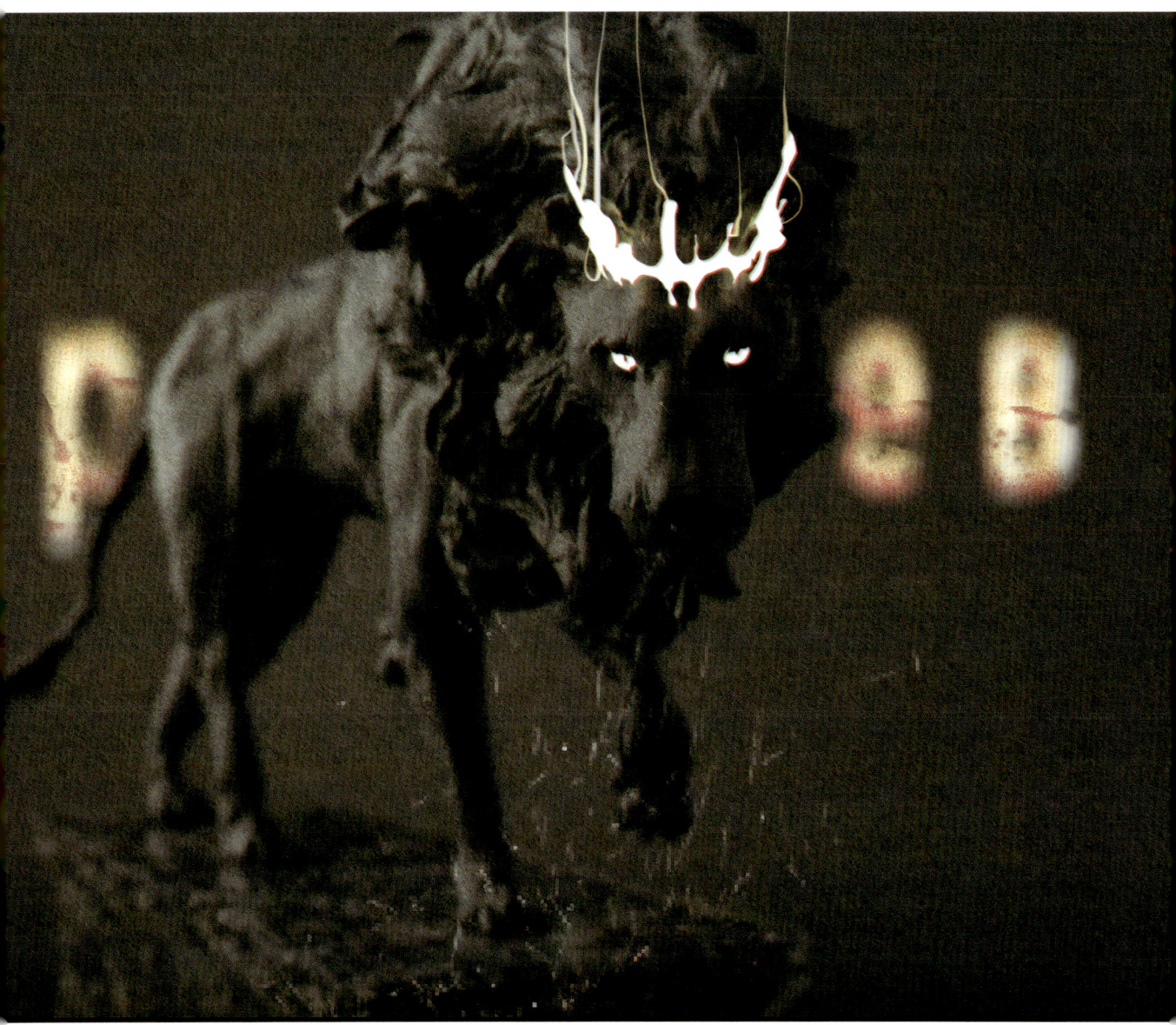

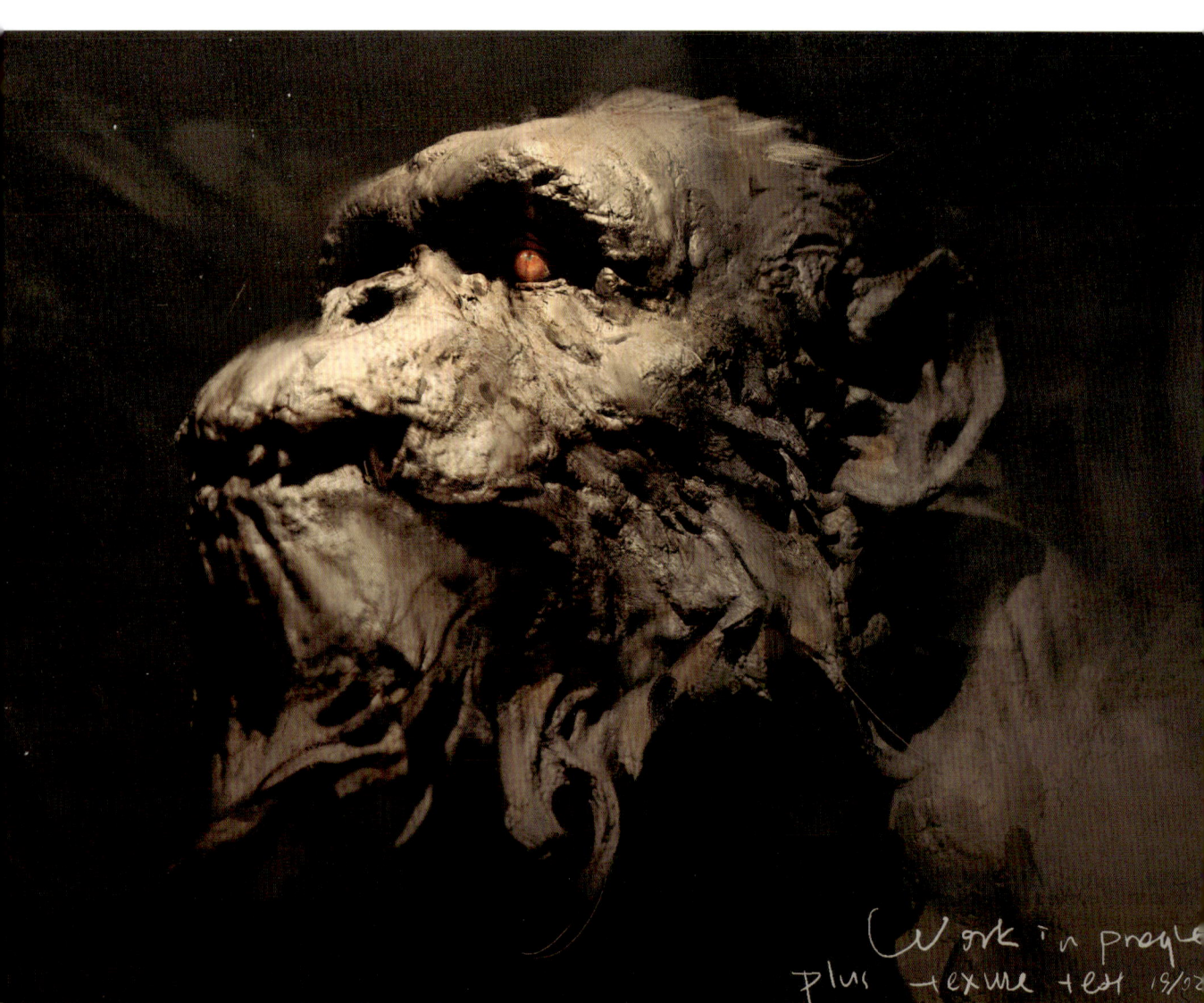

Work in progre
plus texture test 18/02

Ghost Monkey '22

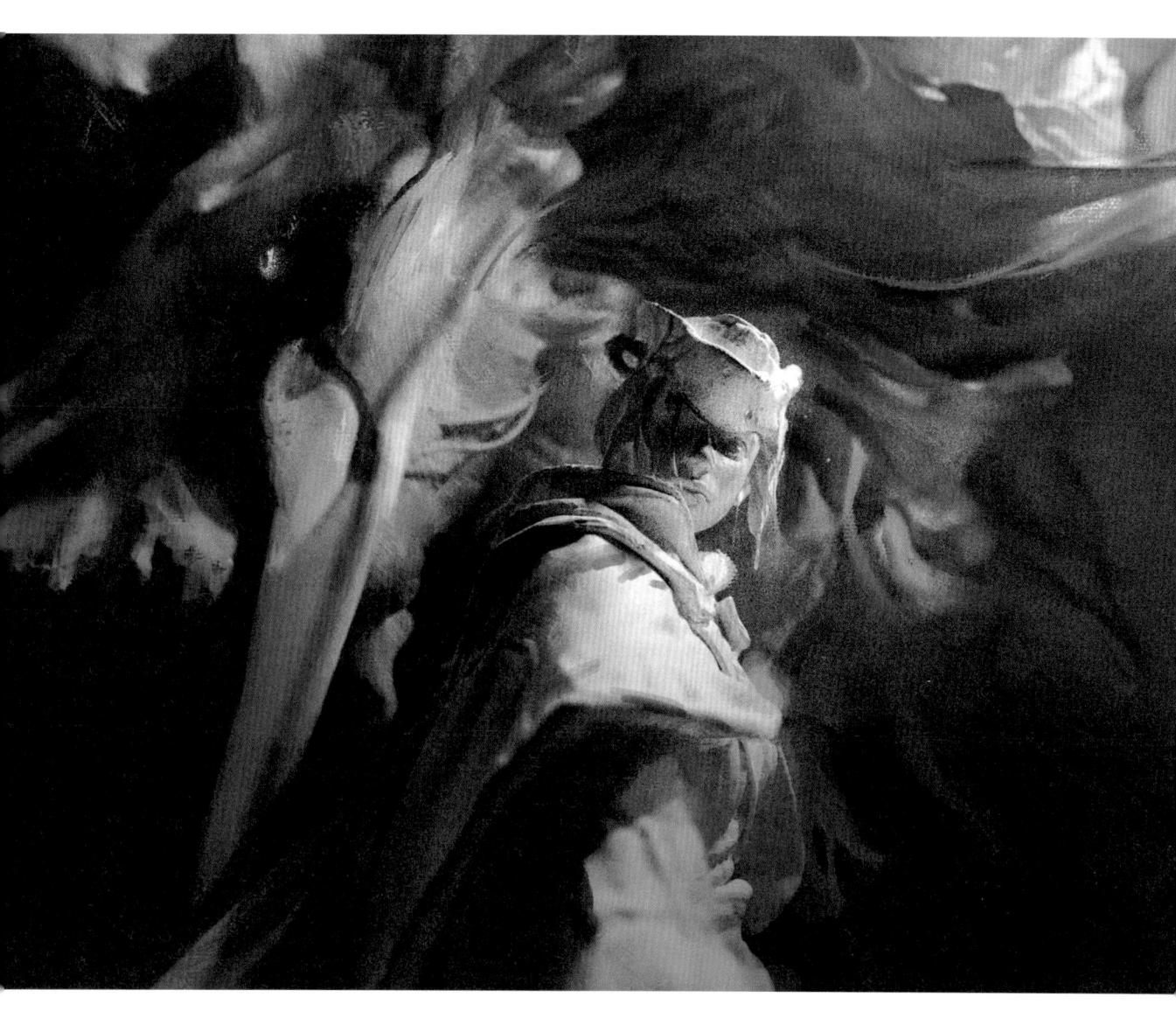

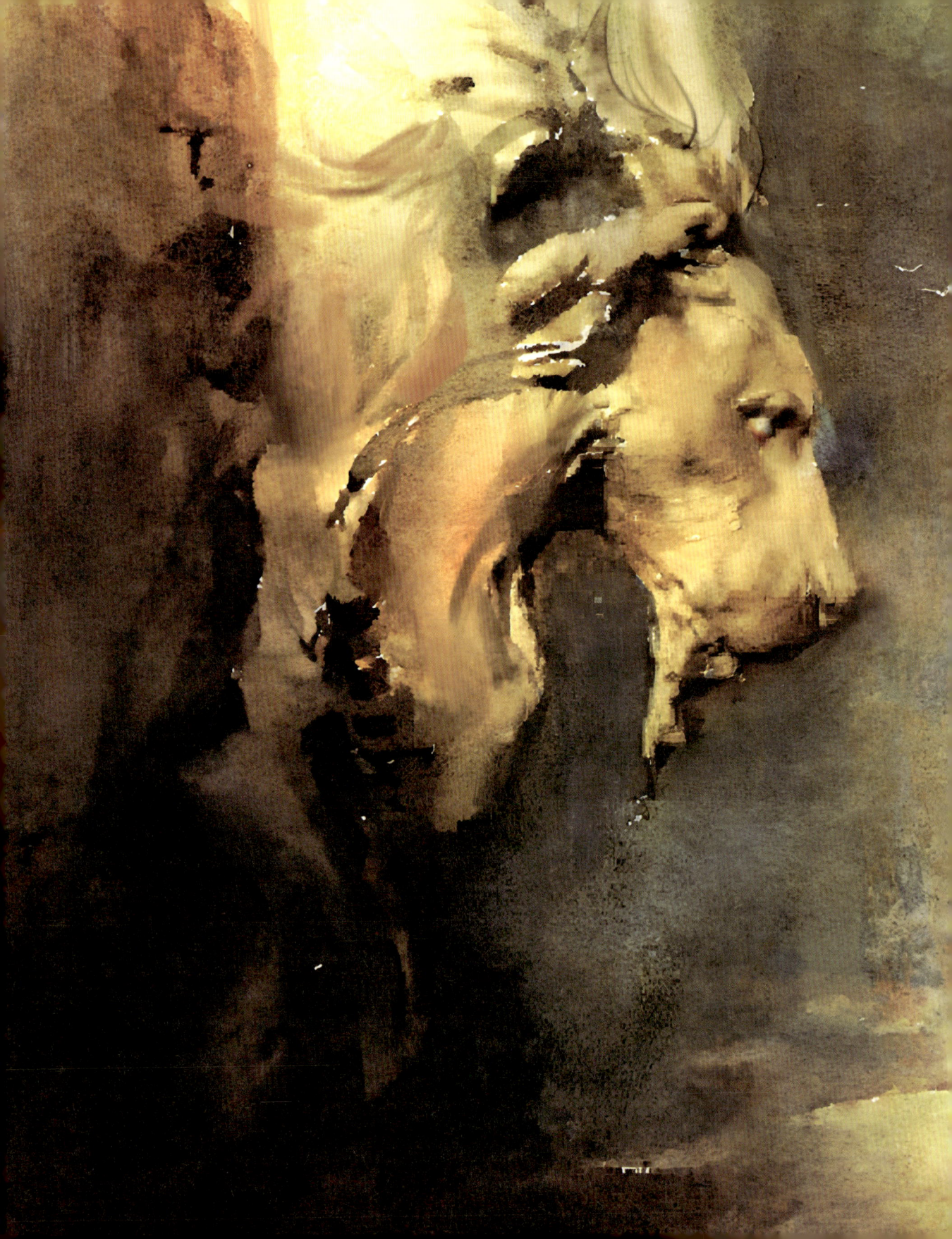

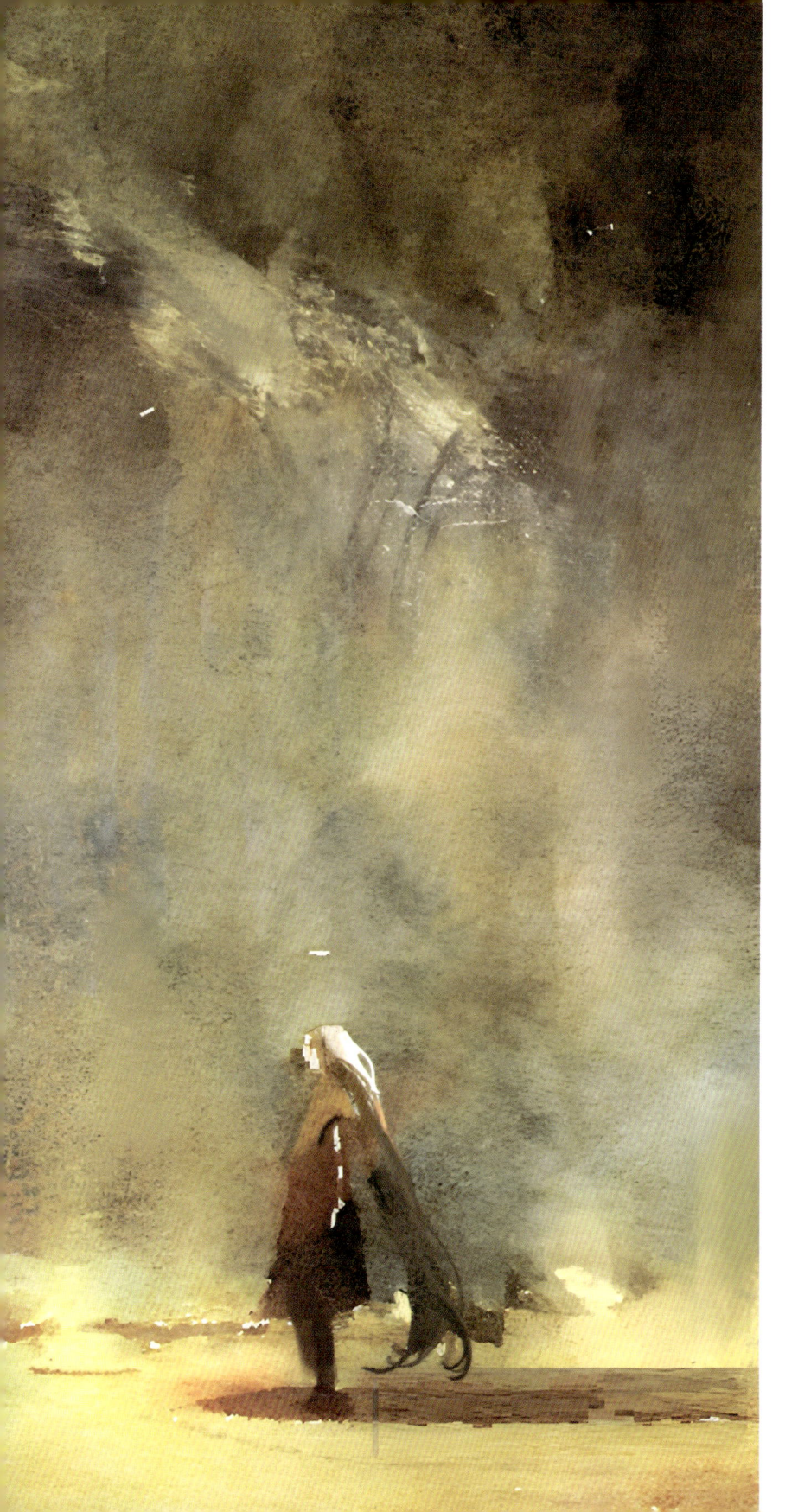

Golden Valley '22

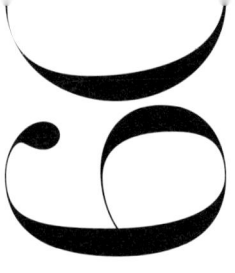

# ETERNITY

In a sense, I consider my sculptures to be three dimensional *sumi e*. The first tool I ever handled, even before a pencil, was a calligraphy brush, so now I hold a pencil in the manner of holding as if it were a brush. The use of *sumi* is deeply ingrained in me, almost like a philosophy. The pace of the strokes, the density of the ink, and the flow of energy on rice paper all these memories influence my later creations across various themes and media.

This chapter represents the Eastern influences in my work. It features historical poets, full of vitality fisherman fishing alone on solitary peaks, and gentle moments of mighty warriors…

私はある意味で、自分の立体作品を三次元の墨絵だと考えています。鉛筆よりも先に手にした最初の道具が書道の筆だった私は、今も筆を握る要領で鉛筆を握ります。墨を用いることは私の中に深く染みついていて、ほとんど哲学のようなものです。筆を運ぶペース、墨の濃さ、わら半紙の上のエネルギーの流れ——こうした記憶のすべてが、多様なテーマとメディアを横断するその後の創作に影響しています。

この章では、私の作品における東洋の影響が表れています。描かれるのは、力強い歴史上の詩人たち、独立してそびえ立つ峰の上でひとり魚を釣る釣り人、逞しい戦士たちの穏やかな瞬間……。

Meditation 228

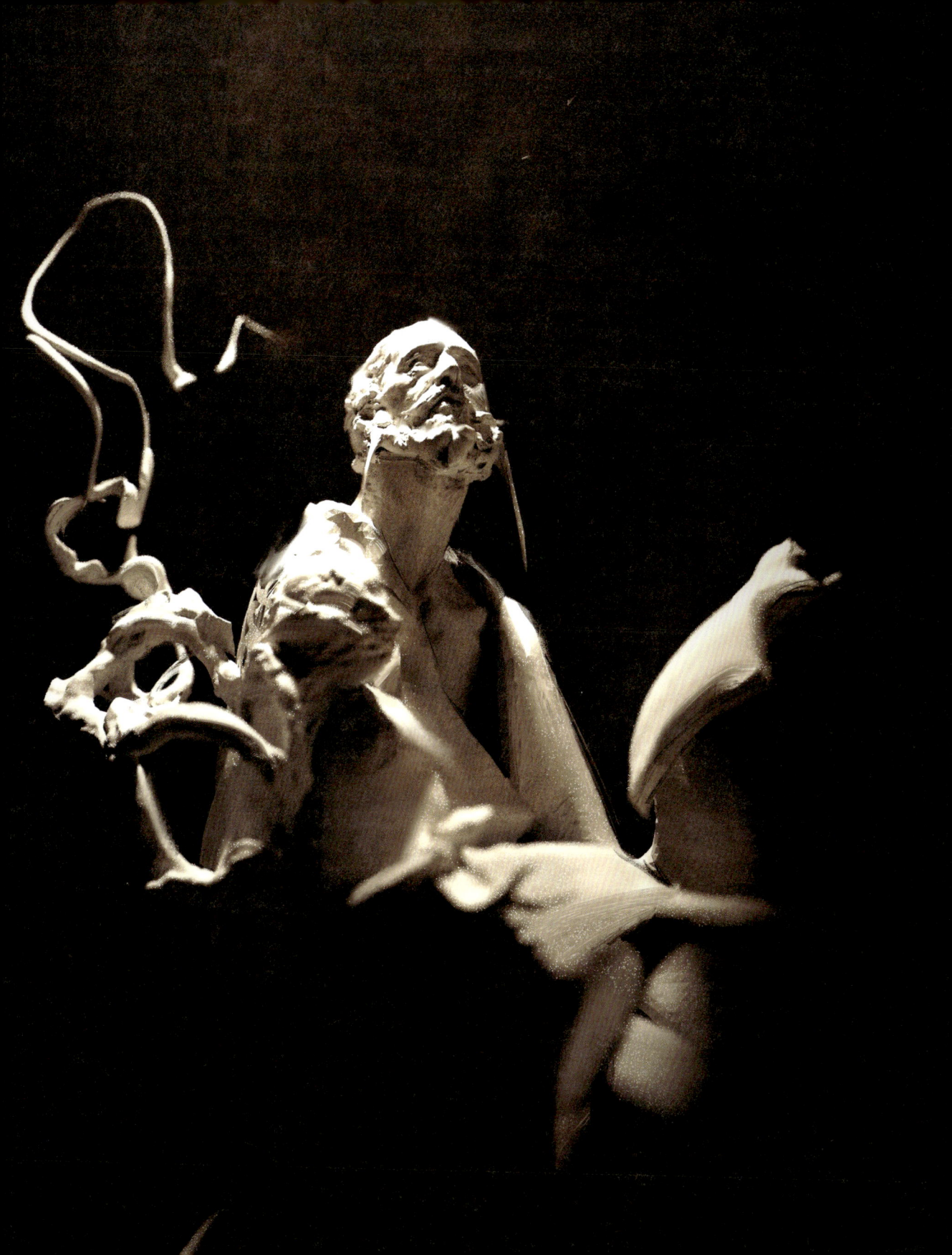

# Fishing 230

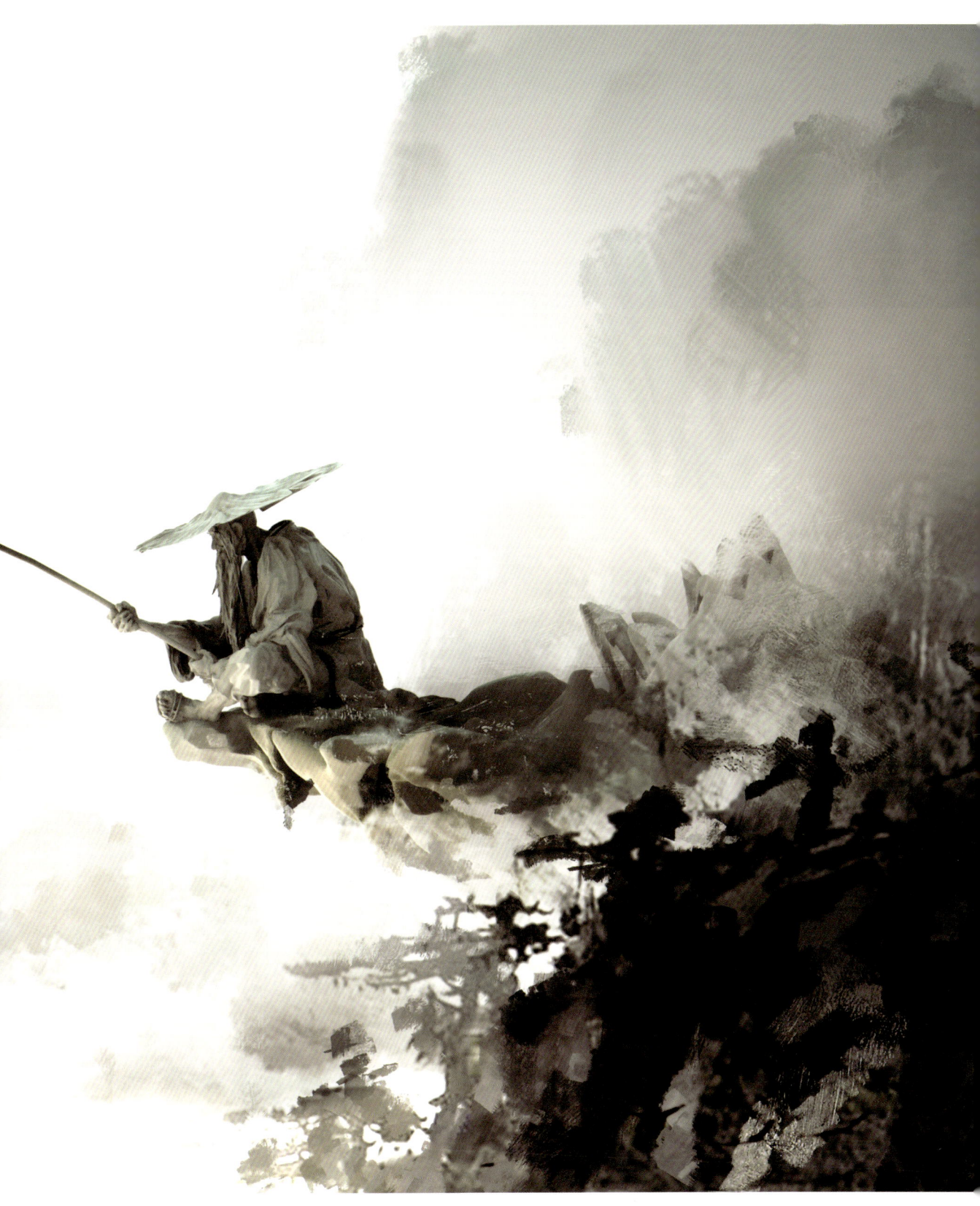

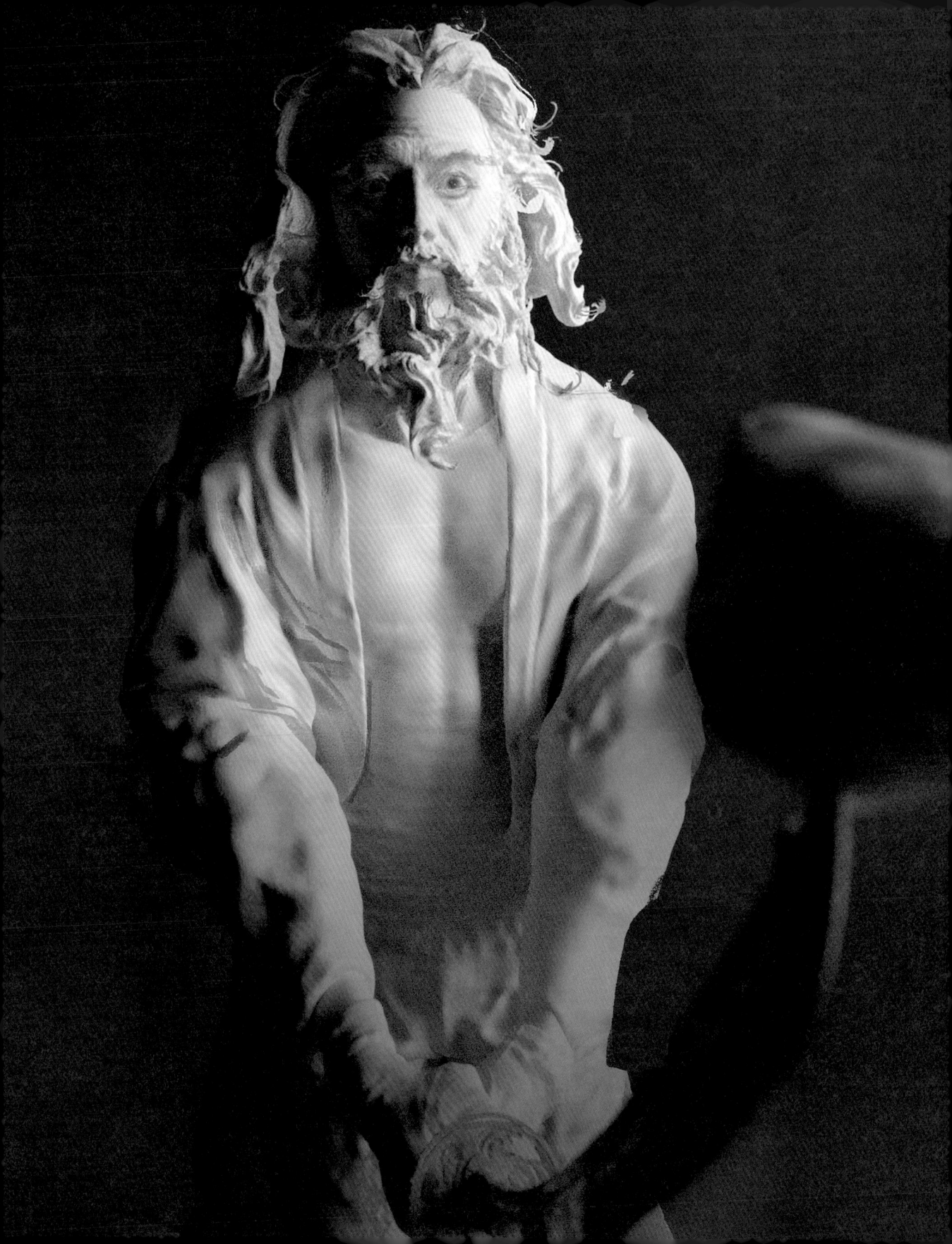

# Samurai & Bird 23?

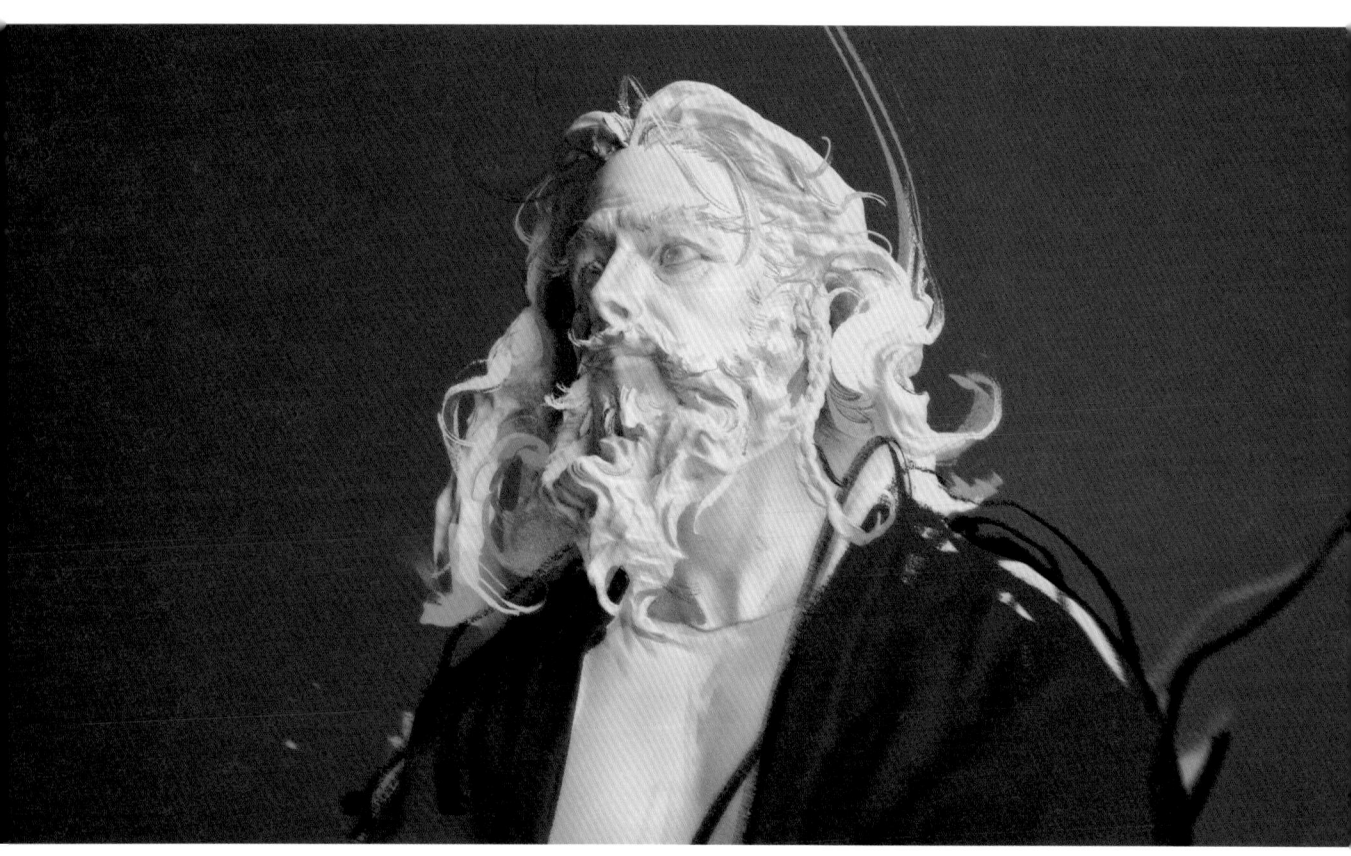

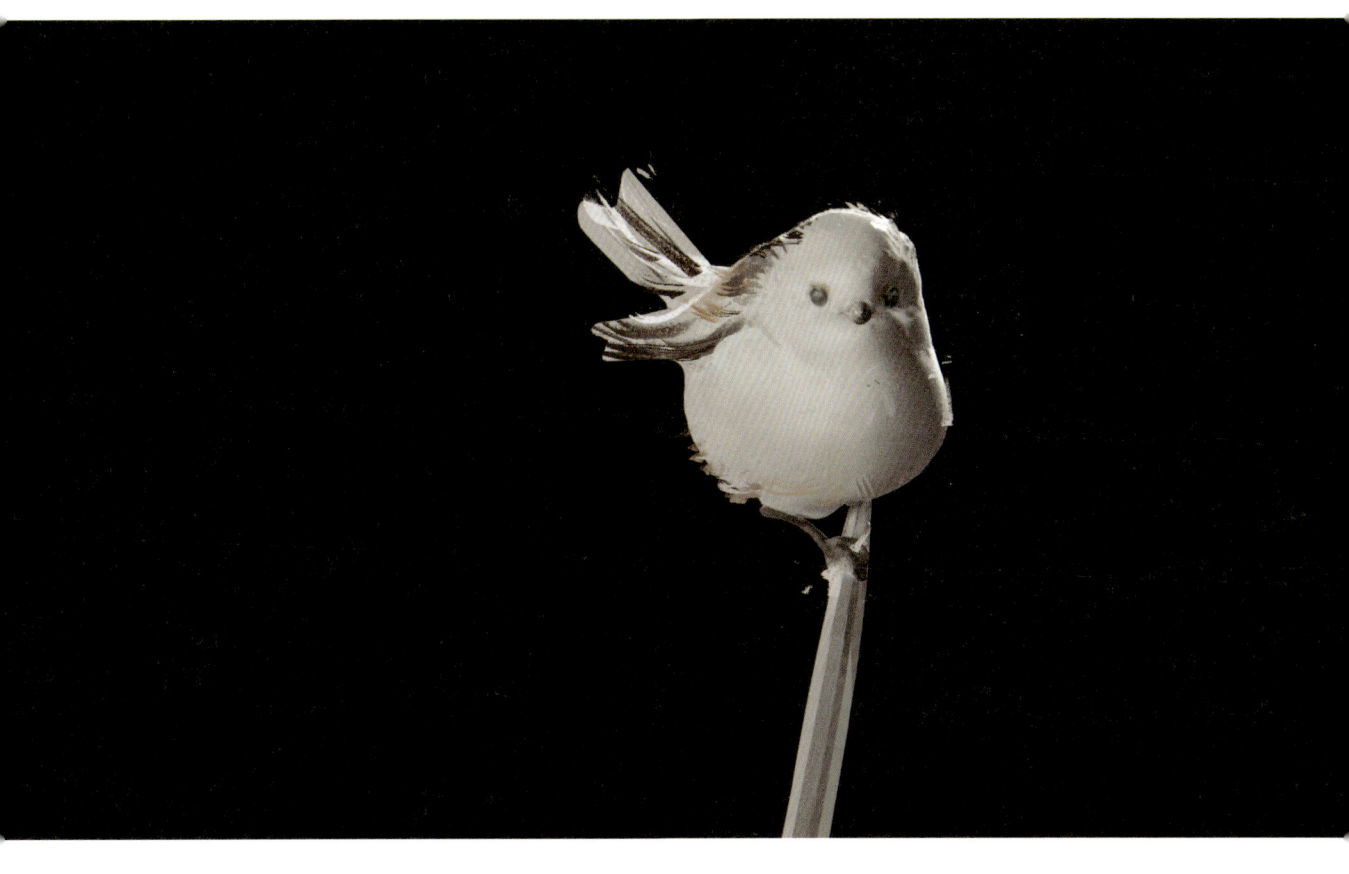

# Buddha 230

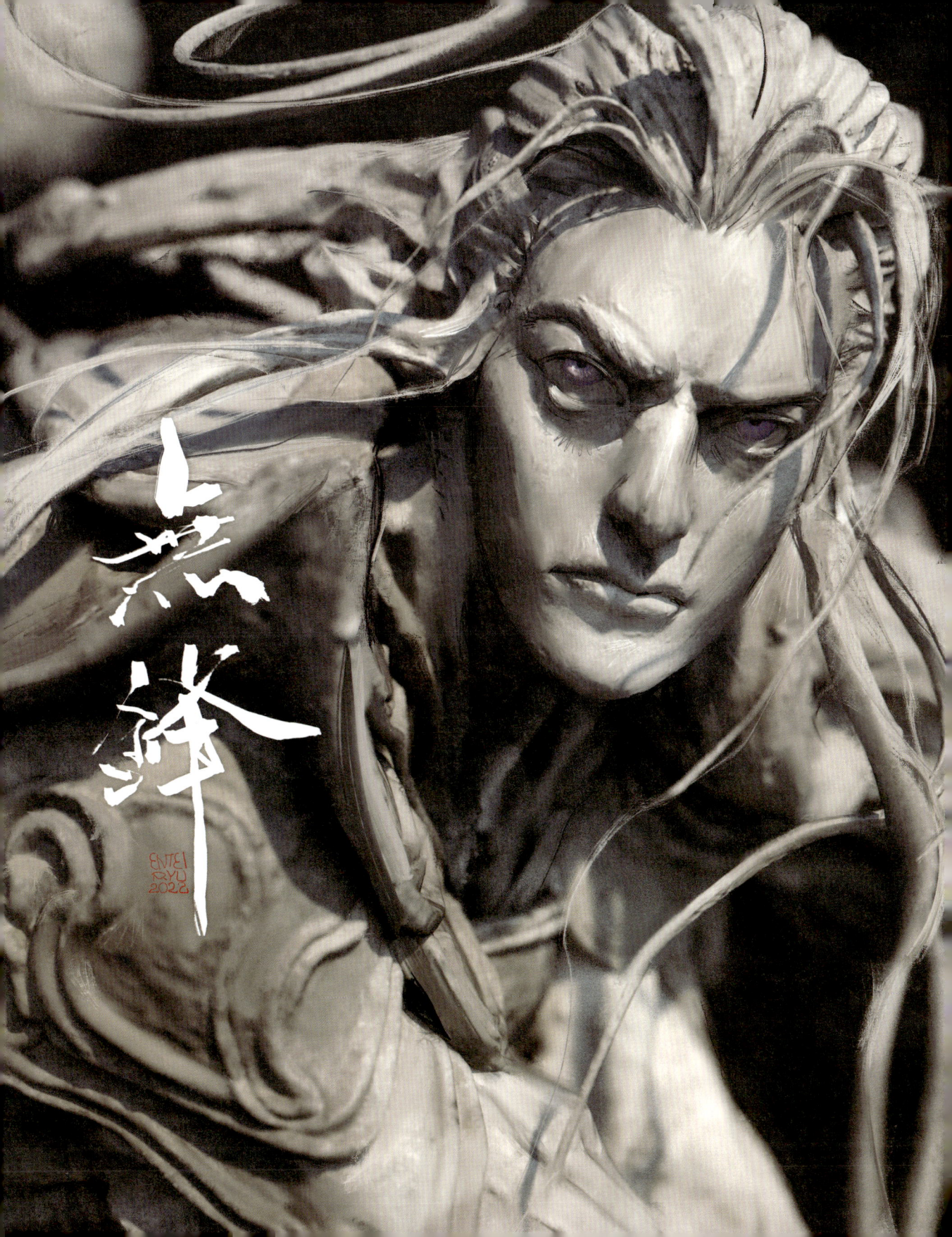

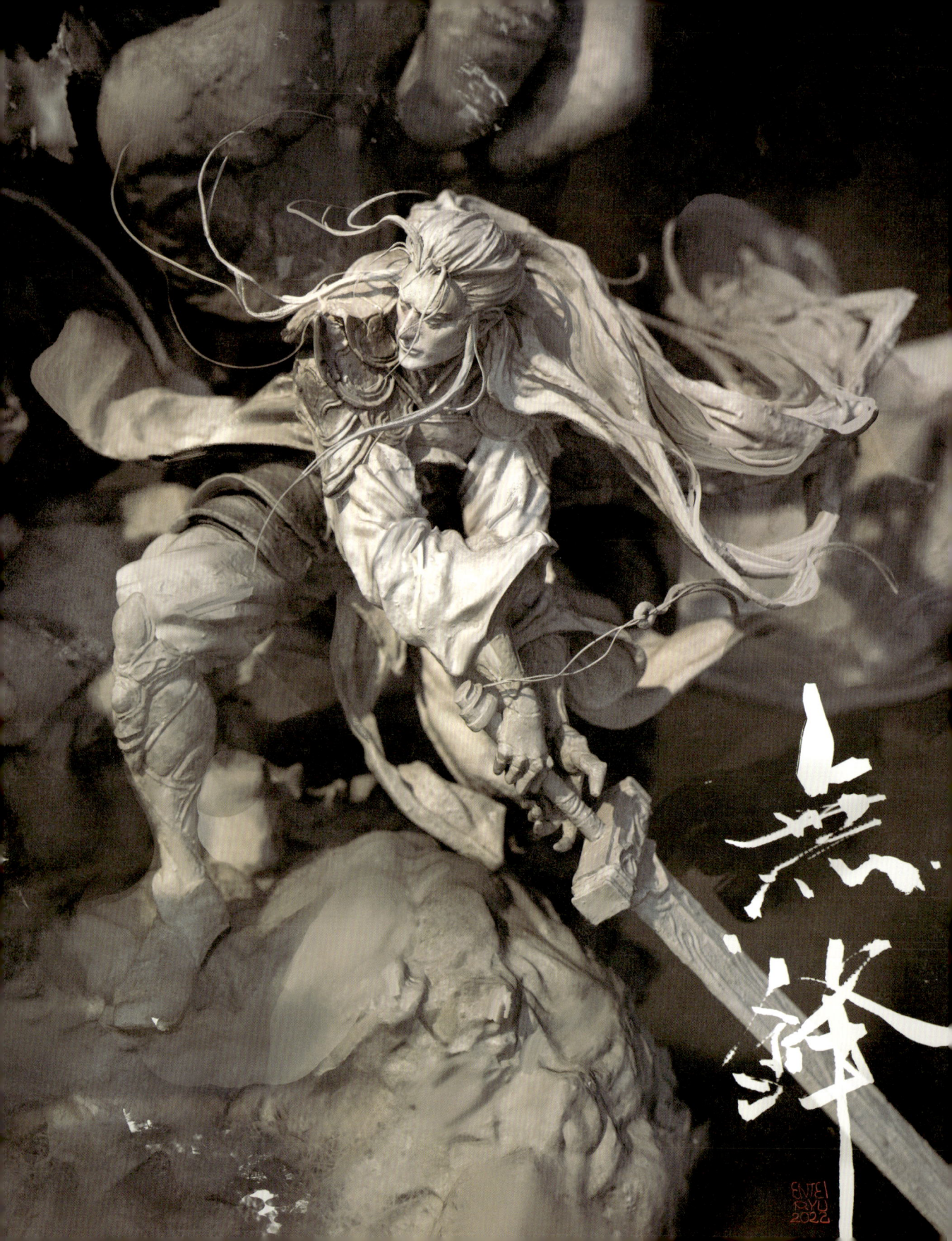

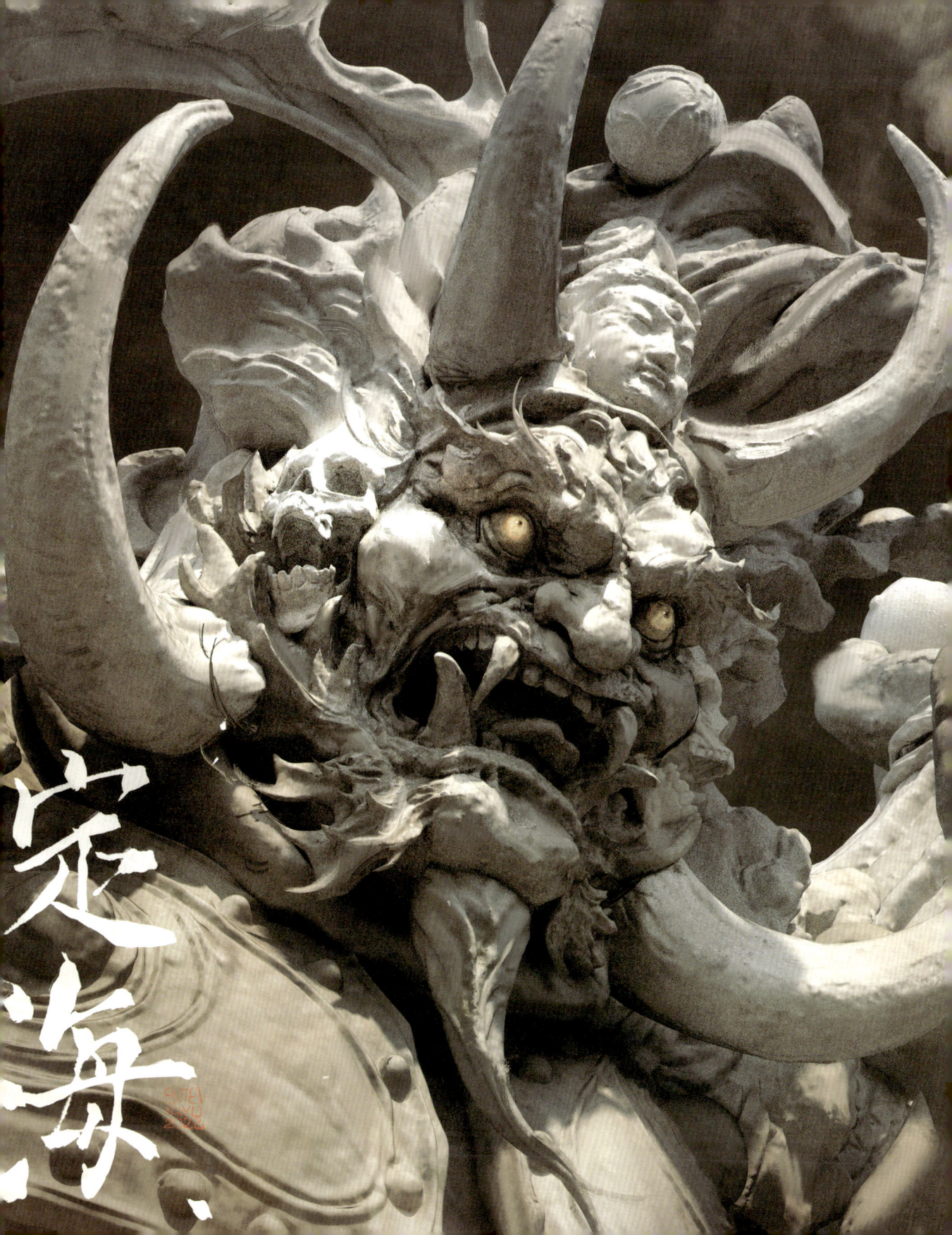

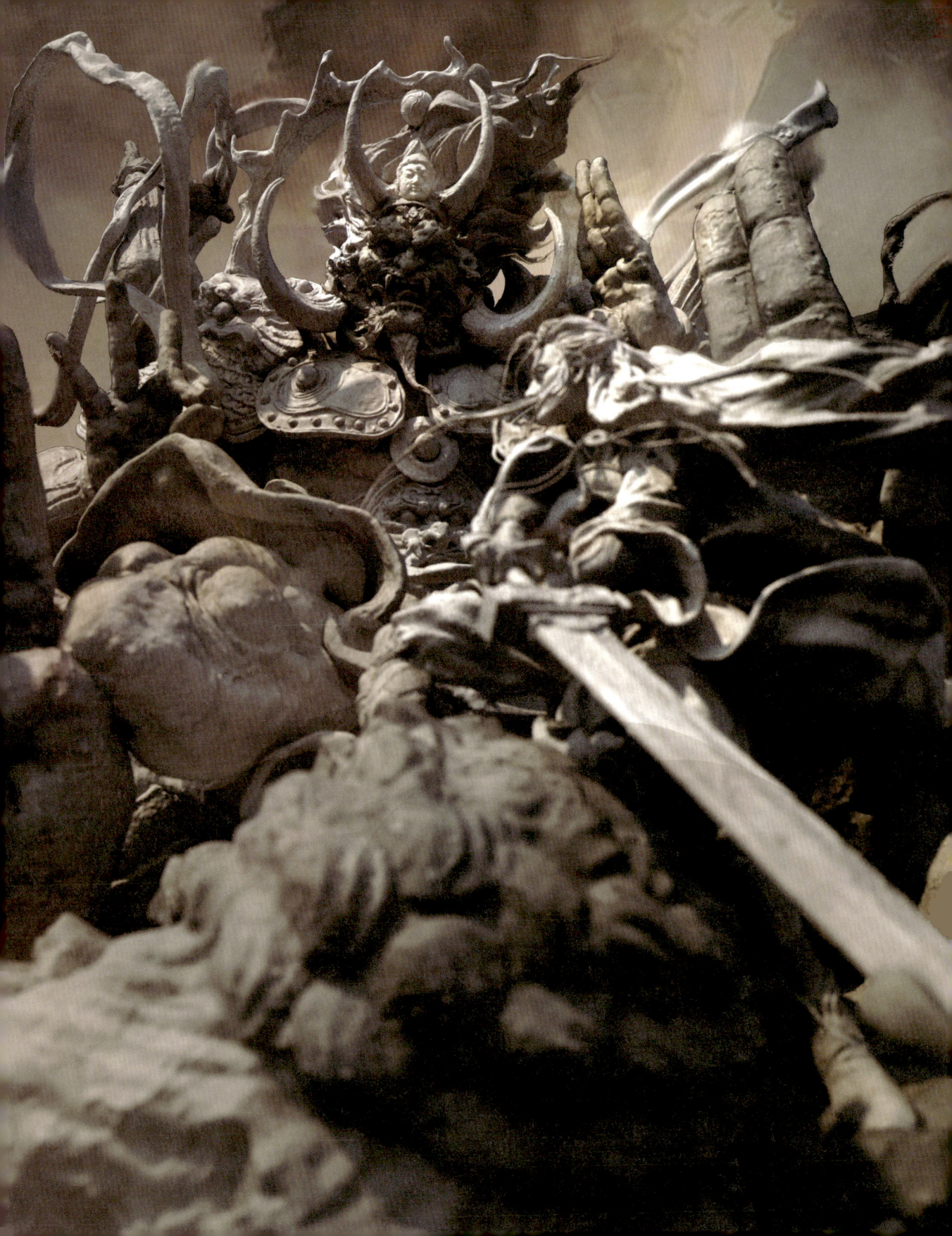

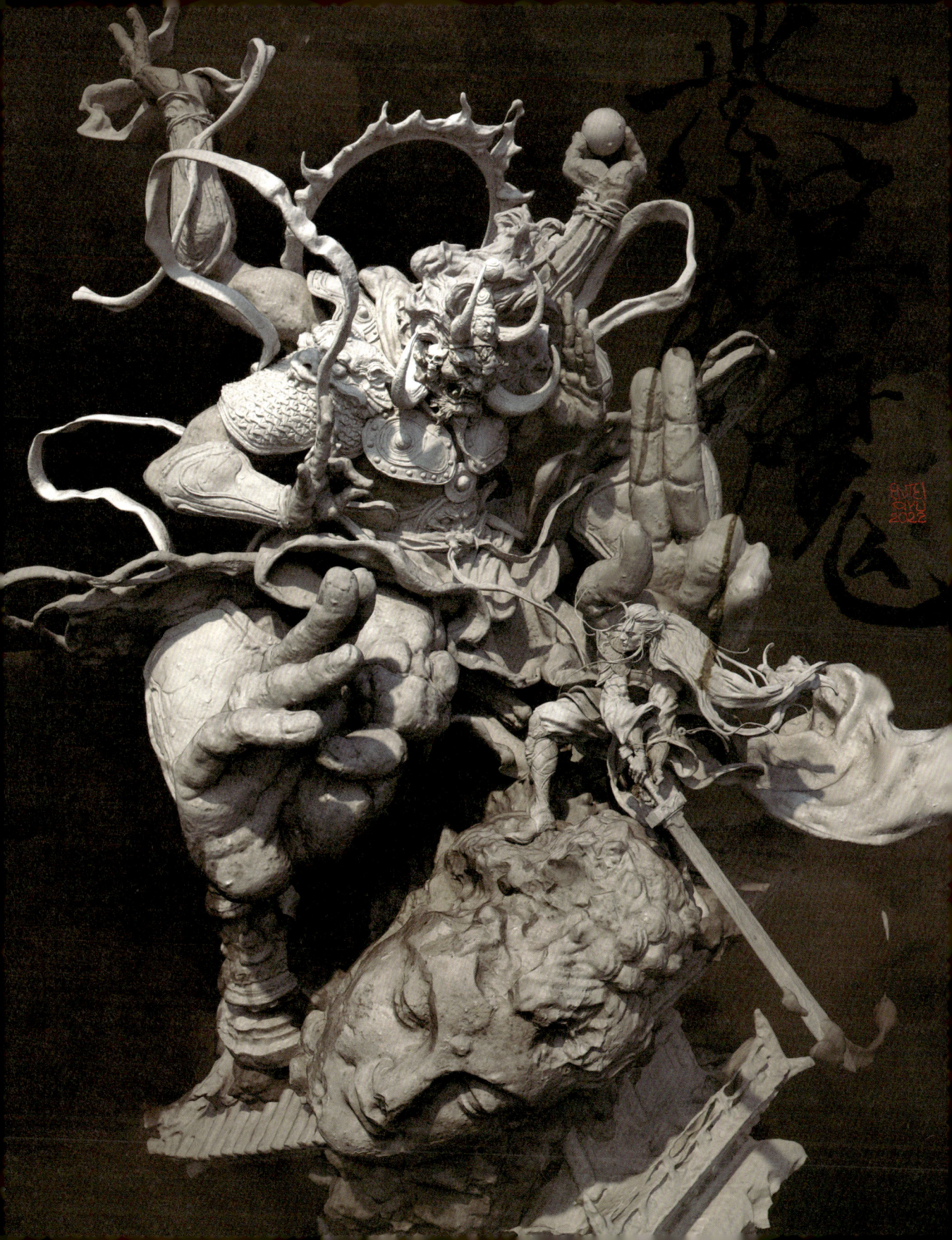

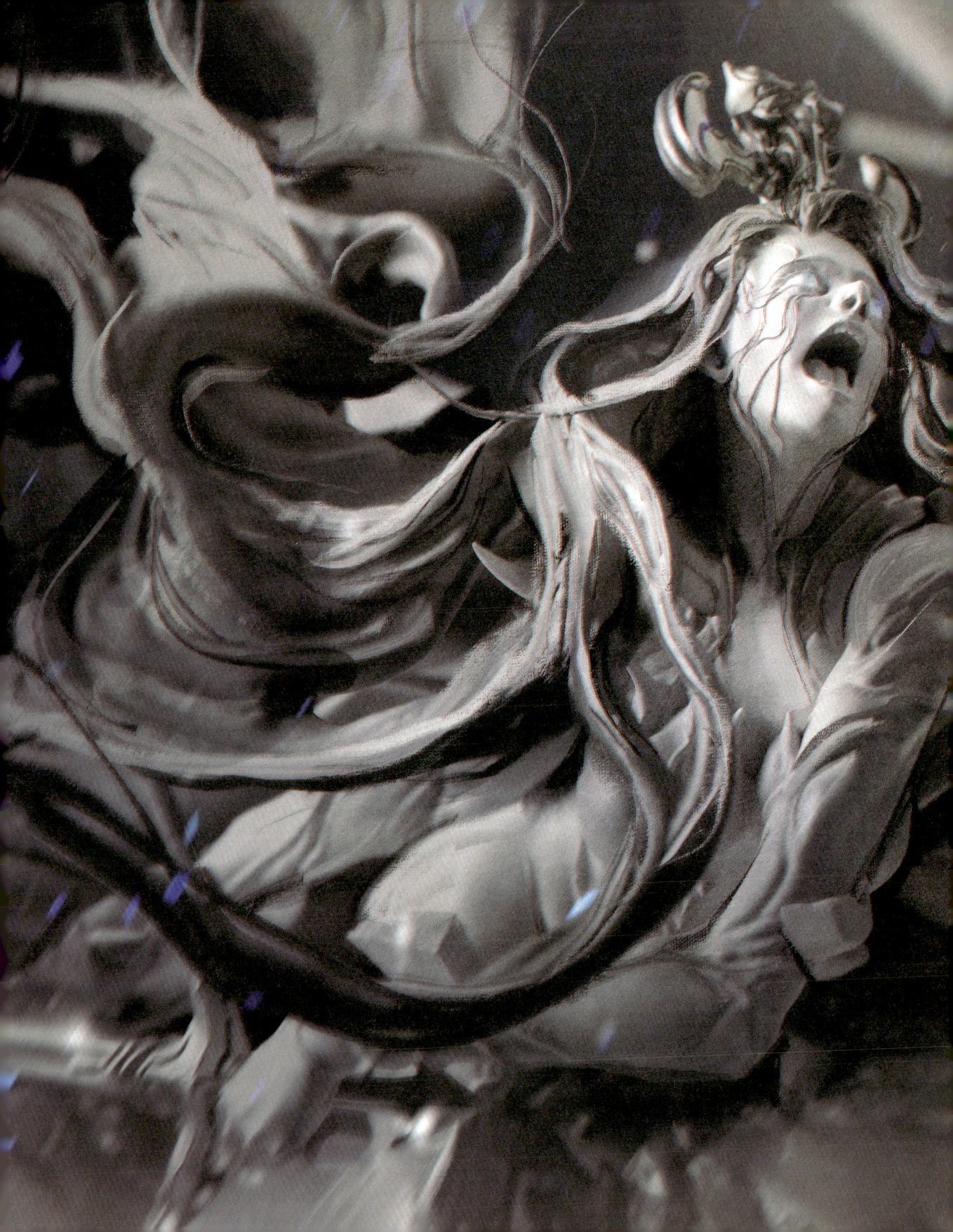

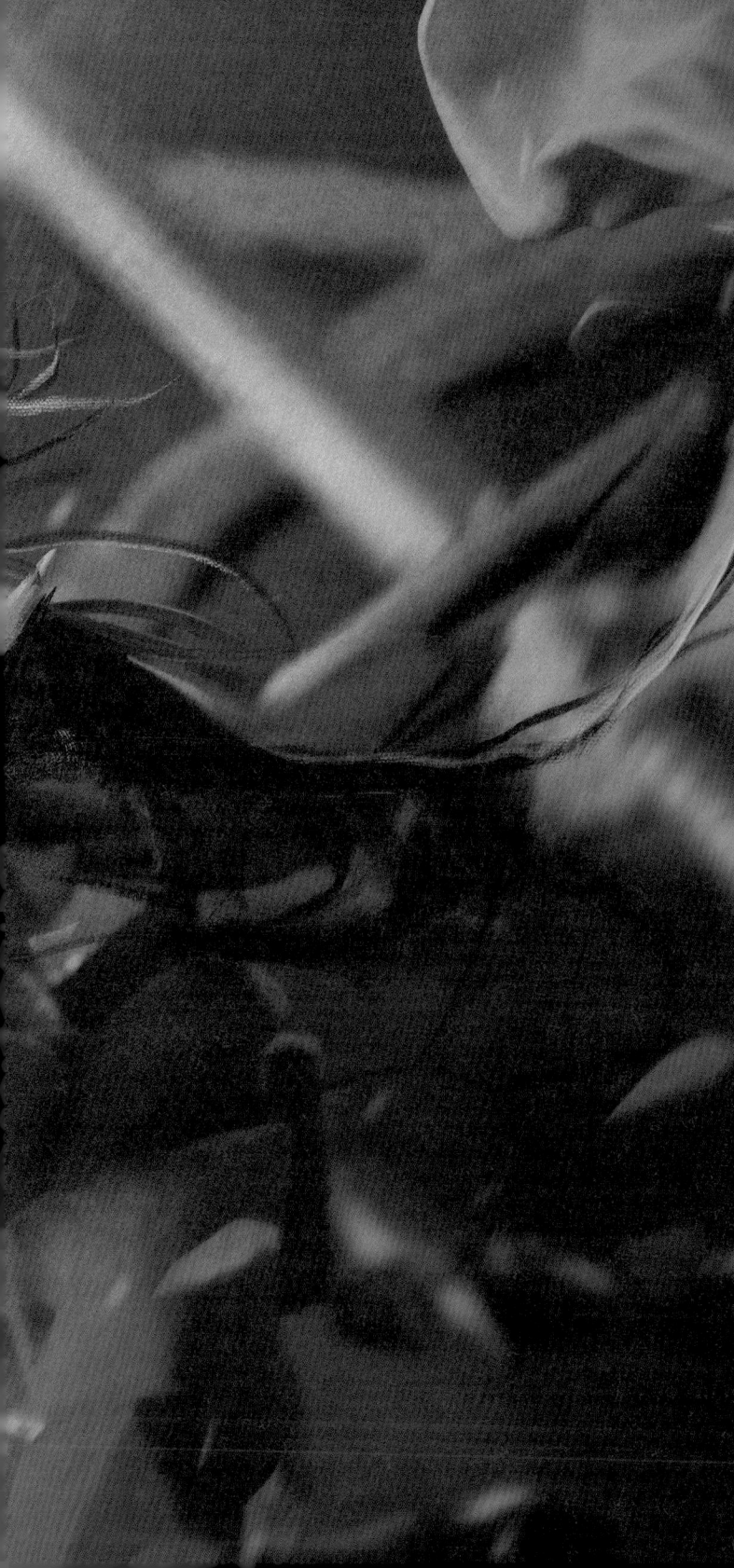

Marionette

09   ETERNITY

Marionette '24?

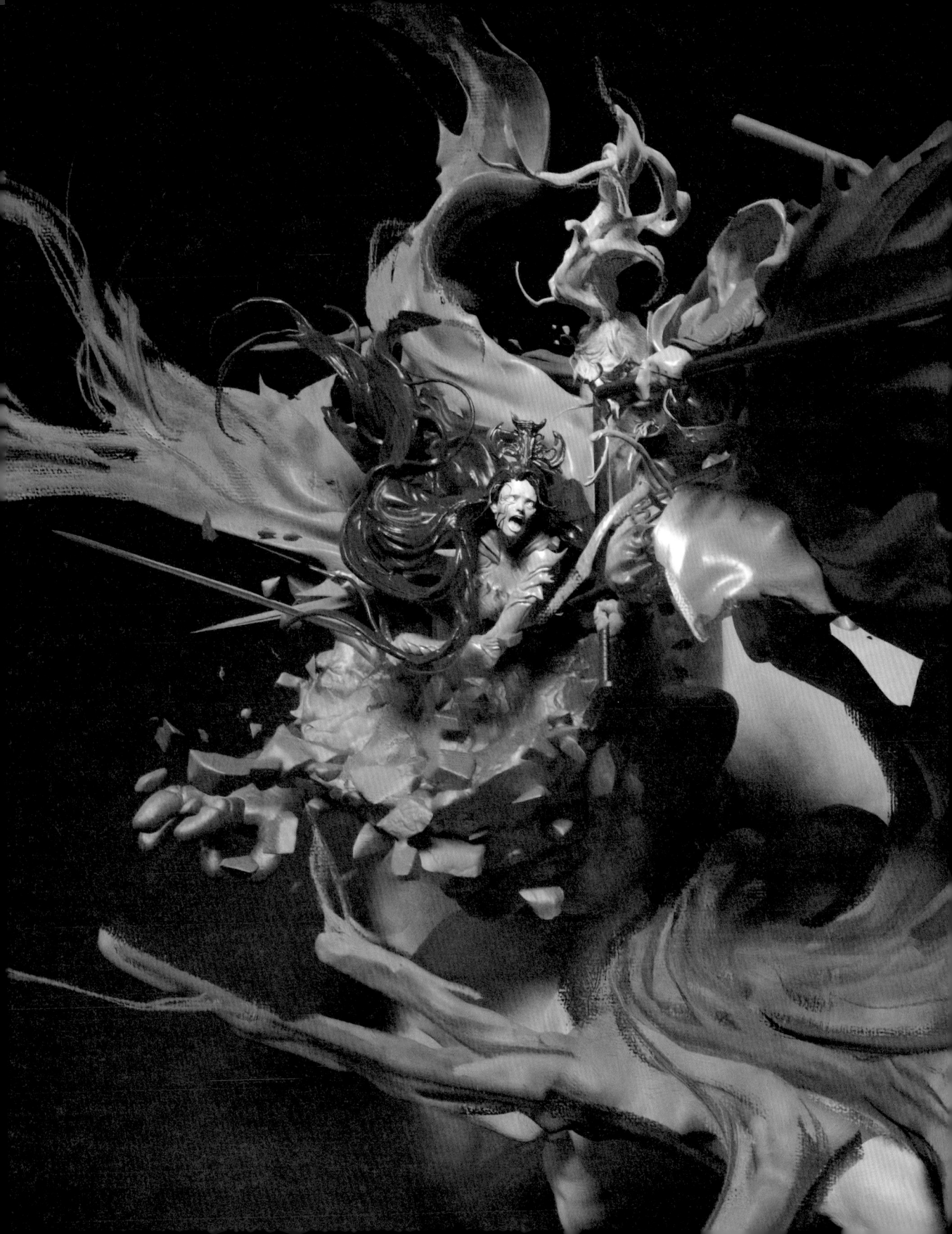

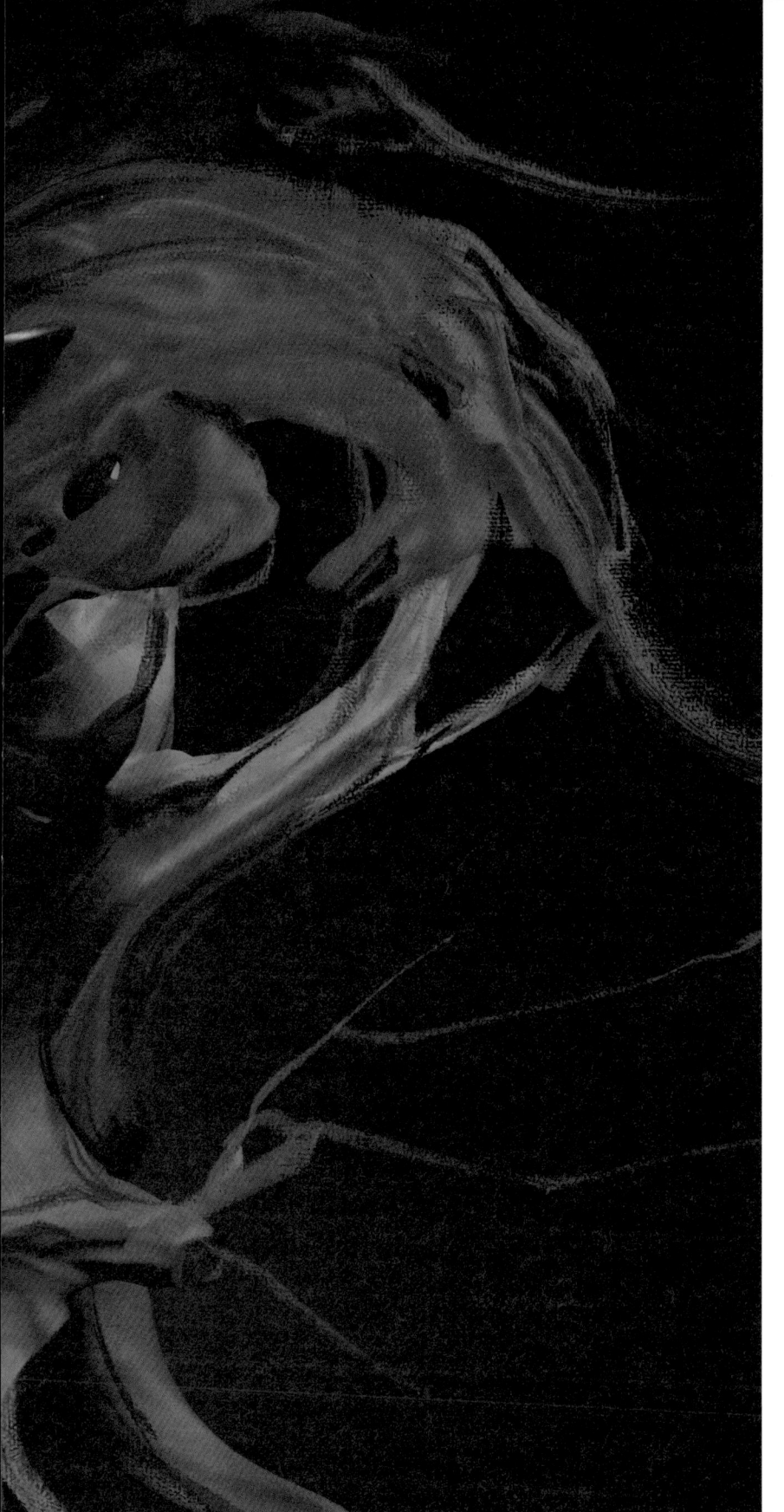

Marionette

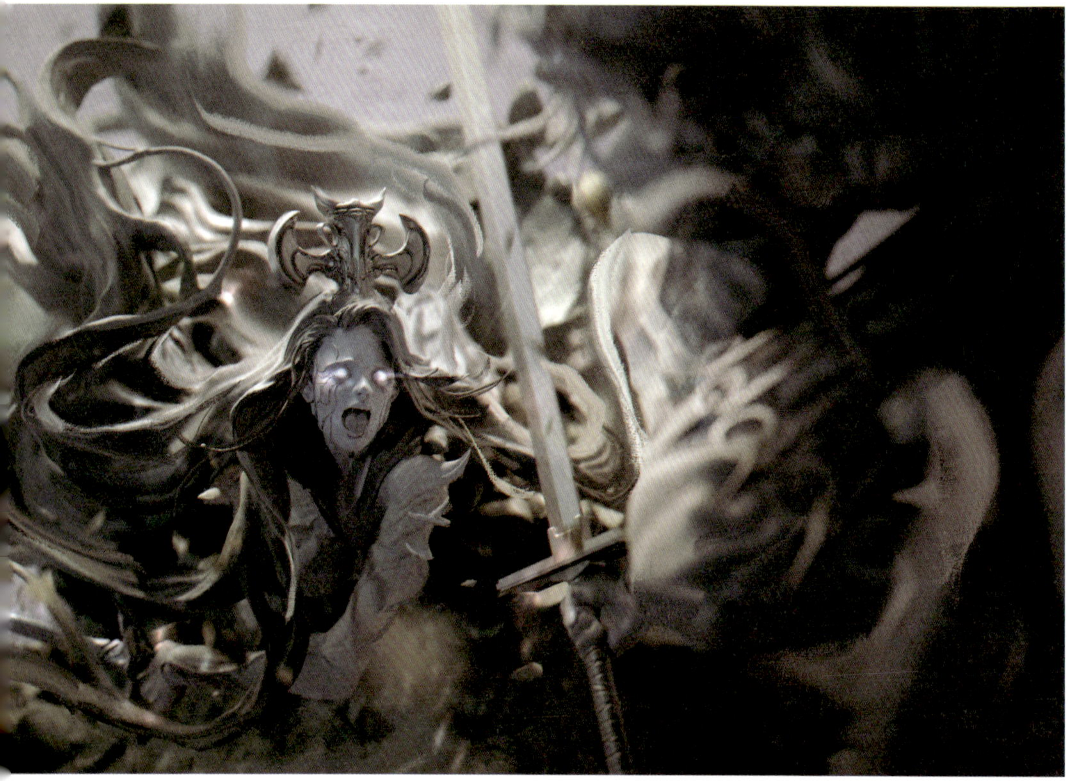

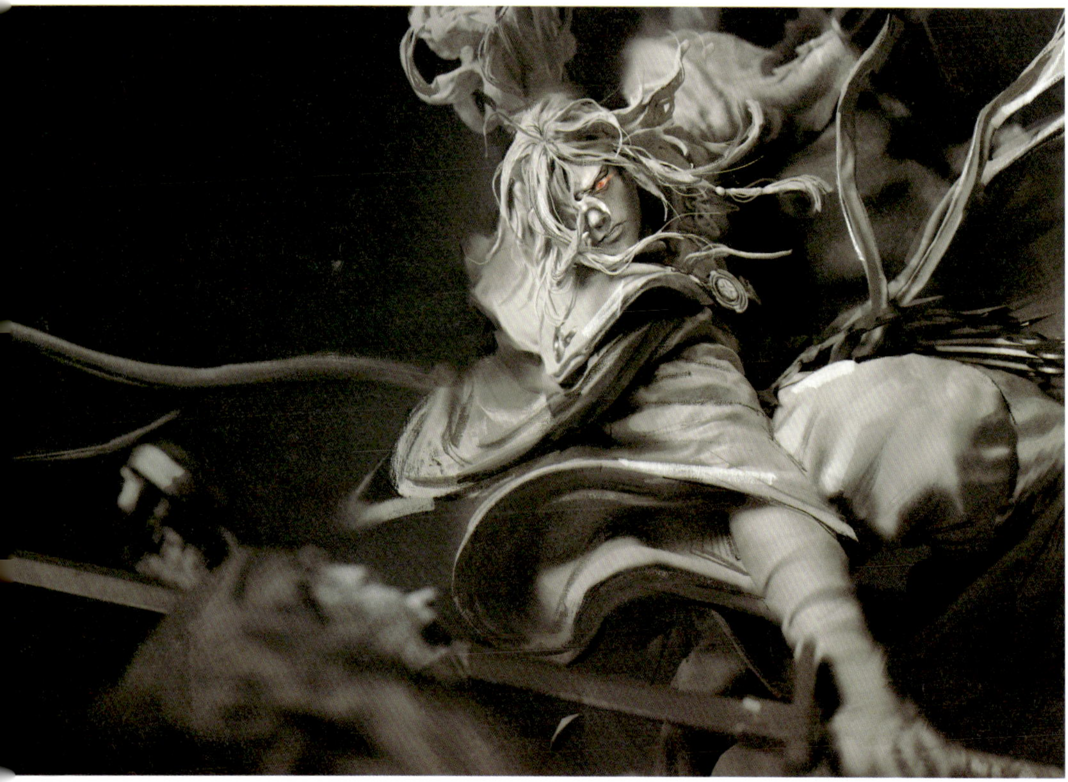

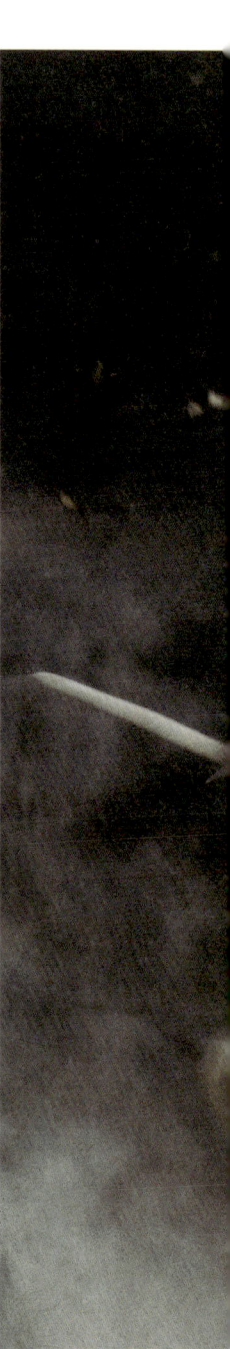

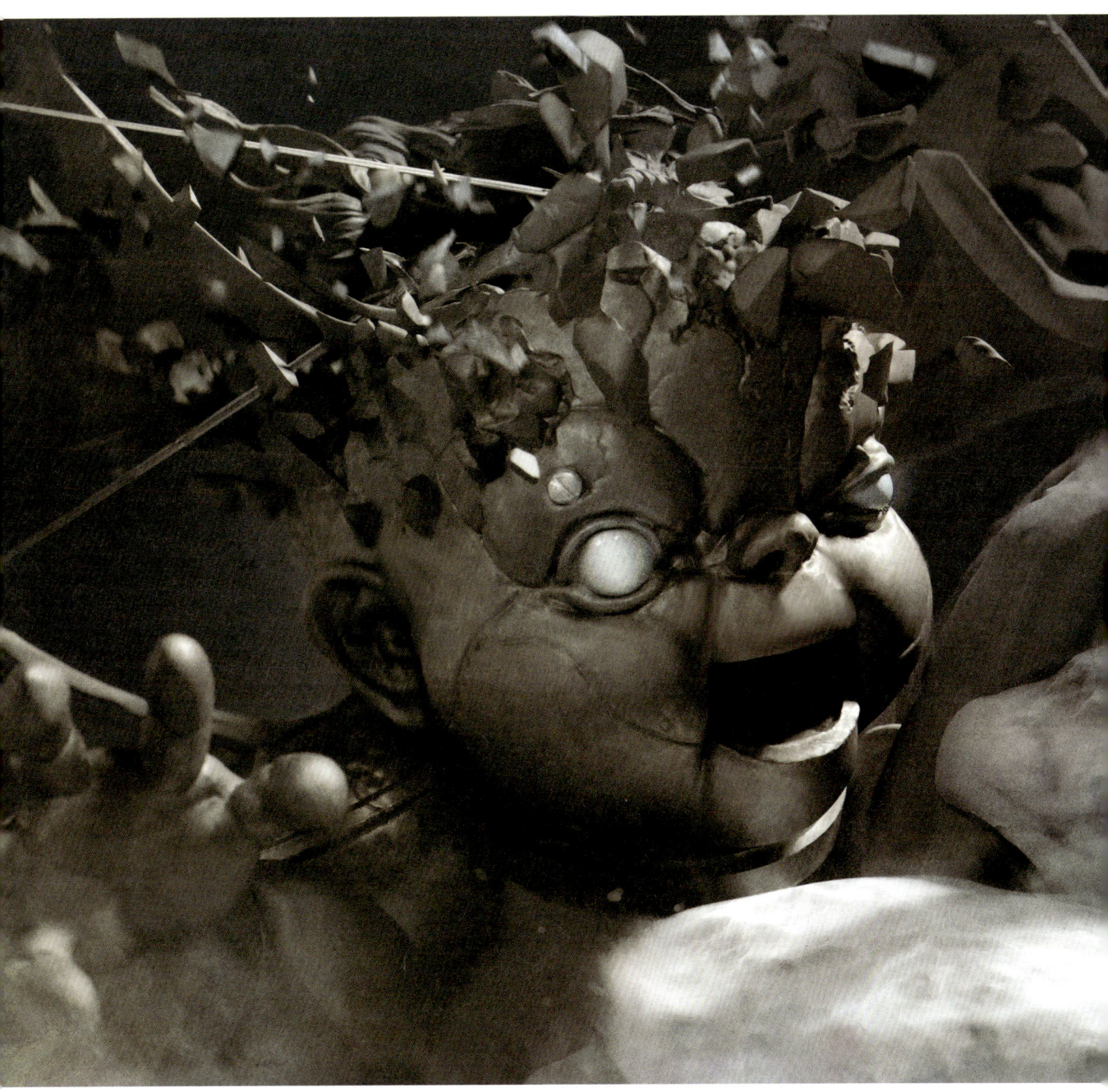

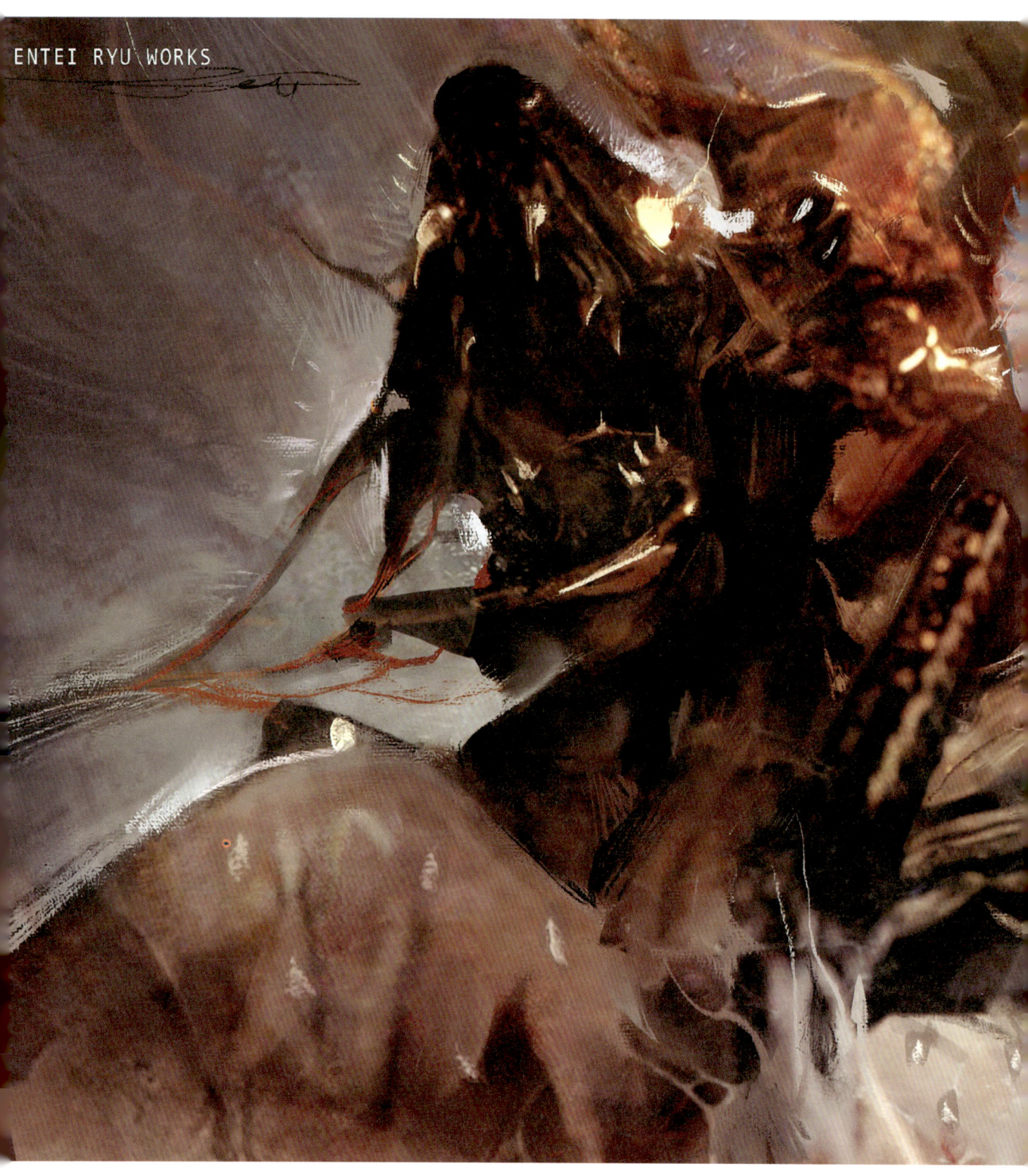

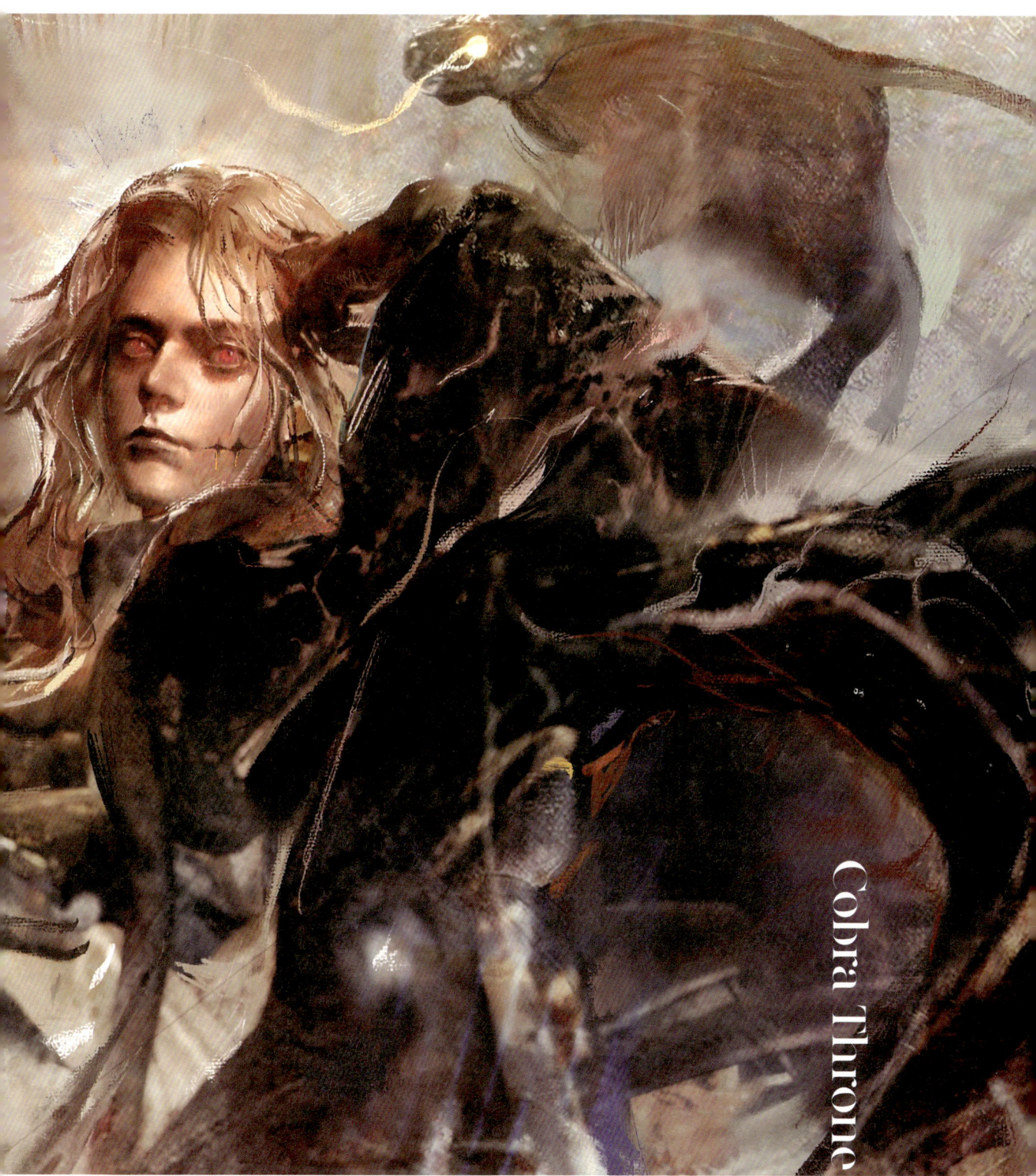

Cobra Throne

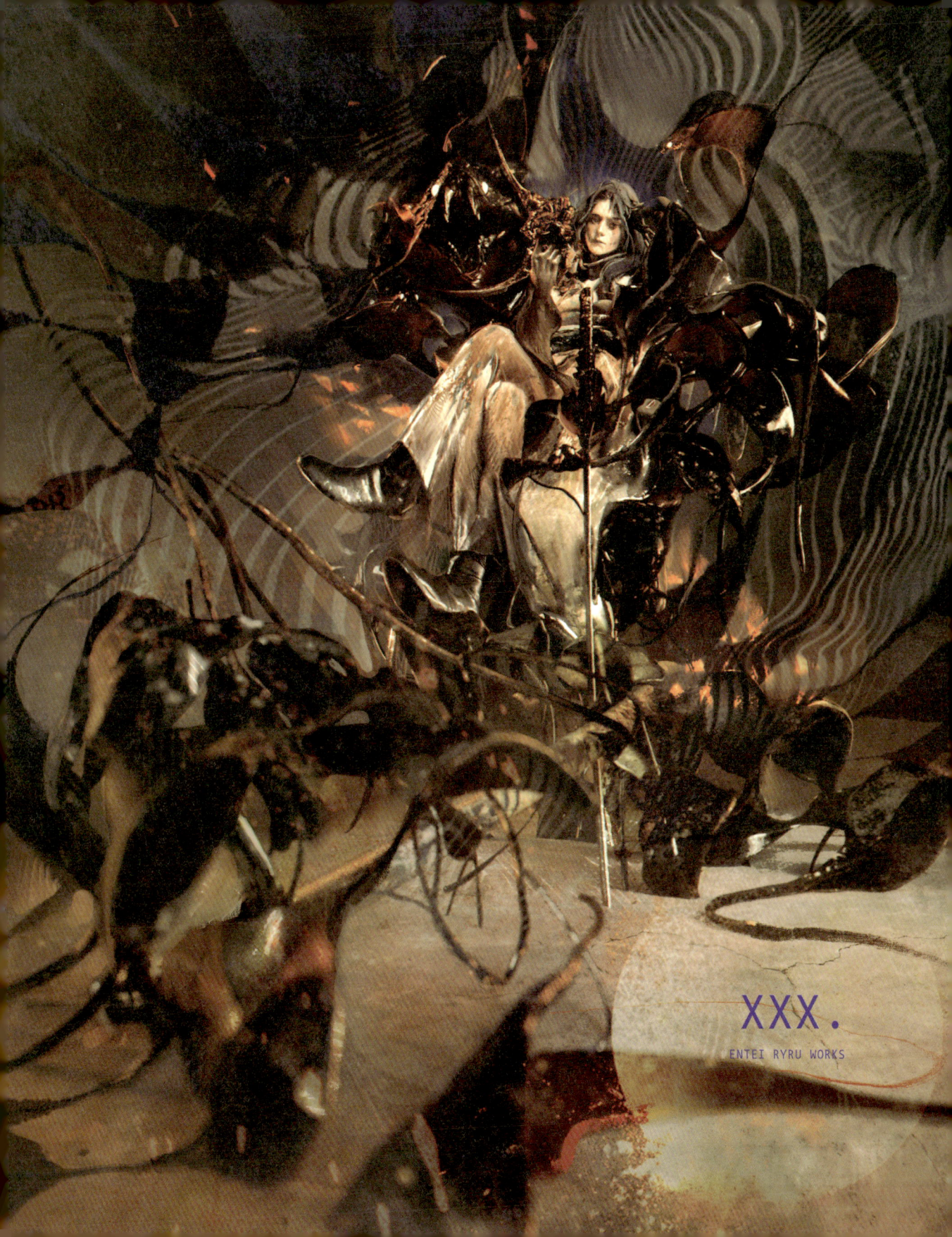

XXX.

ENTEI RYRU WORKS

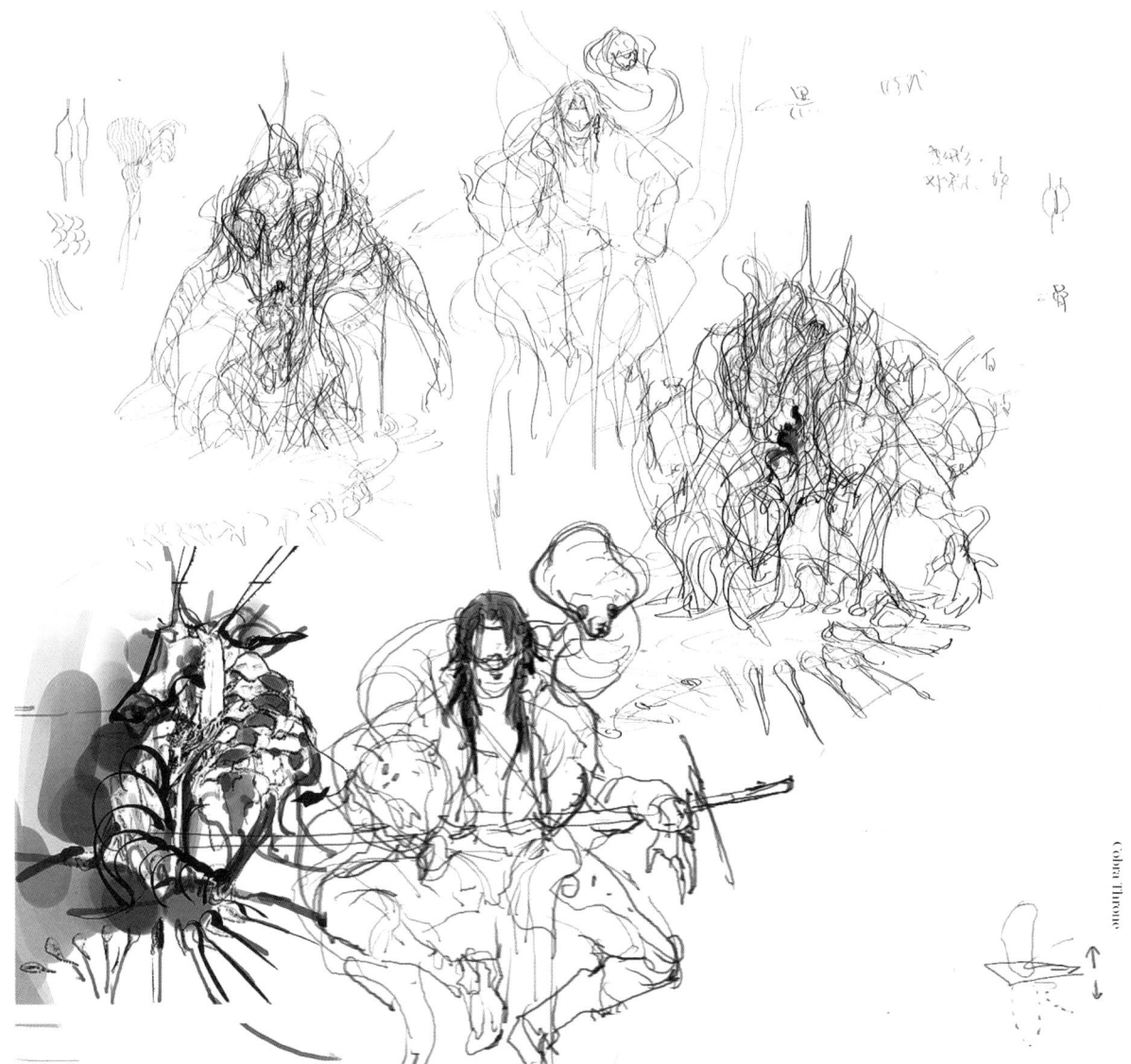

Cobra Throne

09 ETERNITY

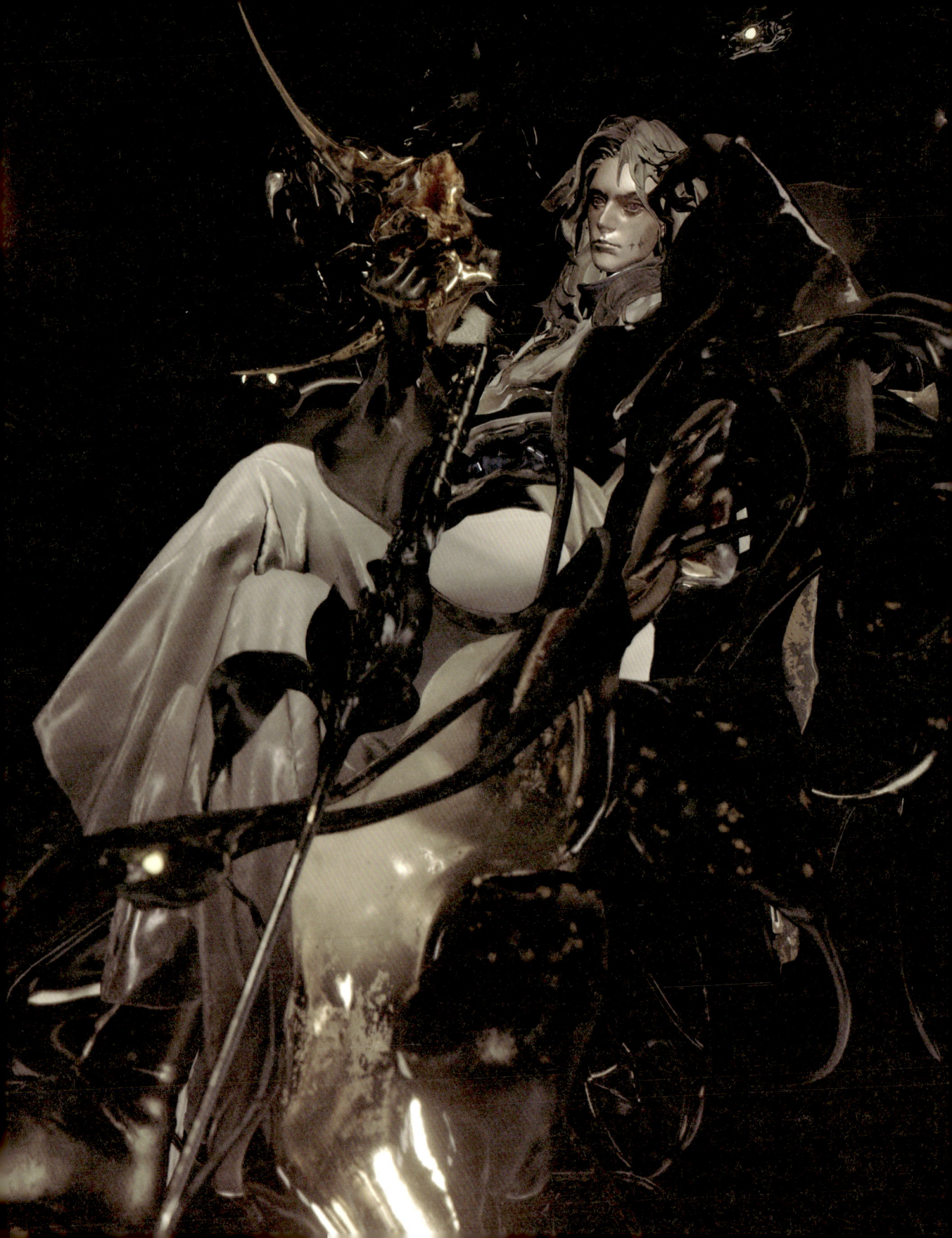

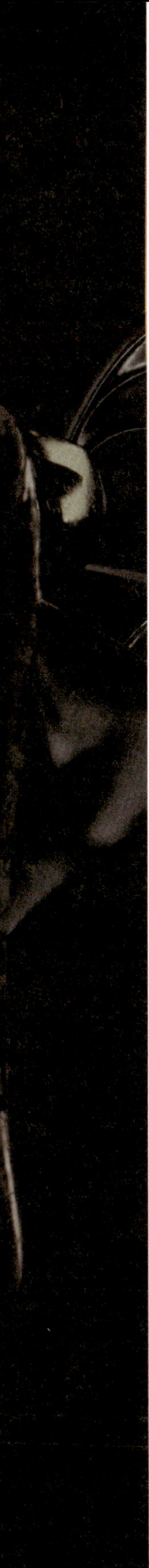

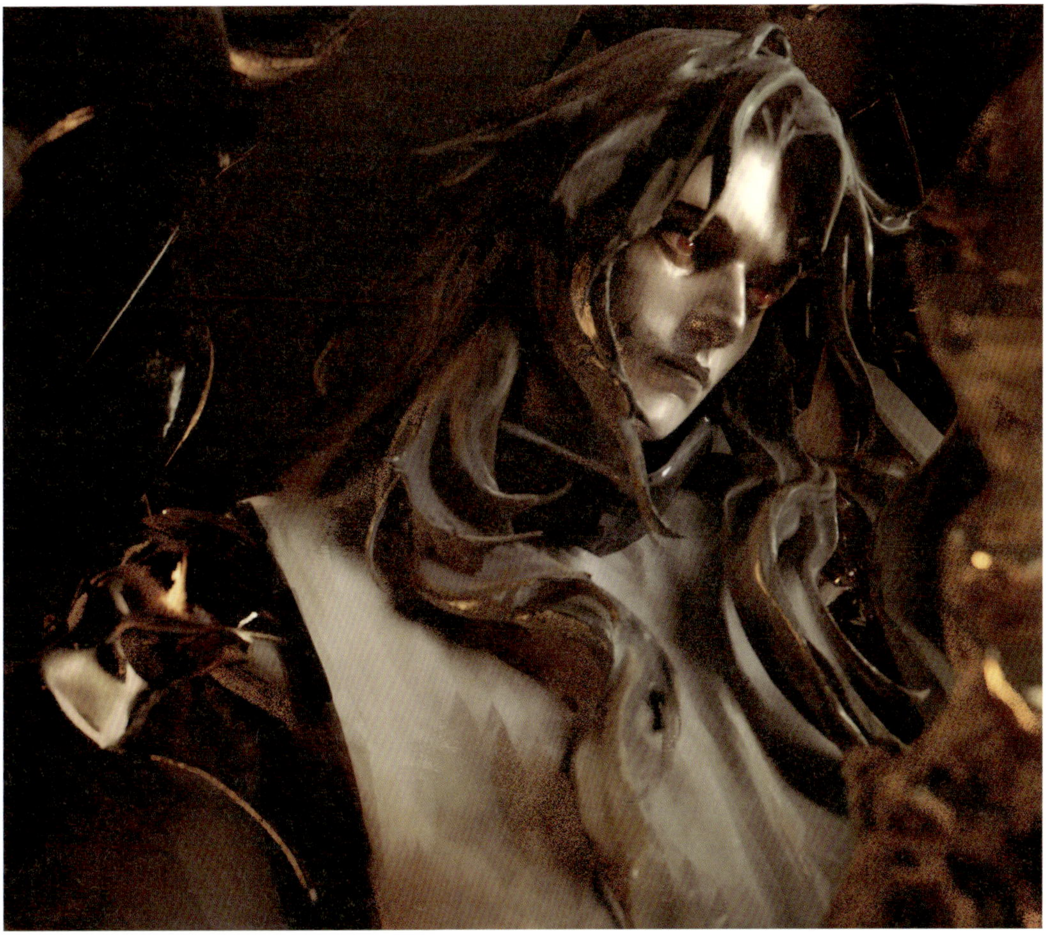

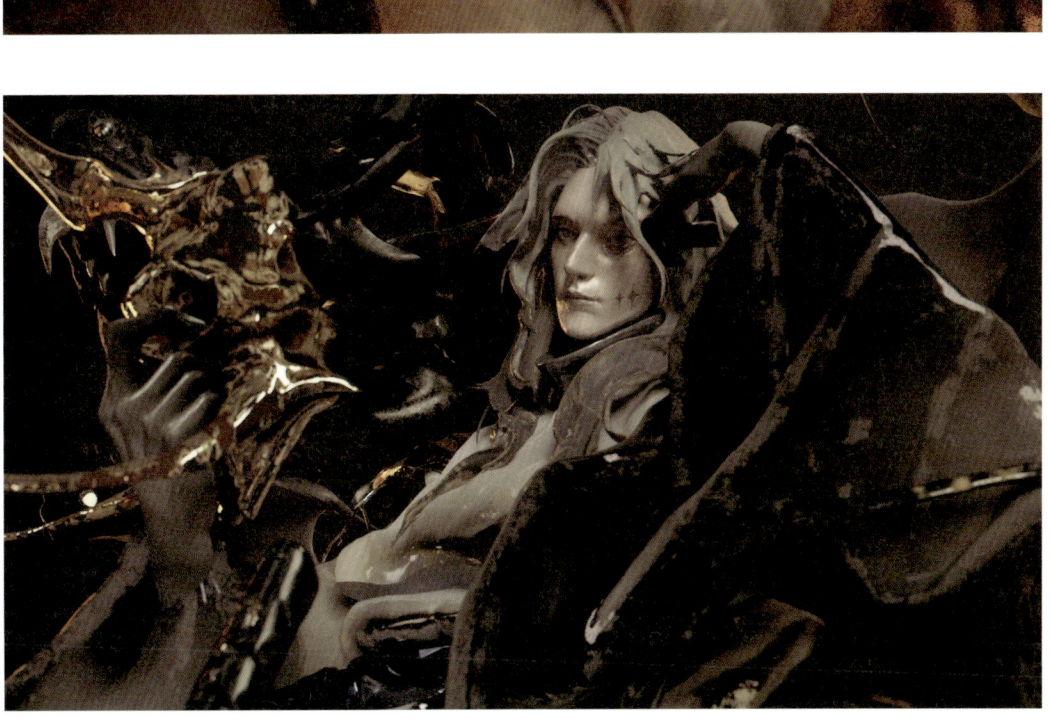

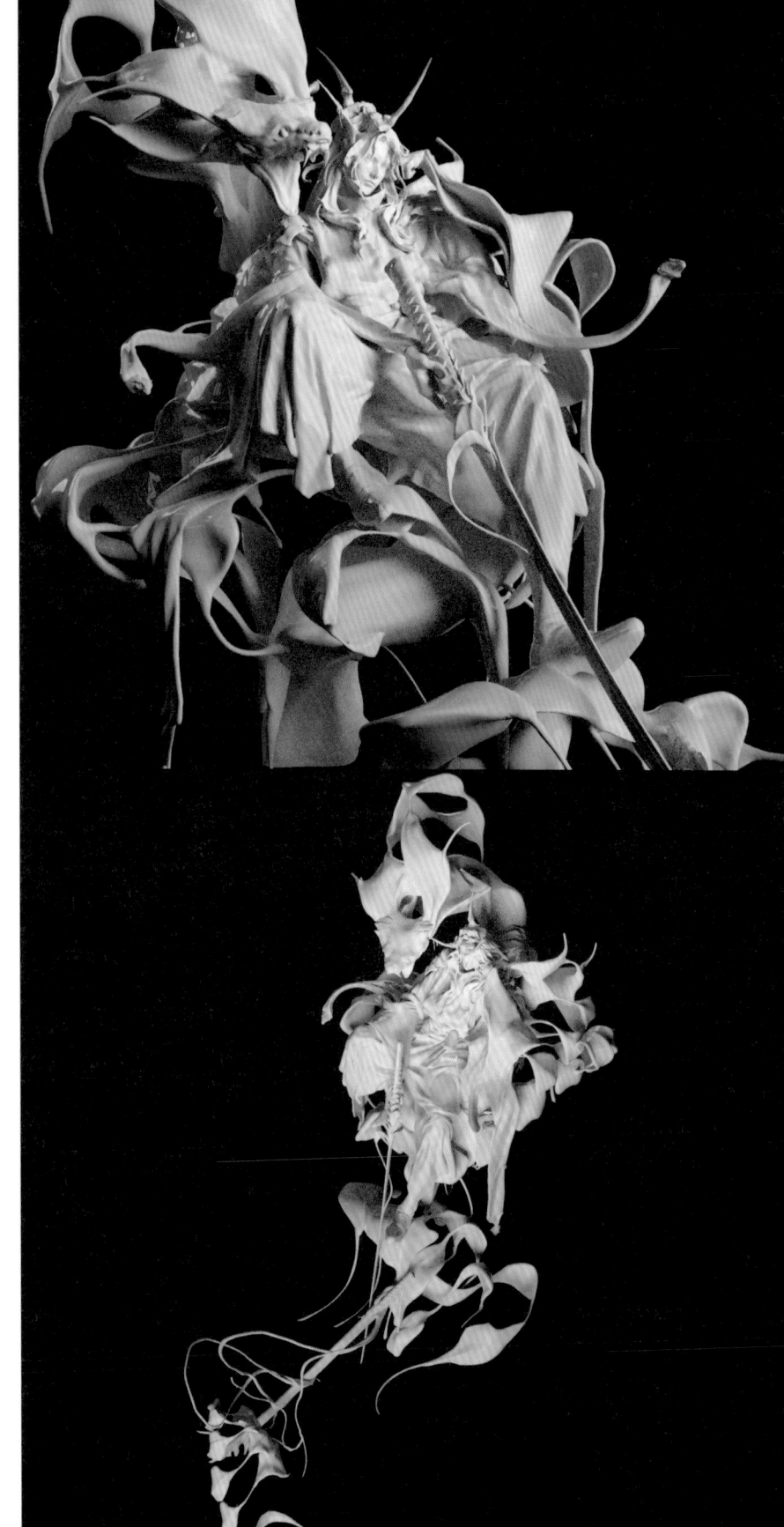

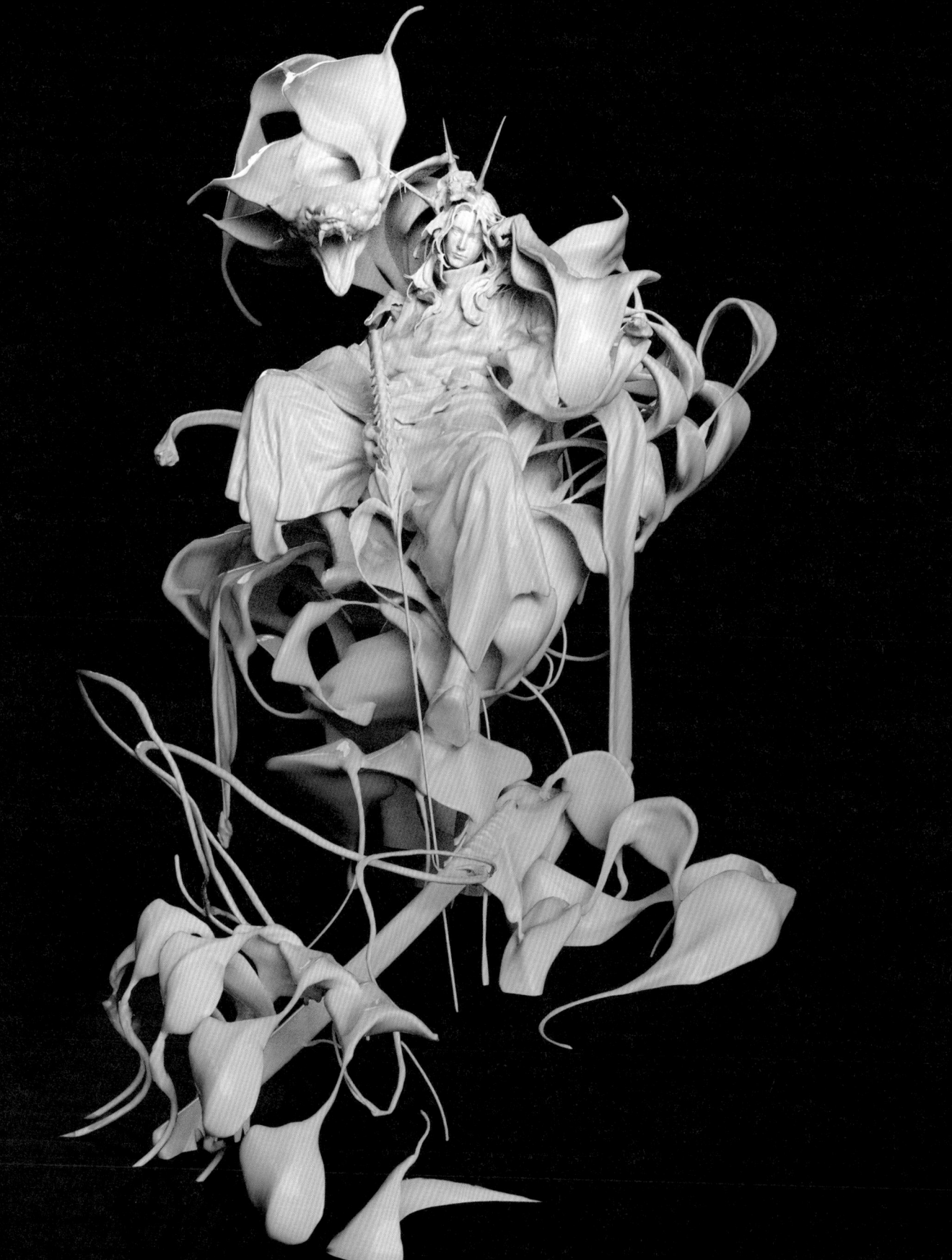

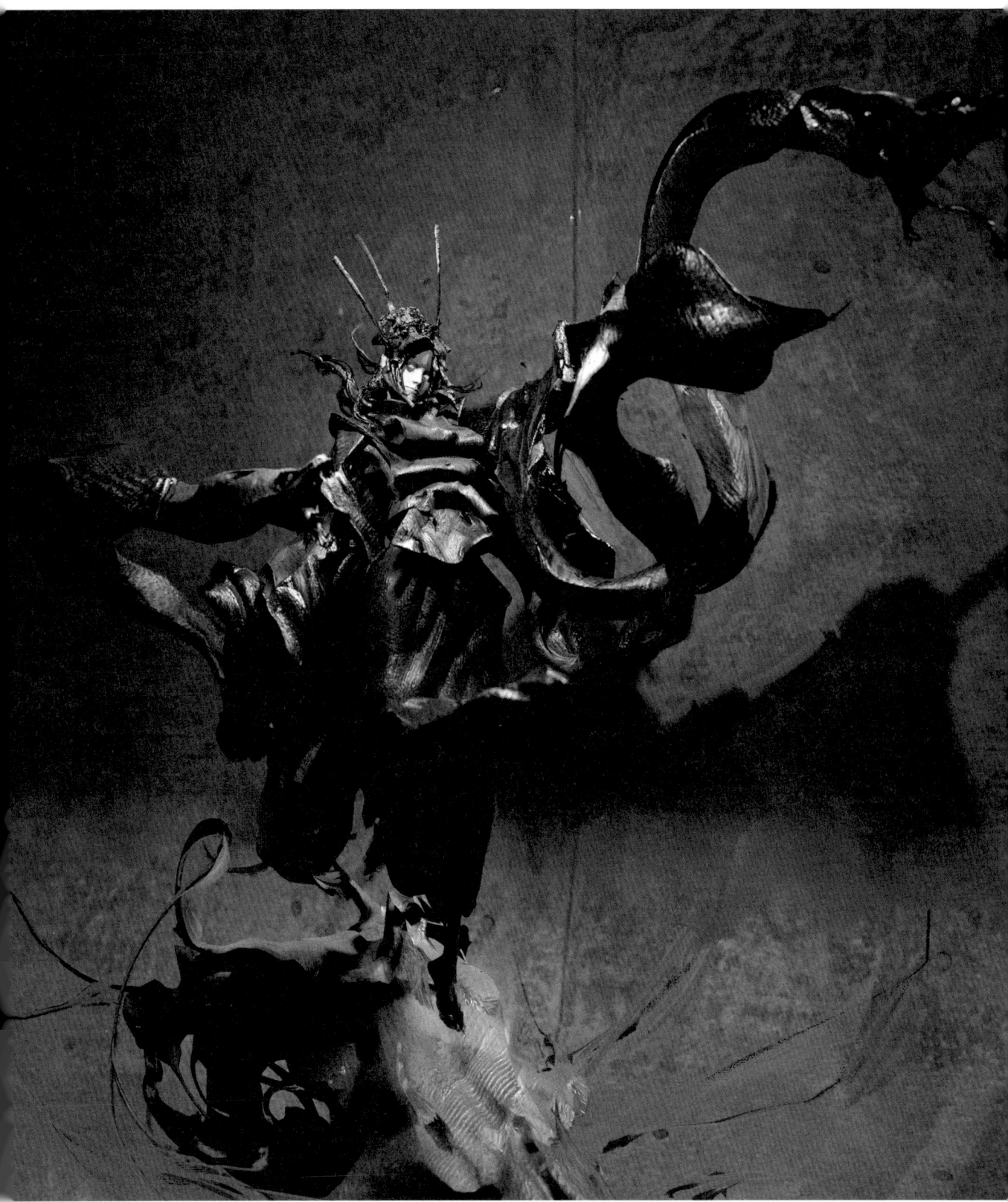

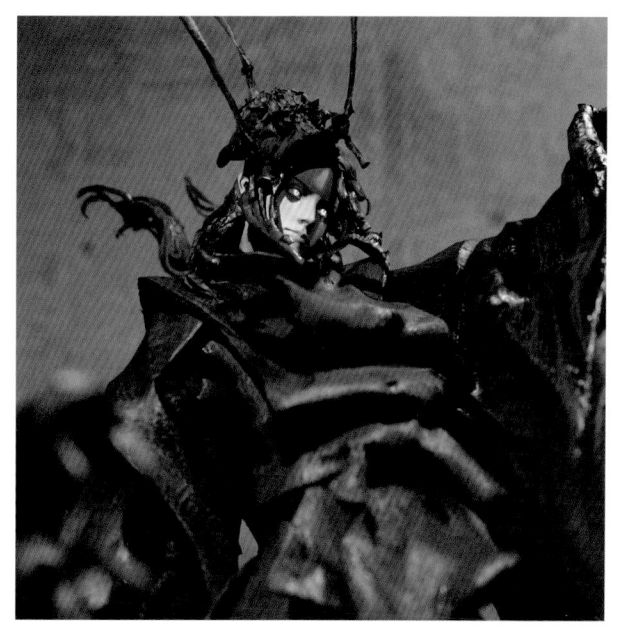
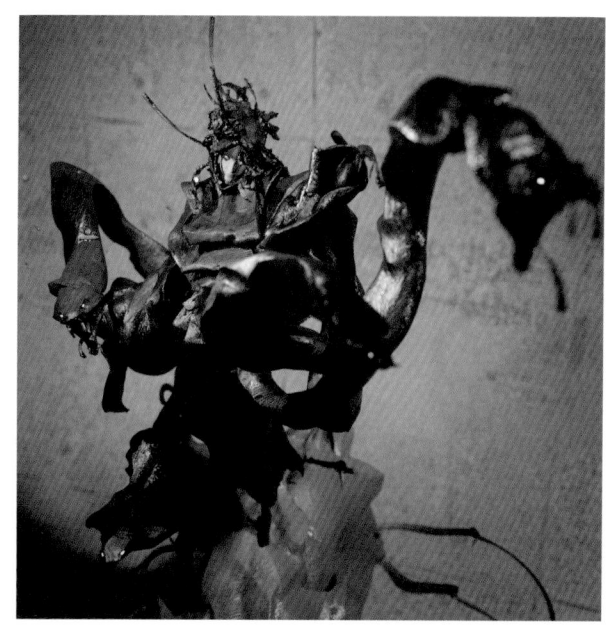
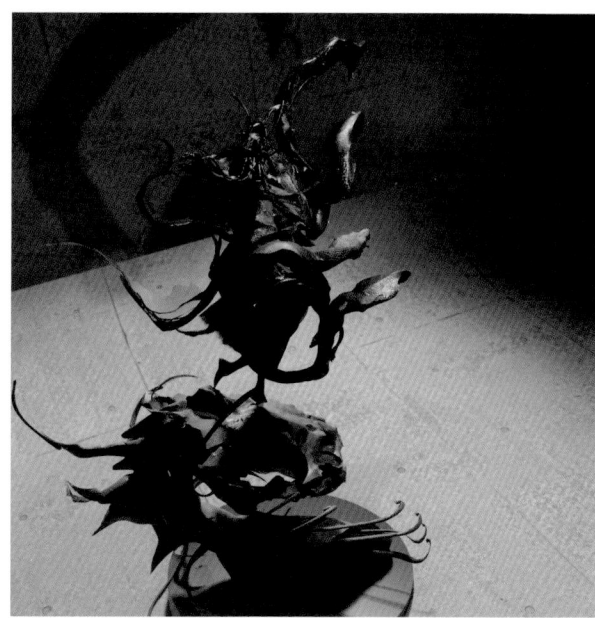
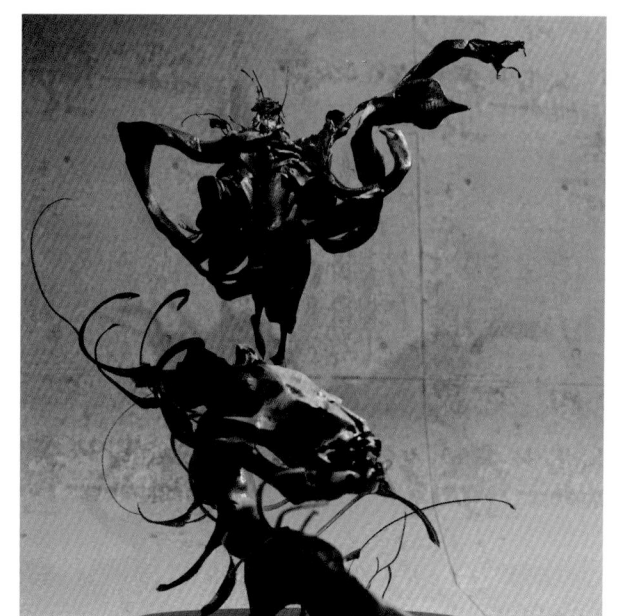

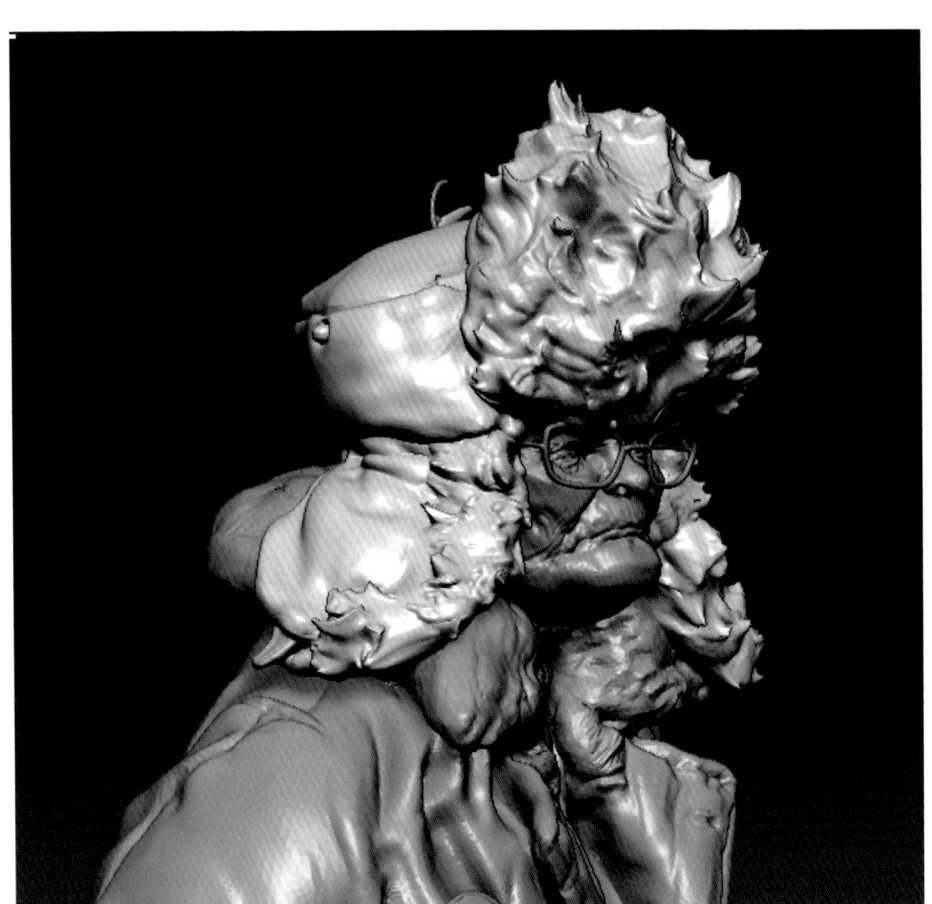

Northern Grandma 25

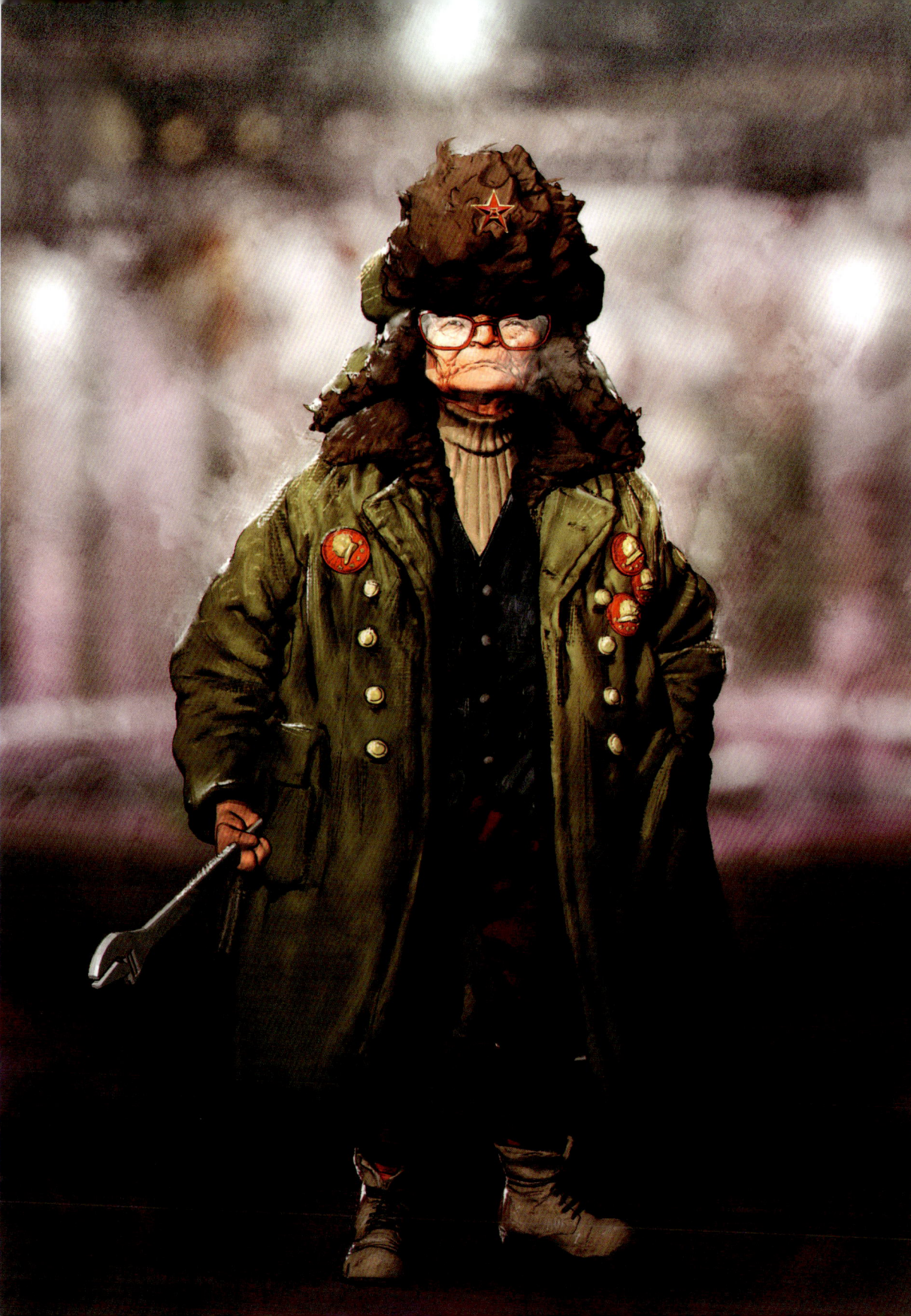

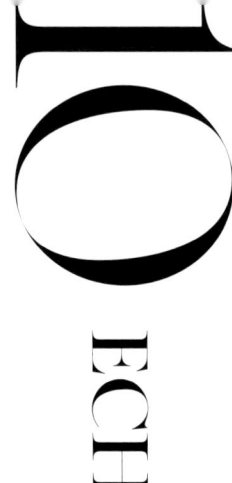

# 10

## ECHO

Aside from creating my own works, I also deeply enjoy making derivative art from artists and pieces I admire, offering fresh interpretations from my own perspective. For an artist with a distinct style, this feels more like a dialogue than mere imitation. It's akin to stepping onto the stages of different worlds, embodying various roles, and allowing one's soul to echo through the emotional spectrum of these characters.

自分自身の作品の制作に留まらず、尊敬するアーティストや作品の二次創作にも取り組んで、私の視点から新鮮な解釈を提供することも、大きな楽しみです。独特のスタイルを持つアーティストにとって、これは単なる模倣というより対話のように感じられます。それは異なる世界の舞台に上がり、多様な役割を体現し、そうしたキャラクターの感情の幅を通して自らの魂を反響させることに似ています。

Tribute Sculpting 26(

From Original illustration by Kim Jung Gi

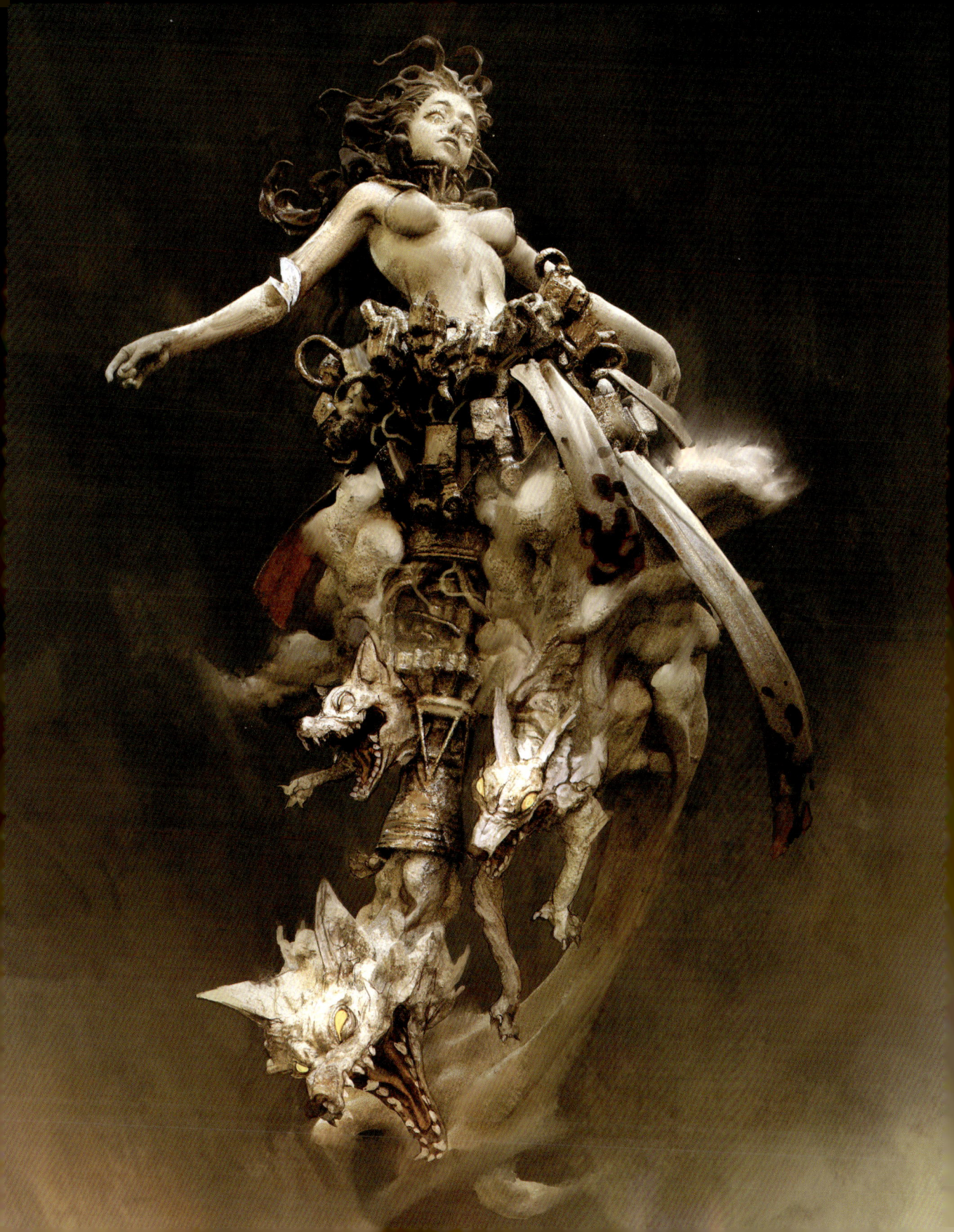

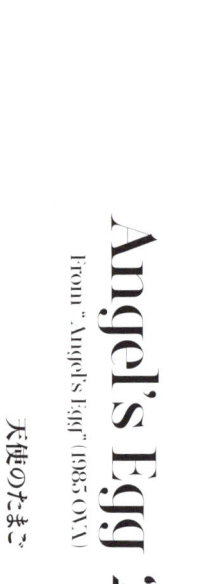

# Angel's Egg 26:

From " Angel's Egg" (1985 OVA)

天使のたまご

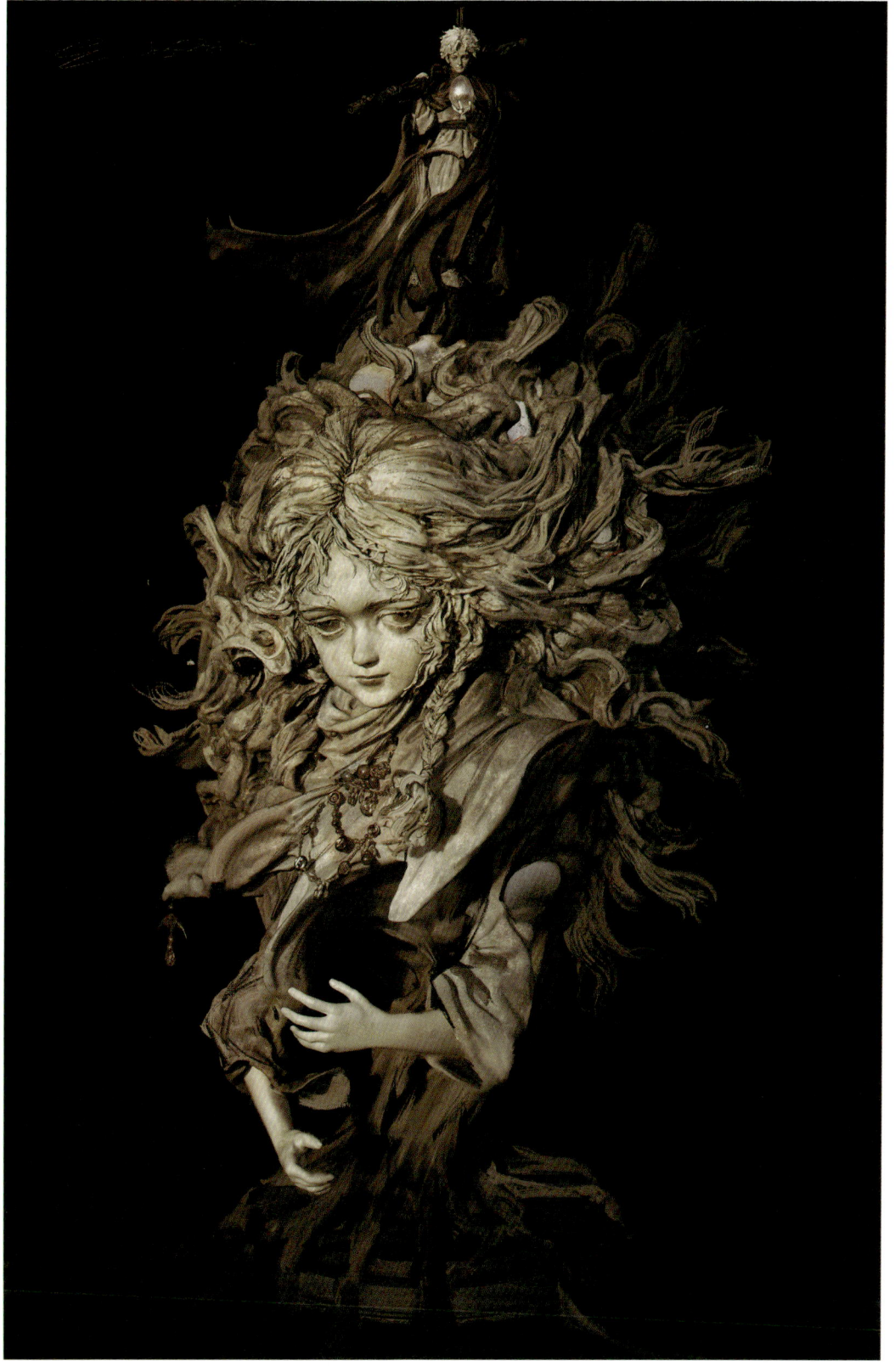

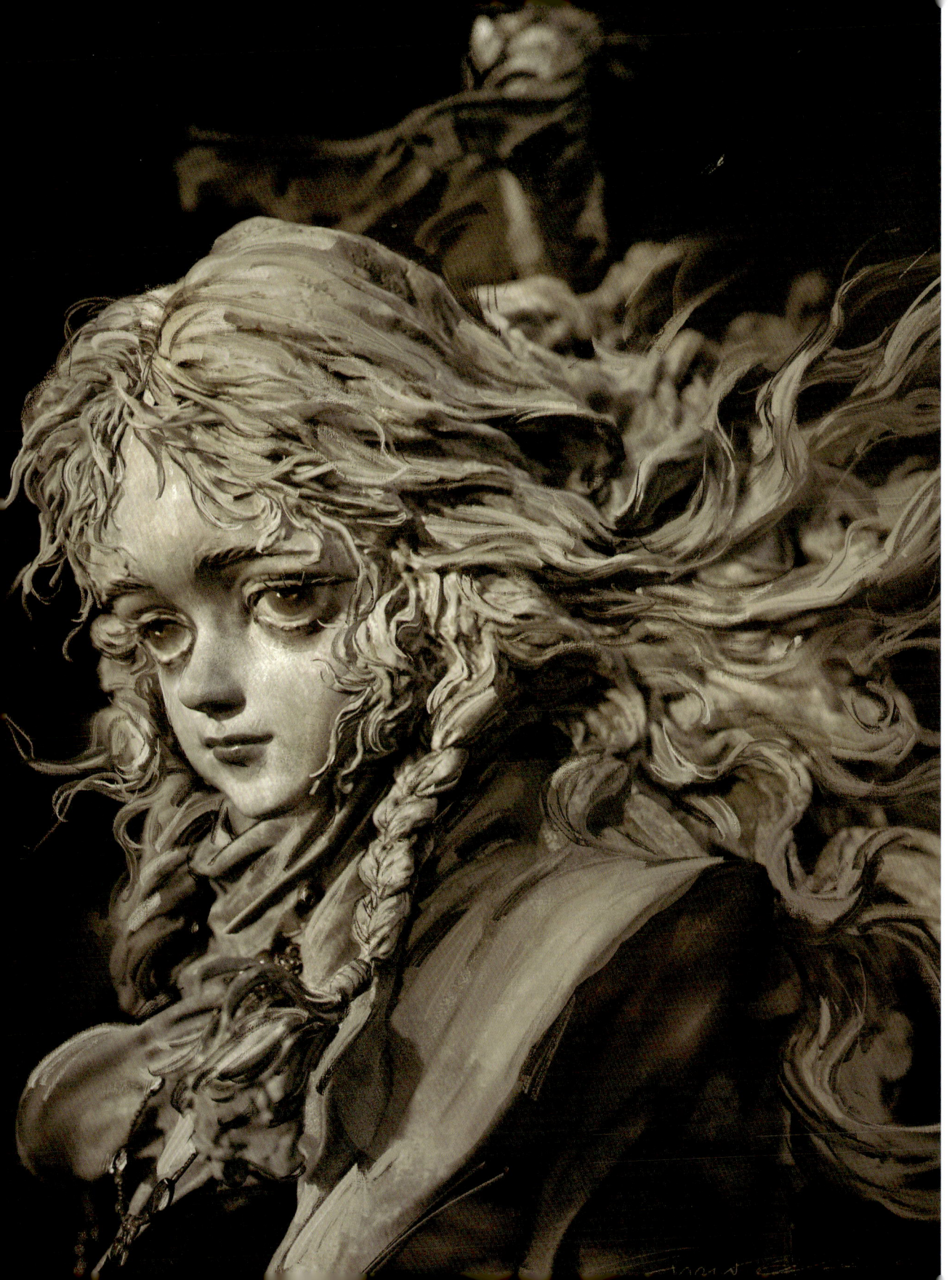

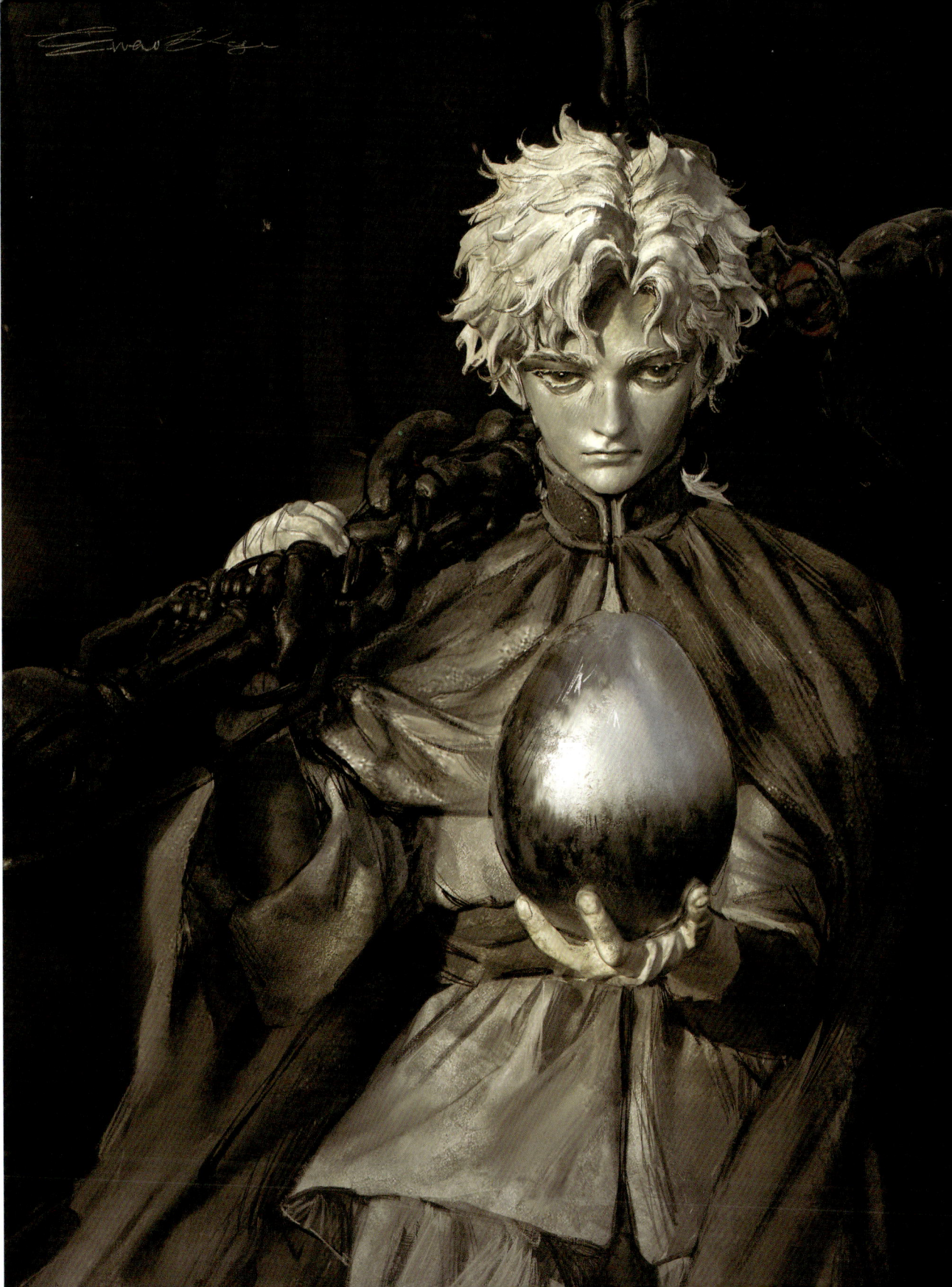

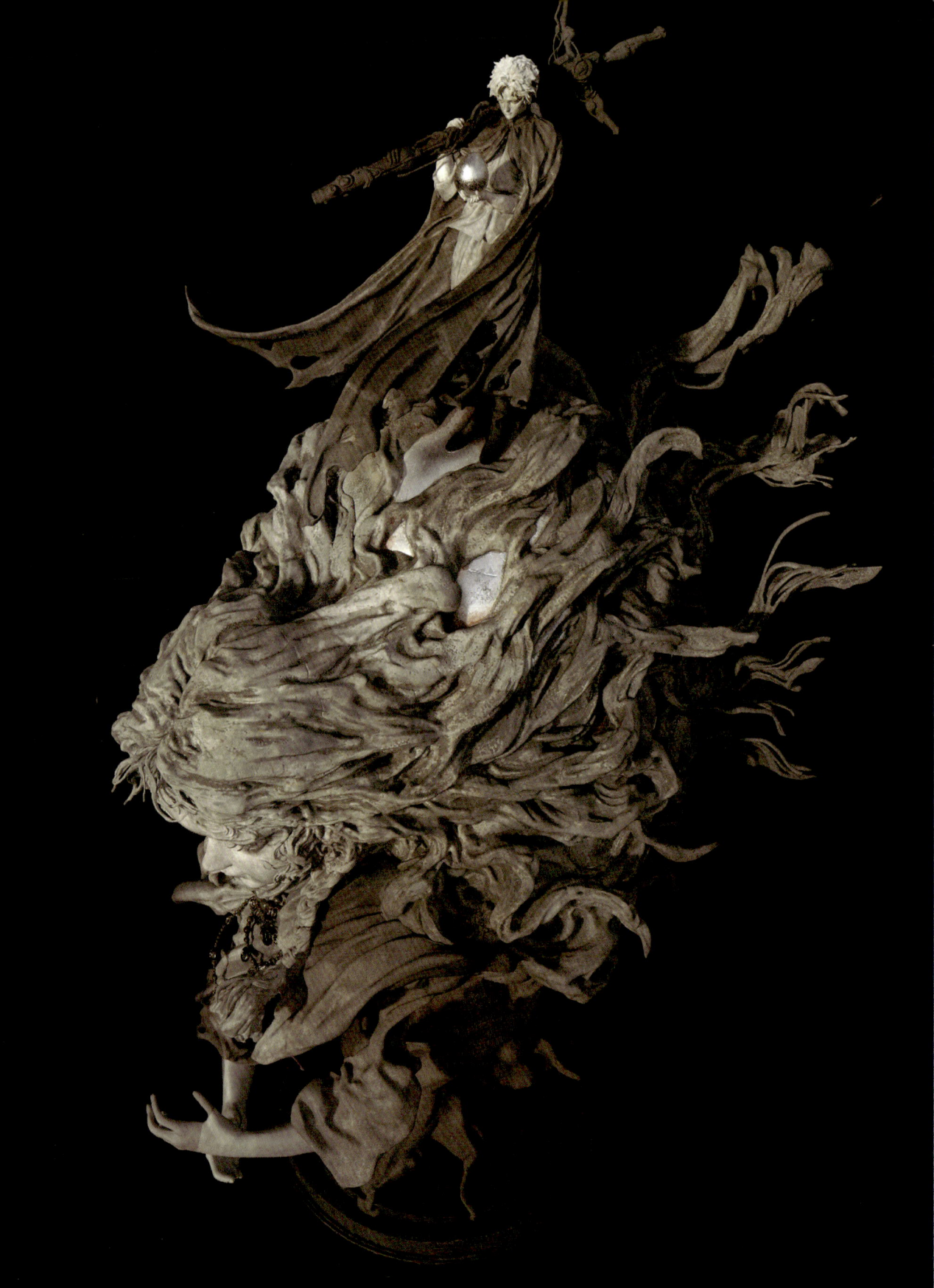

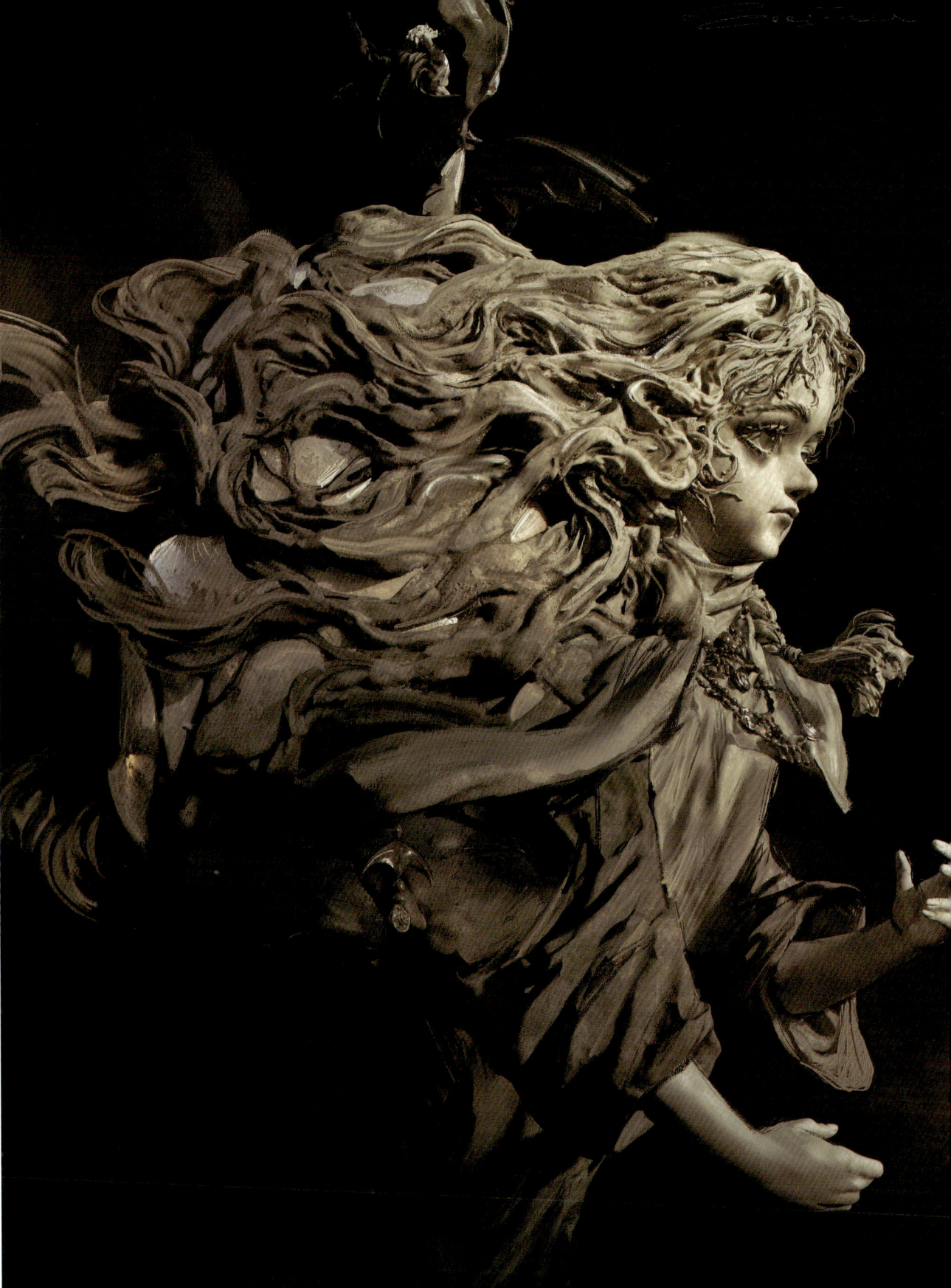

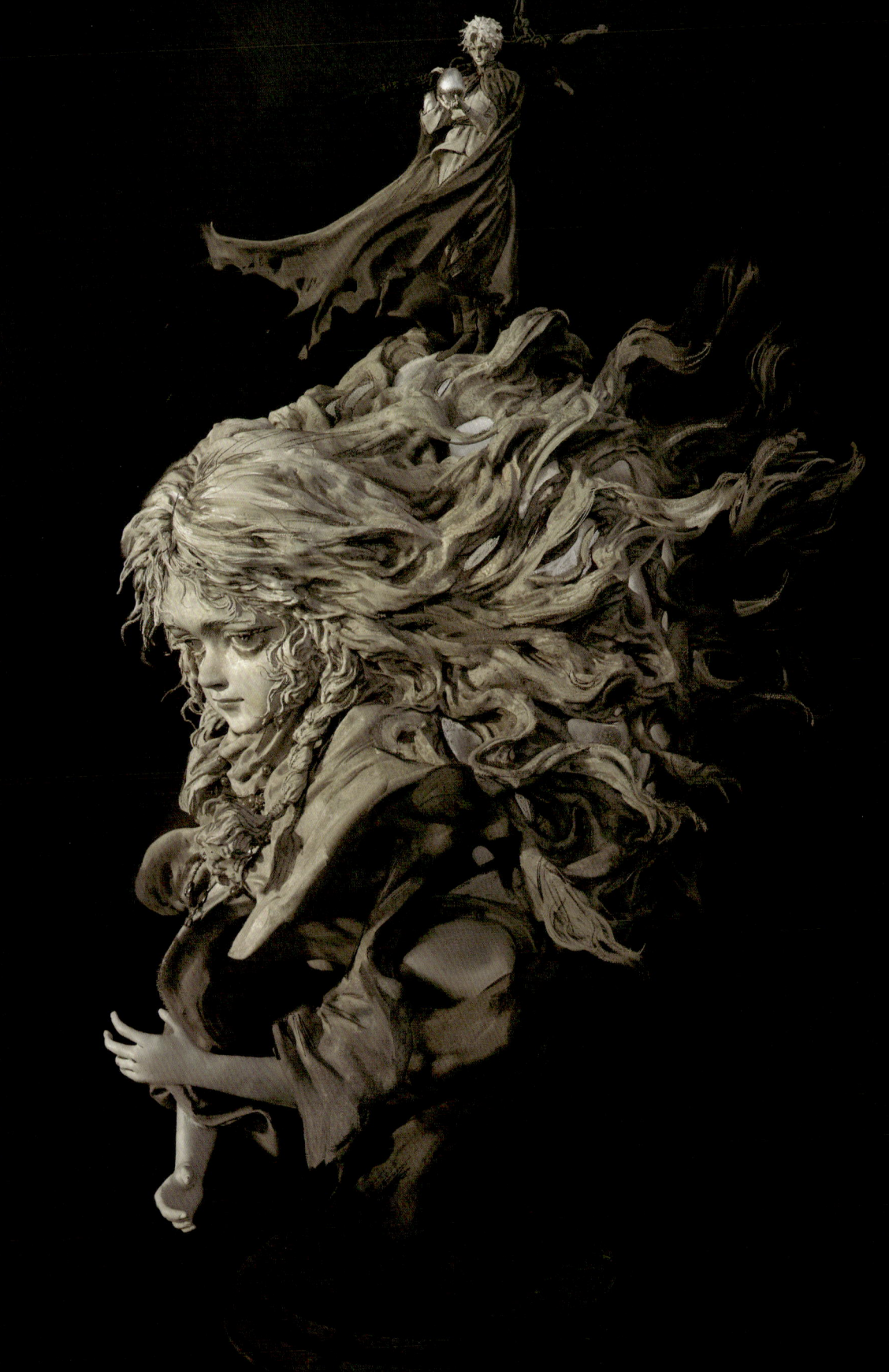

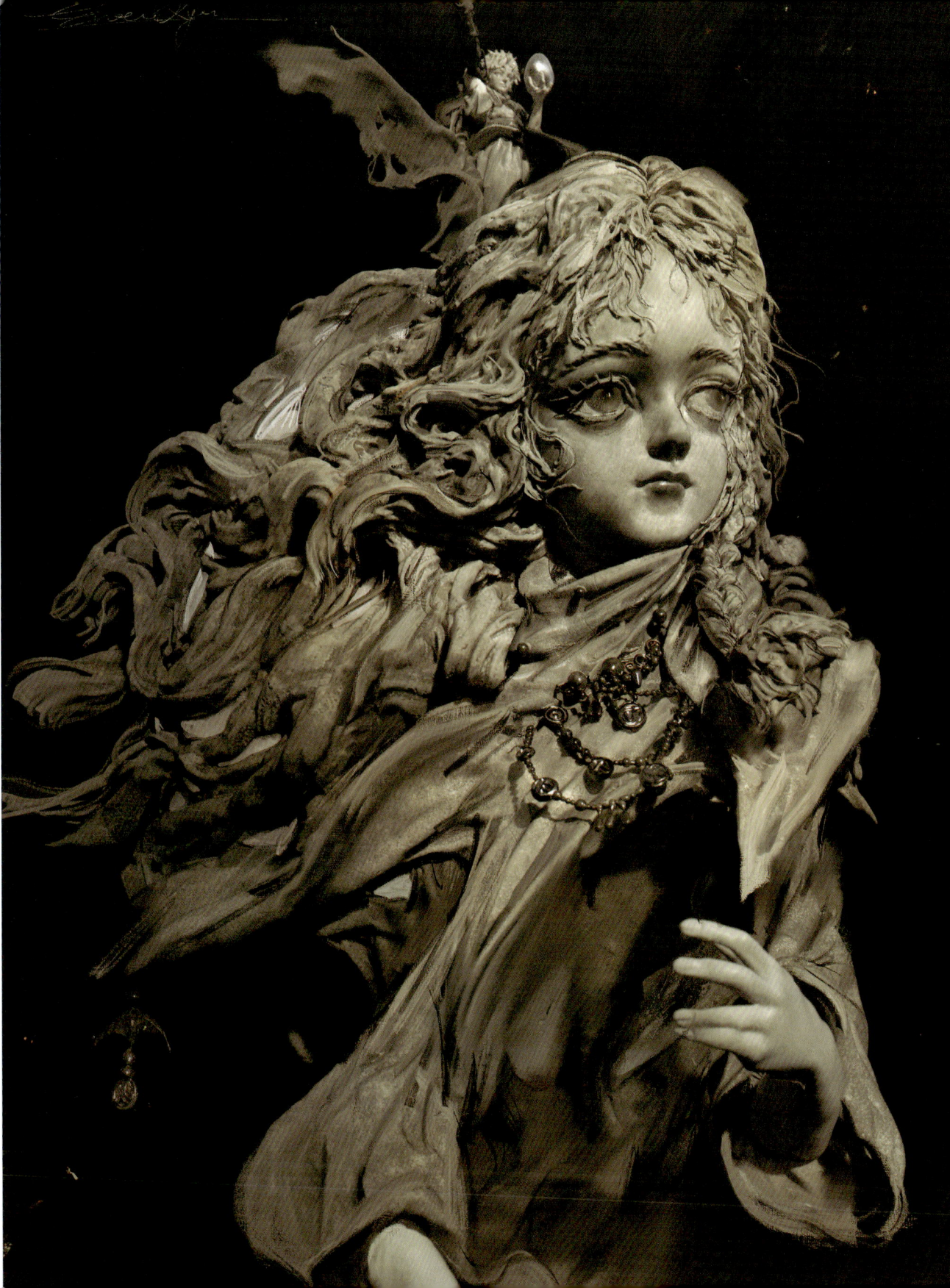

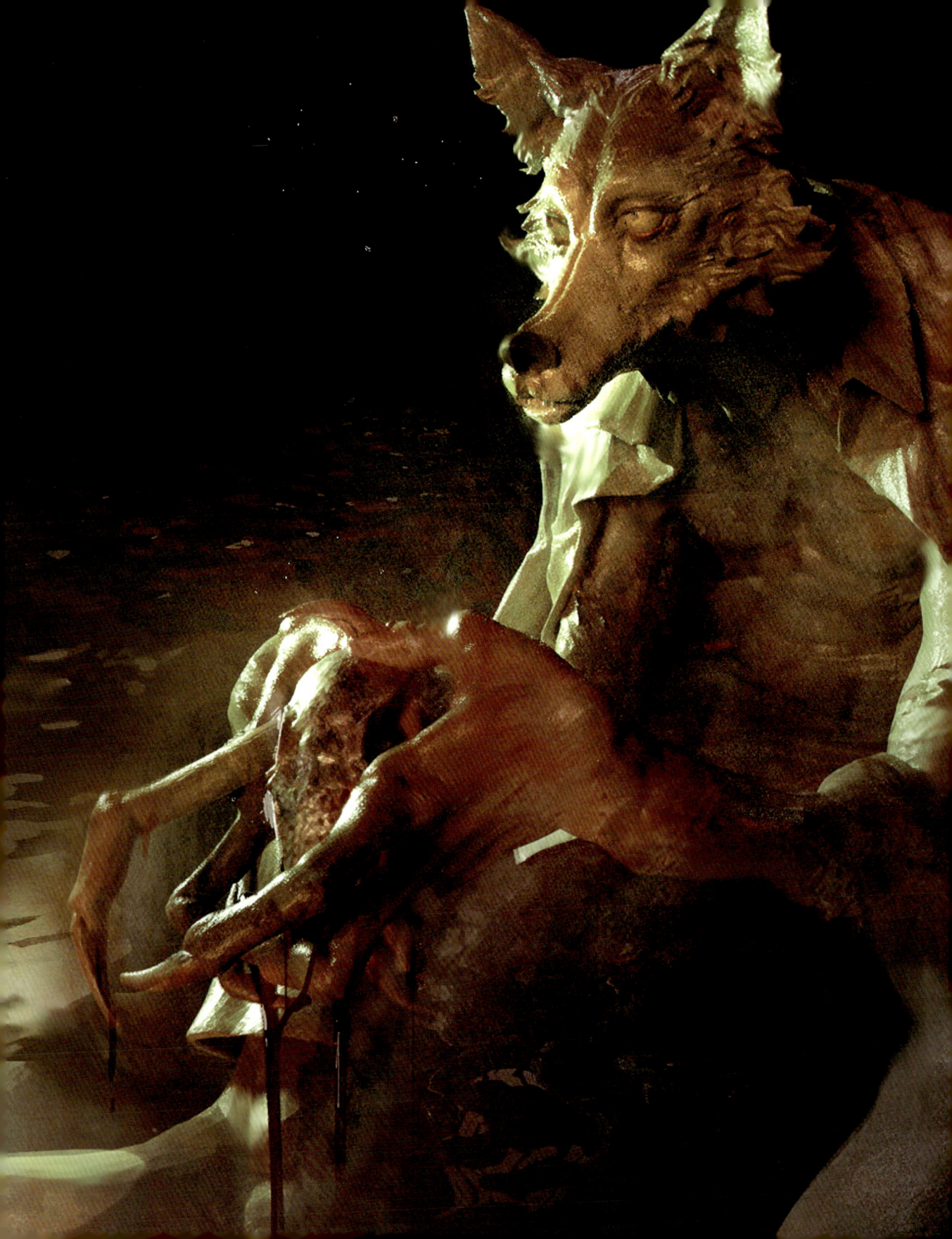

# Legoshi & Rouis

桜庭正樹『BEASTARS』(秋田書店「少年チャンピオン・コミックス」刊)より

Based on the manga "BEASTARS" by Paru Itagaki

originally serialized in the Weekly Shonen Champion published by AKITASHOTEN.

Legoshi & Rouis 271

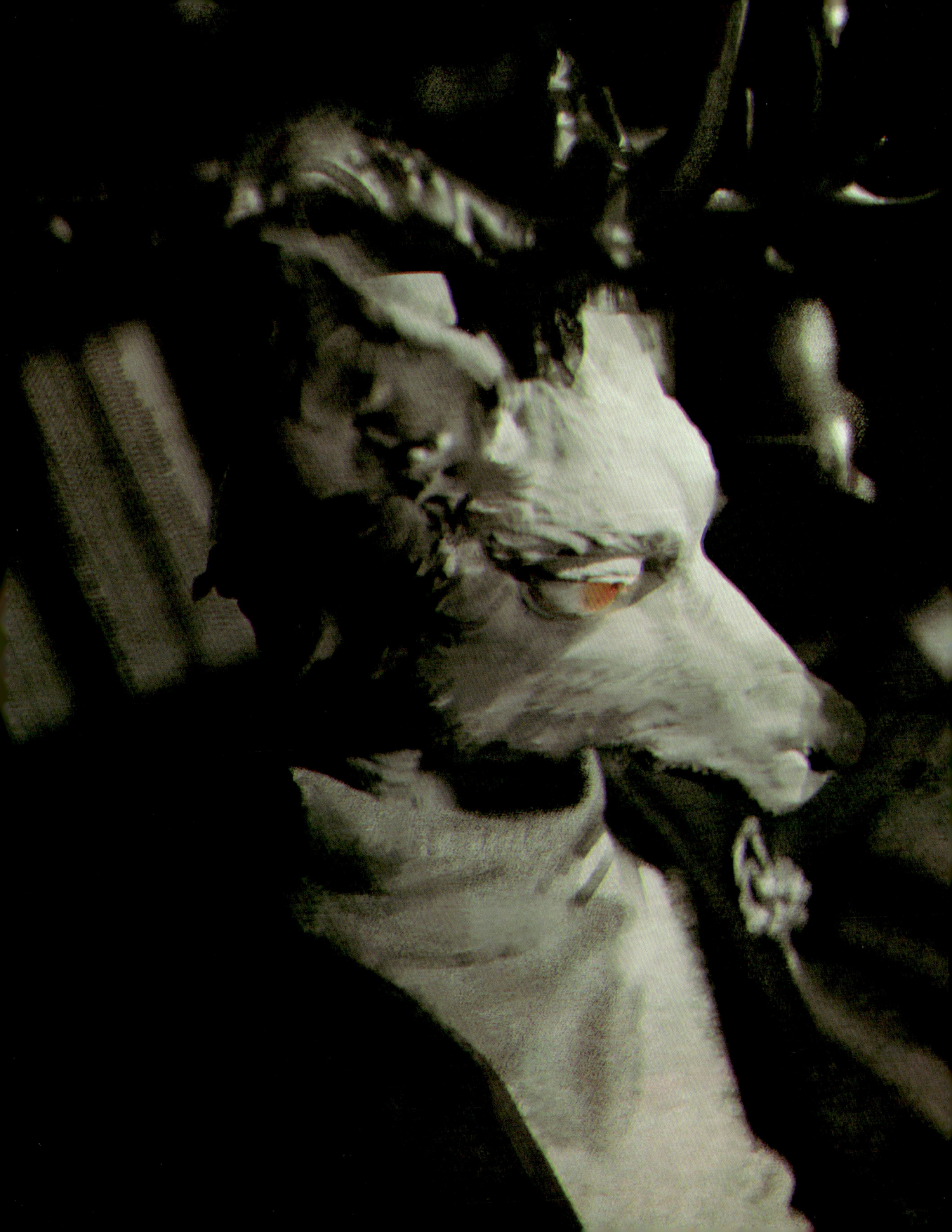

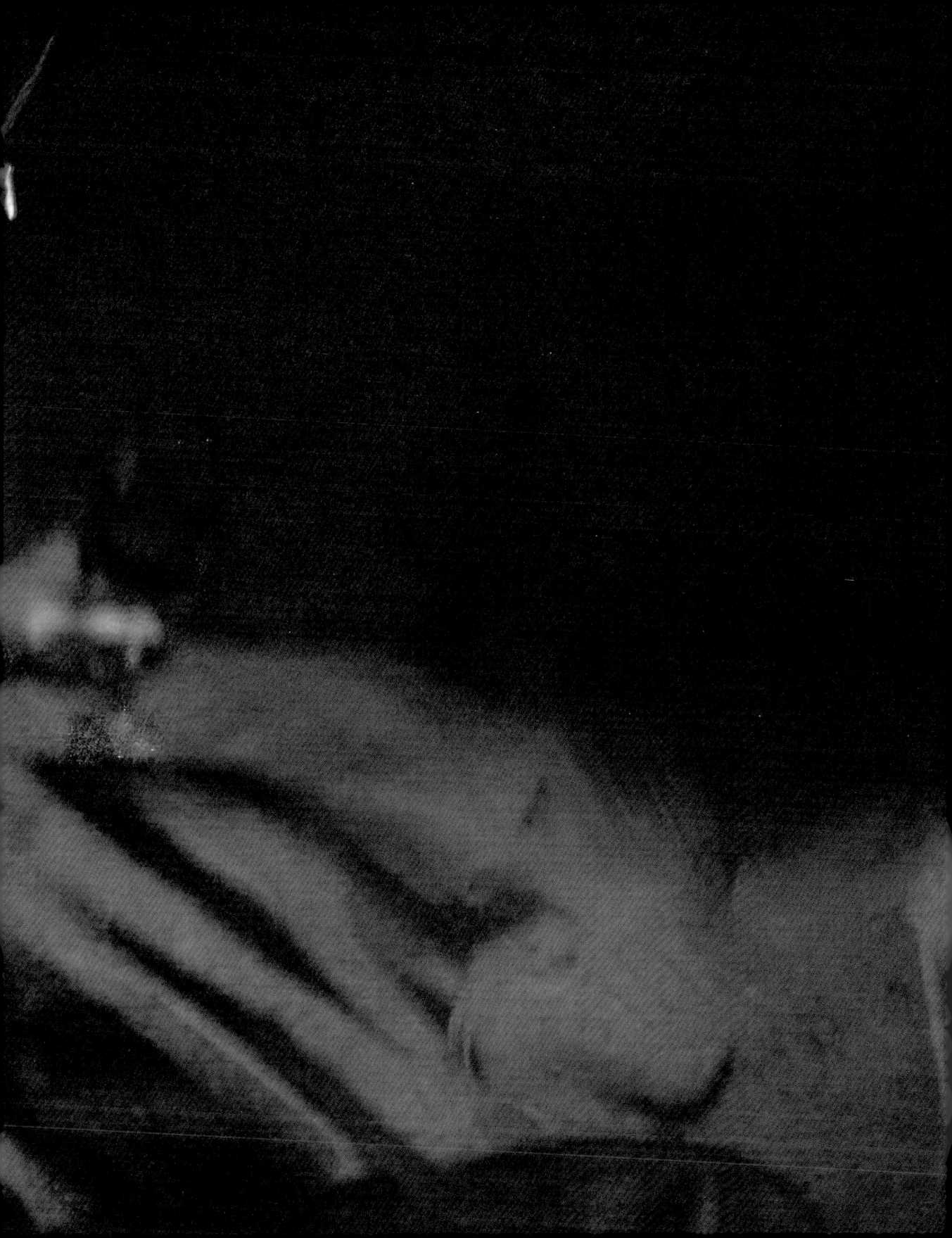

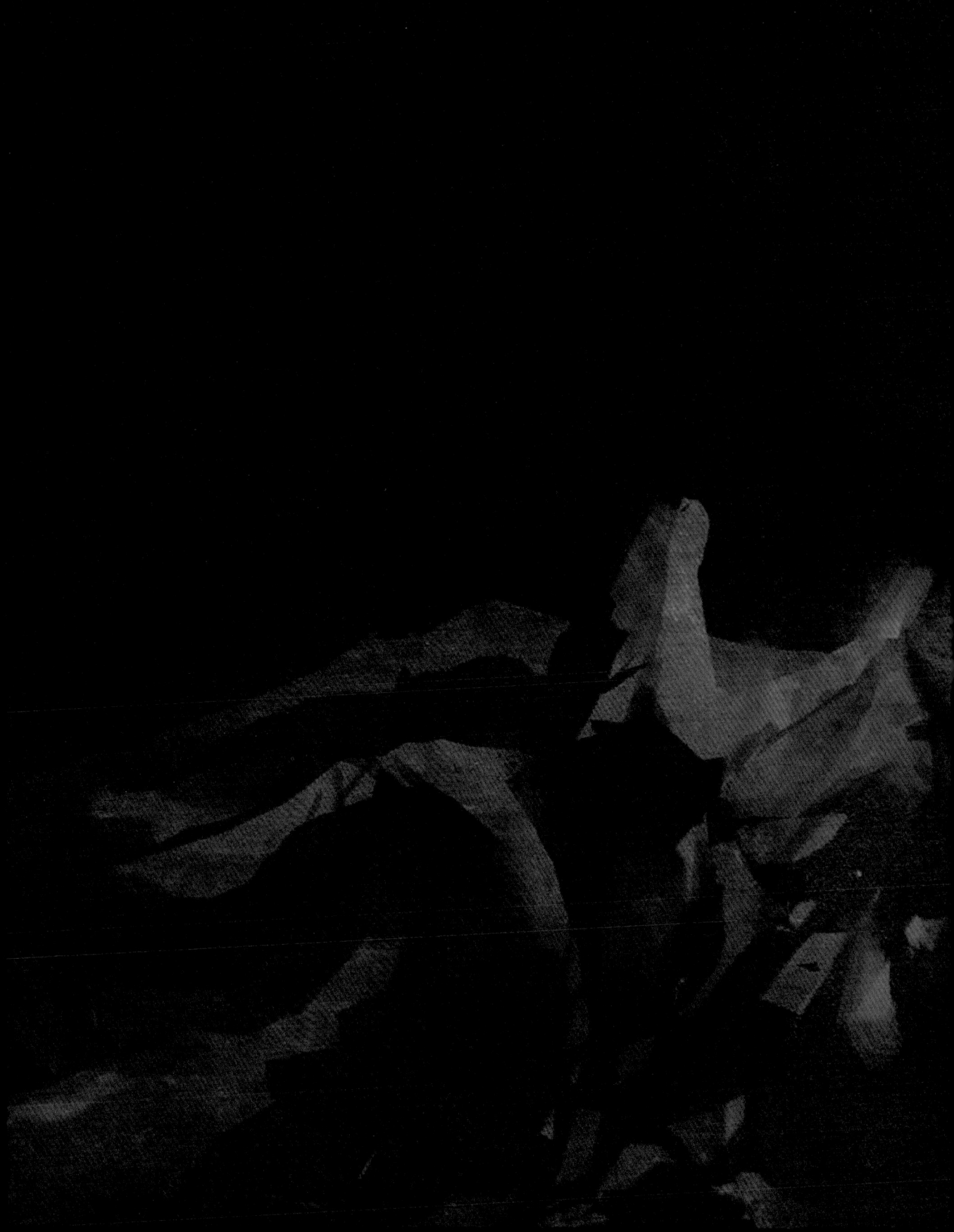

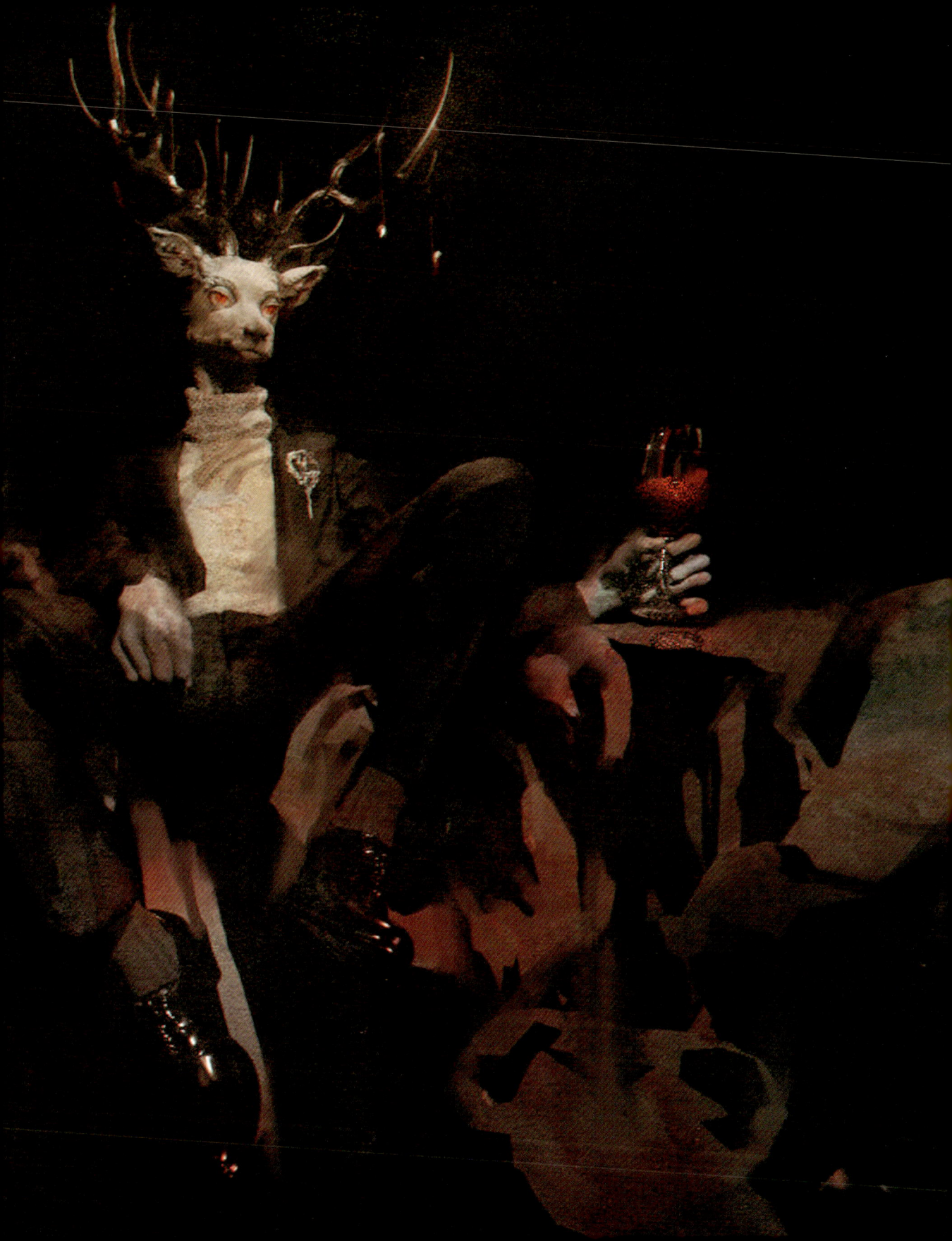

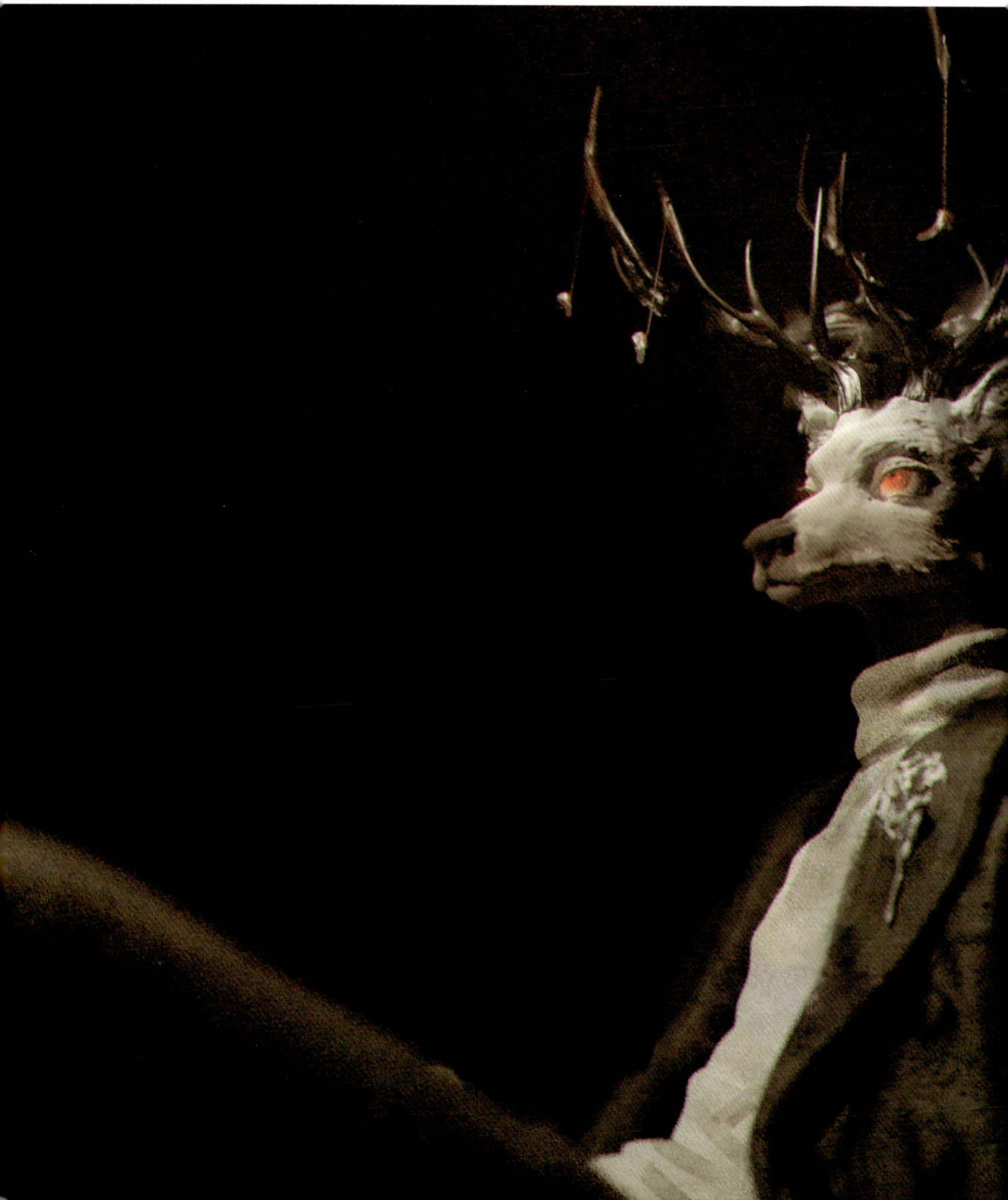

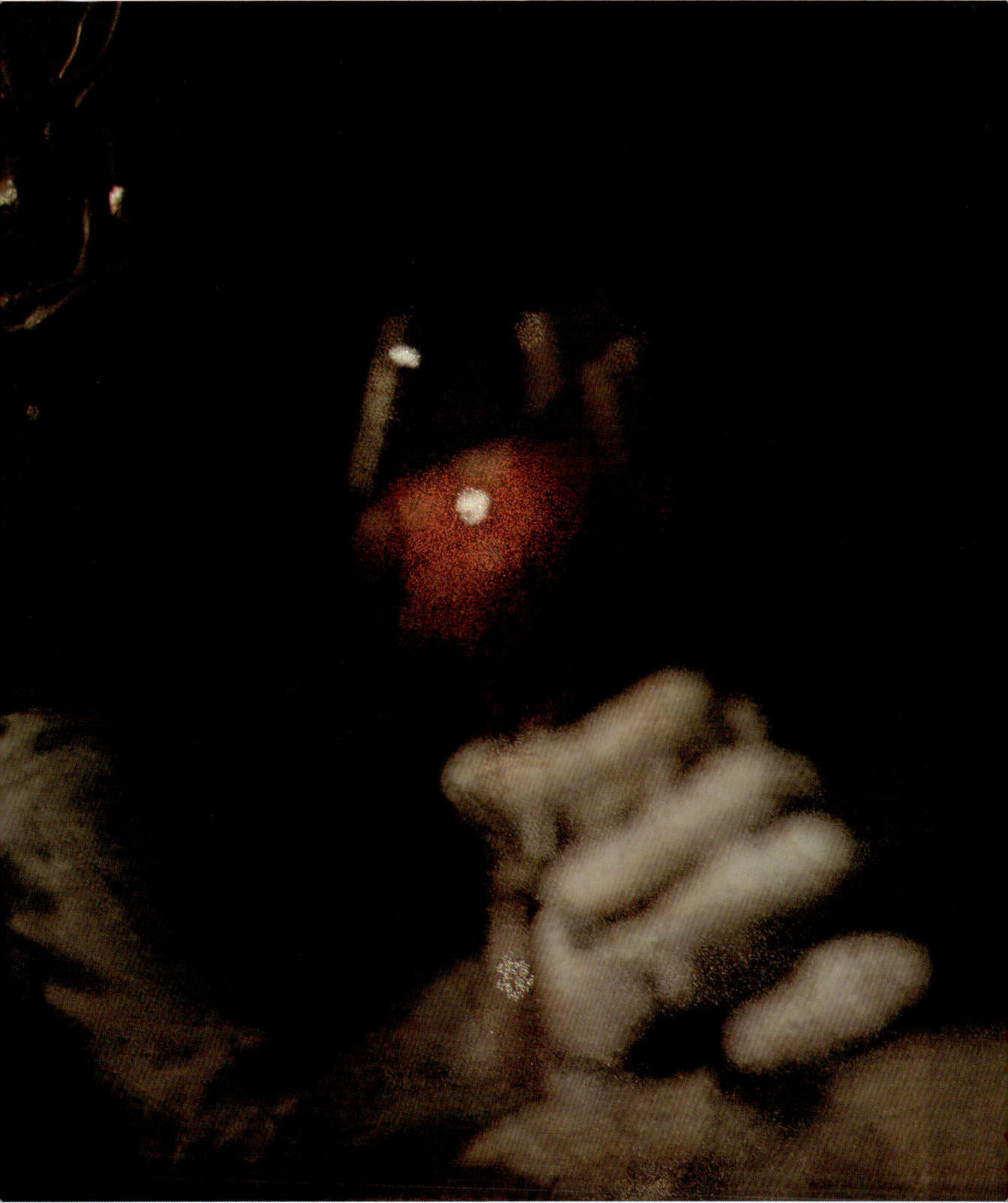

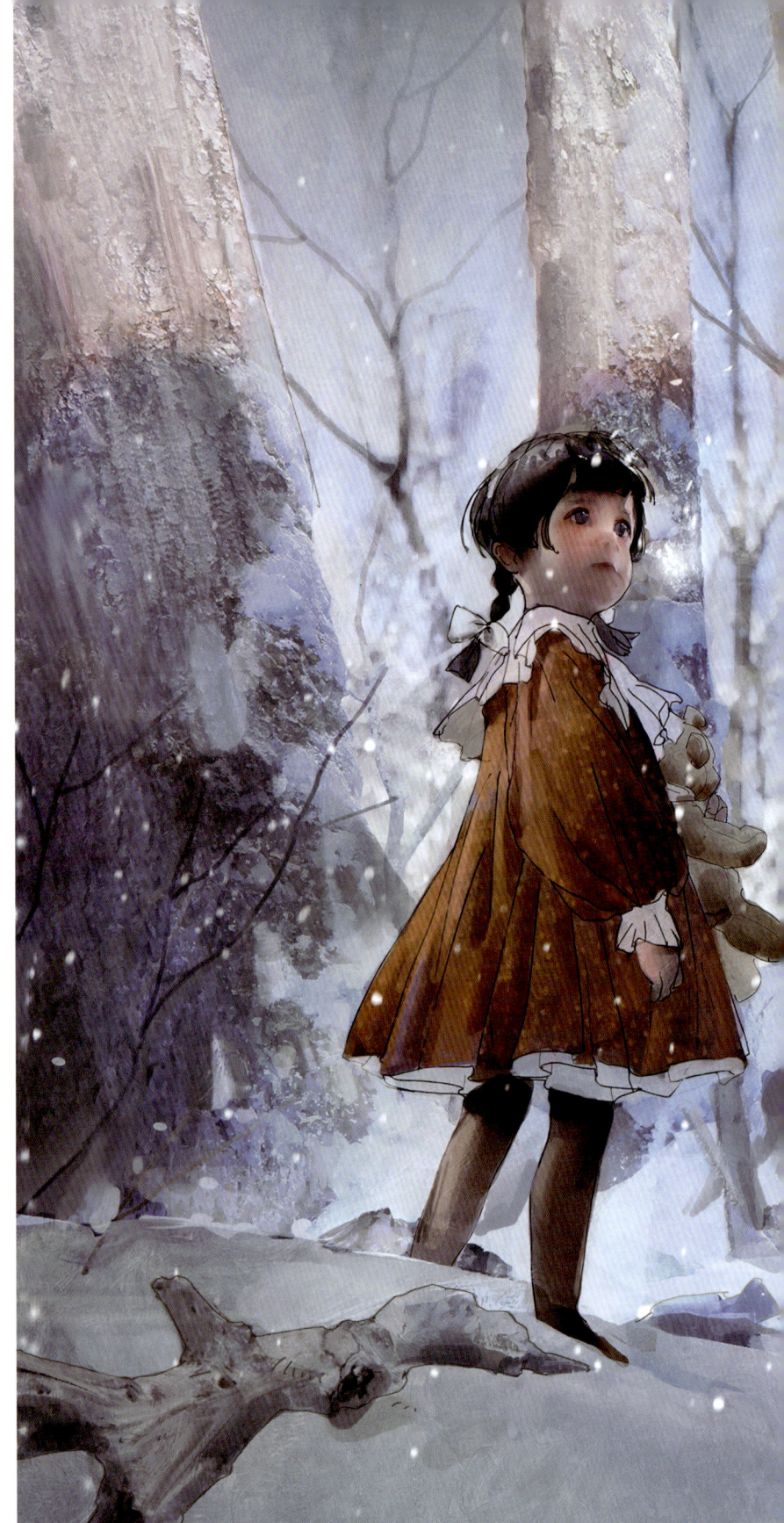

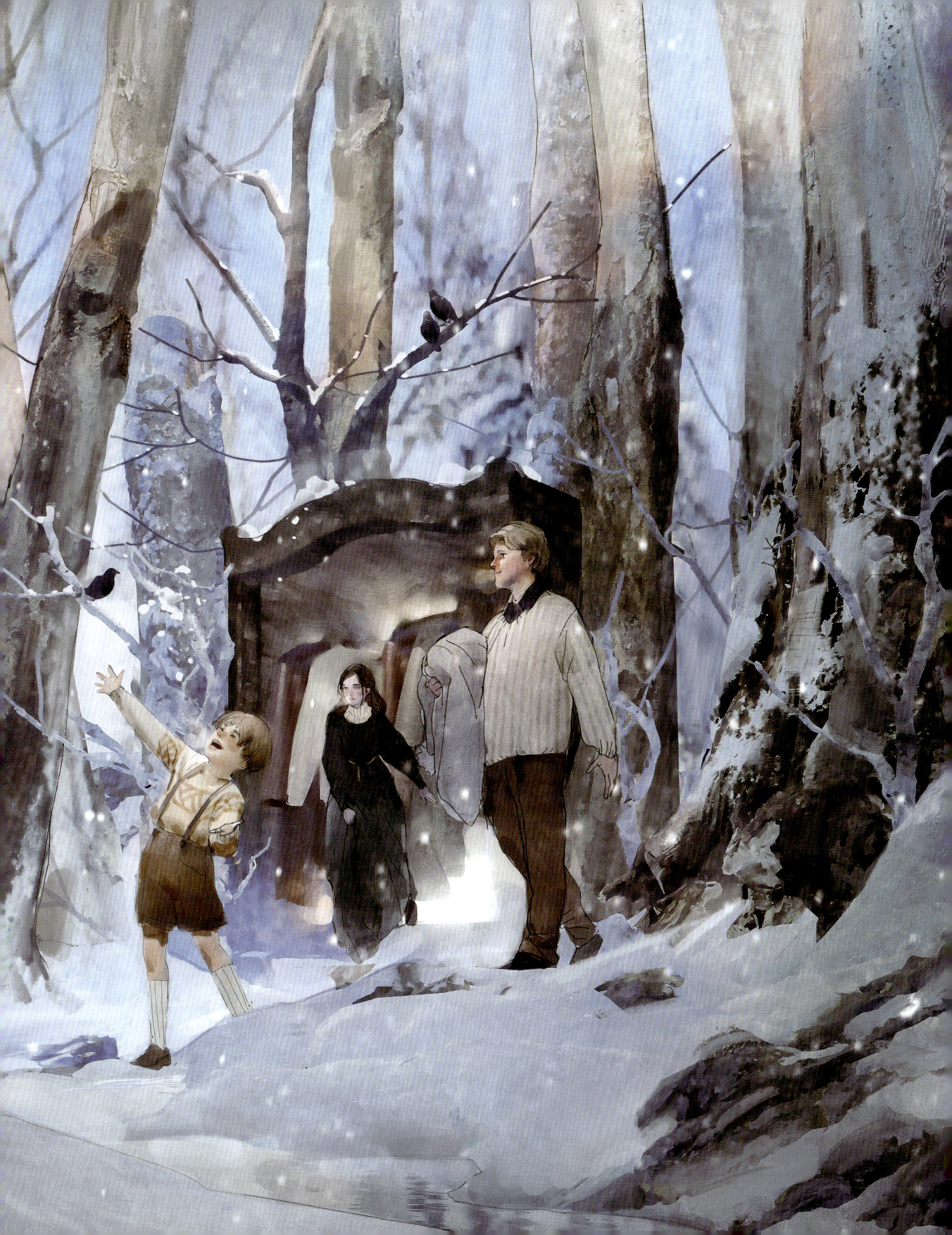

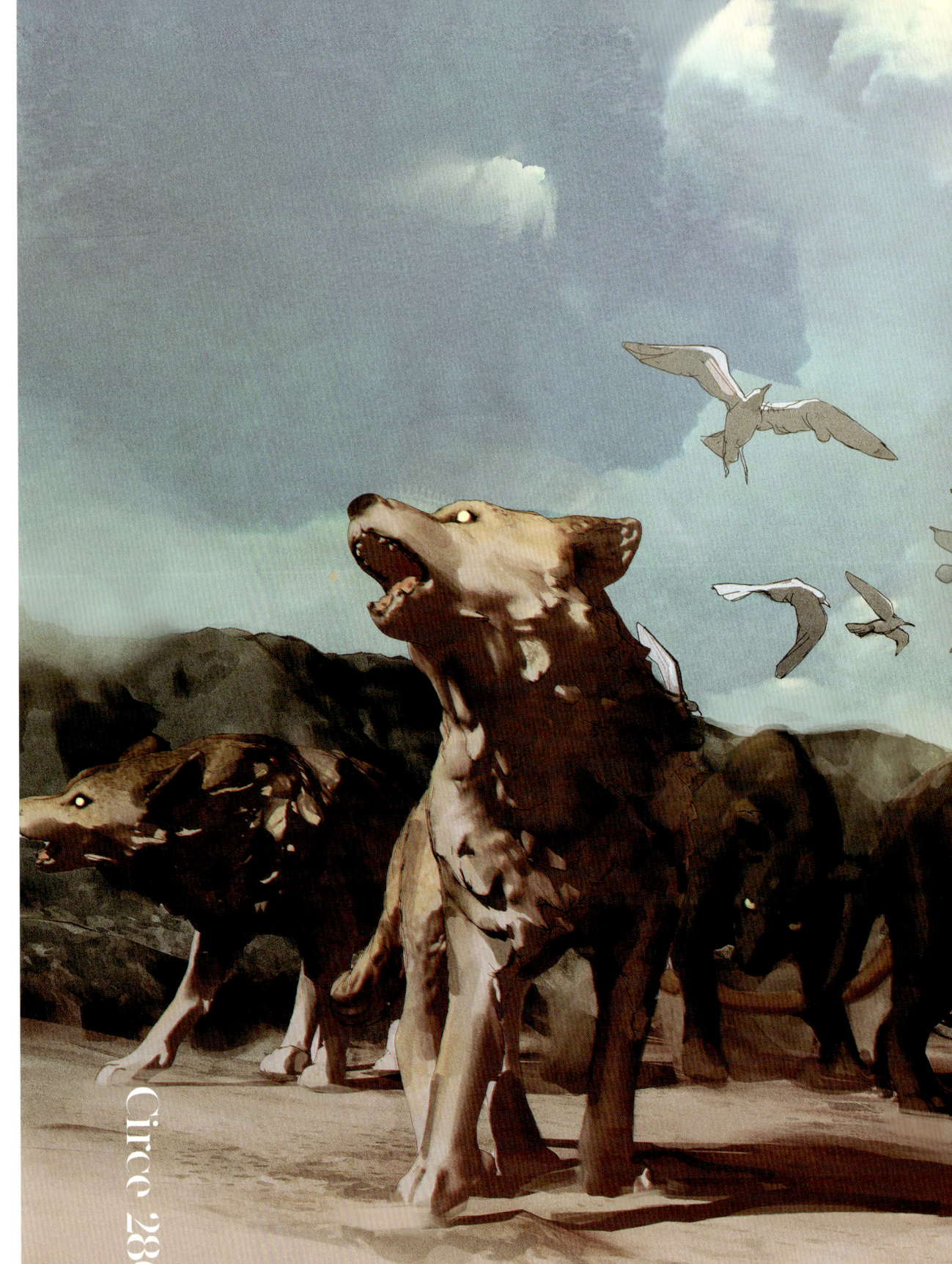

Circe 280

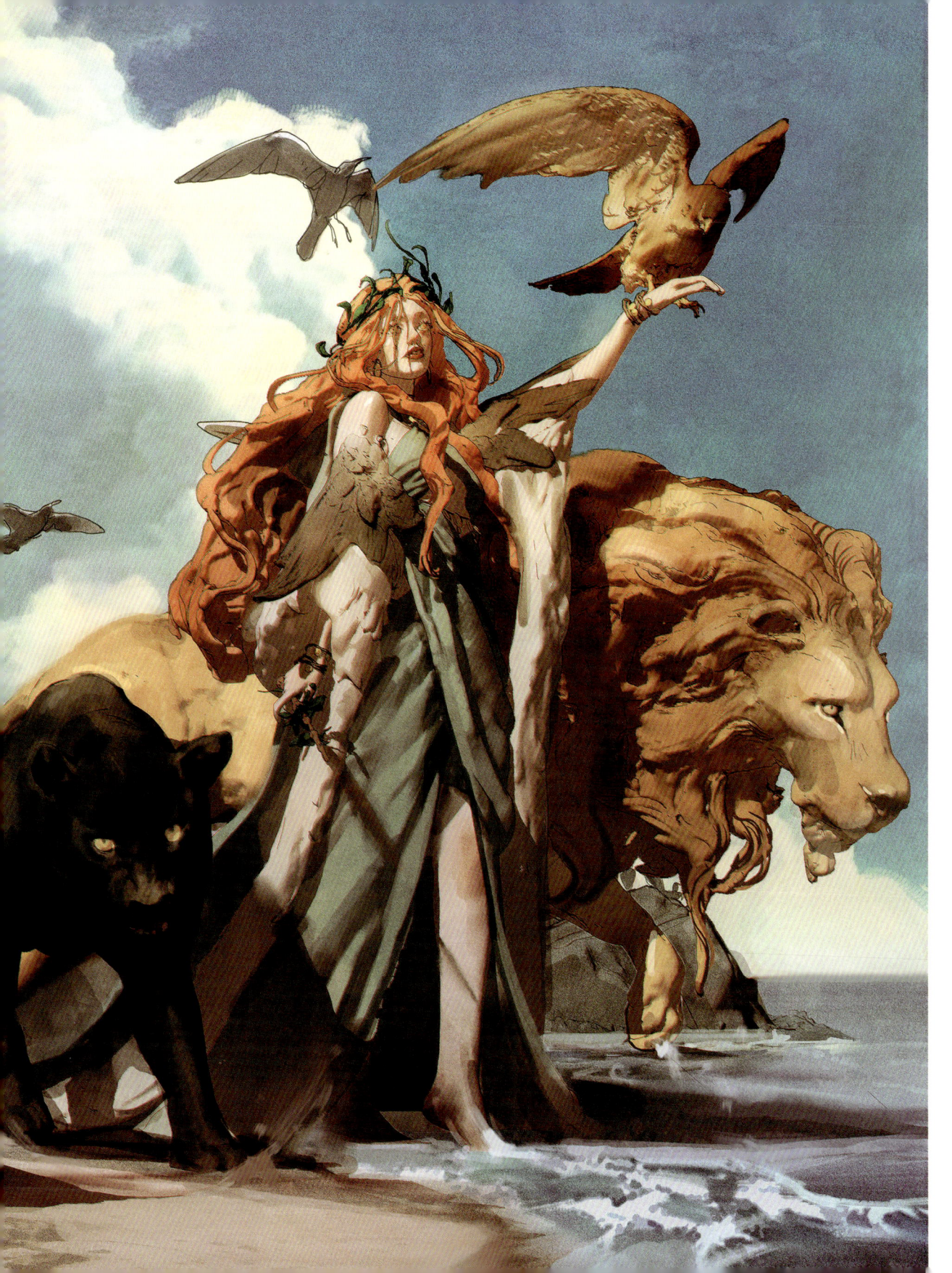

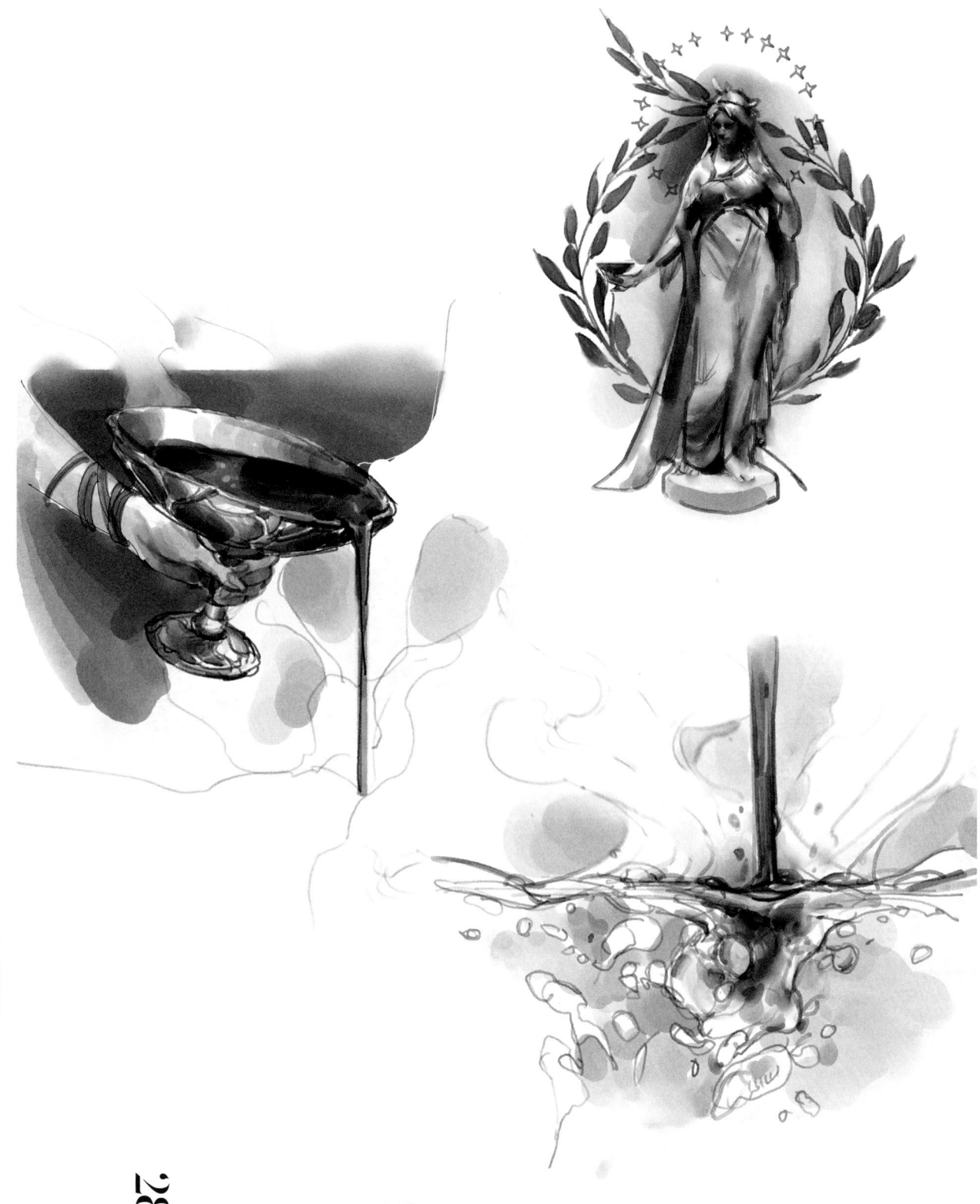

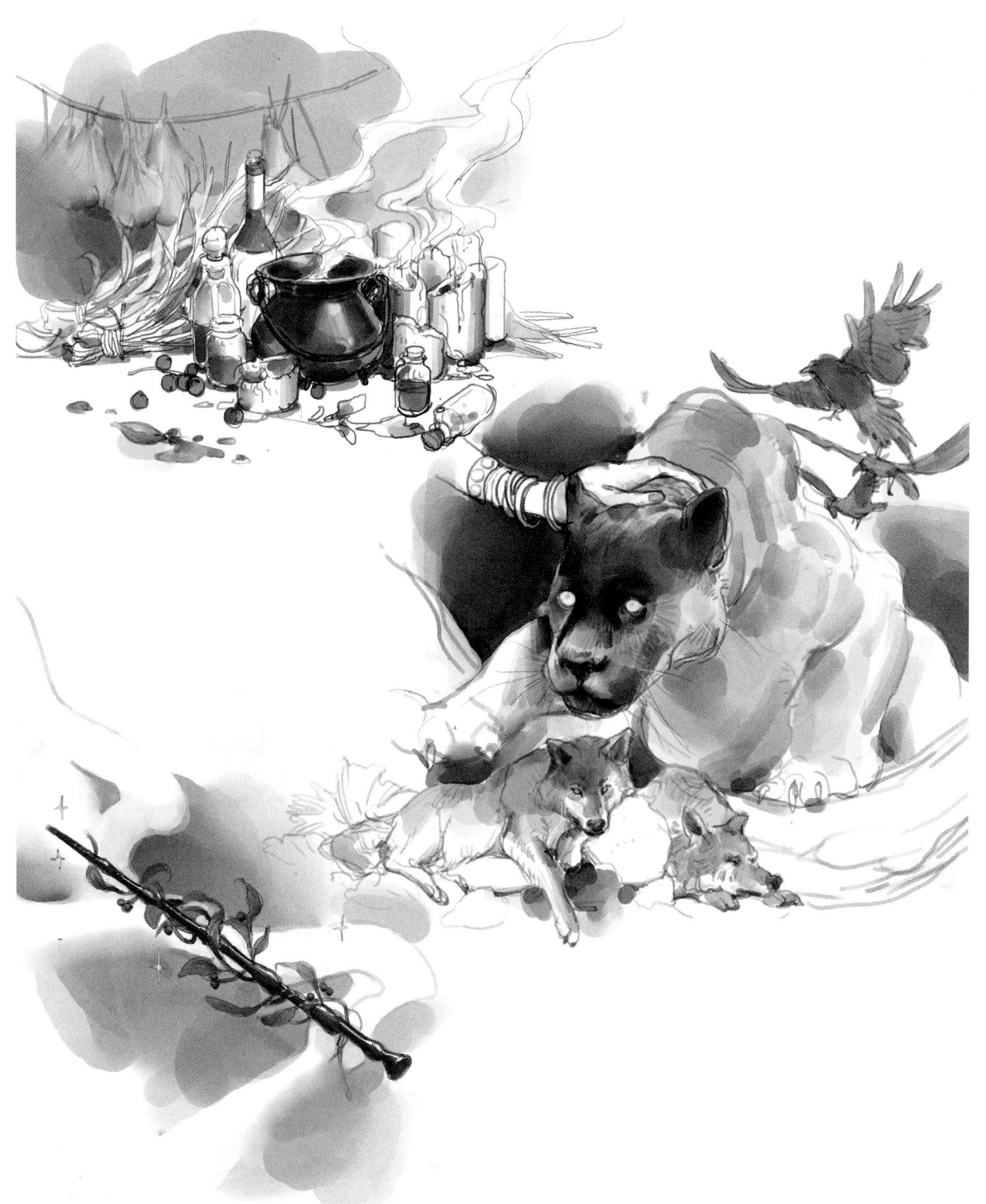

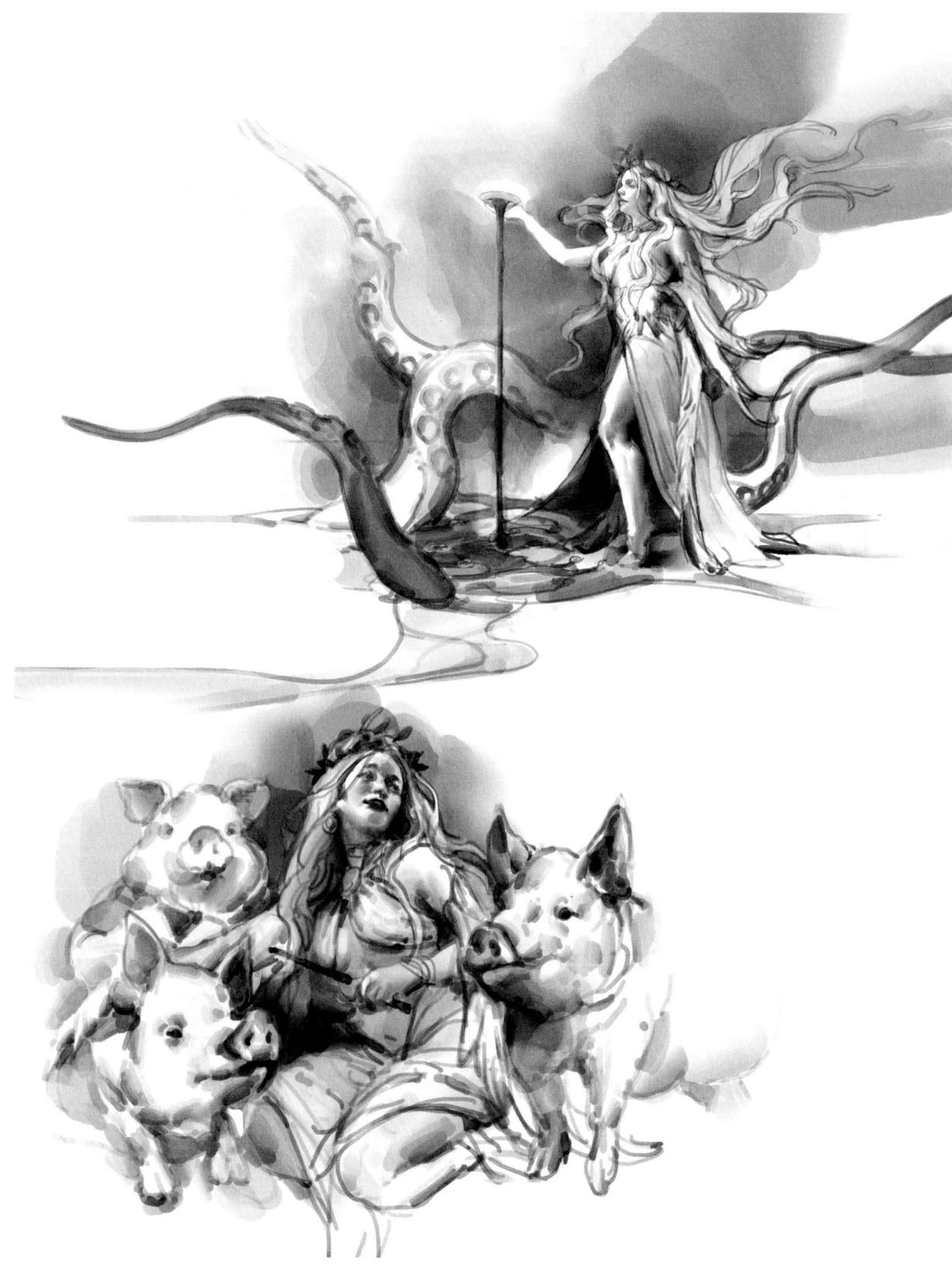

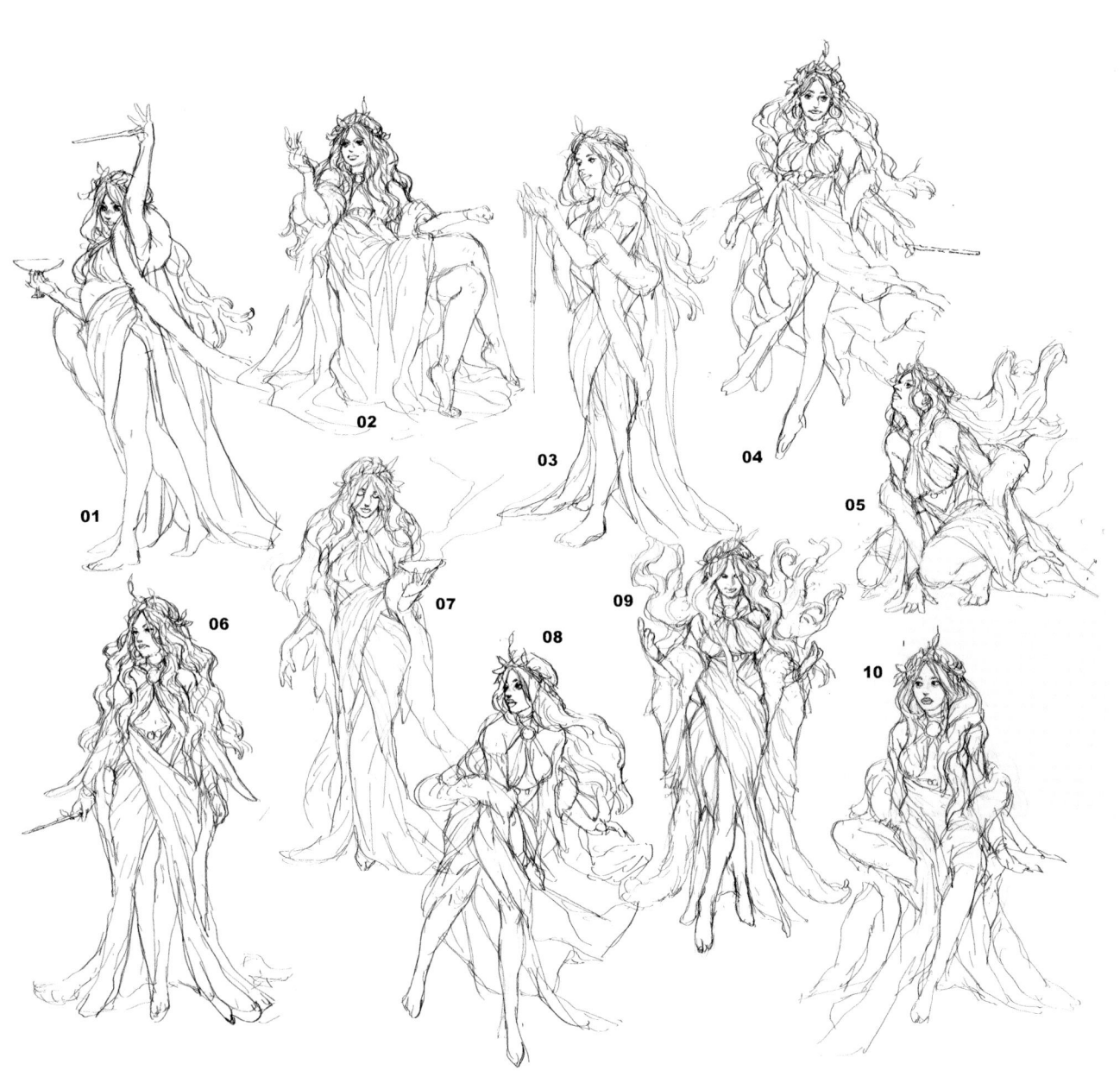

01

02

03

04

05

06

07

08

09

10

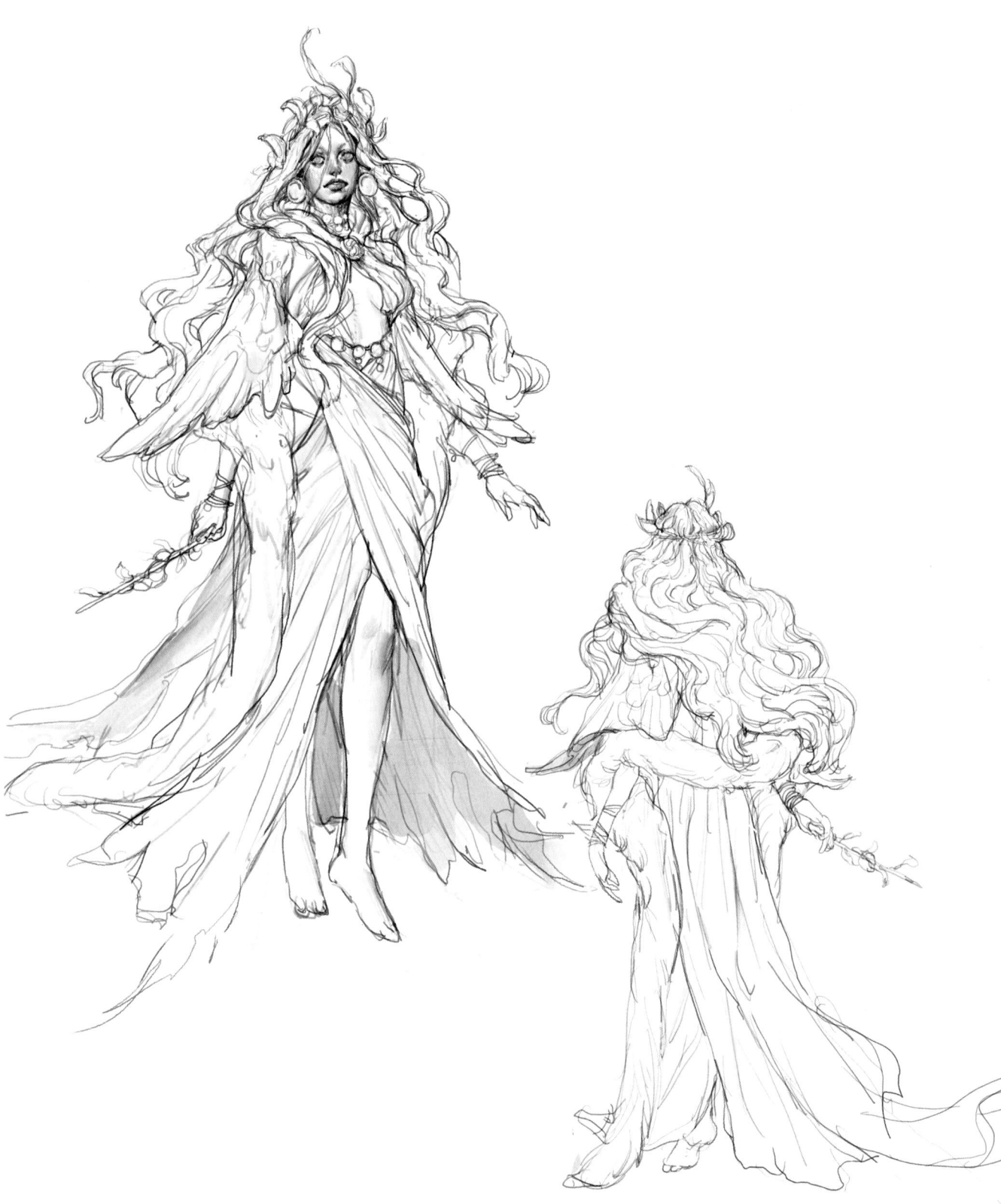

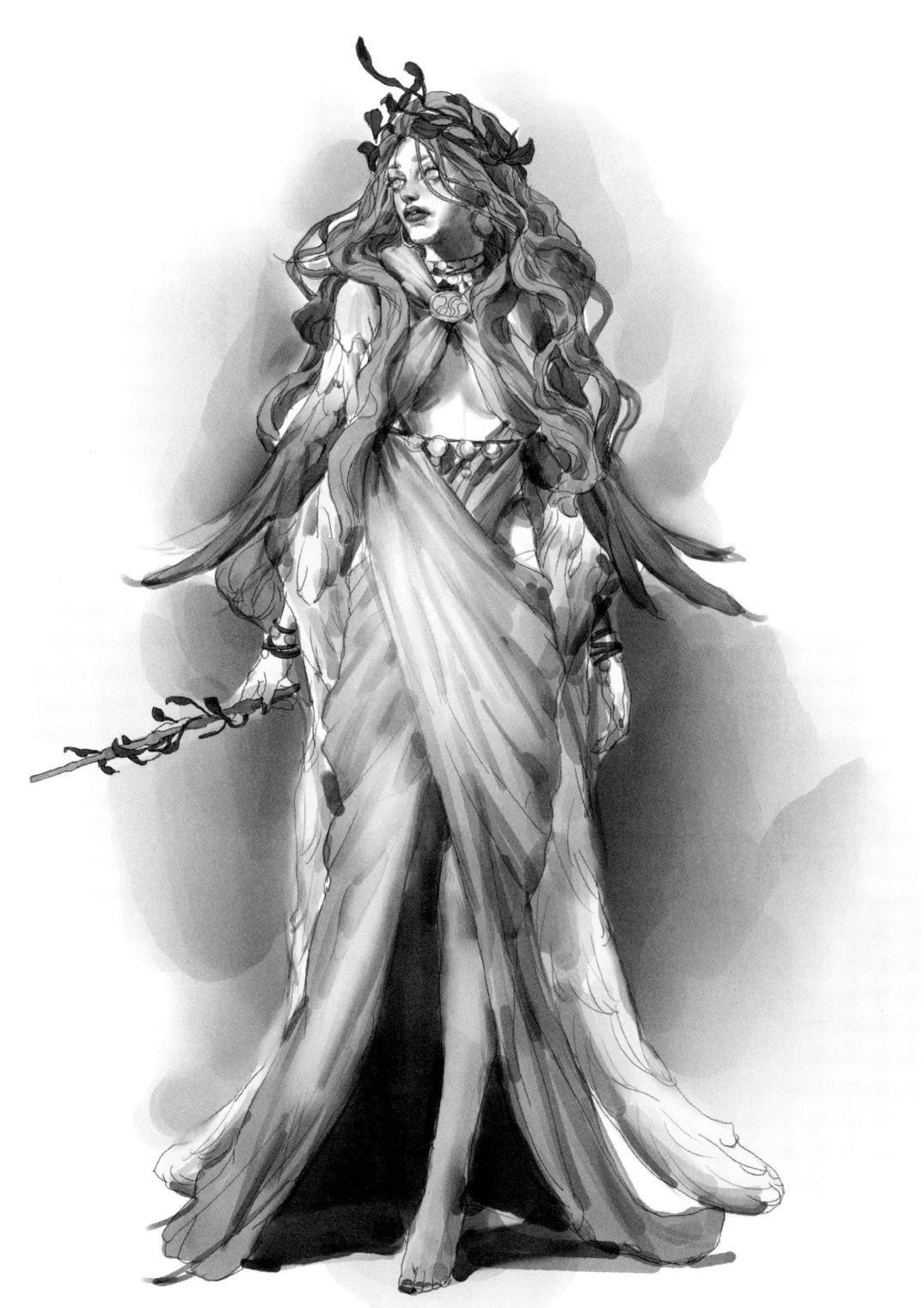

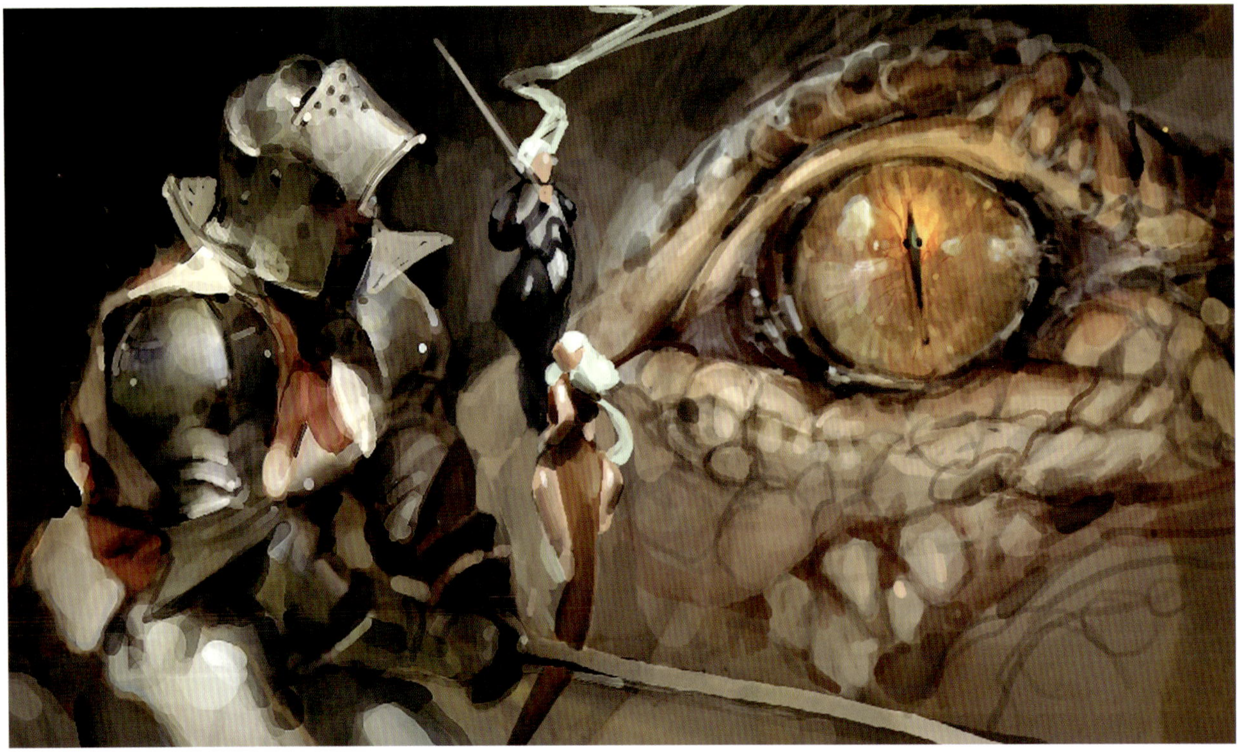

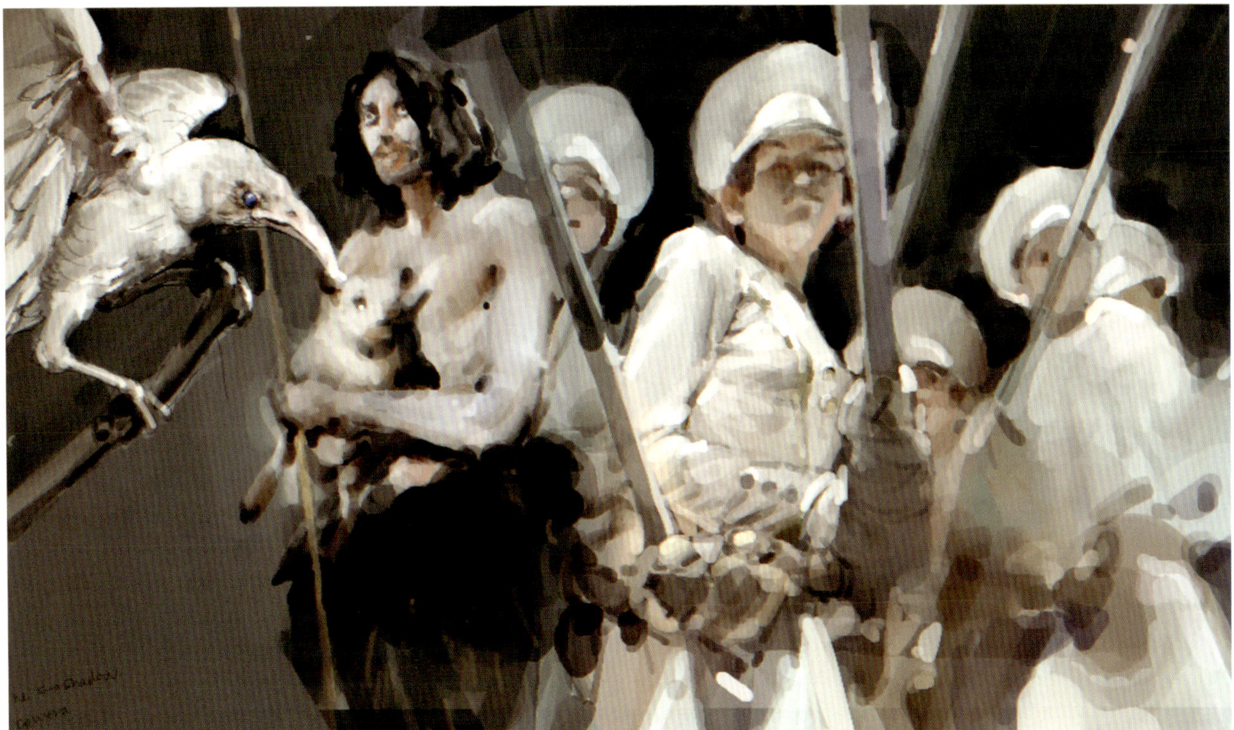

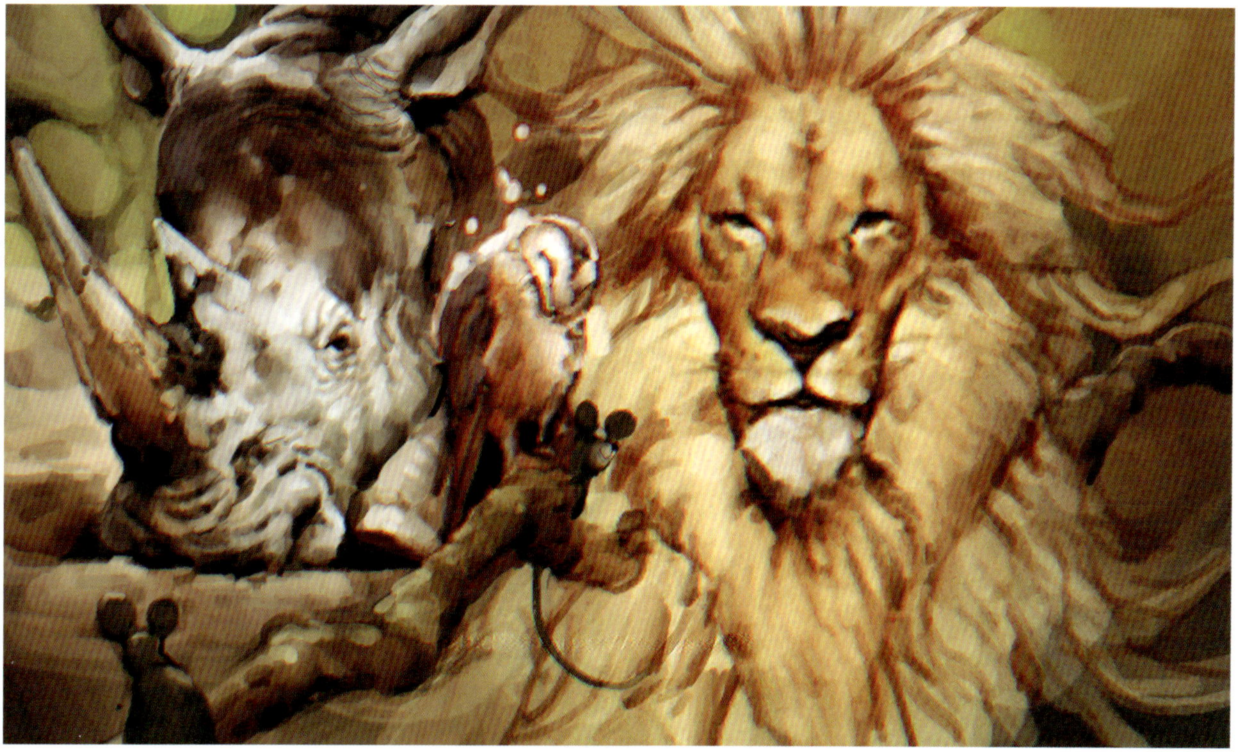

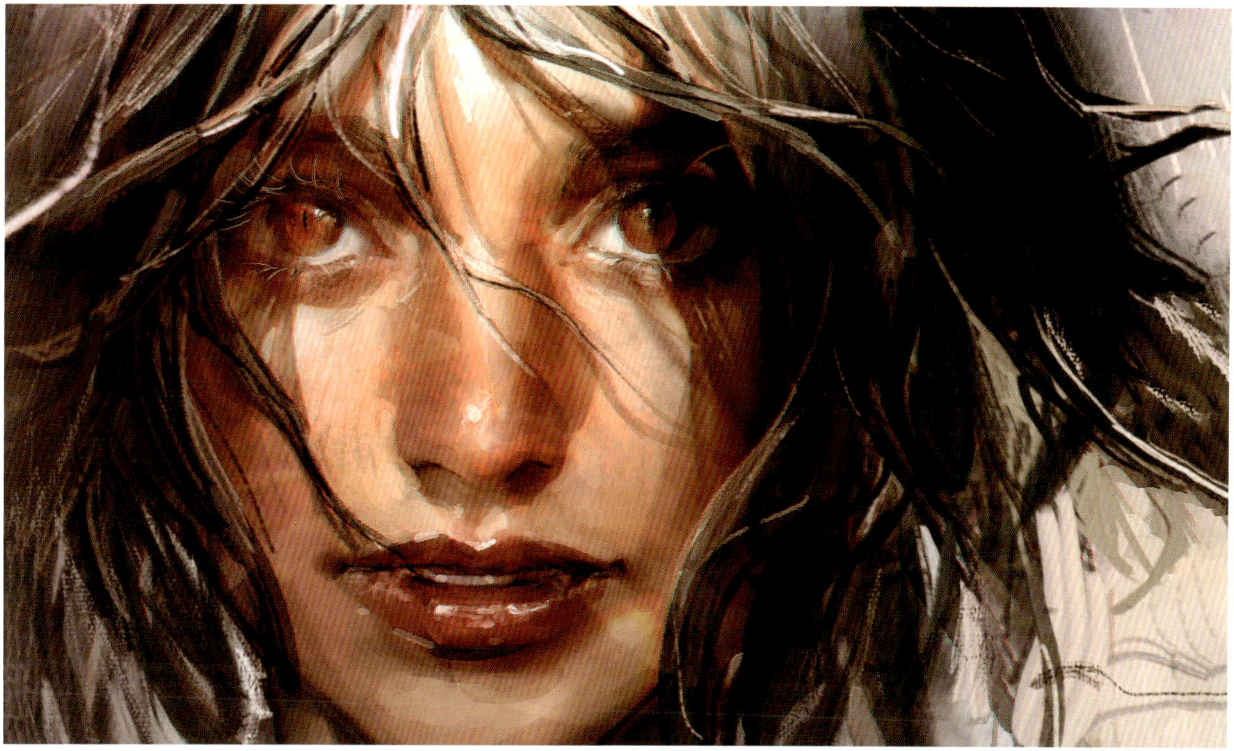

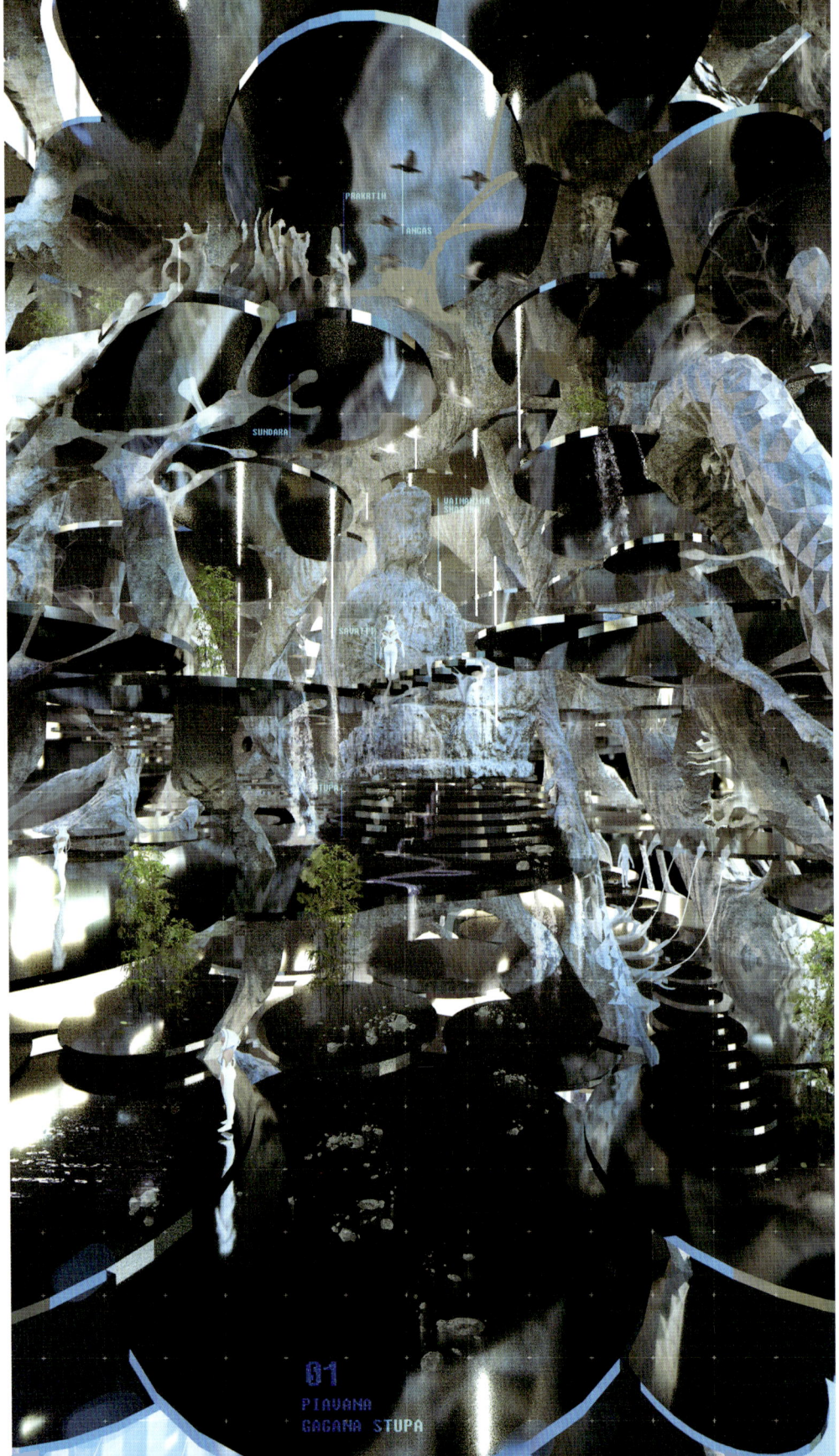

01
PINVANA
GAGANA STUPA

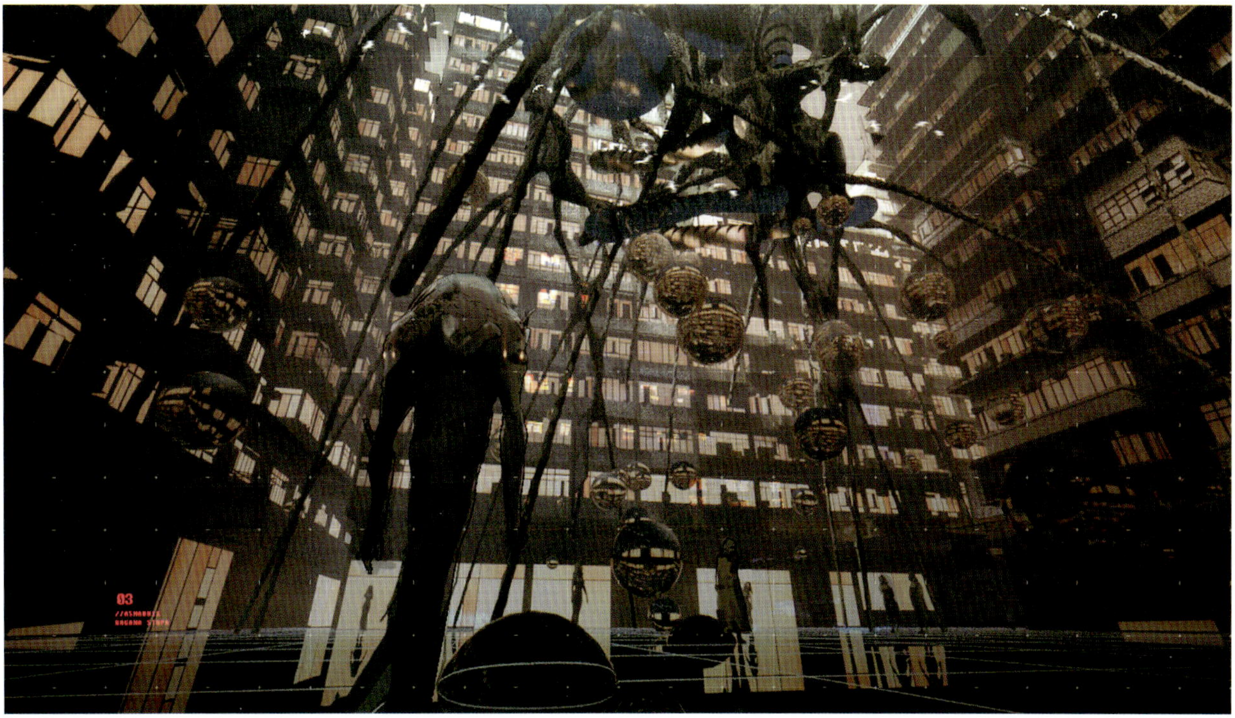

03
//ASHMARGA
GAGANA STUPA

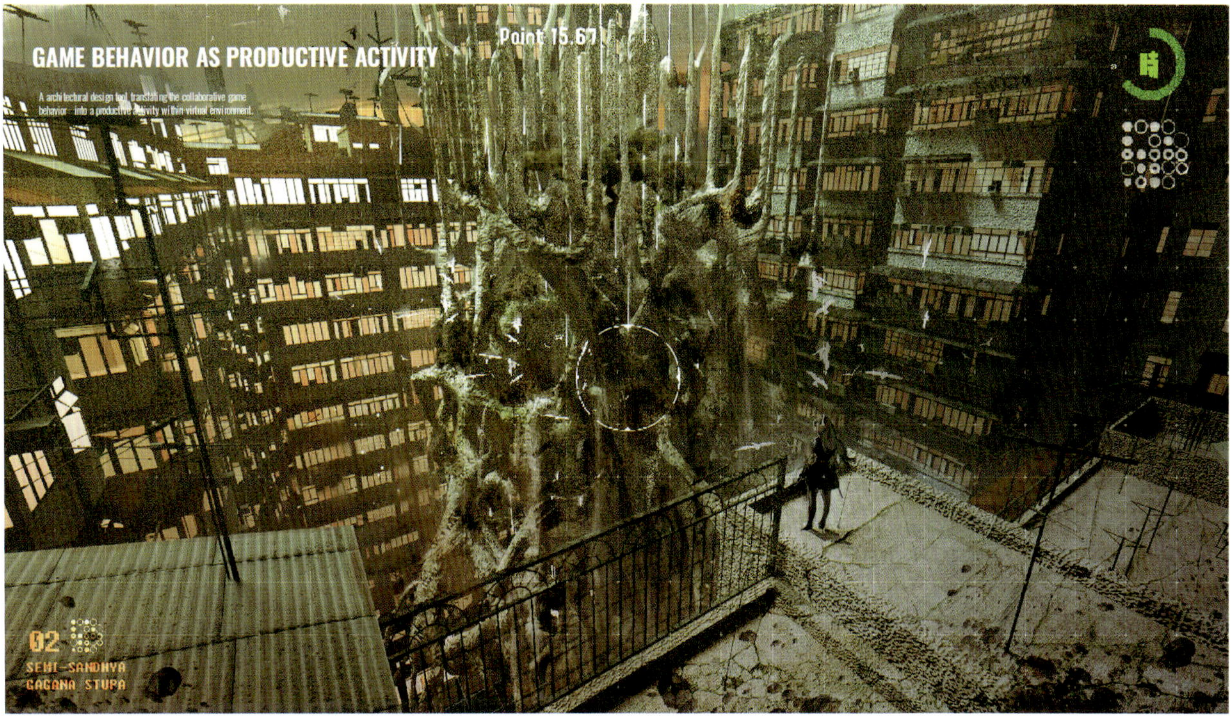

# GAME BEHAVIOR AS PRODUCTIVE ACTIVITY

Point 15.67

A architectural design tool, translating the collaborative game
behavior into a productive activity within virtual environment.

02
SEMI-SANDHYA
GAGANA STUPA

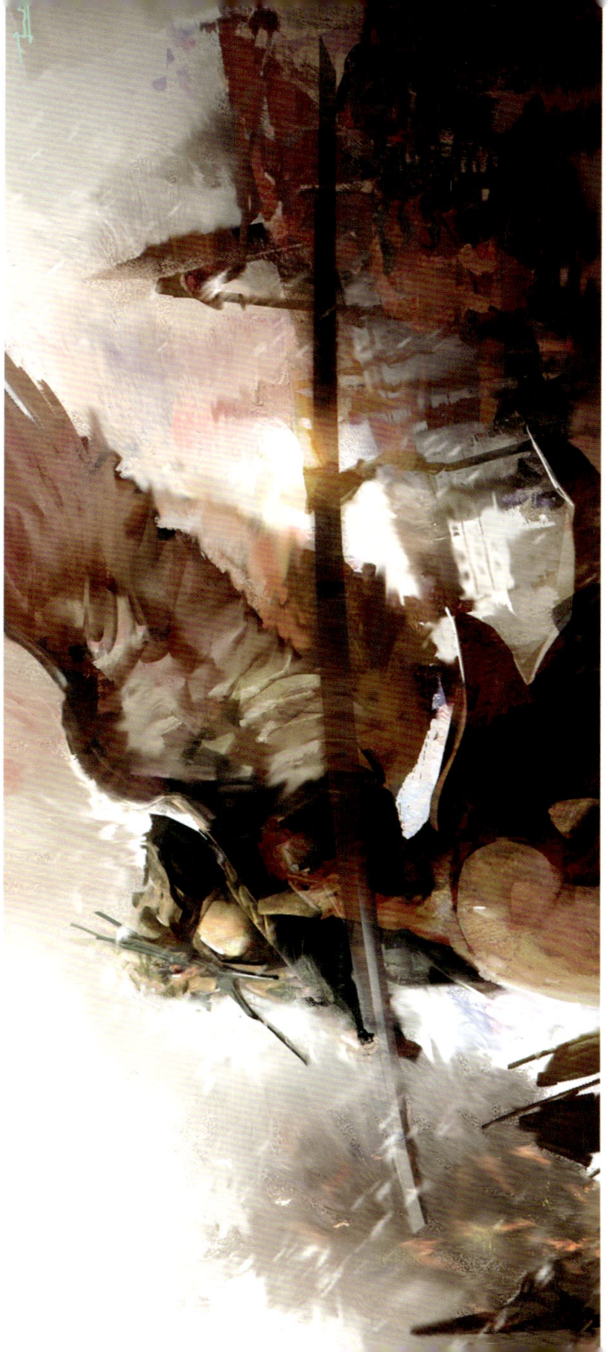

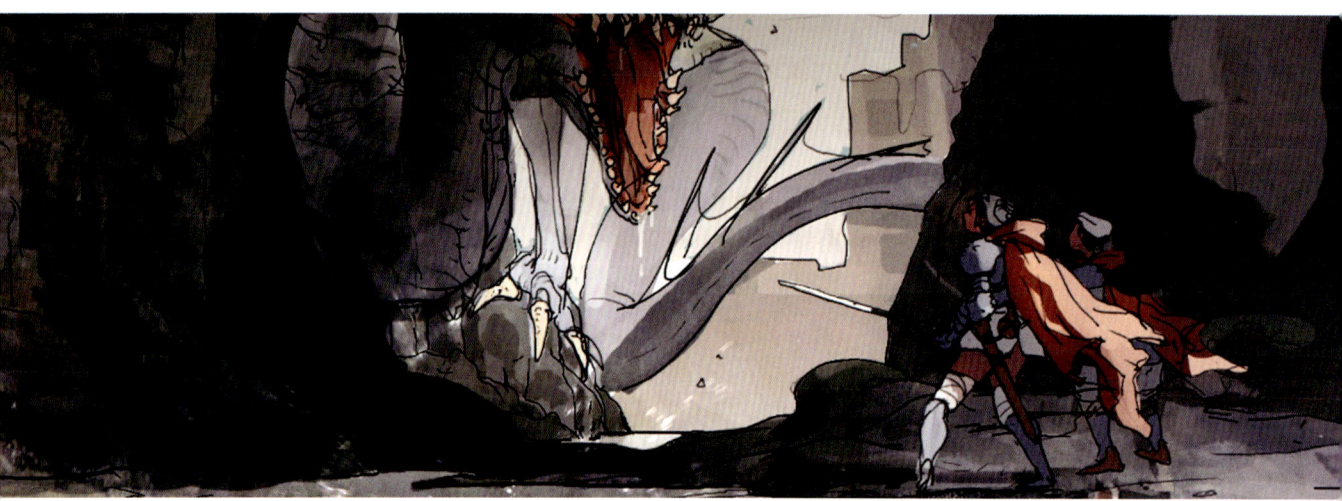

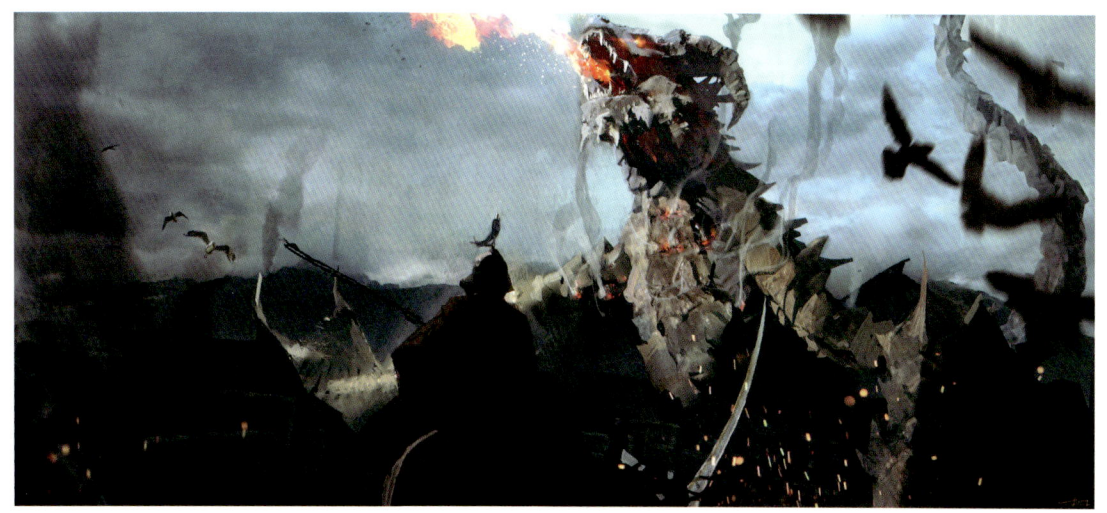

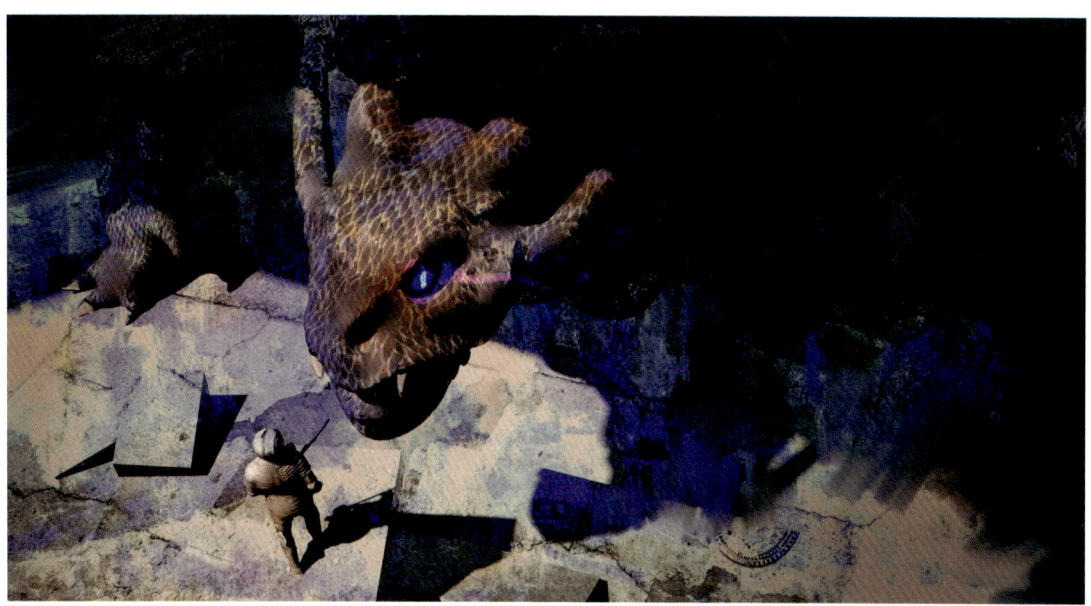

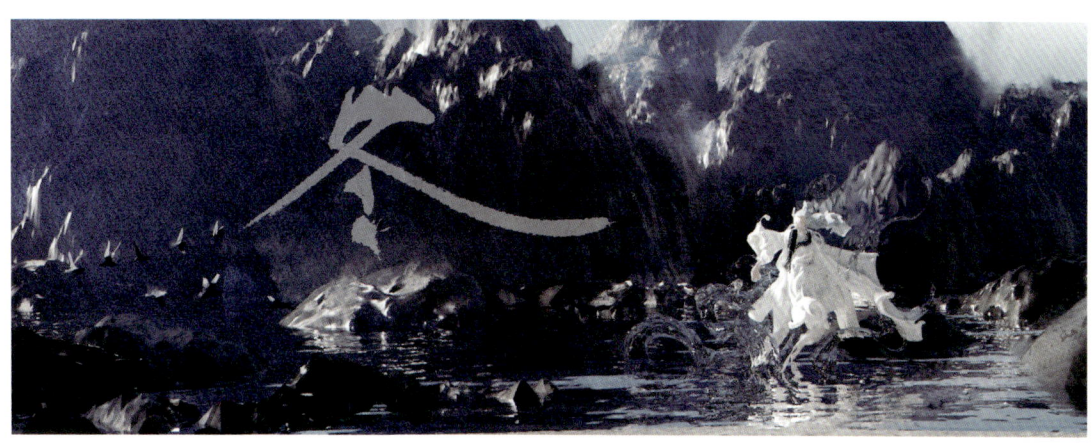

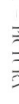

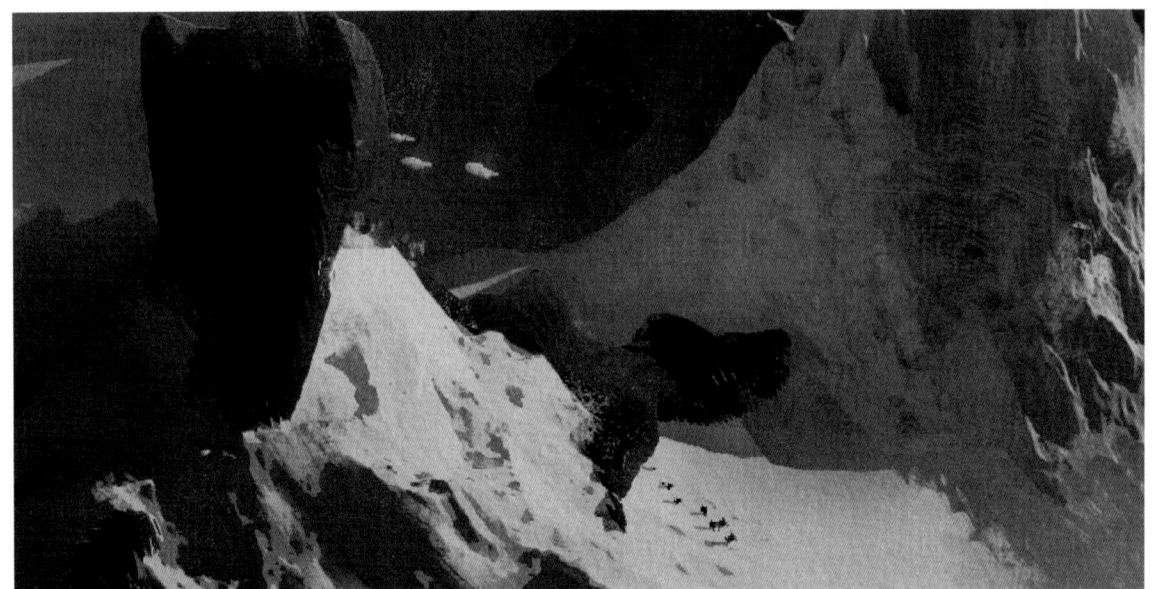

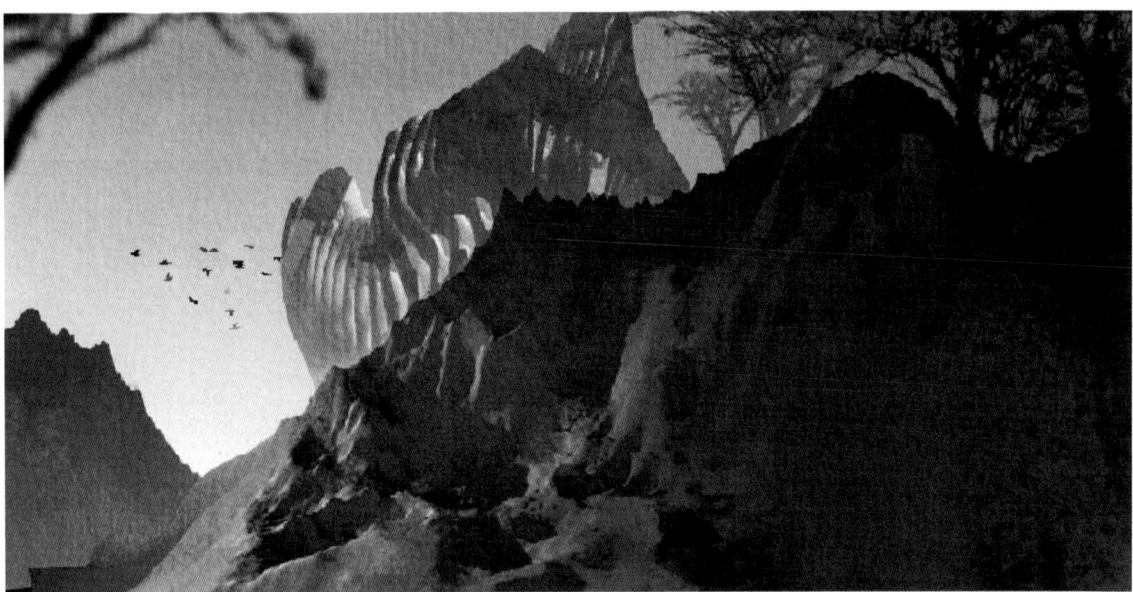

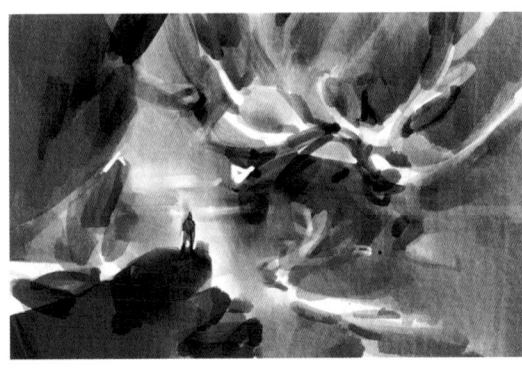
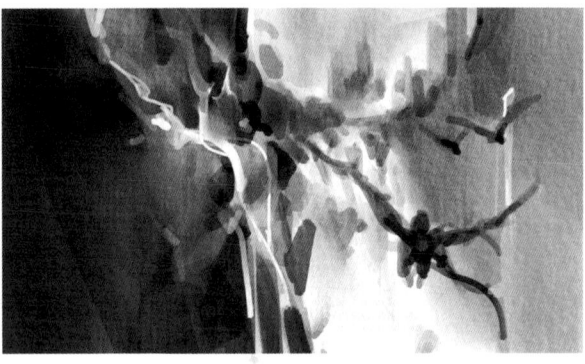

Peeling
From skin

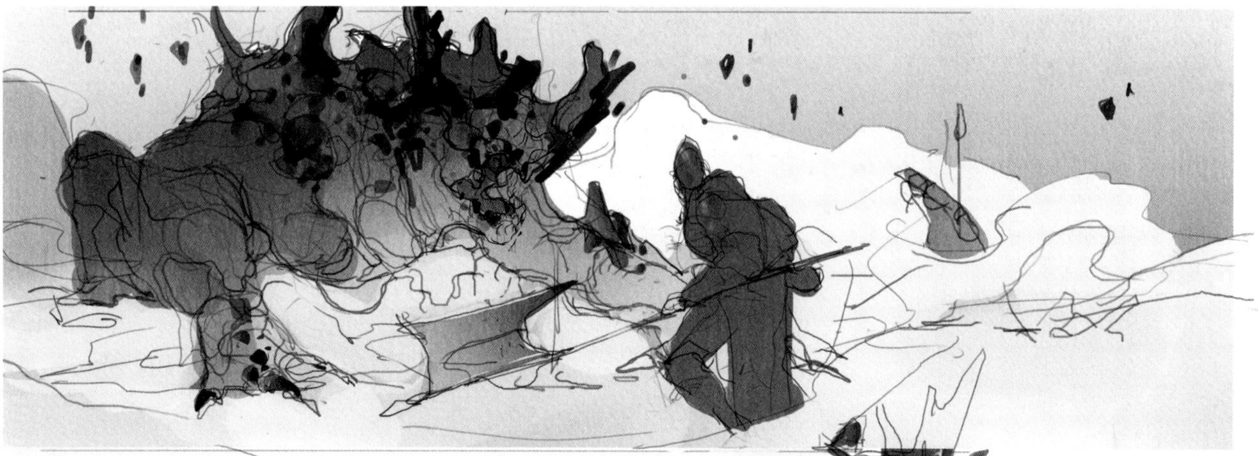

Erosion From Bone

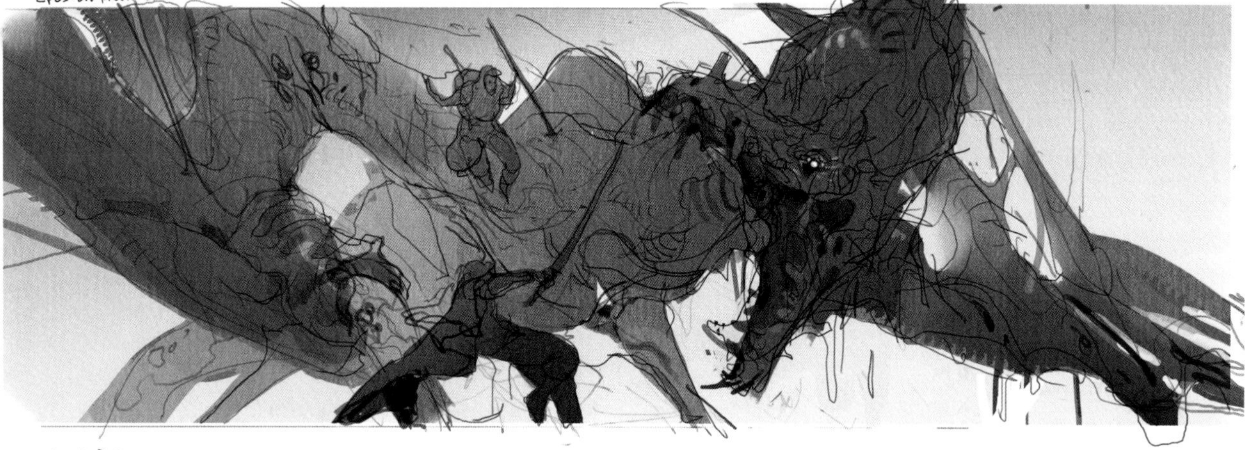

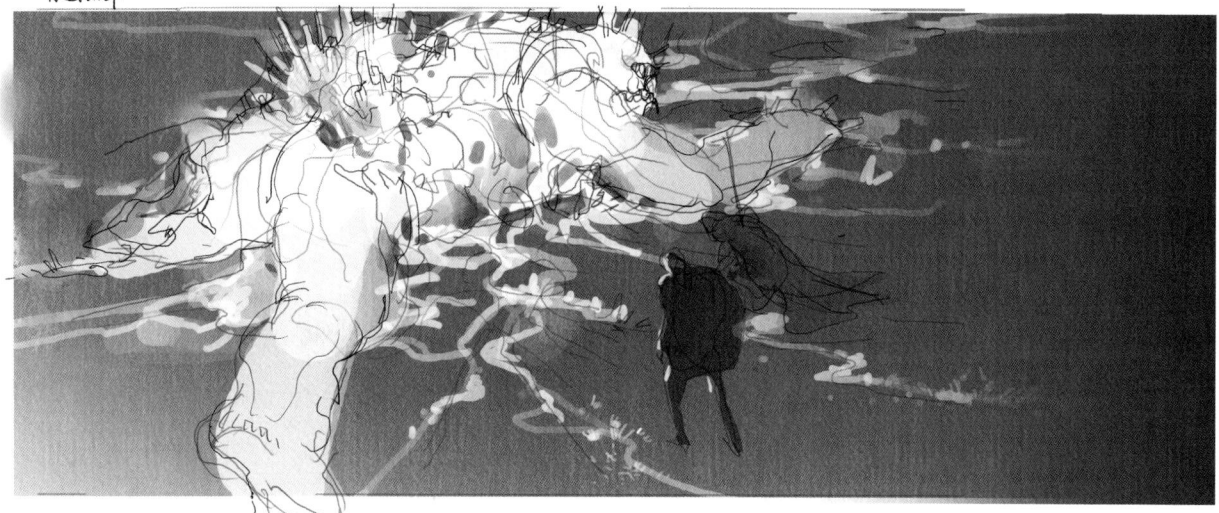

Melting

# 12

## SCHEMATIS

Painting is as natural as breathing.
I always carry a sketchbook or an iPad with me to maintain a habit of drawing in everyday life. Therefore, you will see my drawings of animals I've seen, train passengers, and my favorite movies. There are also concept designs, sculptures, jewelry sketches, and drafts of my other illustration works. I prefer to think through painting rather than having a clear idea before I start; the act of painting itself provides me with answers. My journey with sketching will continue along the path of my life.

絵を描くことは呼吸と同じくらい自然なことです。
日常生活の中で絵を描く習慣を維持するために、いつもスケッチブックやiPadを持ち歩いています。ですから皆さんは、私が見た動物や列車の乗客、好きな映画の絵を目にすることになるわけです。コンセプトデザイン、造形、ジュエリーのスケッチ、他のイラスト作品の下書きもあります。描き始める前に明確なアイデアを用意するよりも、描きながら考える方が好きです。描く行為そのものが答えをくれるからです。私のスケッチの旅は、これからも人生とともに続いてゆくでしょう。

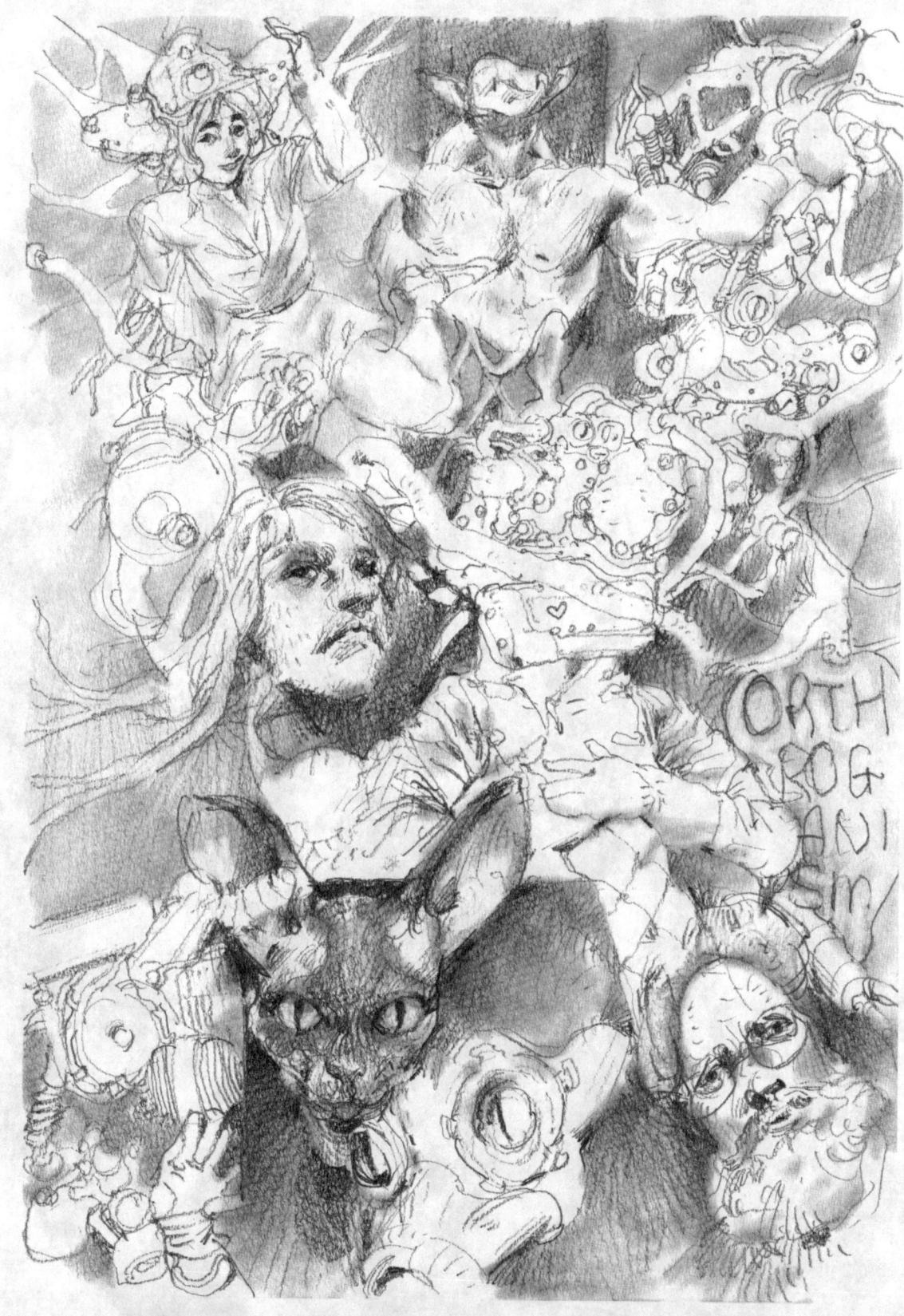

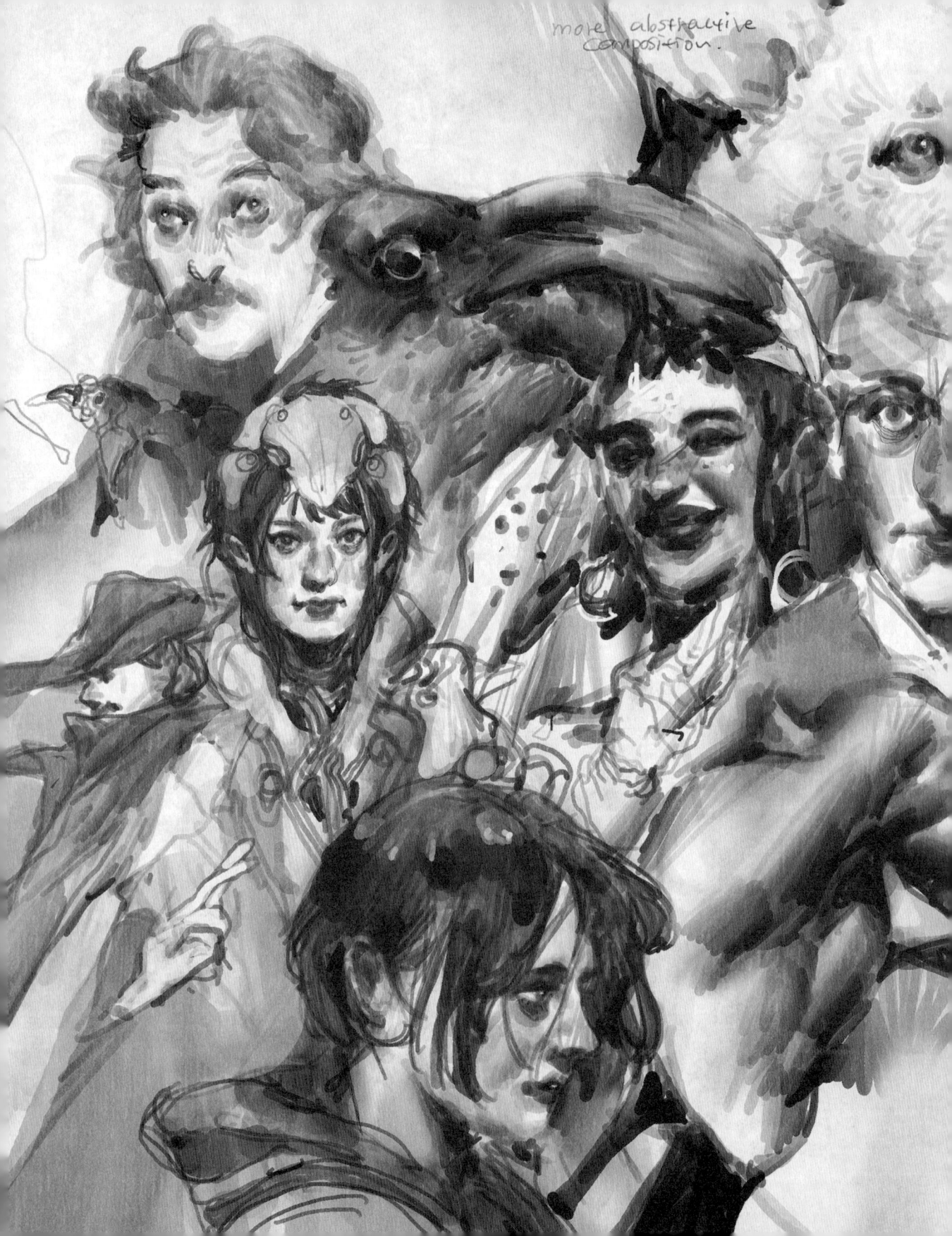

more abstractive
composition.

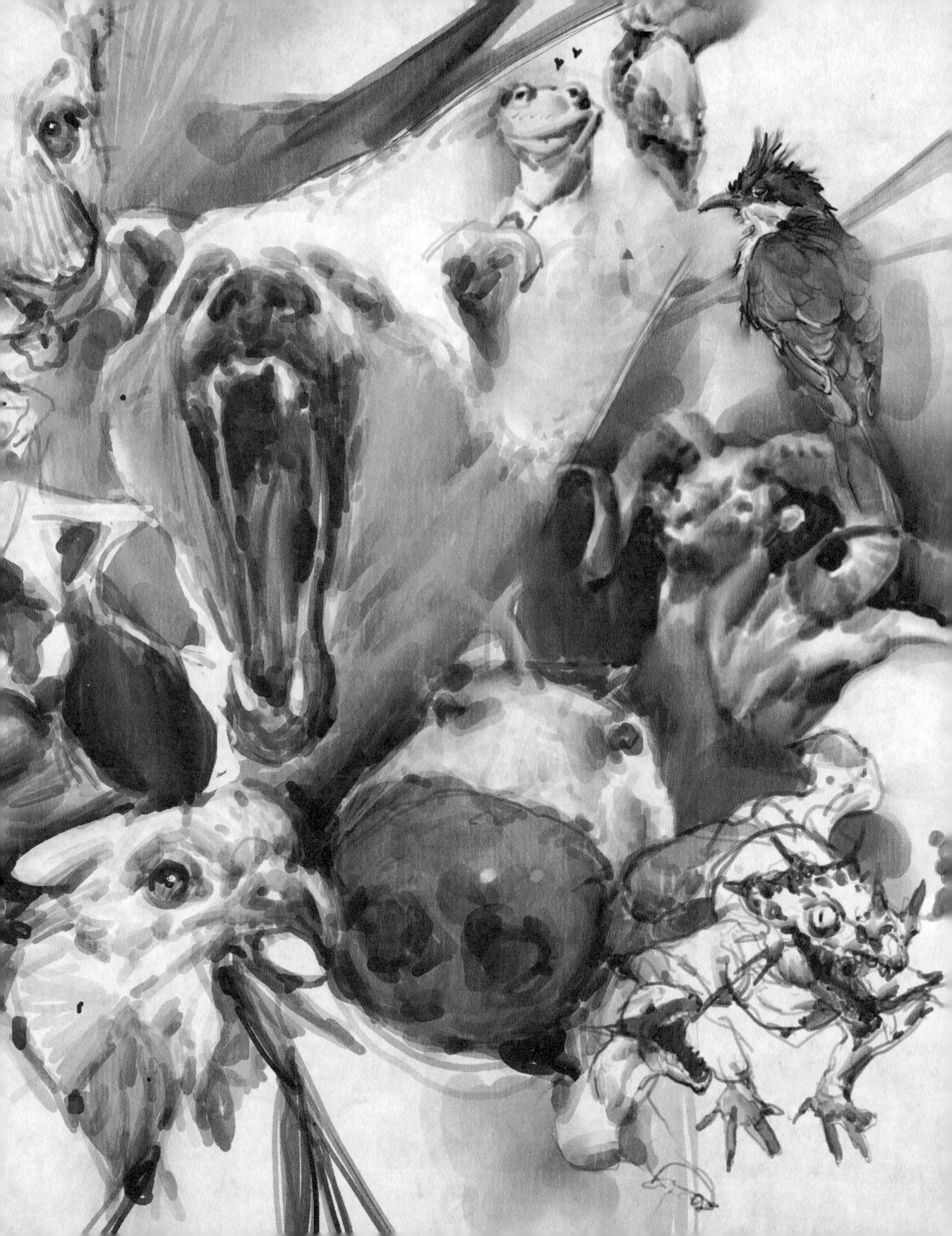

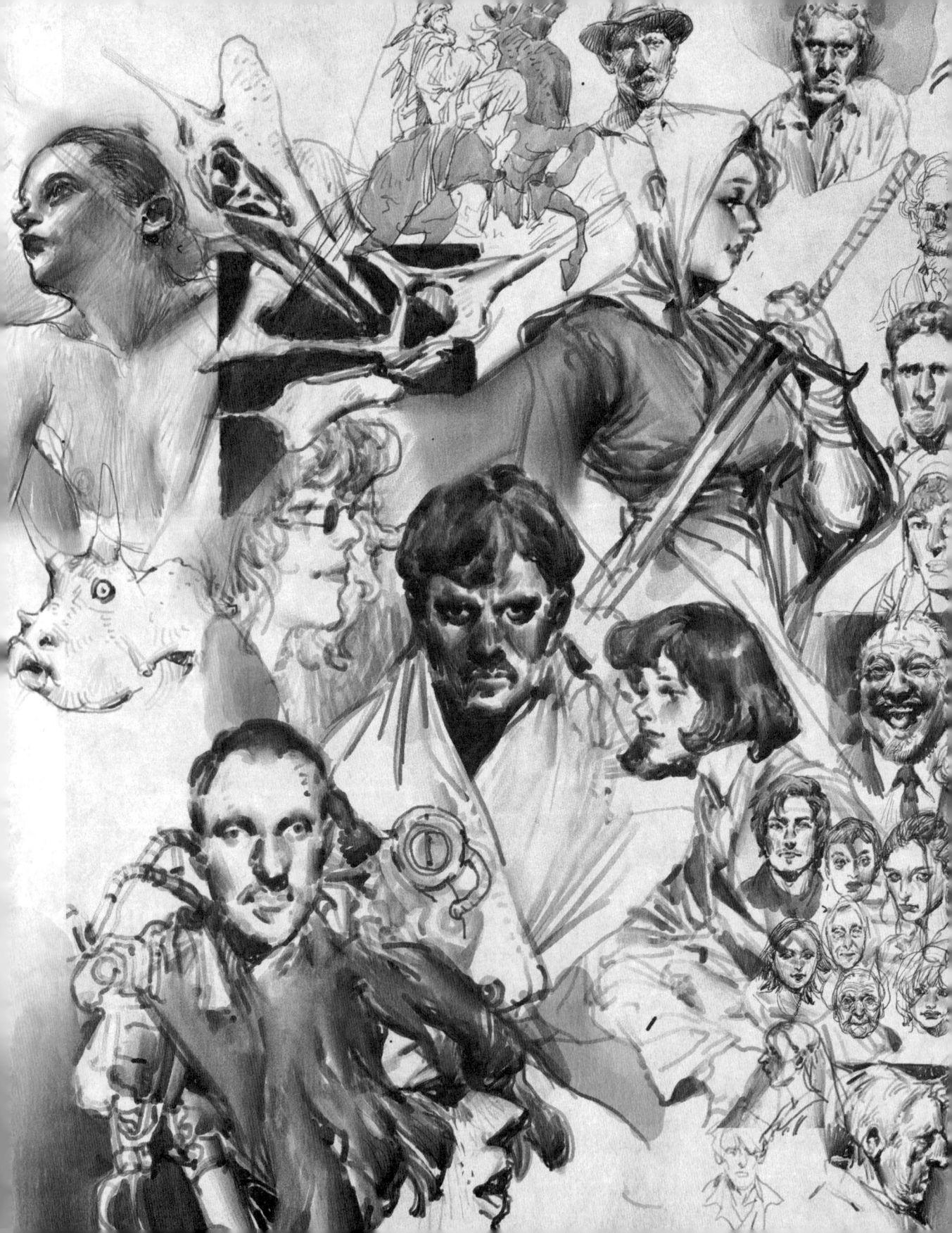

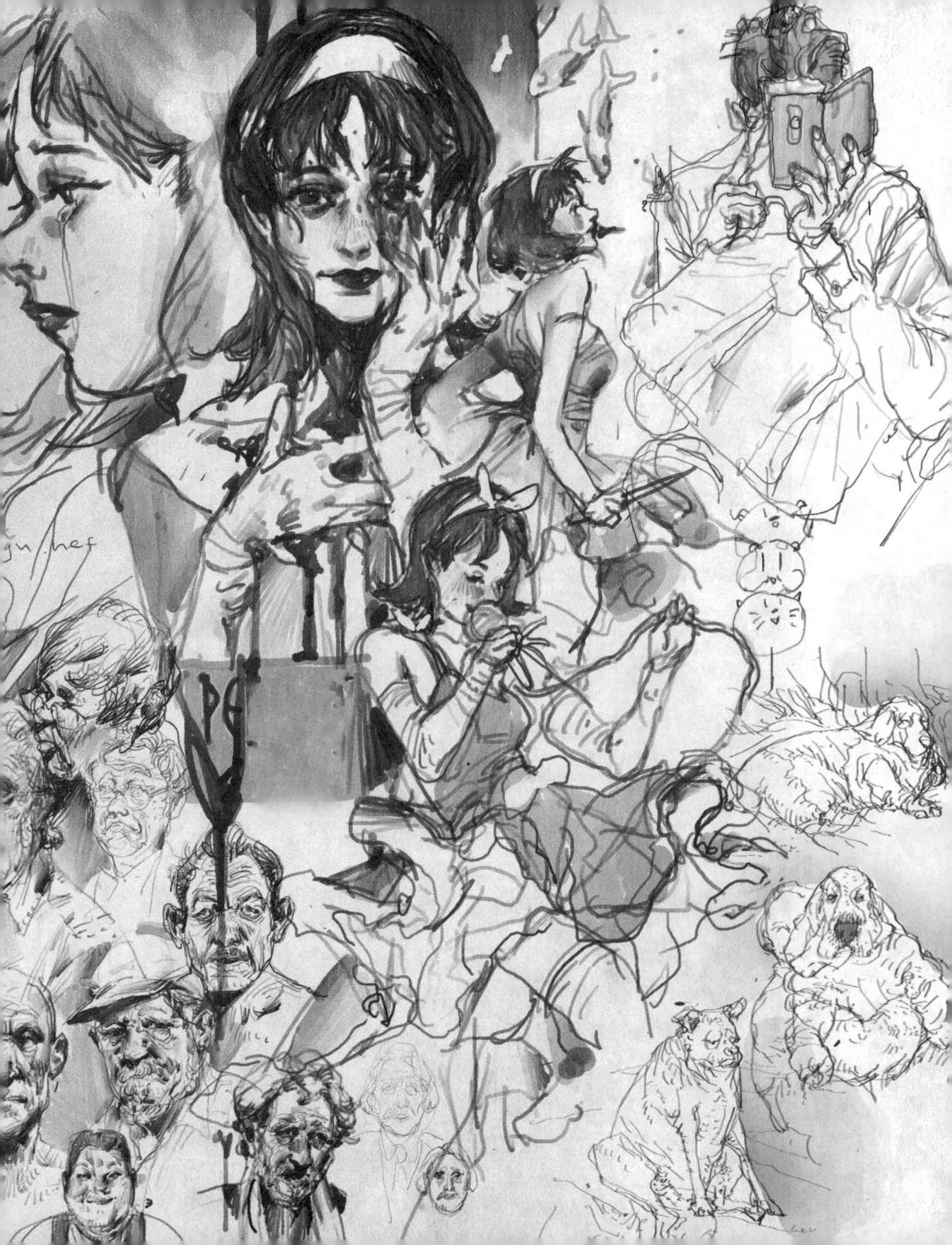

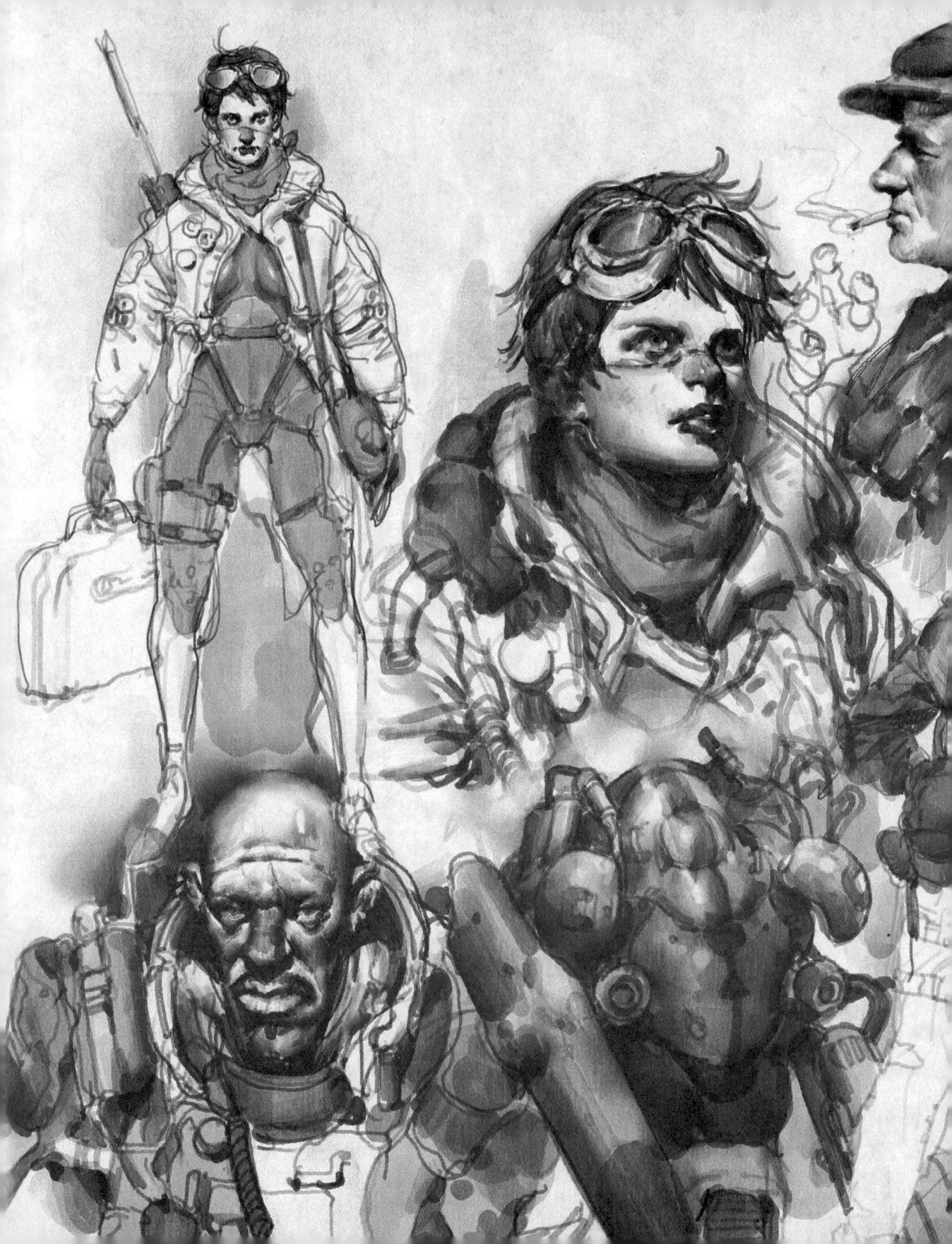

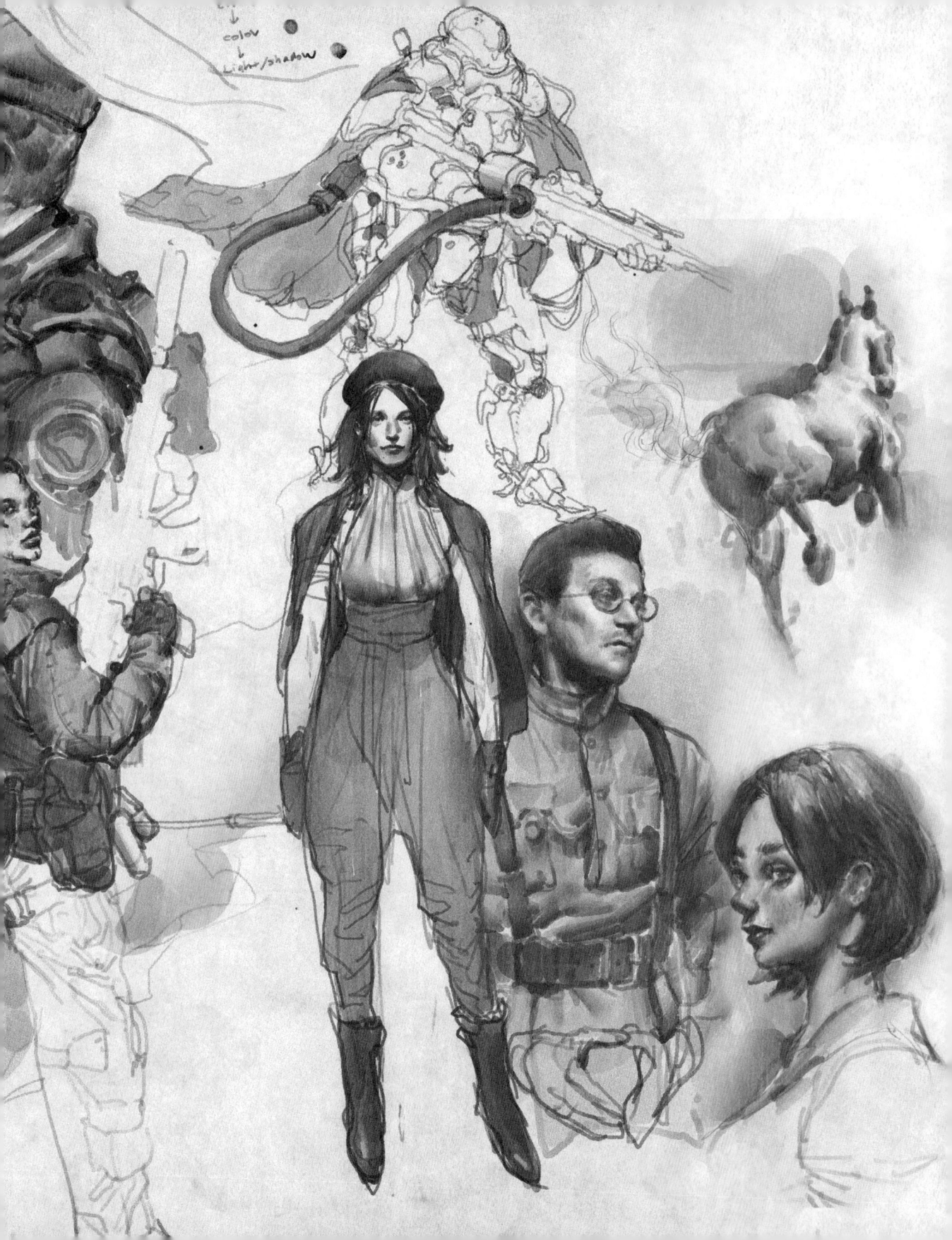

color
Light/shadow

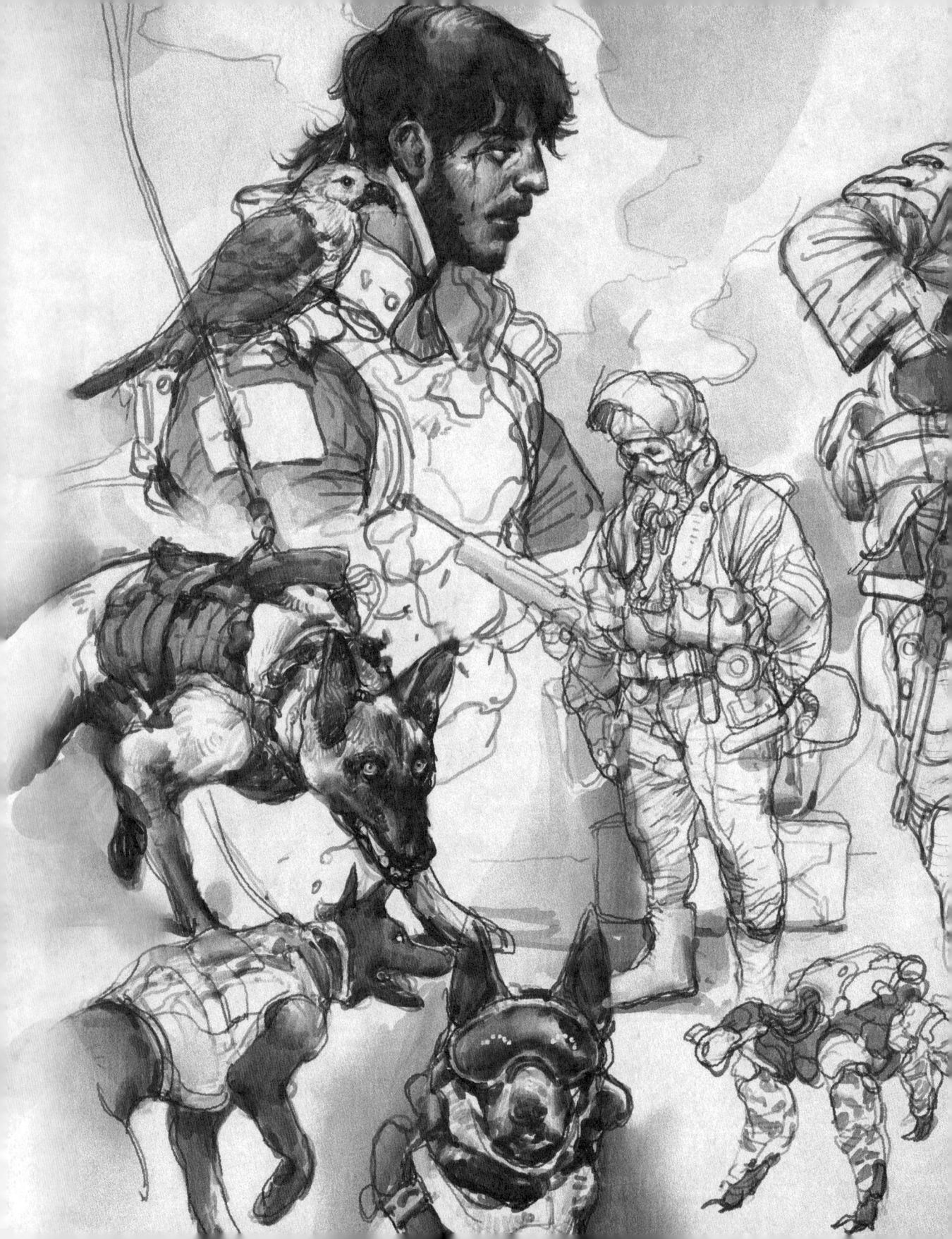

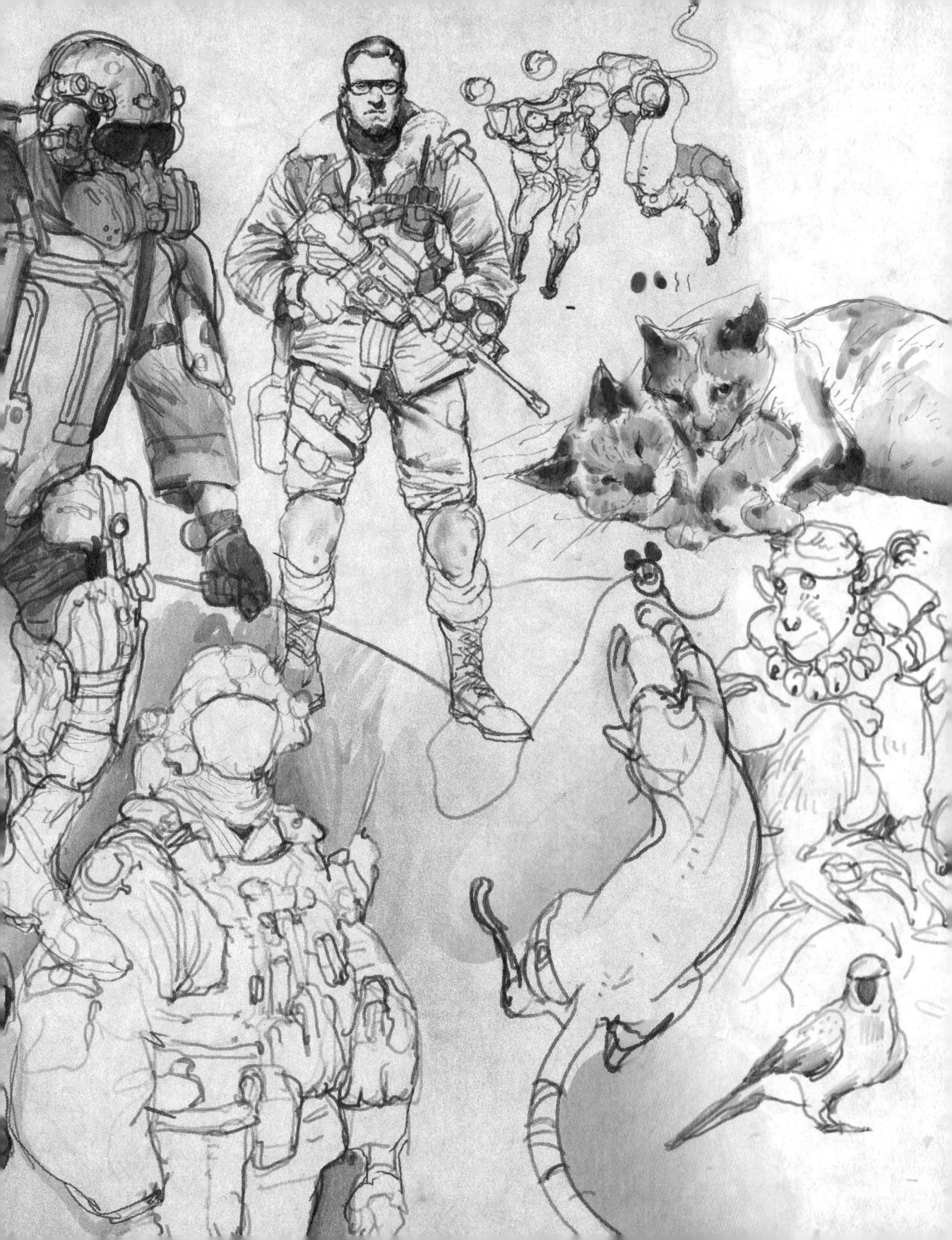

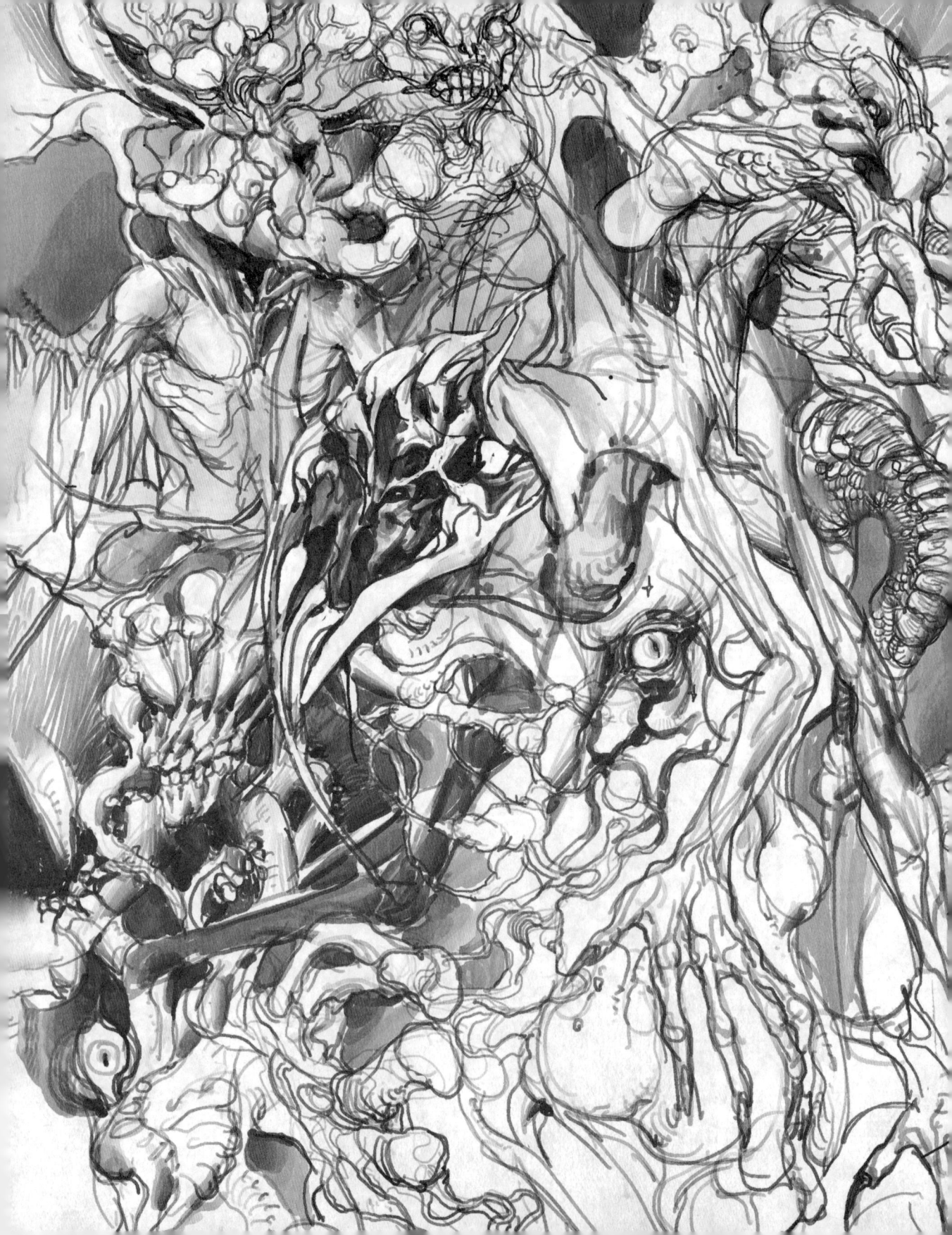

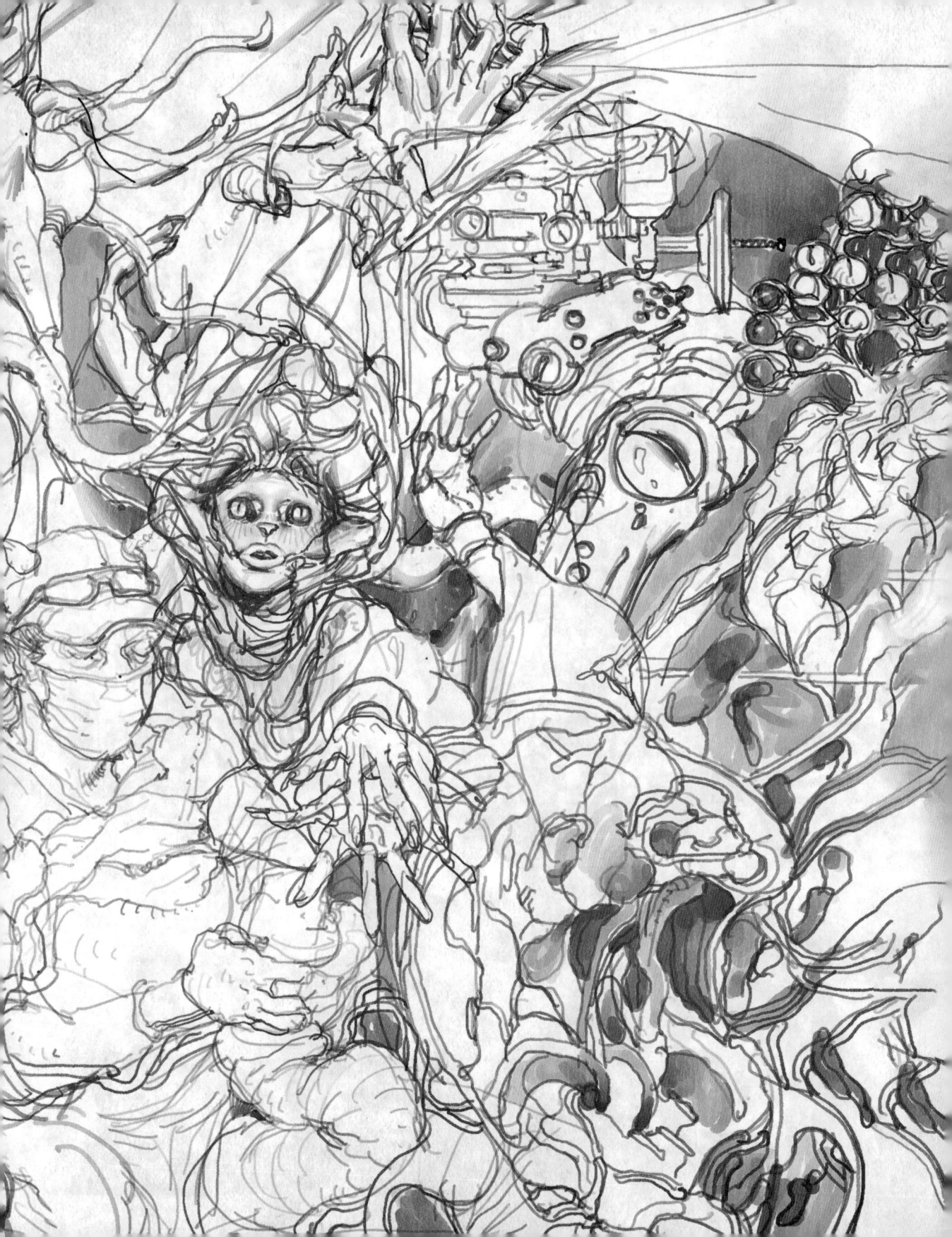

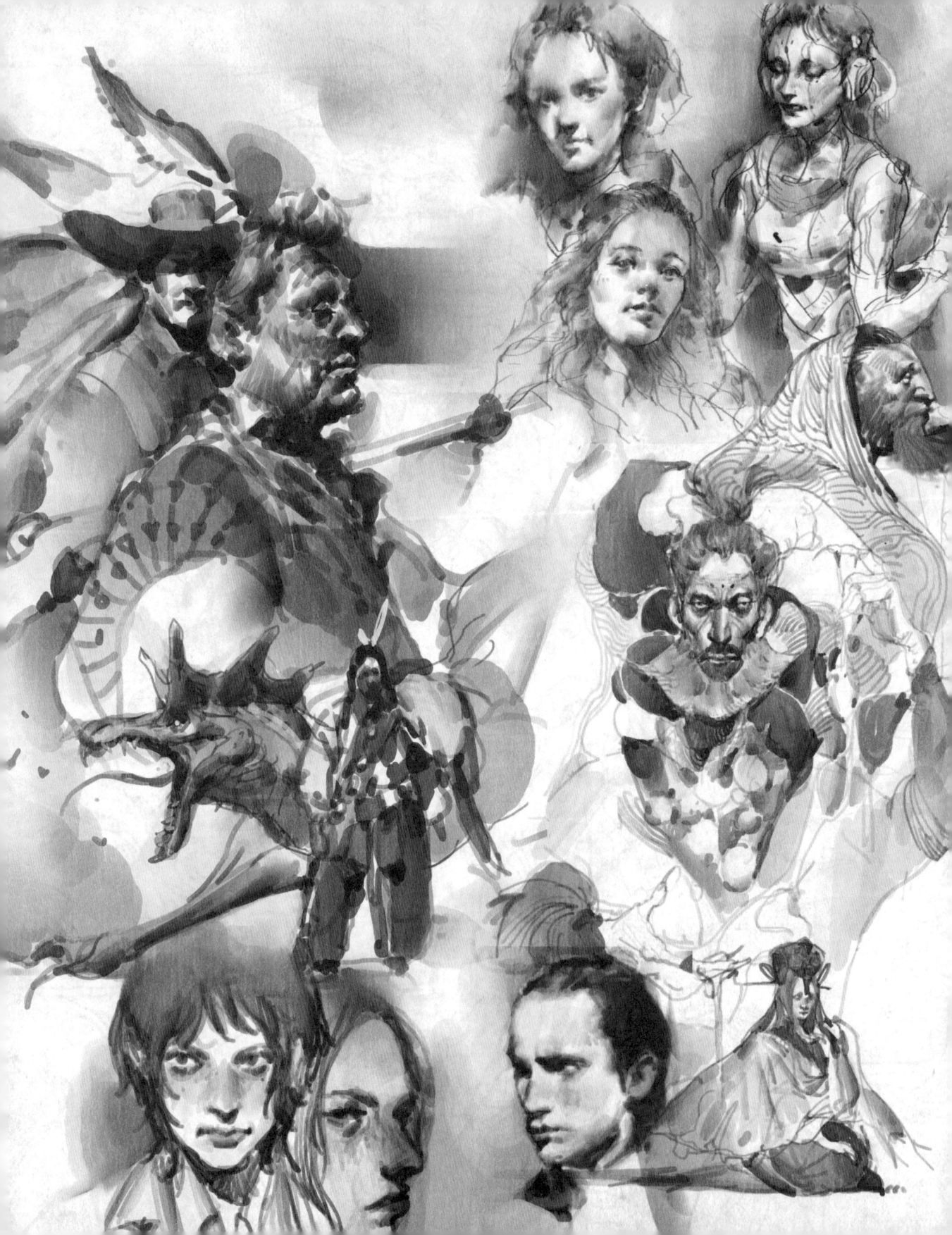

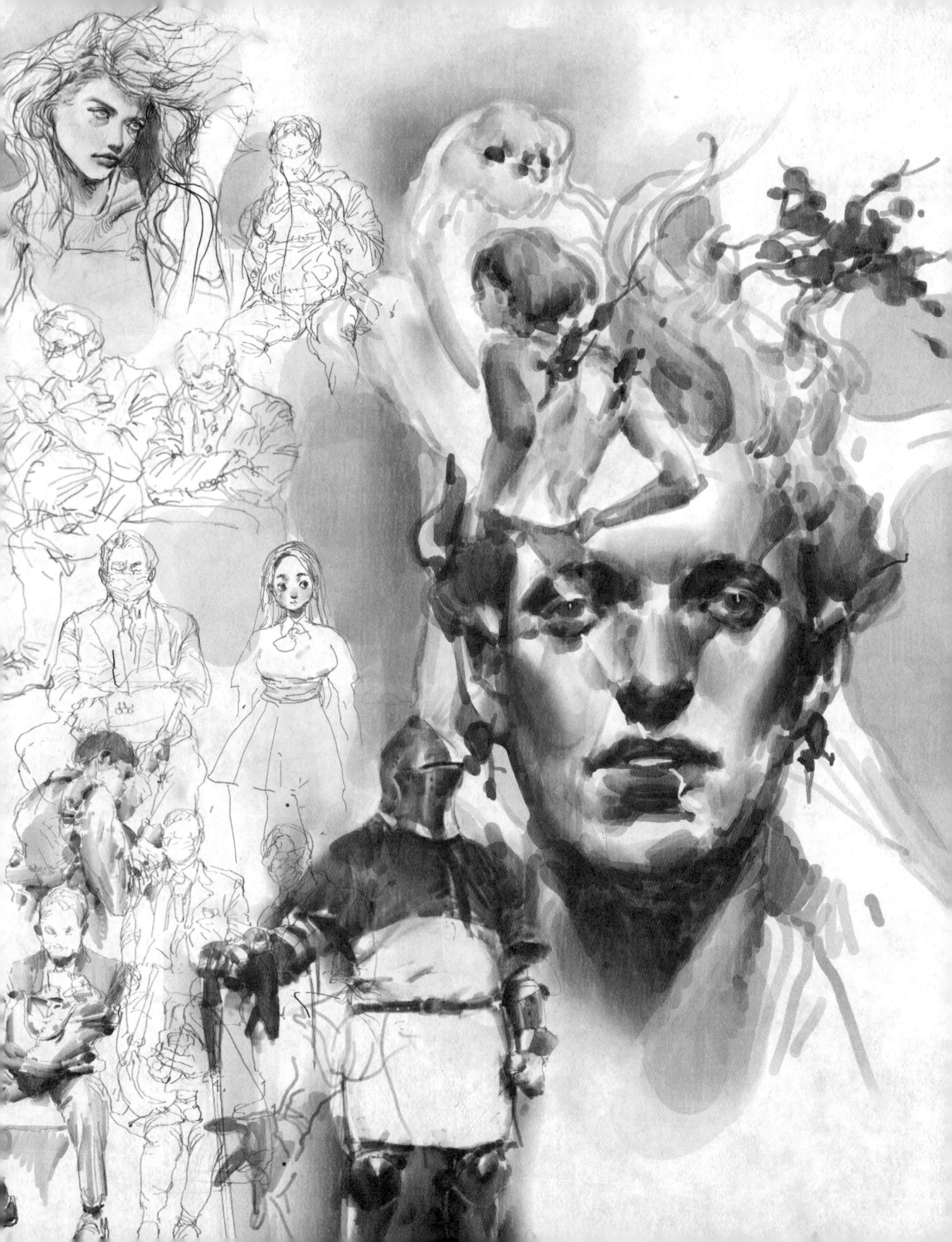

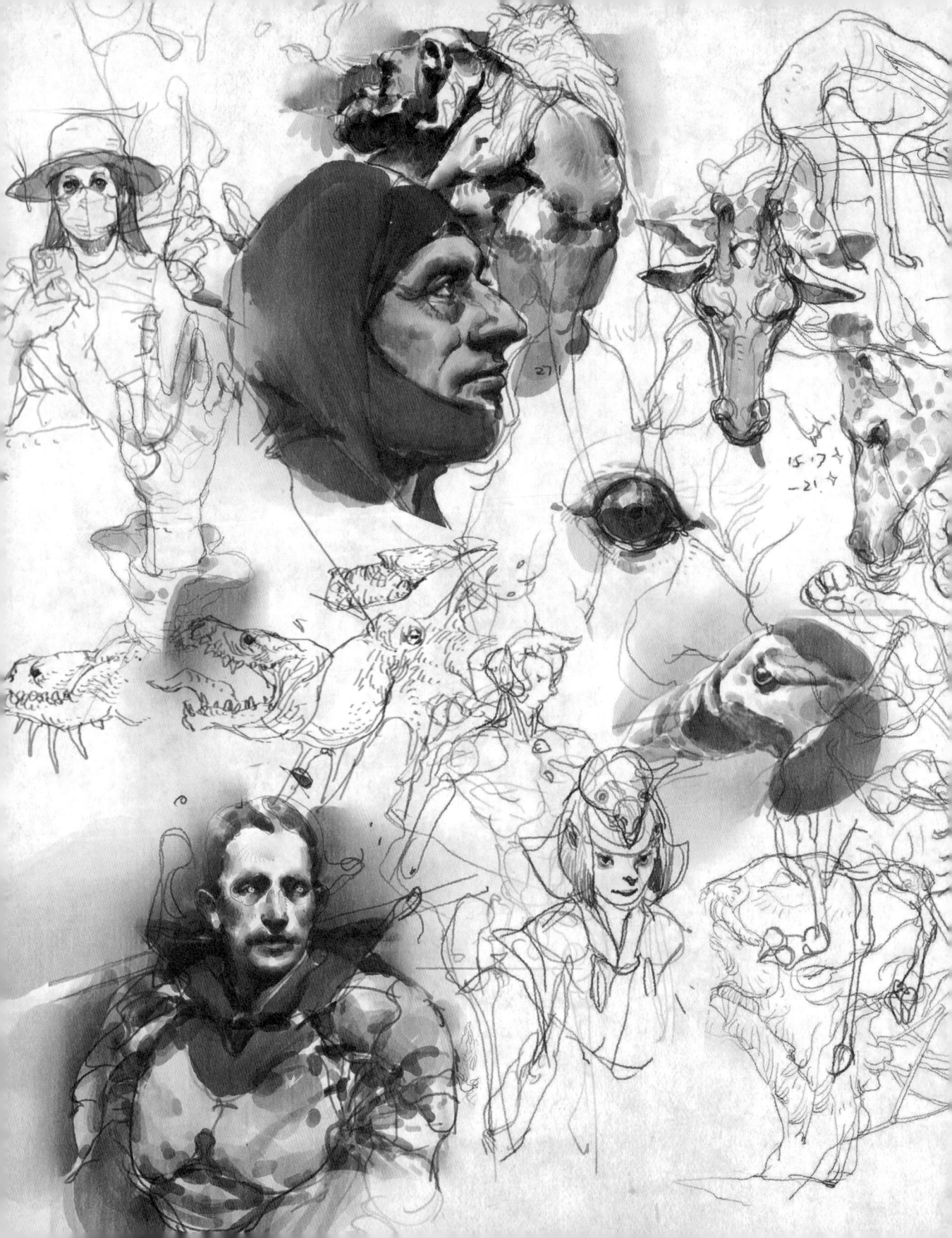

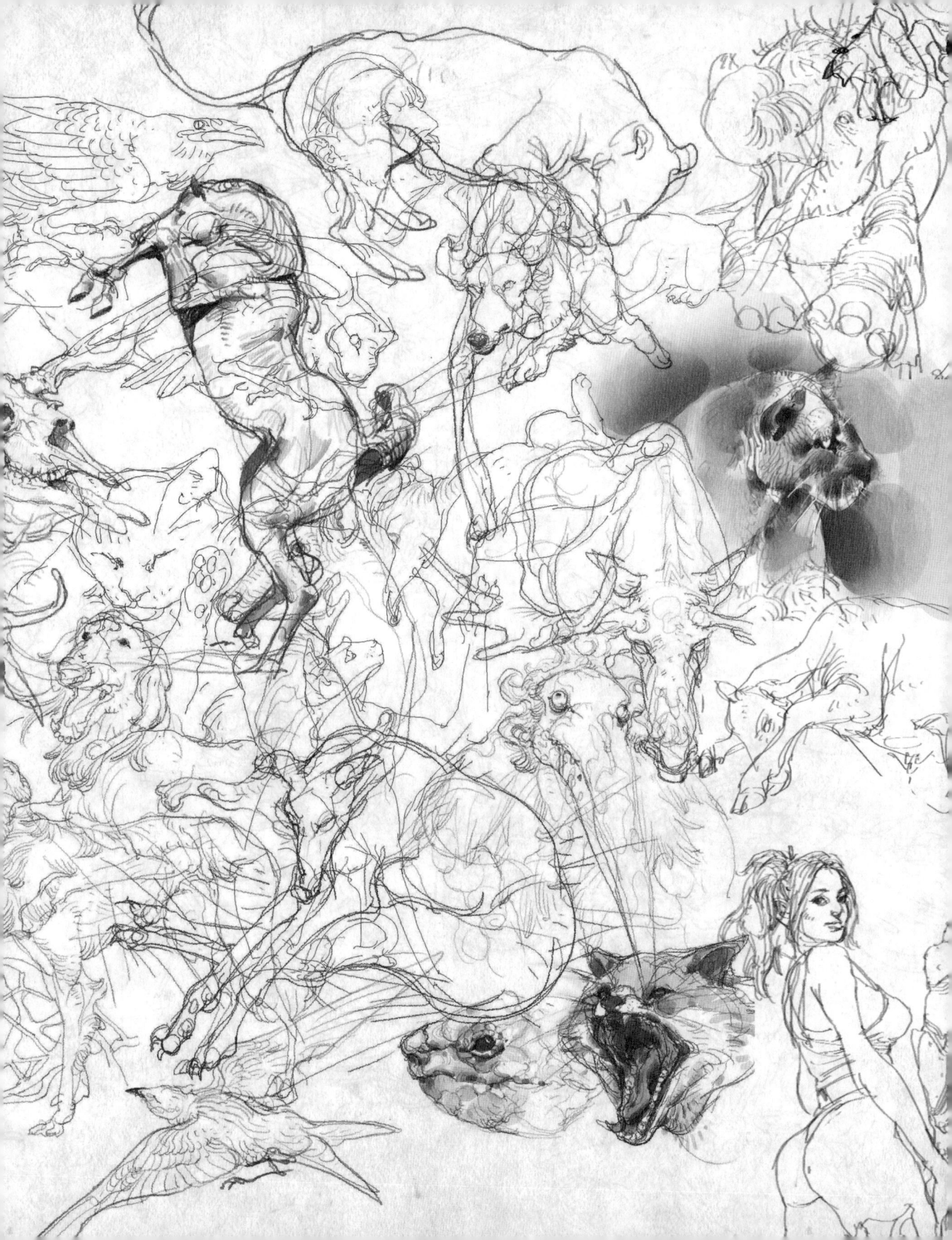

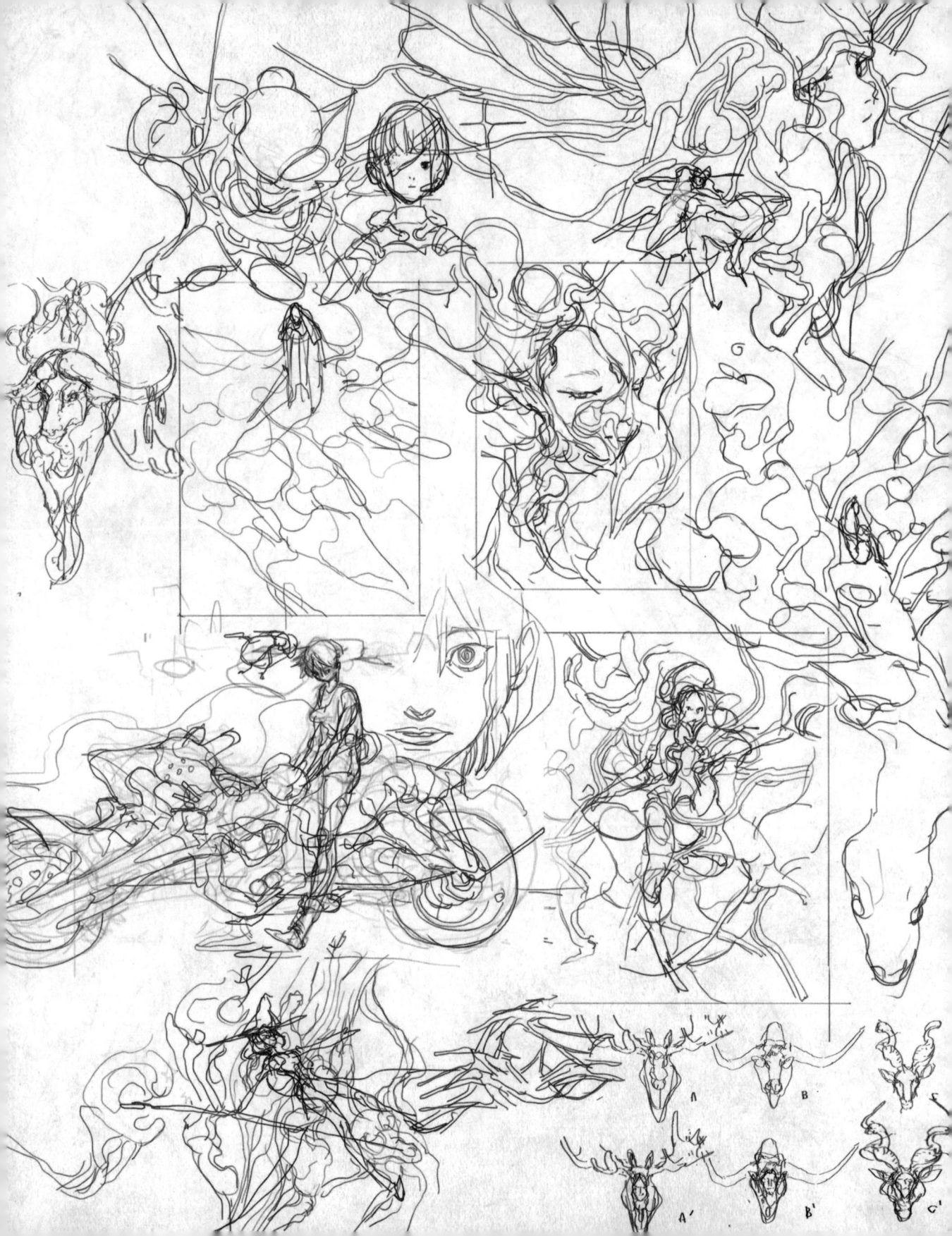

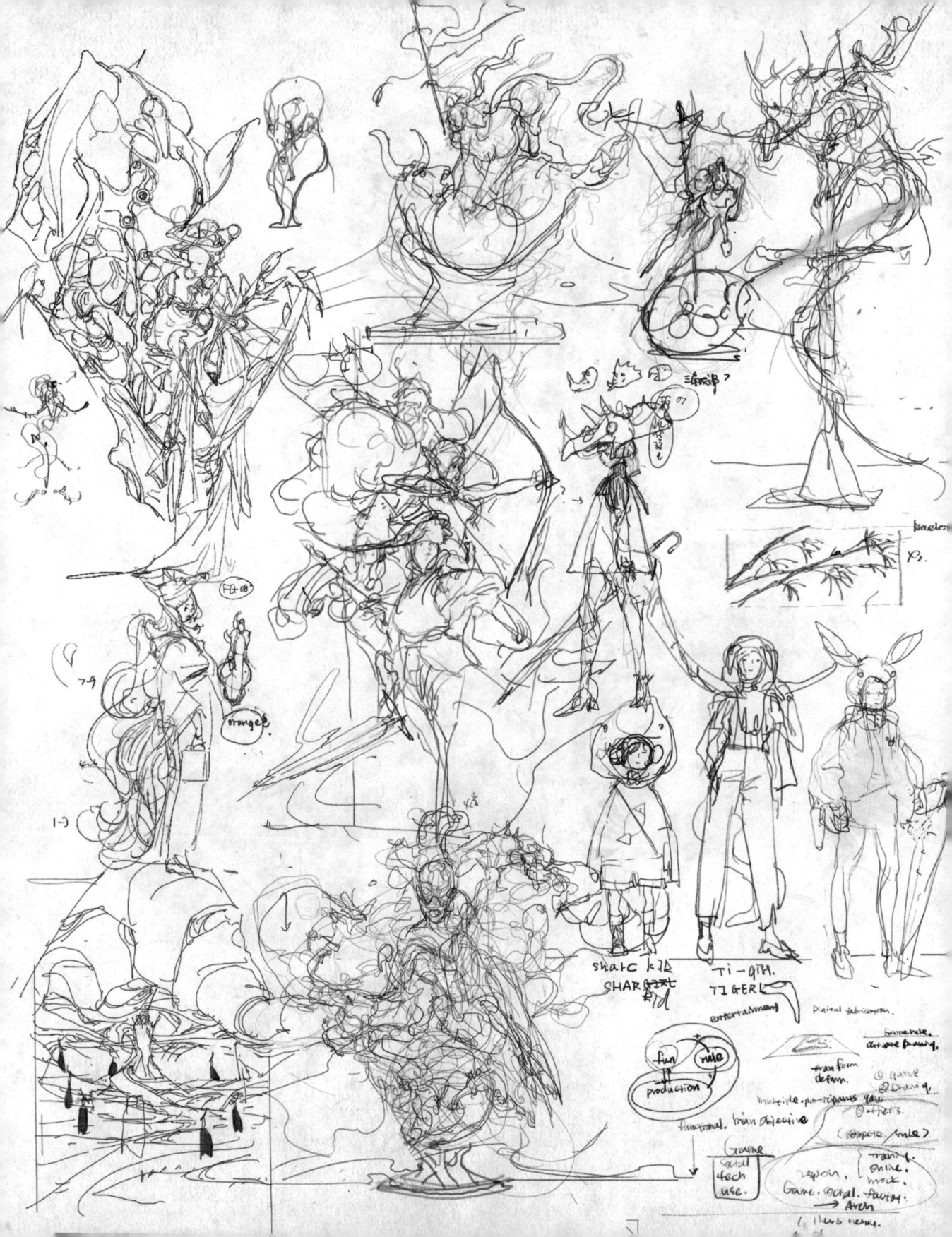

shark kid
SHARK GIRL
RIO

TI-GTH.
TIGERL

orange.

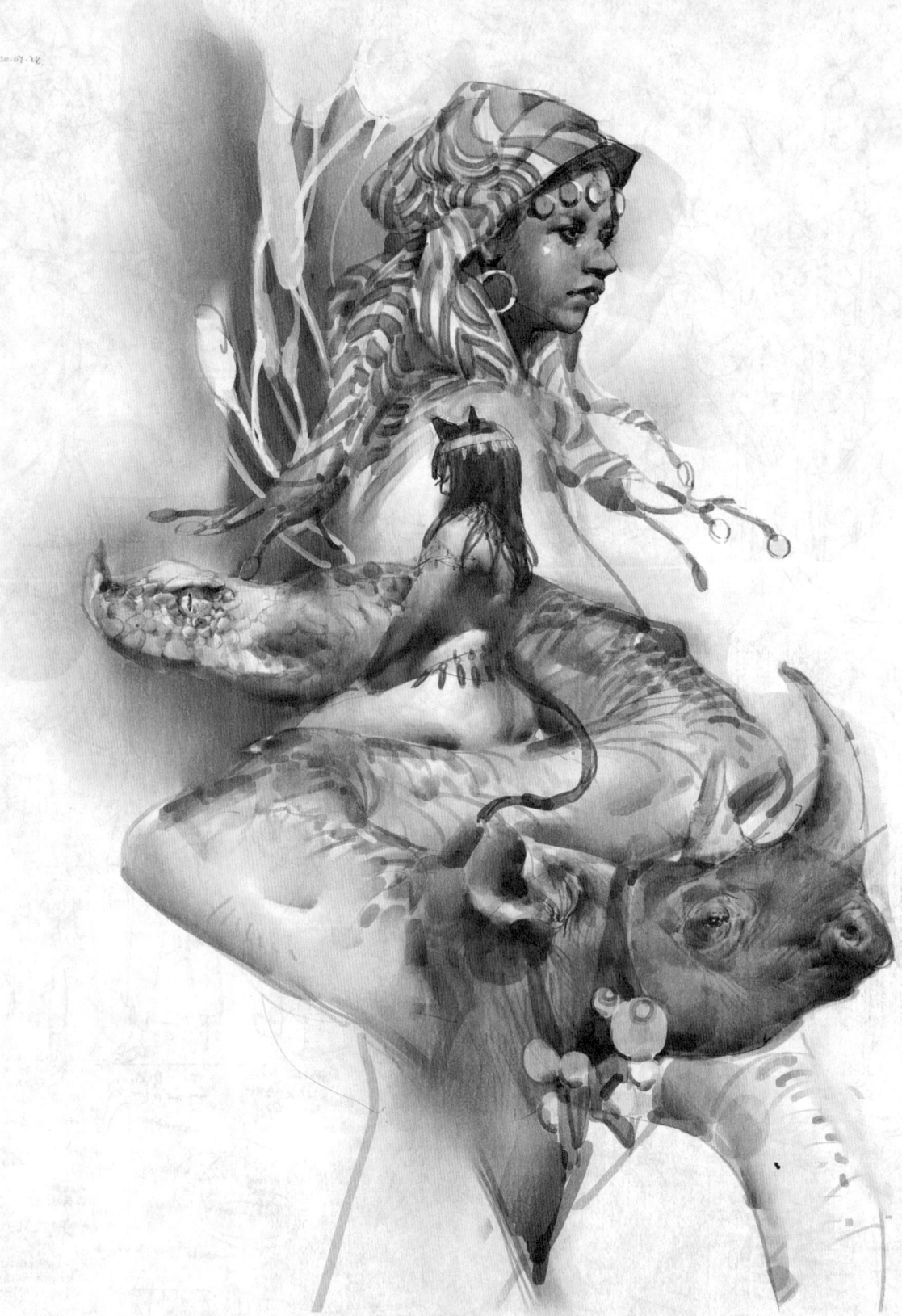

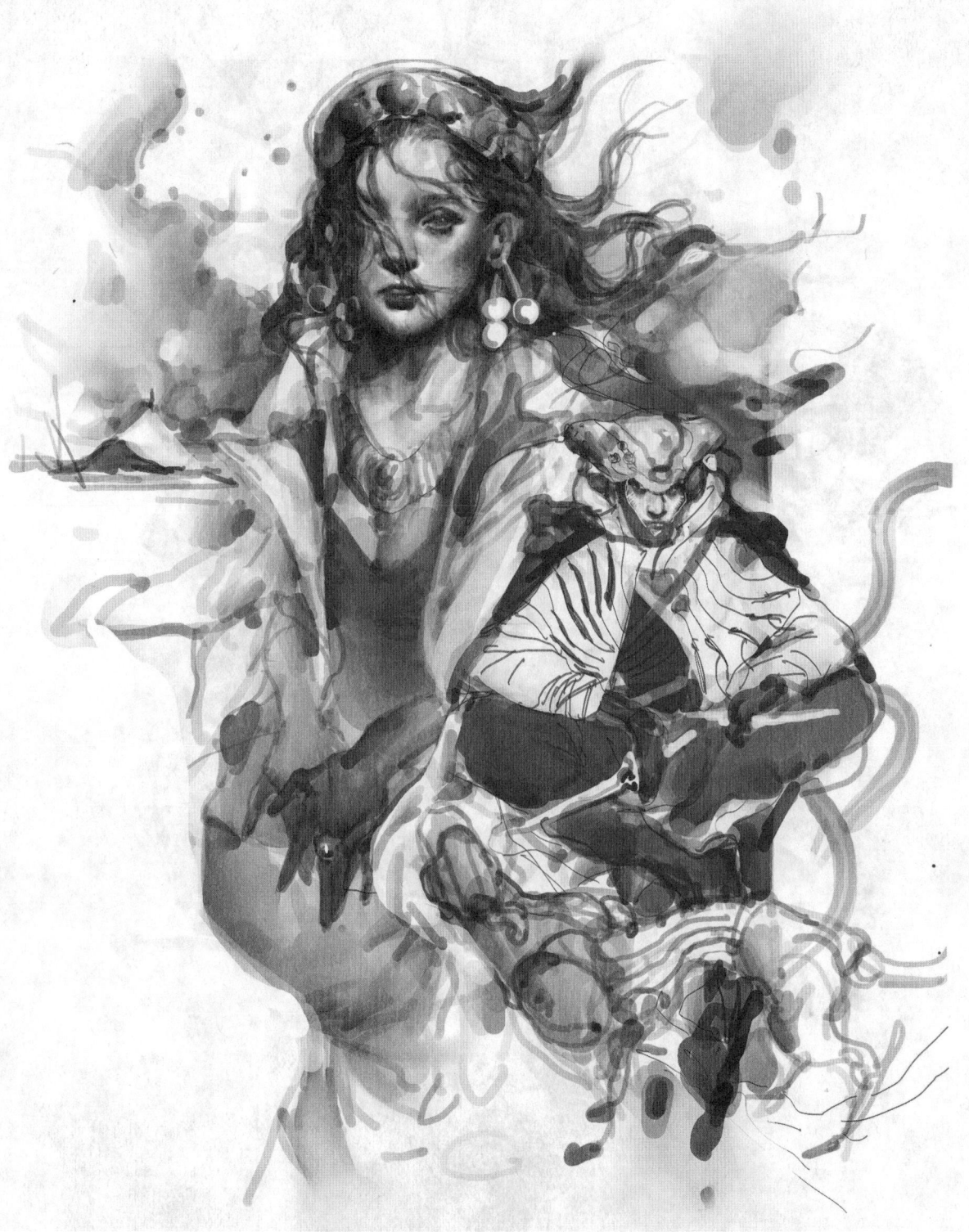

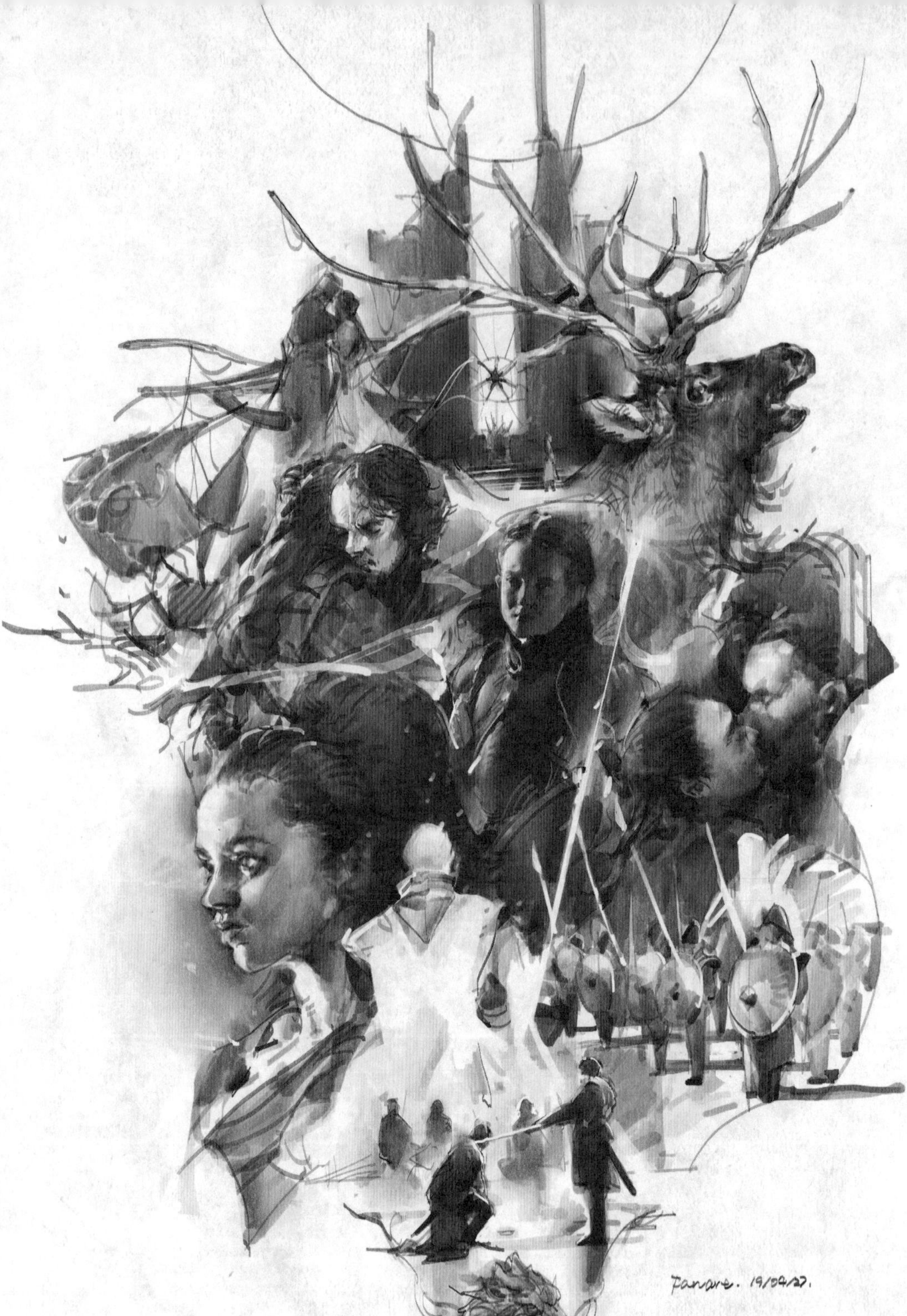

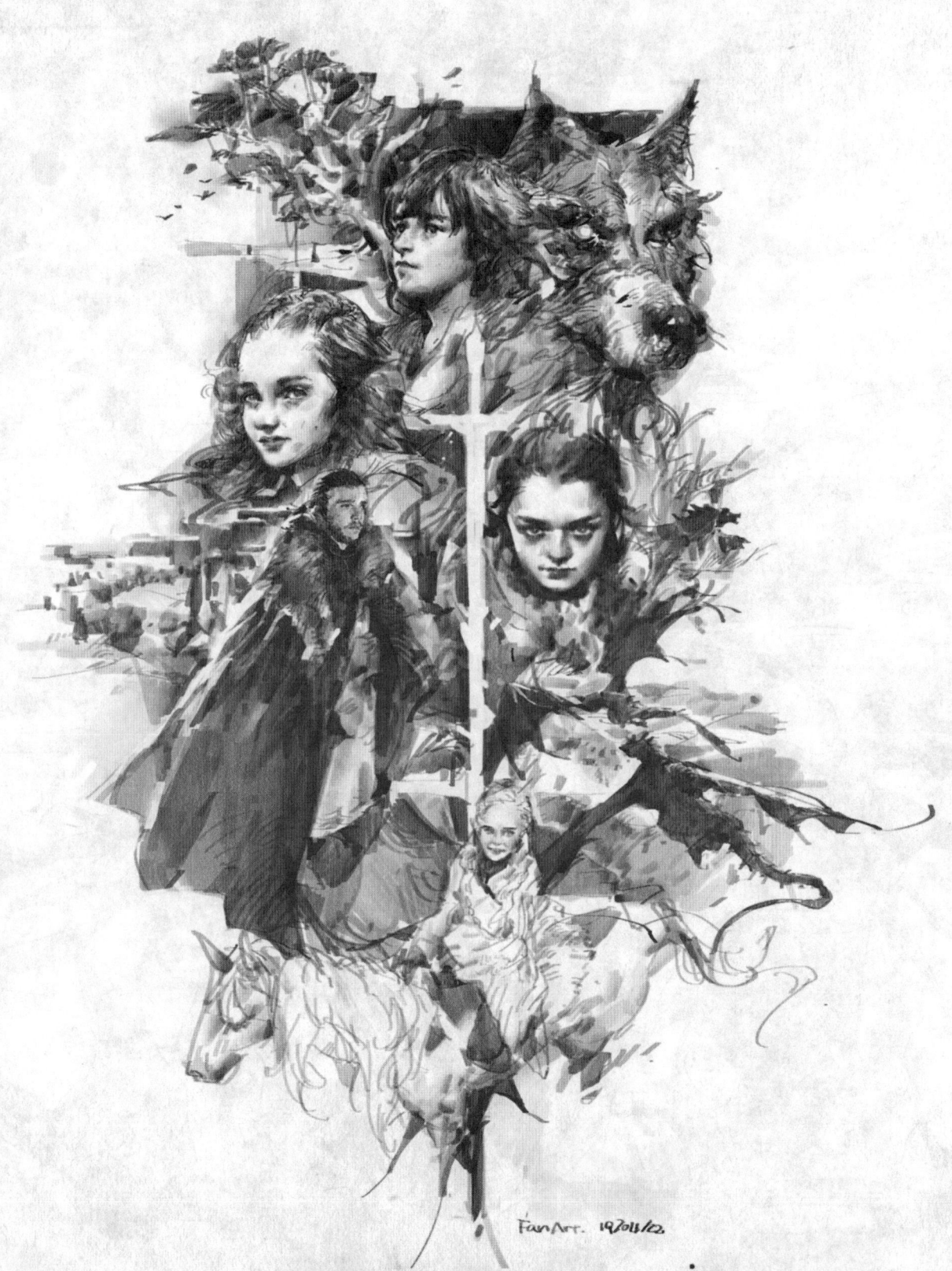

FanArt. 19/01/12.

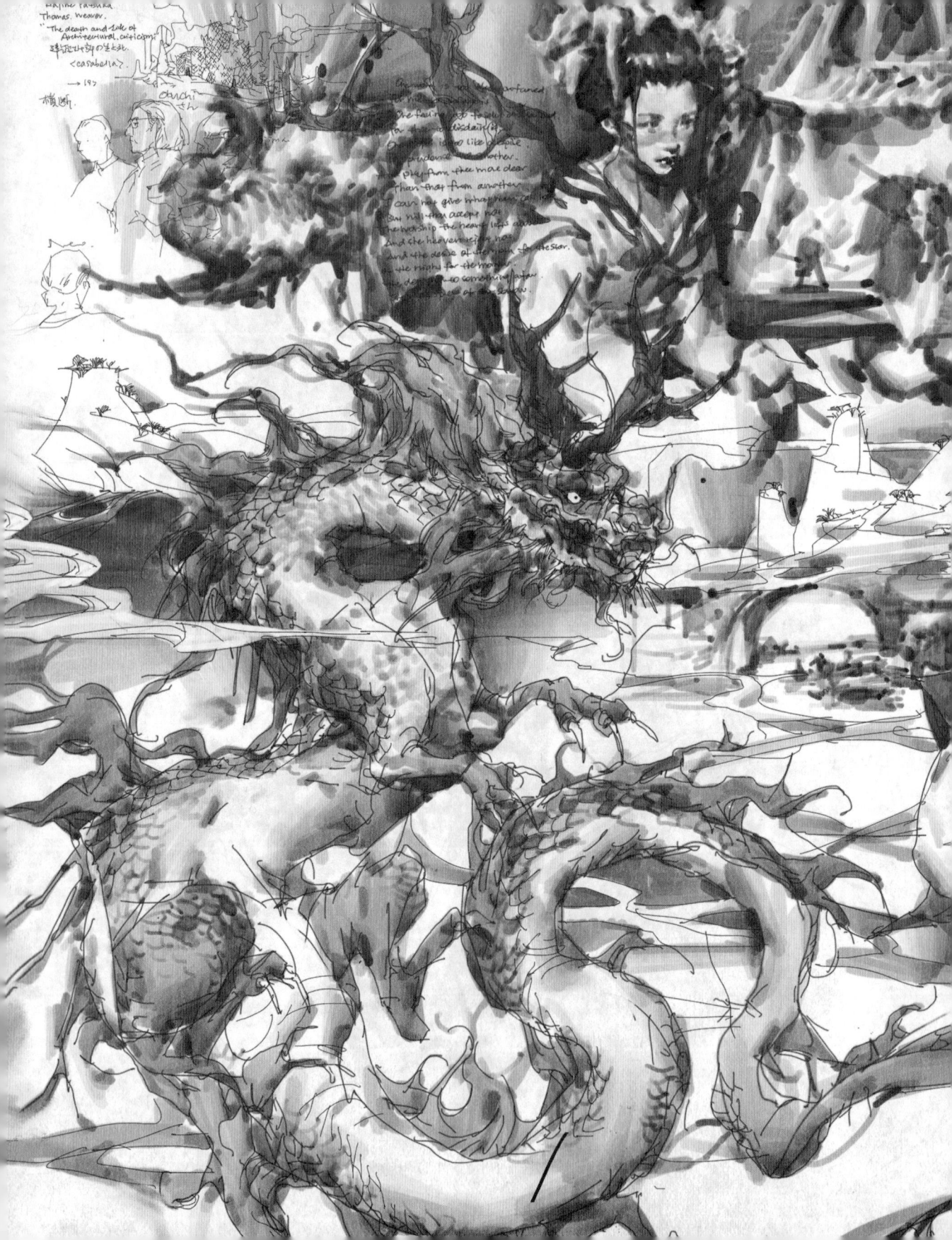

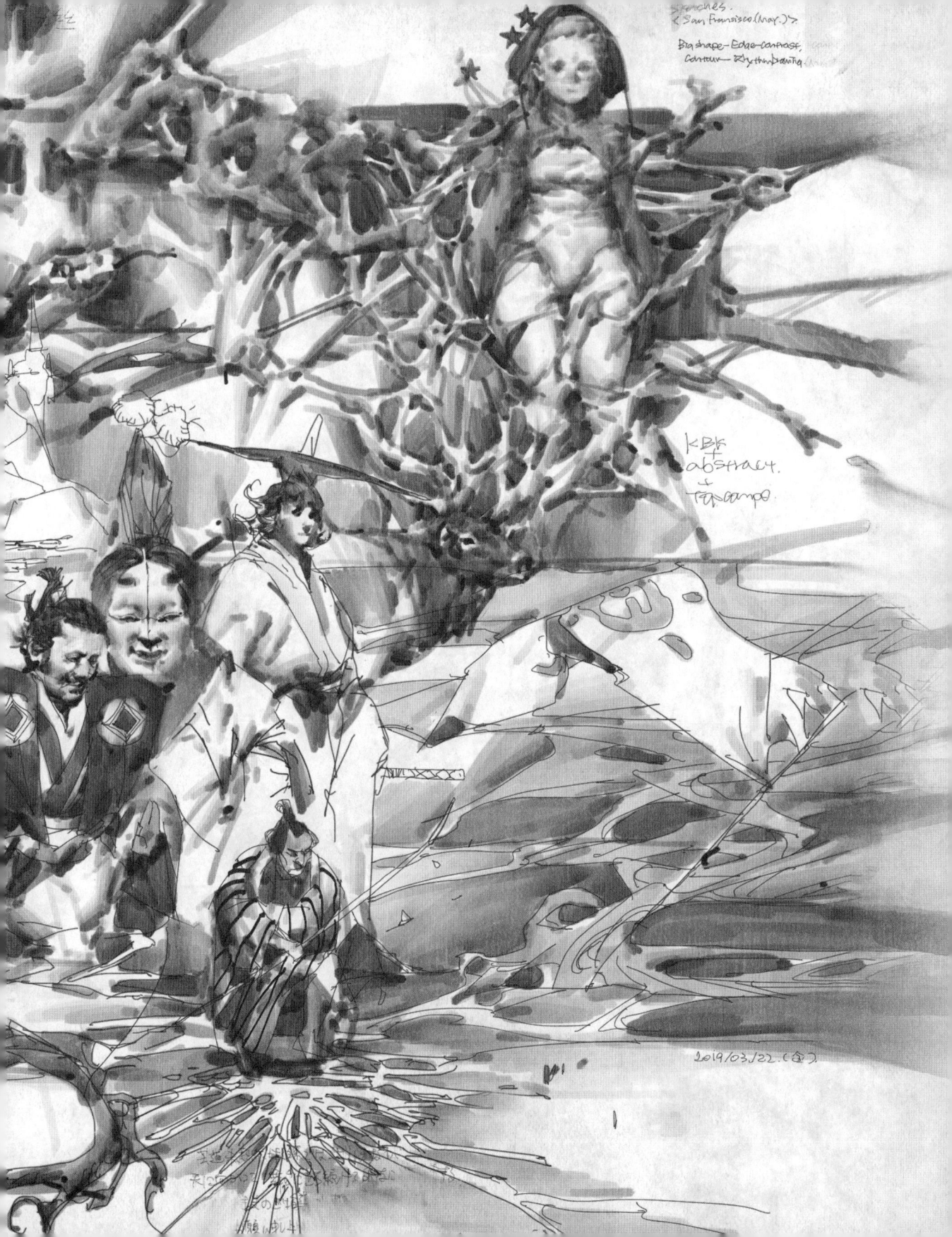

KBK
+
abstract.
+
Tasgampo.

2019/03/22. (金)

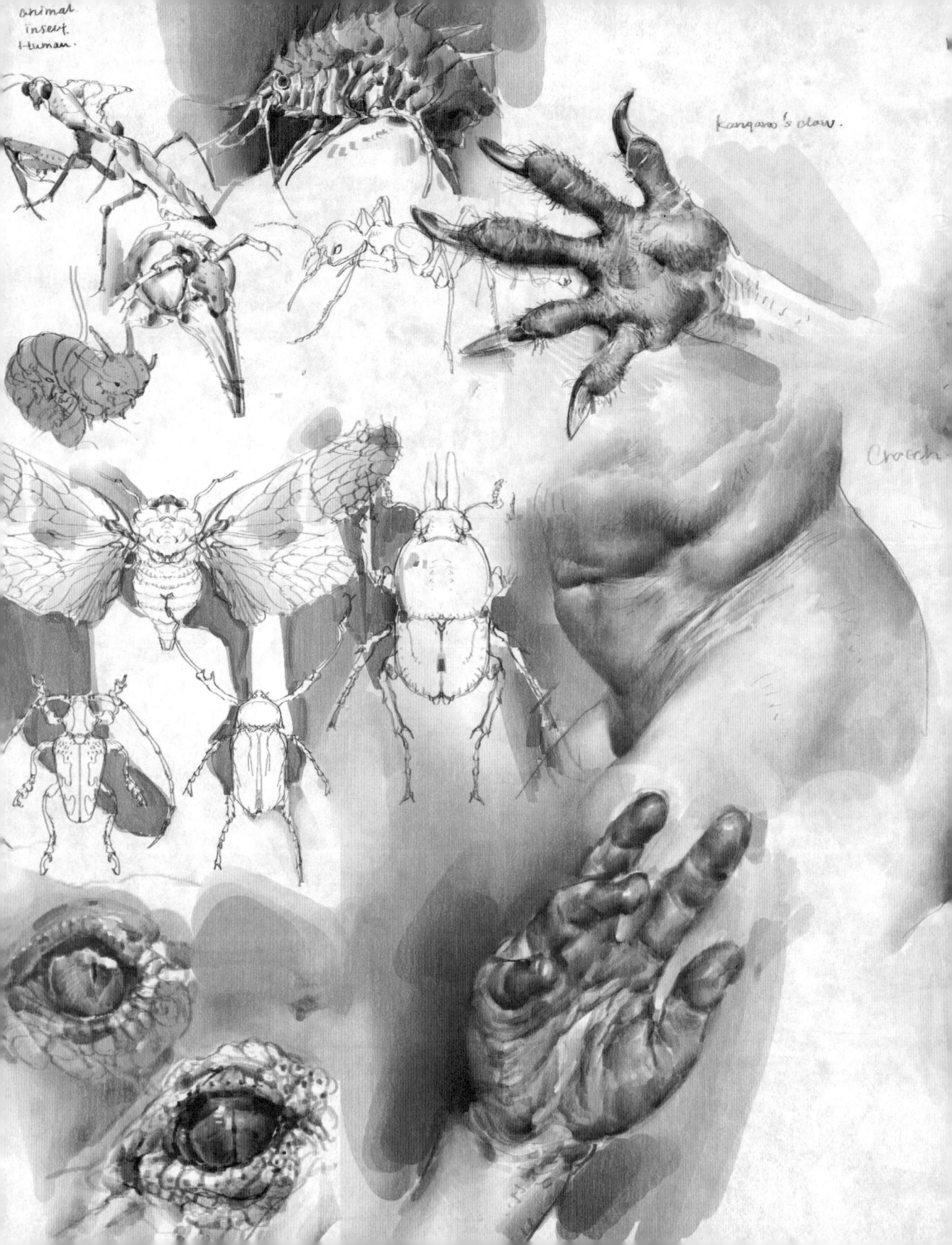

animal.
insect.
Human.

Kangaroo's claw.

Crotch.

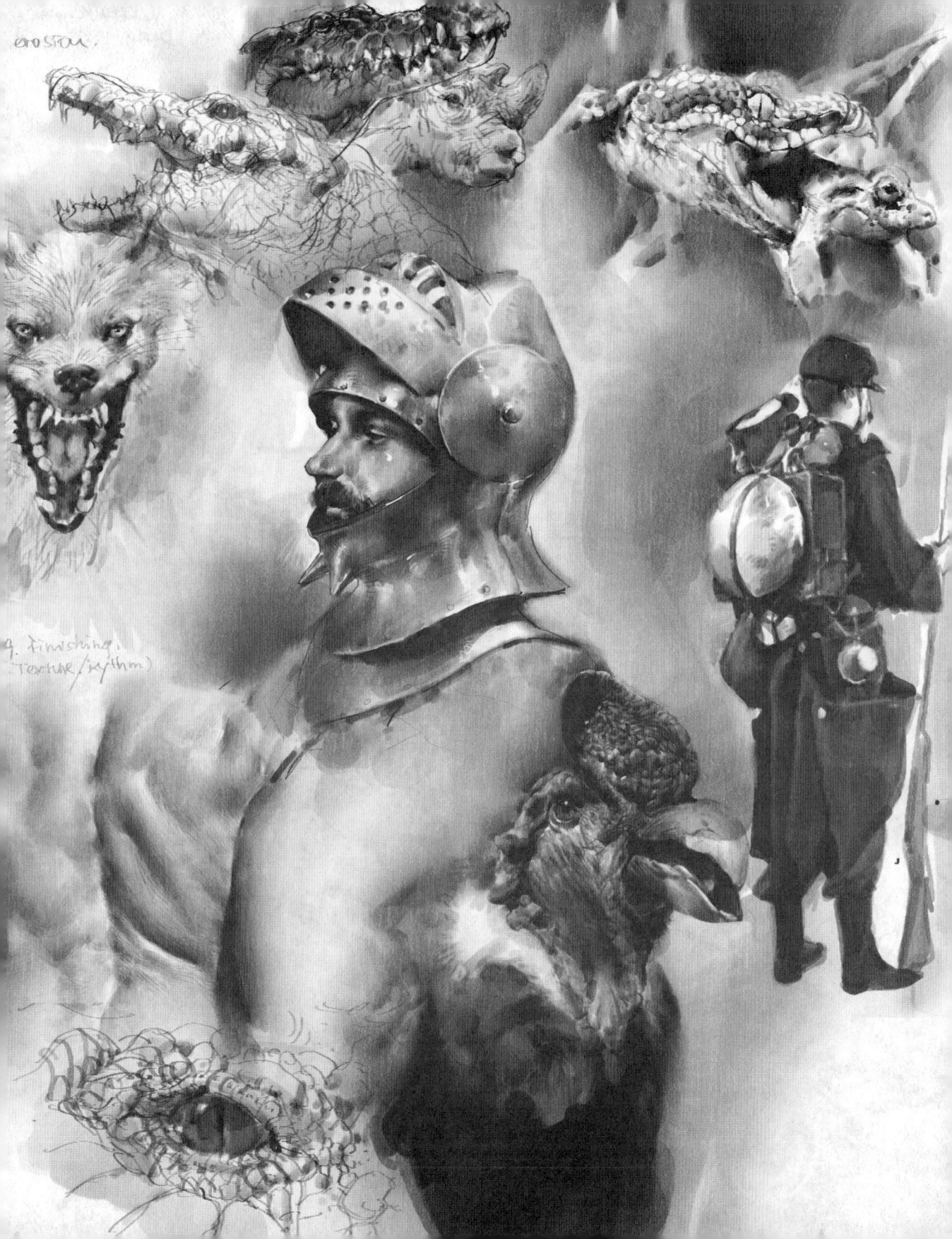

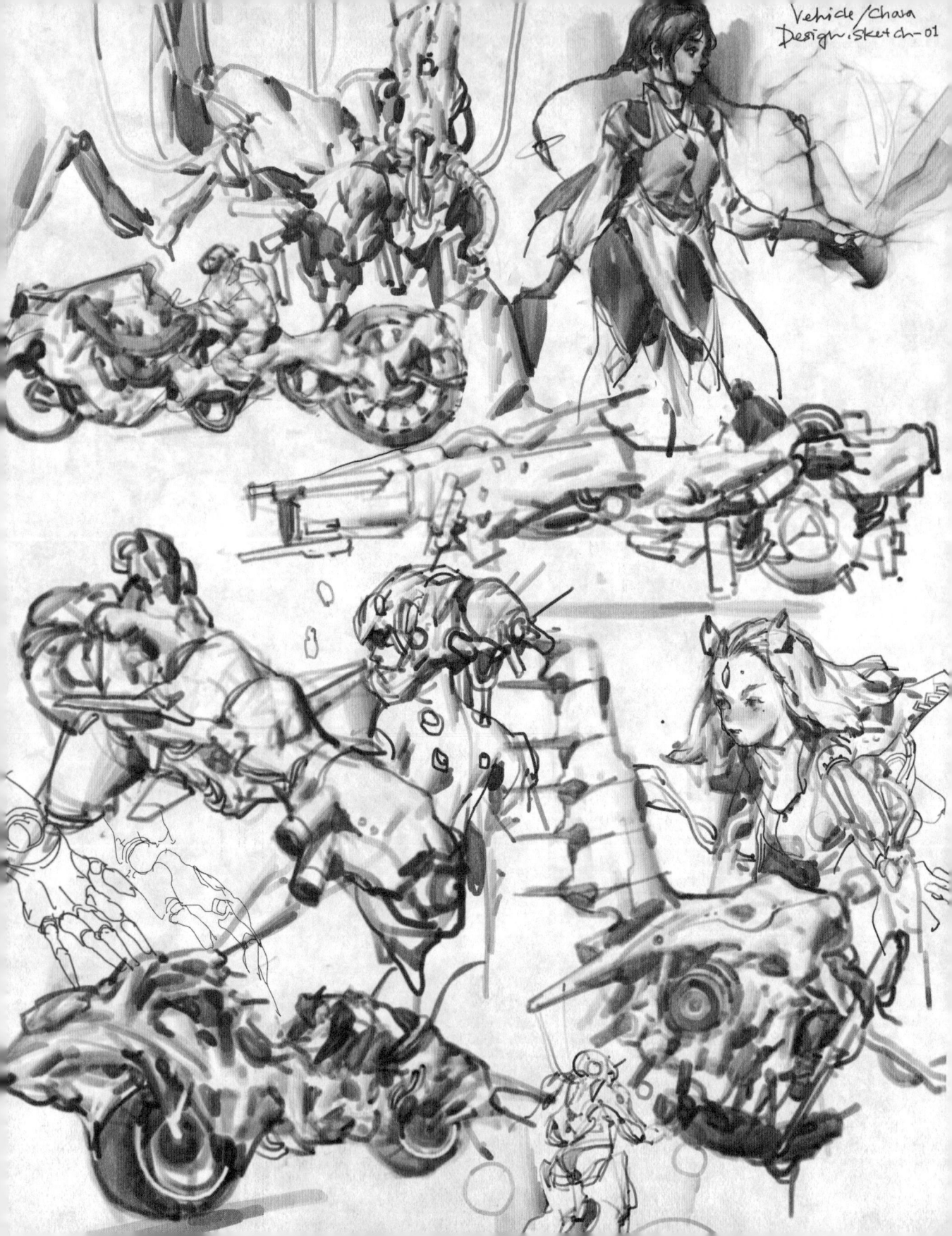

Vehicle/Chara
Design. Sketch-01

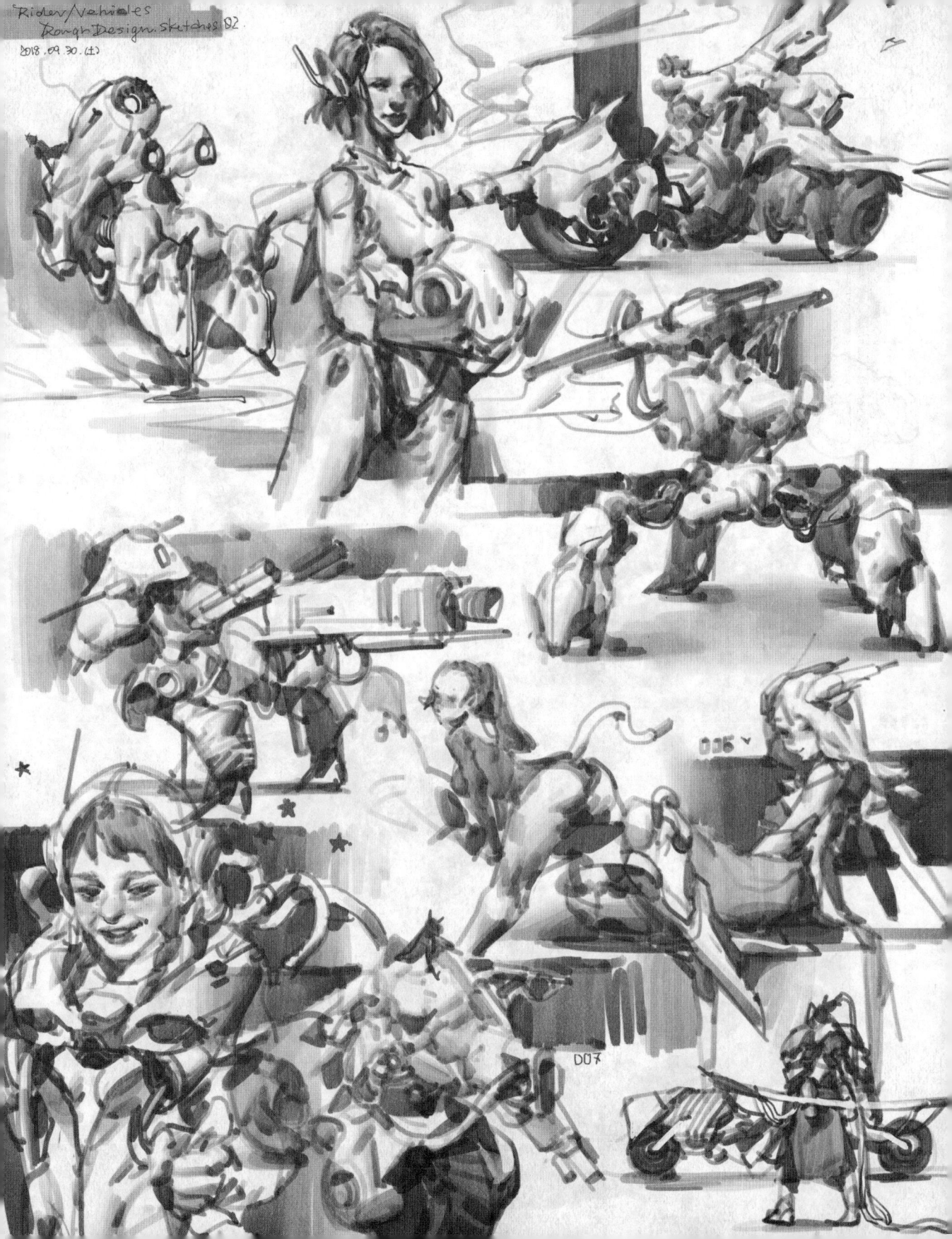

Rider/Vehicles
Rough Design sketches 02
2018.09.30.(土)

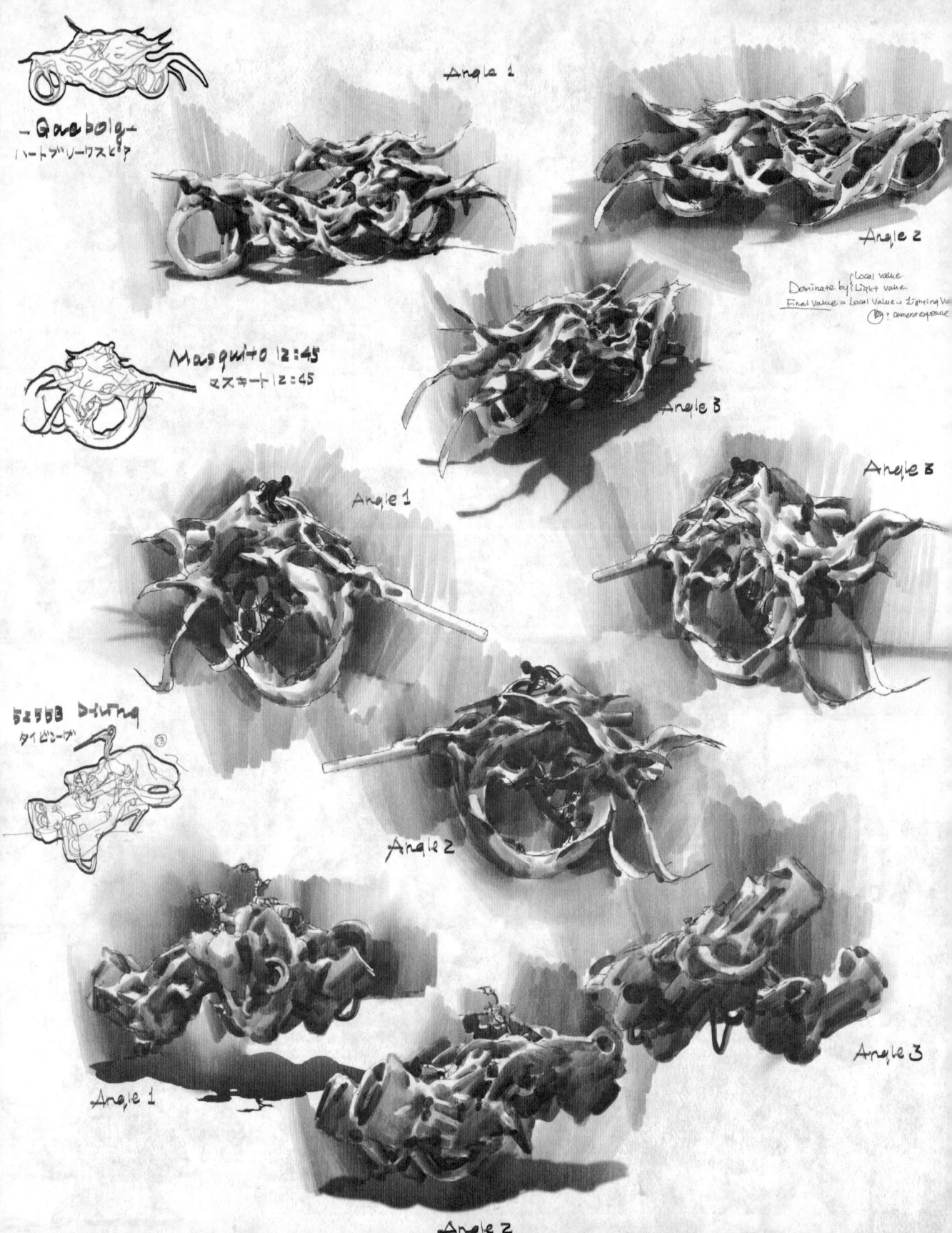

- Qaebolg -
ハートブリークスピア

Angle 1

Angle 2

Angle 3

{ Local value
Dominate by { Light value.
Final value = Local Value + Lighting Va
⊗ = camera exposure

Masquito 12:45
マスキート 12:45

Angle 3

Angle 1

Angle 3

タイビシーブ Diving
タイビシーブ

Angle 2

Angle 1

Angle 3

Angle 2

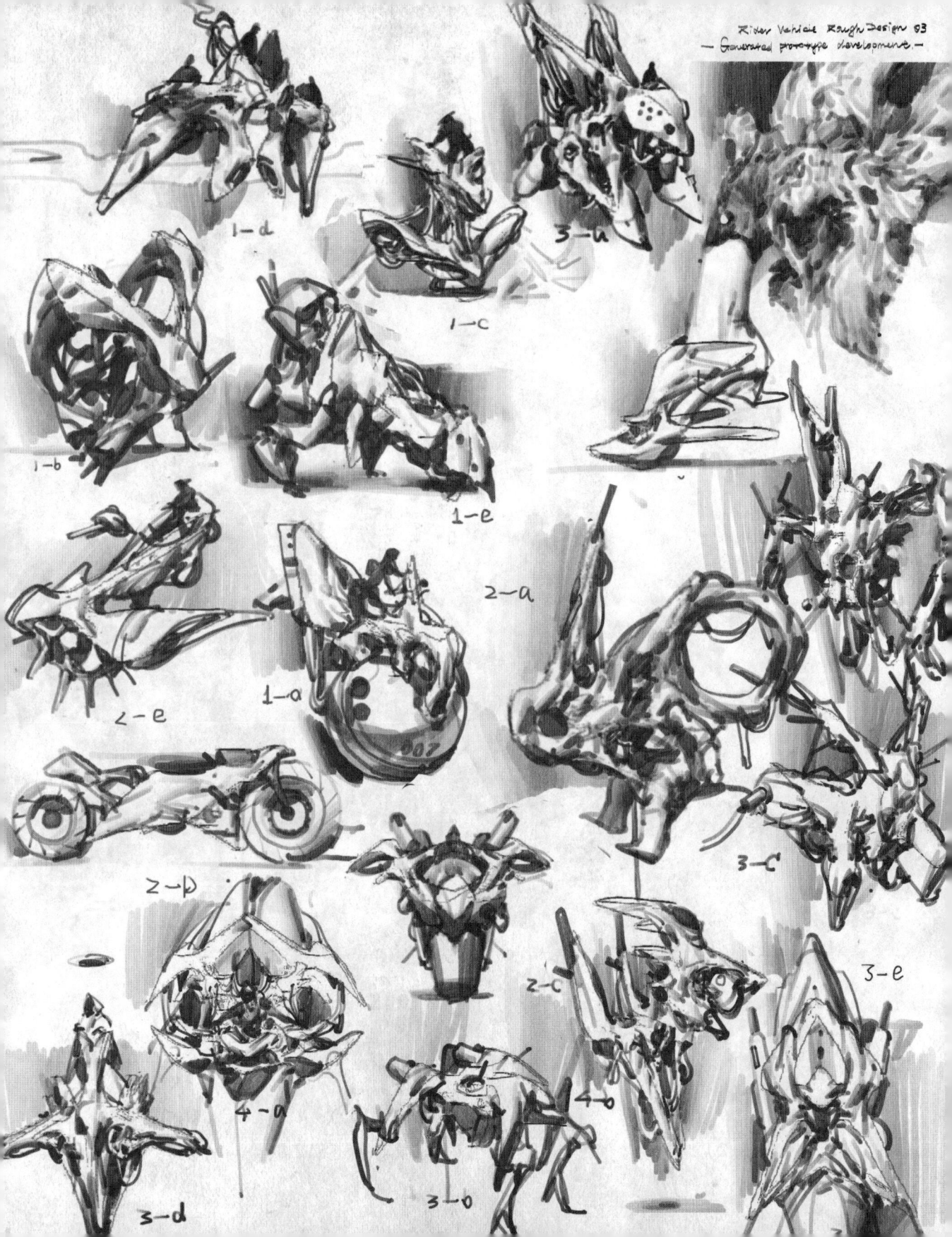

1-d

1-c

3-u

1-b

1-e

2-a

2-e

1-a

3-c

2-b

2-c

3-e

4-a

4-b

3-d

3-b

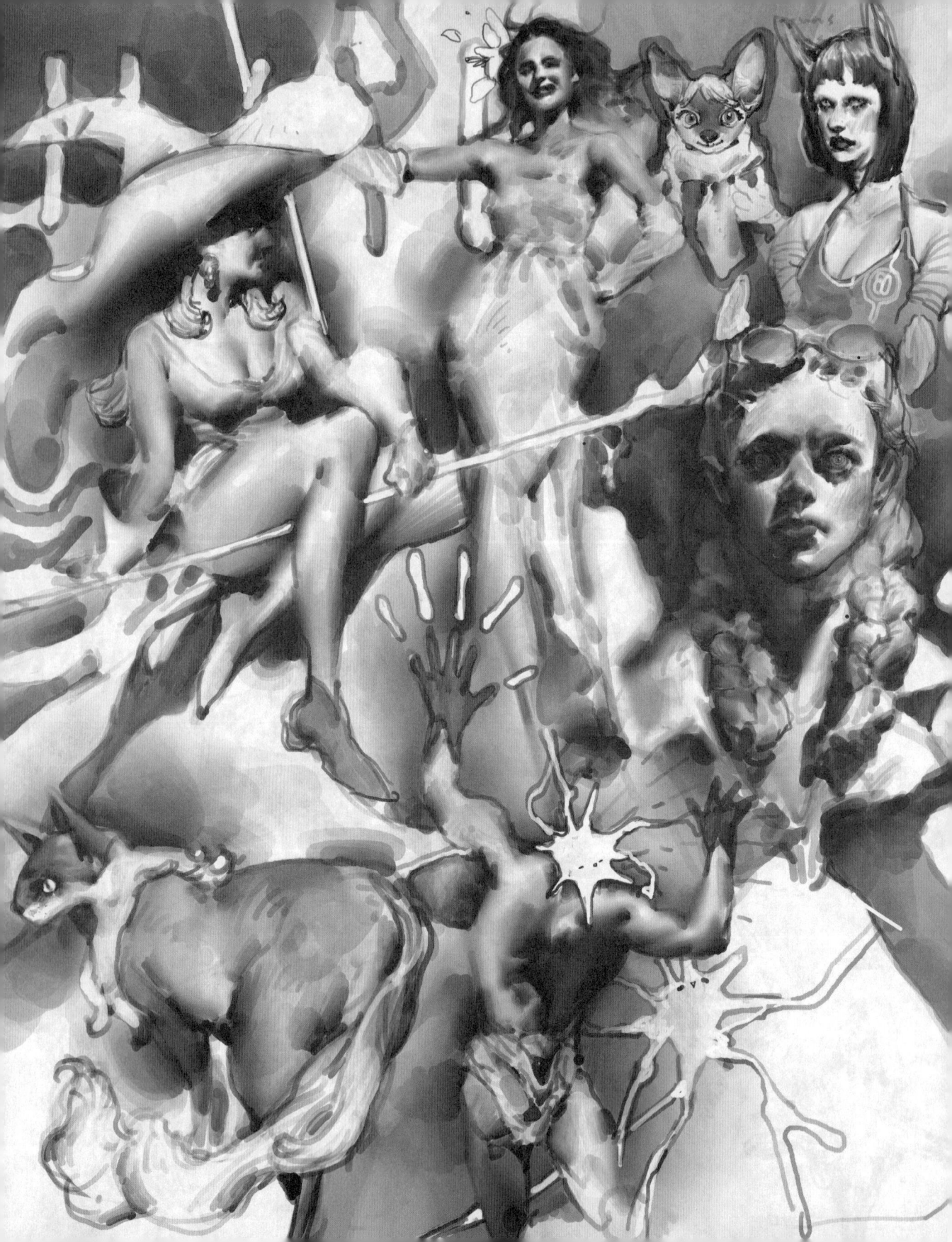

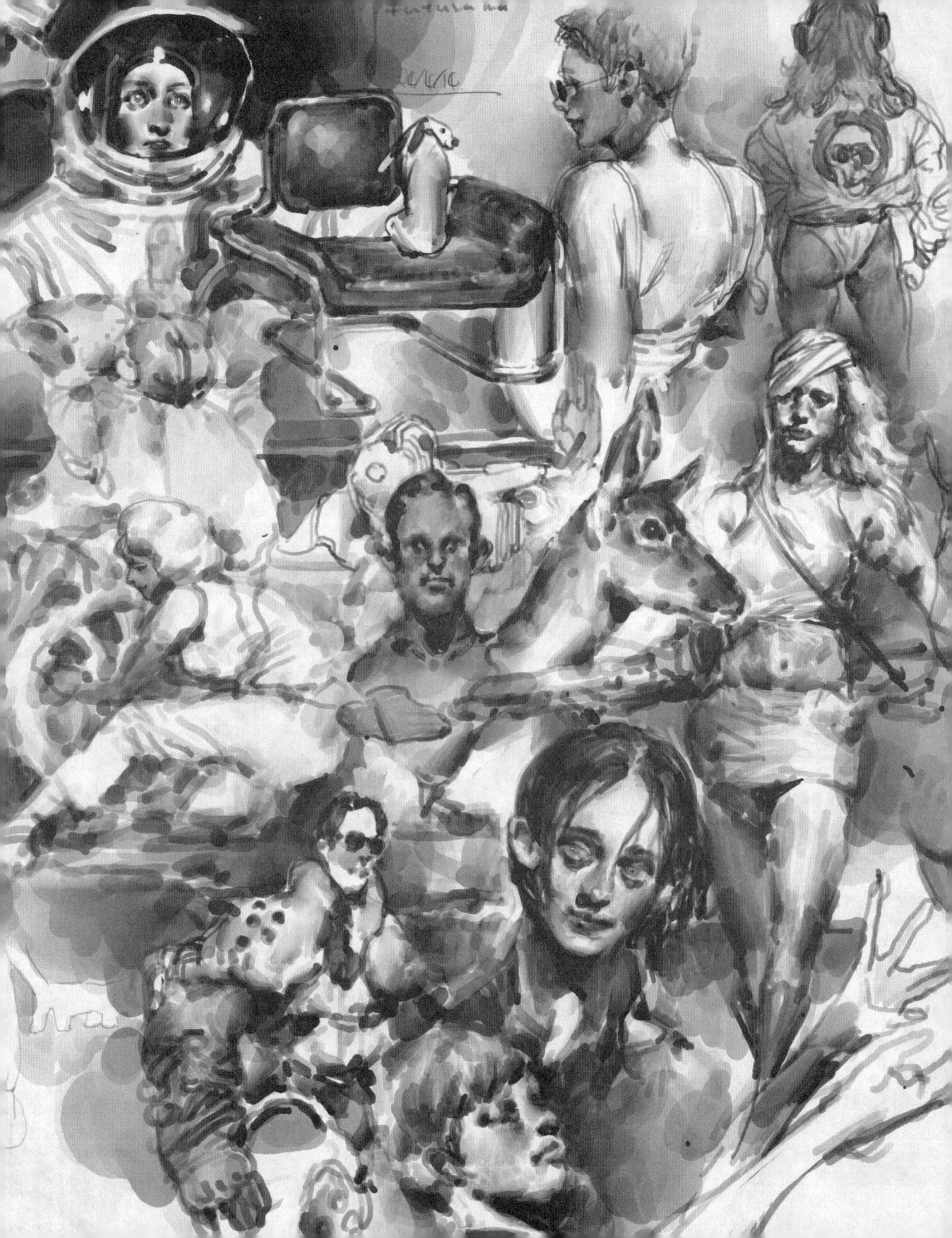

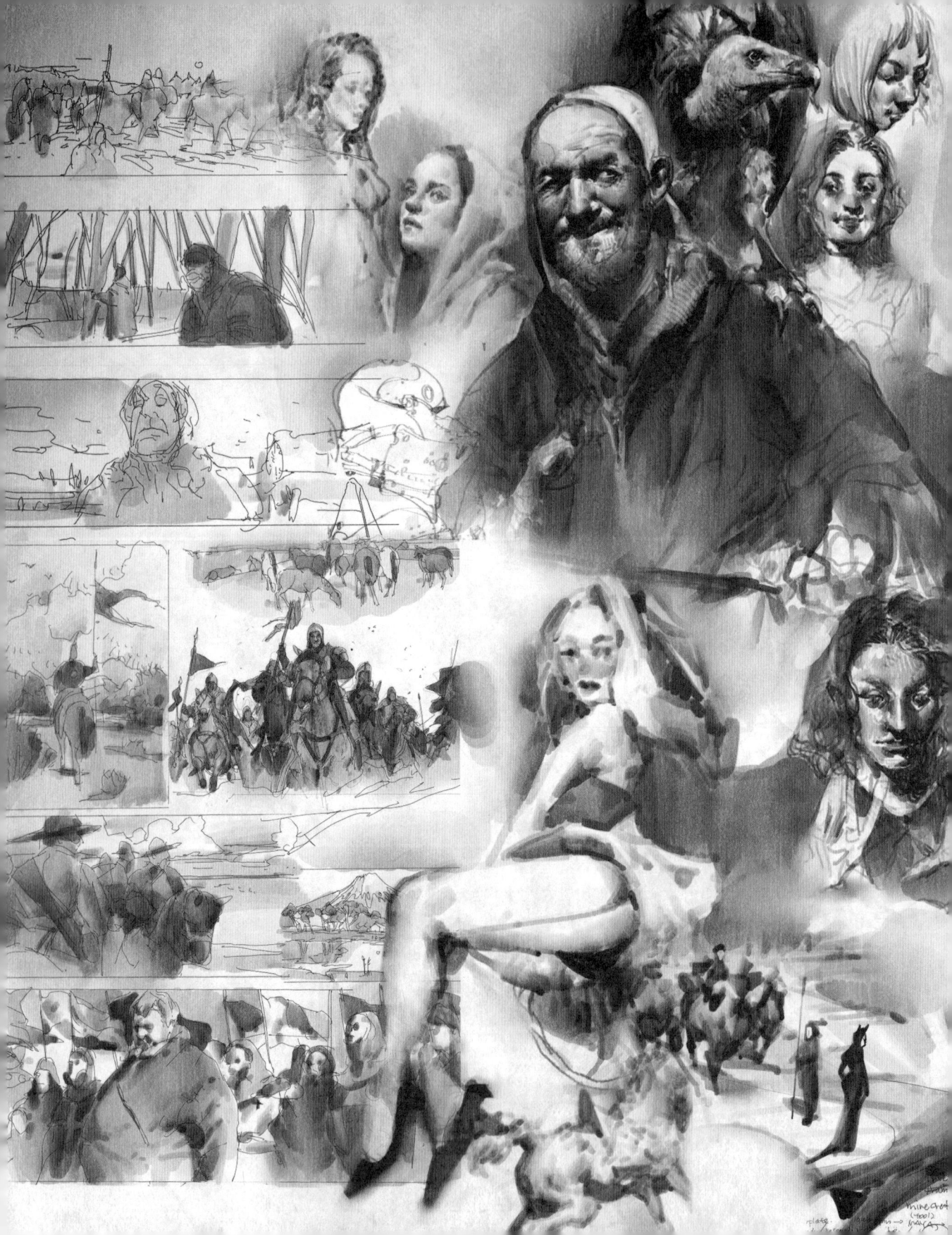

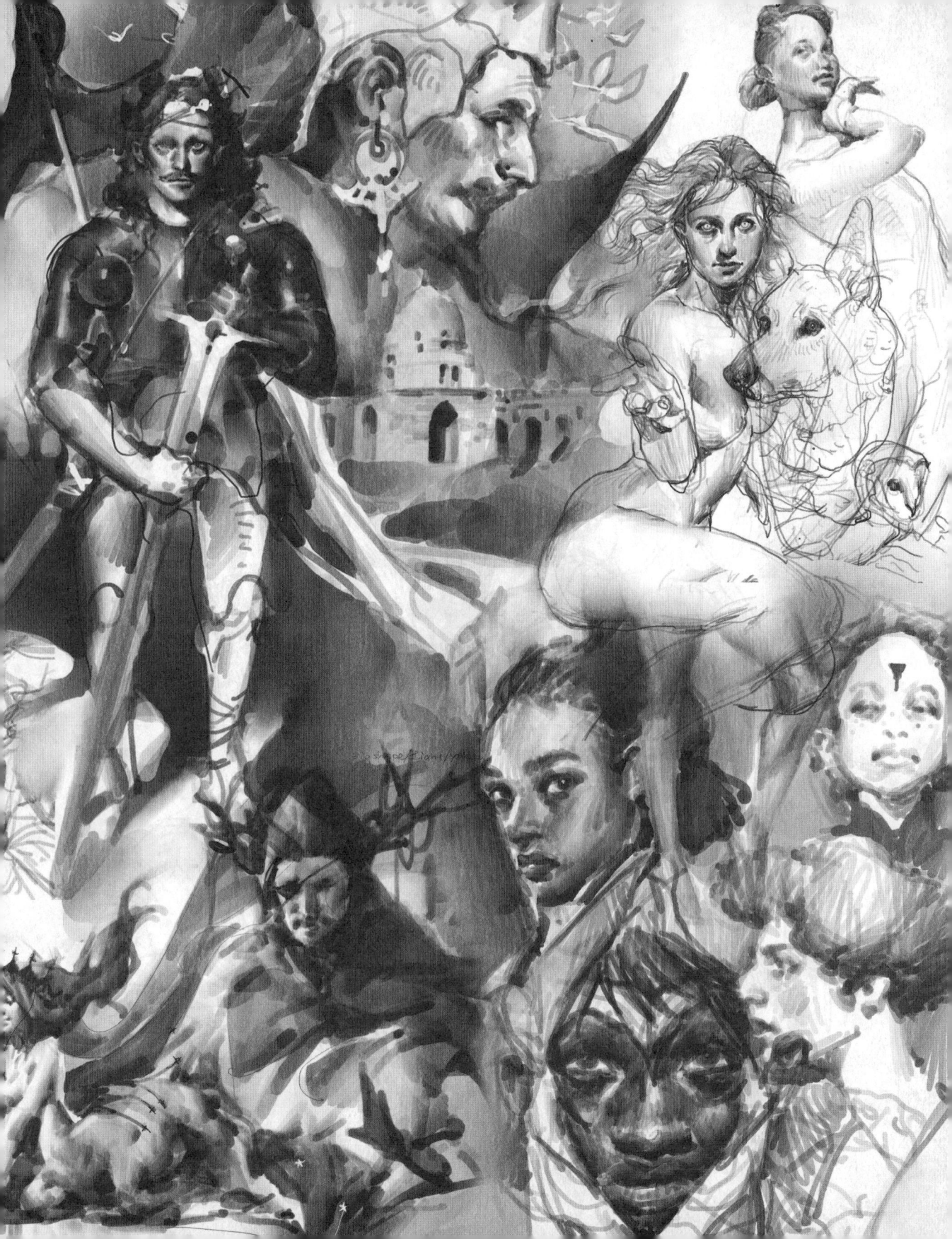

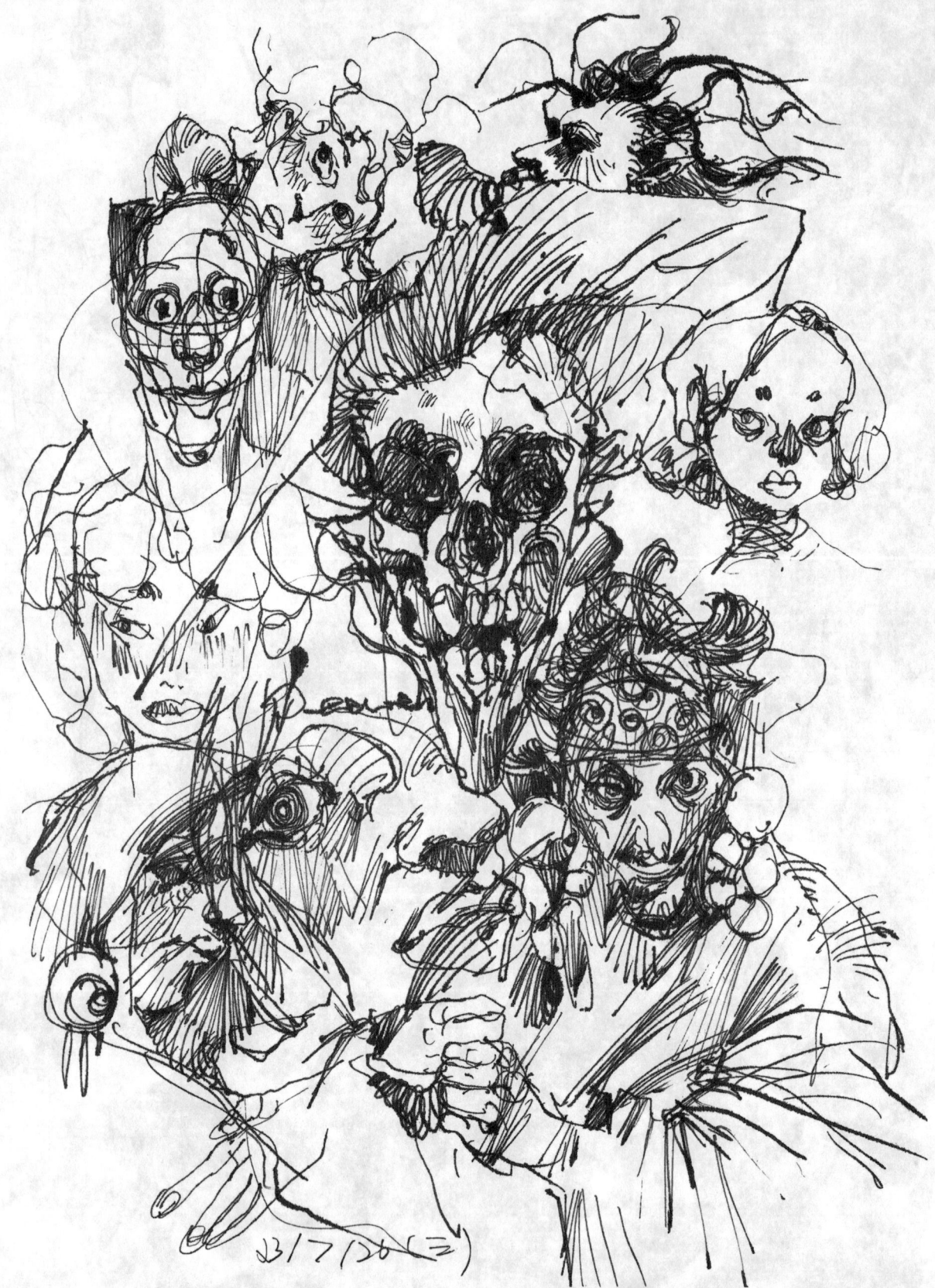

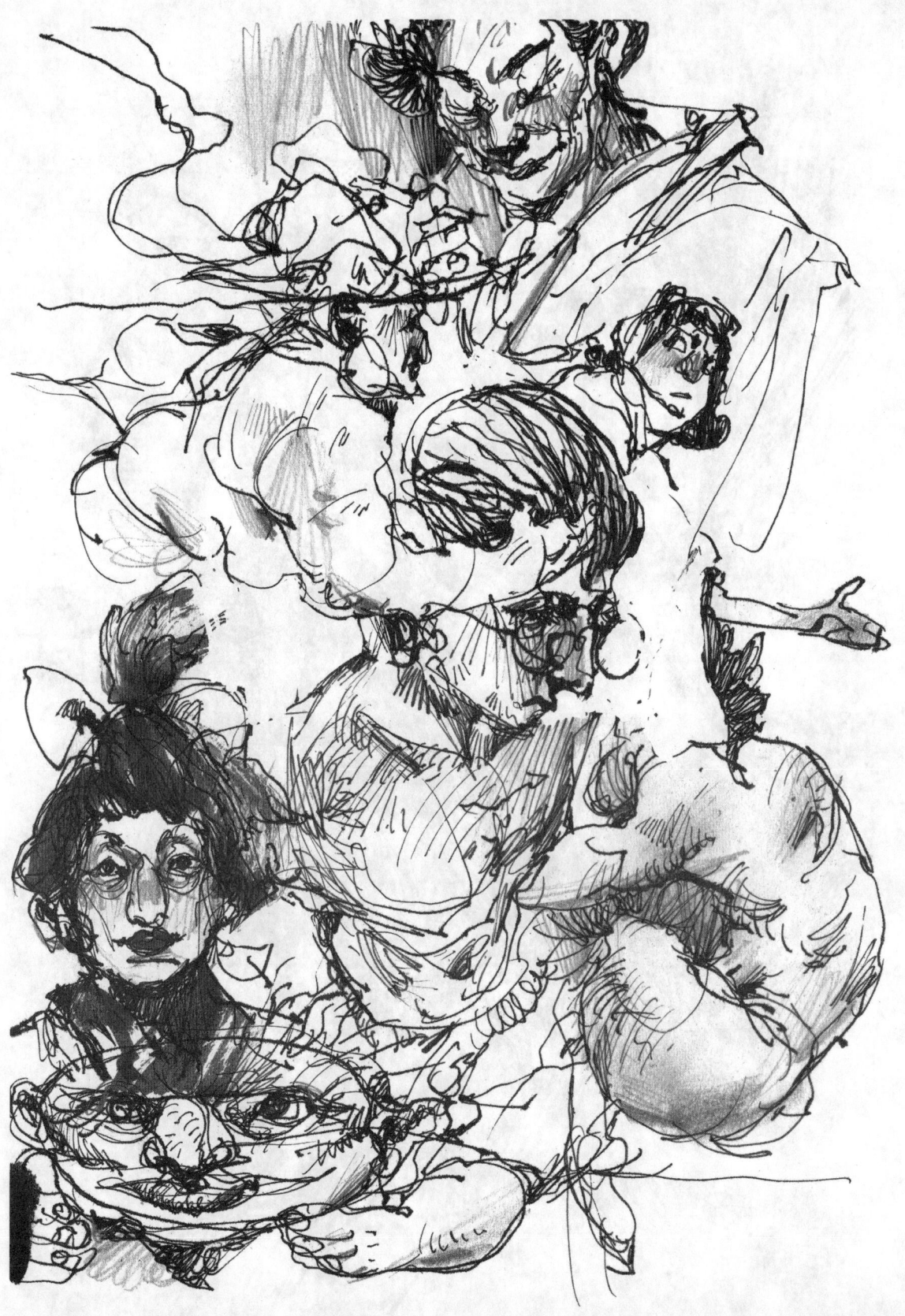

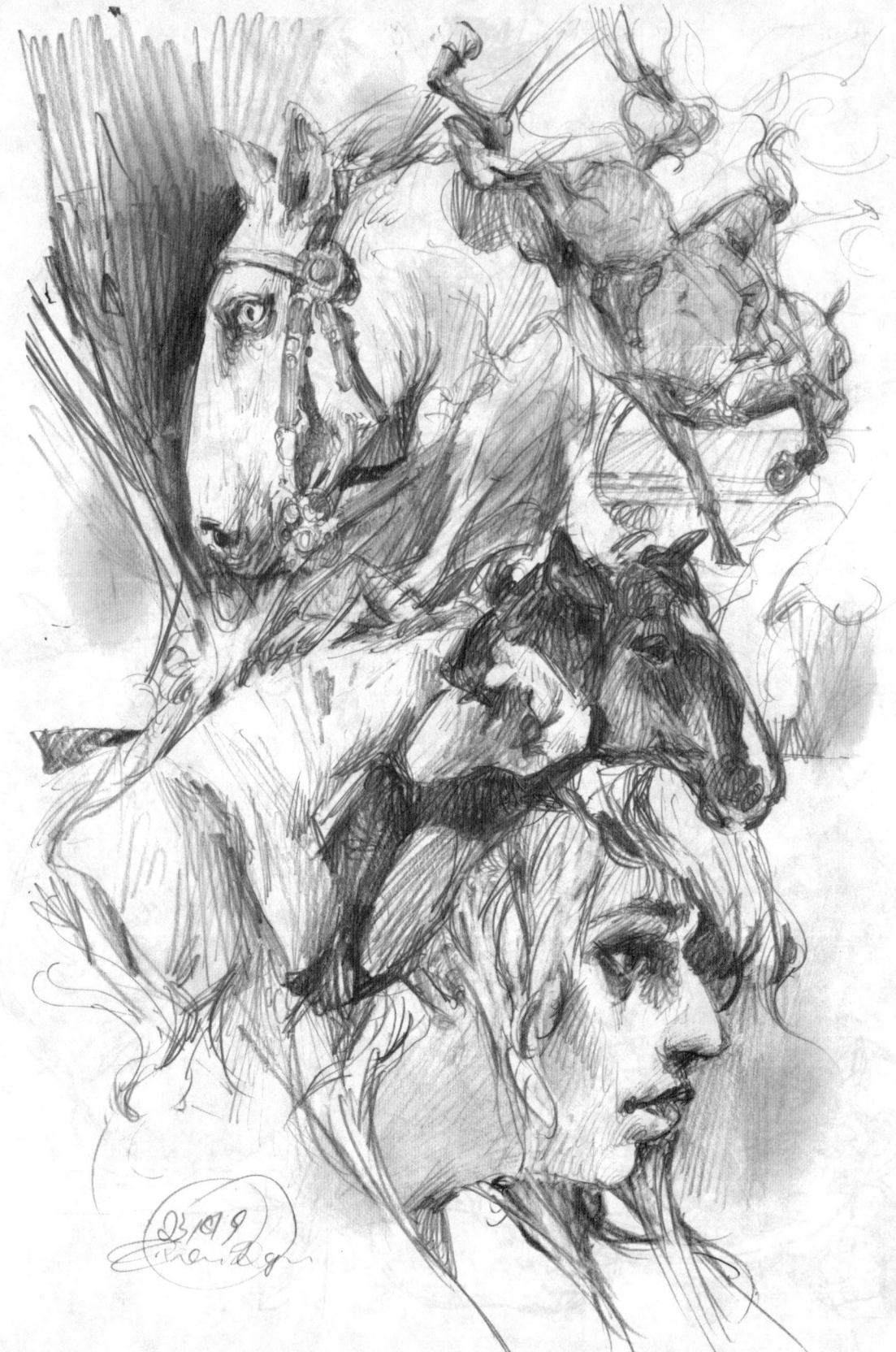

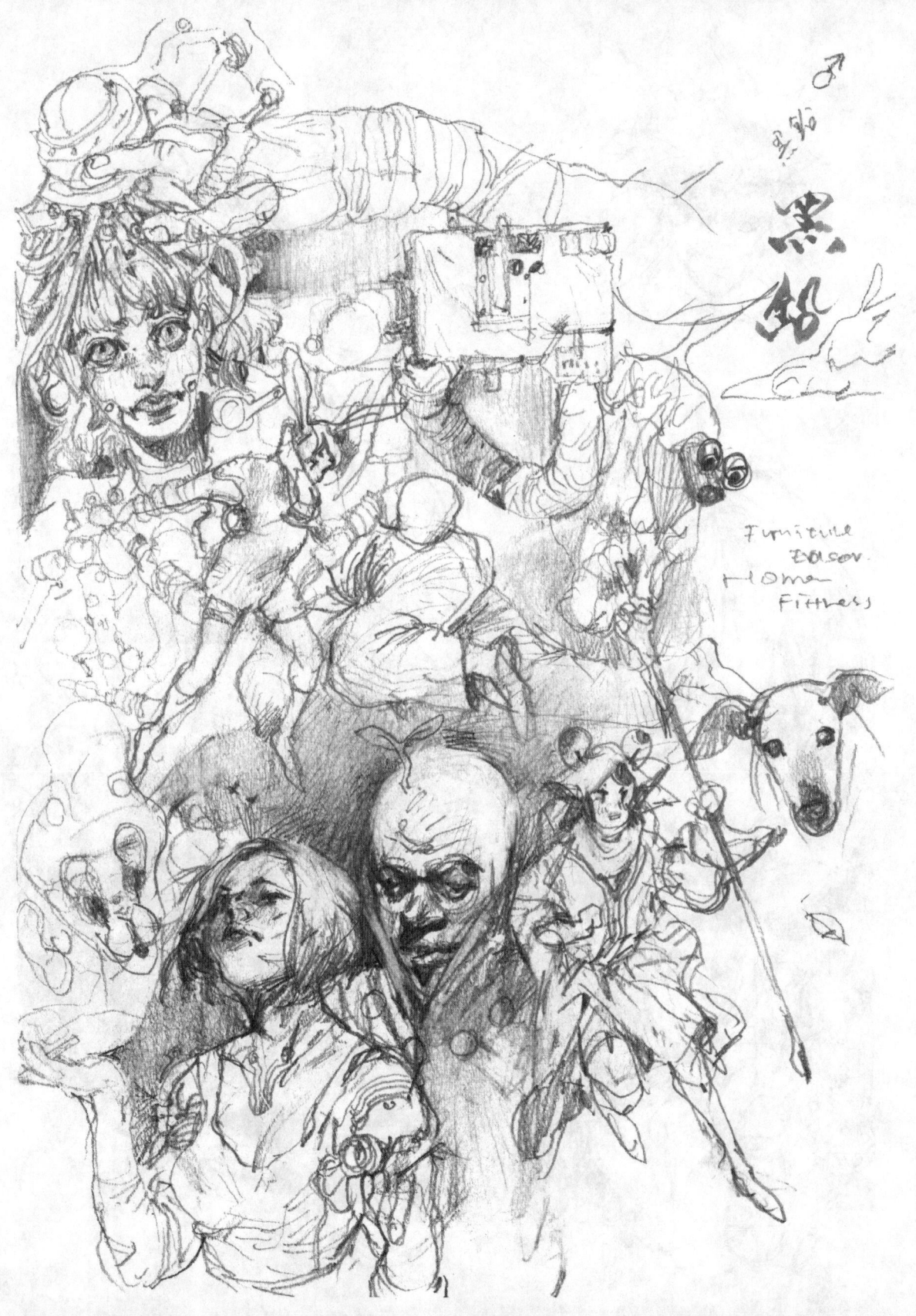

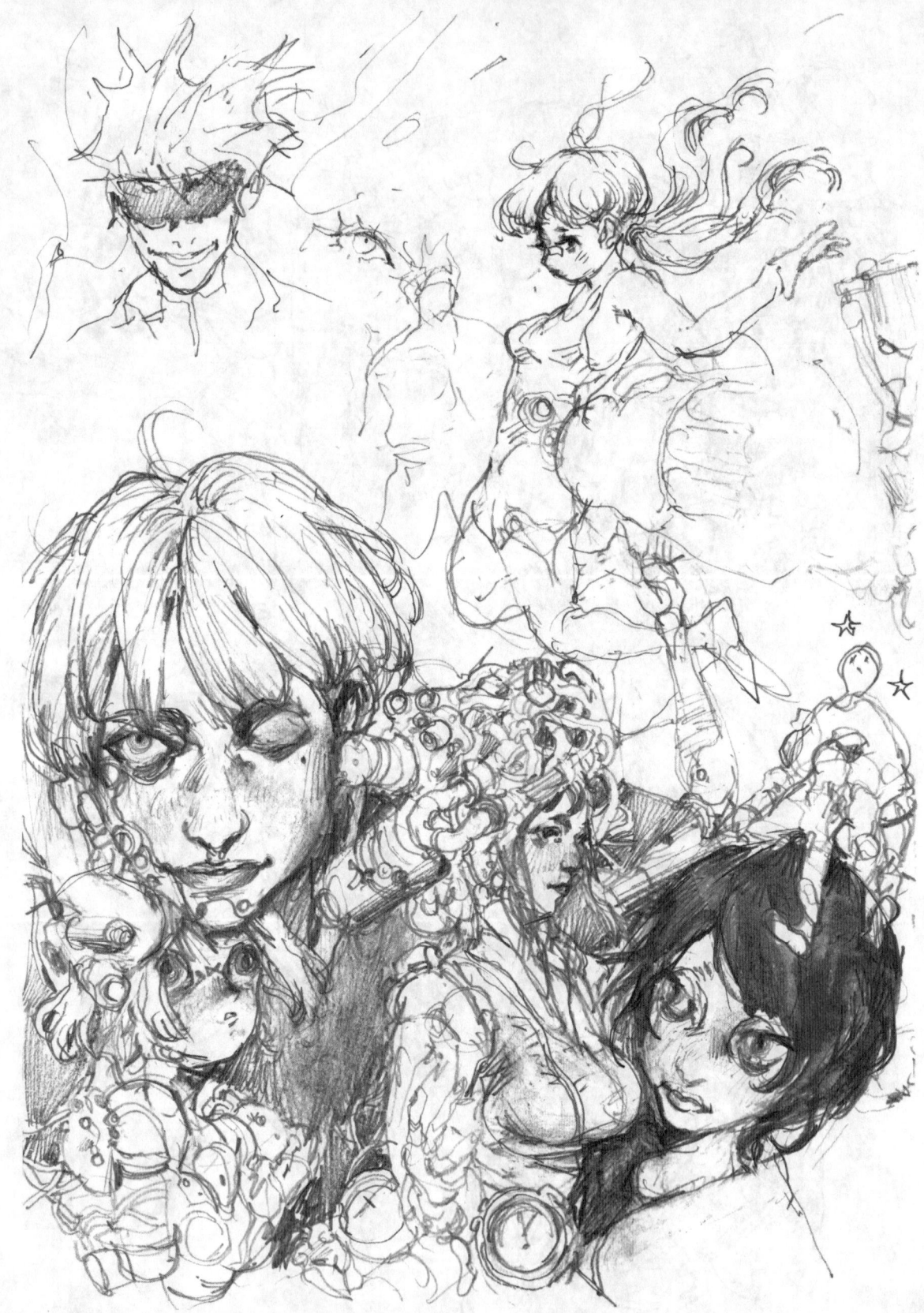

# AFTERWORD あと書き

Nice to meet you. I'm Enrei Ryu.

Art is something I've decide to dedicate my life to, much
like setting out on a long and winding journey

This book is my first published art collection
featuring works from the past decade.

It is a chronicle of the time I've spent and the paths I've spent
and the paths I've traversed.

I am grateful to my family and friends who have always
supported and accompanied me along this journey.

I am also deeply thankful to everyone who contributed to
the publication of this book.

And to you, the reader, thank you for holding this book
in your hands. I am delighted to share the vistas of
my journey with you and invite you to join me as we embark
on future adventures together.

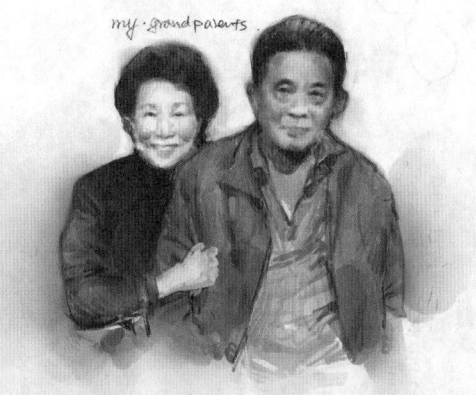

my · grandparents

エンテイ・リュウ です。はじめまして。

アートとは、私が一生を捧げると決意したものであり、

長く曲がりくねった旅に出ることによく似ています。

本書は過去10年間の作品を集めた、私の初めてのアートコレクションです。

過ごしてきた時間と、旅してきた道のりの記録です。

この旅路をいつも支え、ともに歩んでくれる家族と友人に感謝します。

また、本書の出版にご尽力くださったすべての方々に深く感謝を捧げます。

そして読者の皆様、この本を手にしてくれてありがとうございます。

私の旅の景色を皆様と分かち合えることができて、とても嬉しいです。

そしてこれから続く冒険に、ともに旅立っていきたいです。

Enrei Ryu.
2024.6

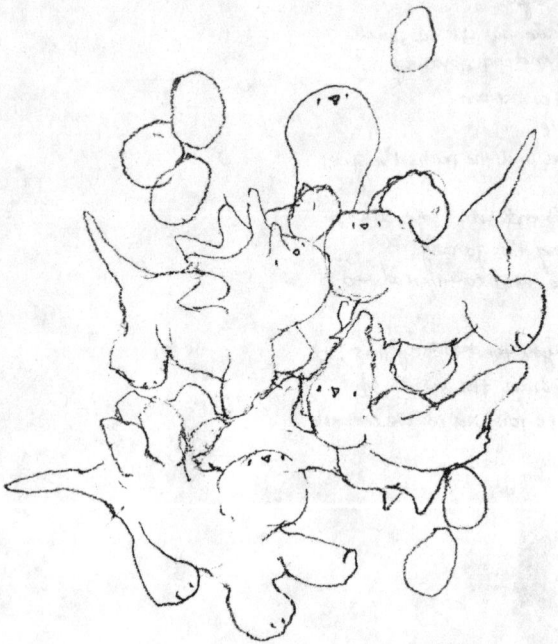

**Entei Ryu**（えんてい・りゅう）
コンセプトアーティスト／造形作家
東京大学大学院で建築を学び、工学系研究科修士号を取得。現在はゲーム・映画のキャラクターデザインを中心に活動。ほかにフィギュア原型やイラストレーション、ファッションなどの分野にも活動の場を広げている。携わったタイトルは「ASSASSIN'S CREED」、「FINAL FANTASY」シリーズなど。
X: @BadZrlwt
Instagram: @badzrliuwt

エンテイ・リュウ アートワークス キメラ
2024年7月11日　初版第1刷発行
2024年11月14日　第2刷発行

著者　Entei Ryu

進行協力　株式会社 W's
デザイン　眞々田稔（ホンマチ組版）
編集　大場義行
編集協力　田中里奈（ヒヨコ舎）
協力　株式会社コジマプロダクション
　　　Underverse Co.,Ltd.
　　　Superani
　　　株式会社徳間書店
　　　株式会社秋田書店
発行人　三芳寛要
発行元　株式会社パイインターナショナル
〒170-0005 東京都豊島区南大塚2 32 4
TEL:03-3814-3883 FAX:03-5395-4830
sales@pie.co.jp
印刷・製本　株式会社シナノ印刷

©2021 Entei Ryu / PIE International
ISBN 978-4-7562-5902-8 C0079
Printed in Japan

Entei Ryu ARTWORKS CHIMERA キメラ